AF773736

Energy and Sustainability

Energy and Sustainability

Edited by **Lucas Collins**

Callisto Reference

New York

Published by Callisto Reference,
106 Park Avenue, Suite 200,
New York, NY 10016, USA
www.callistoreference.com

Energy and Sustainability
Edited by Lucas Collins

© 2016 Callisto Reference

International Standard Book Number: 978-1-63239-710-2 (Hardback)

This book contains information obtained from authentic and highly regarded sources. Copyright for all individual chapters remain with the respective authors as indicated. All chapters are published with permission under the Creative Commons Attribution License or equivalent. A wide variety of references are listed. Permission and sources are indicated; for detailed attributions, please refer to the permissions page and list of contributors. Reasonable efforts have been made to publish reliable data and information, but the authors, editors and publisher cannot assume any responsibility for the validity of all materials or the consequences of their use.

The publisher's policy is to use permanent paper from mills that operate a sustainable forestry policy. Furthermore, the publisher ensures that the text paper and cover boards used have met acceptable environmental accreditation standards.

Trademark Notice: Registered trademark of products or corporate names are used only for explanation and identification without intent to infringe.

Printed in China.

Contents

Preface

This book explores the field of energy and its sustainable use. Energy is an essential requirement of today's world. But this dire requirement results in unstable usage and wastage. Thus, using energy like hydroelectricity, solar energy, wind energy, etc. which is sustainable is a wise choice. This book covers in detail some existent theories and innovative concepts revolving around sustainable energy. It aims to present path-breaking techniques that transformed this field. From theories to research to practical applications, case studies related to all the contemporary topics of relevance to this field have been included in this book. It will provide comprehensive knowledge to the readers and prove to be an essential guide for both academicians and experts.

The information contained in this book is the result of intensive hard work done by researchers in this field. All due efforts have been made to make this book serve as a complete guiding source for students and researchers. The topics in this book have been comprehensively explained to help readers understand the growing trends in the field.

I would like to thank the entire group of writers who made sincere efforts in this book and my family who supported me in my efforts of working on this book. I take this opportunity to thank all those who have been a guiding force throughout my life.

Editor

Smart grid terminology development—crossing the boundaries of terminology standardization

Susanne Arndt[1*], Tatyana Sheveleva[2] and Christoph Goeker[3]

Abstract

Background: Standardization is concerned with ongoing terminology standardization activities. Activities are rather complex in divergent topics and current themes of interest. The article is concerned with terminology standardization activities in Germany and international standardization activities for smart grids and smart-grid-related topics like smart metering systems, smart homes, and electromobility. Even though standardization topics are very clearly organized by standardization road maps, and responsibilities are distributed among working groups, there are still conceptual overlaps between activities of different groups that will result in inconsistencies and ambiguities in their respective glossaries. These glossaries, however, undergo only a limited process of synchronization during their development, especially on the level of single concepts and terms. The application of inconsistent and ambiguous terminology in standards may later on reduce their internal and external consistency, readability, and understandability. To create high-quality standards, conceptual consistency needs to be guaranteed. To do this, terminologies under development should be made more openly available to standardization working groups in the development phase (and not only after completion). Furthermore, additional synchronization tasks on the conceptual level are needed to generate consistent and clear conceptualizations of new technologies.

Methods: A number of glossaries have been analyzed for overlaps as described by this article.

Results: The article gives an overview of issues found in the respective glossaries, and the process can lead to proposals which may be put to vote among domain experts.

Conclusions: Overall, domain experts working on terminologies should be given more assistance as regards terminological and linguistic knowledge and methodology to assure linguistic and terminological next to technical quality of their terminologies. Future work will be dedicated to issue resolution and assistance for issue identification.

Background

Energy transition—challenges and aims

With the emerging trend to create an environmentally friendly future, the need for development and dissemination of innovative environmentally friendly technologies is growing. One of these technologies currently being under intensive development is a fully automated intelligent energy systems based on future *smart grids* [1]. The projected merging of smart grids with grid-related domains such as electric mobility, smart home, and smart metering will give rise to a huge "energy organism". Its development requires a multidisciplinary approach that unifies the knowledge of many different scientific areas like automation engineering, electrical engineering, information technology, automotive engineering, or even architecture.

Successful integration of such a complex technology requires standardization of knowledge. Co-participating players of those different domains need to develop national and international technical standards with the aim of harmonization of construction, production, and use of smart grids. Standardized knowledge, however, is among other factors based on linguistic clarity which can only be achieved when experts try to standardize the smart grid terminology within their groups. To provide

* Correspondence: arndt@iva.ing.tu-bs.de
[1]Institut für Verkehrssicherheit und Automatisierungstechnik, Technische Universität Braunschweig, Hermann-Blenk-Straße 42, 38108 Braunschweig, Germany
Full list of author information is available at the end of the article

an overview about the whole existing smart grid terminology, the technical standardization bodies continually collect these terminological units to use them as lexicological basis for standardization activities aiming at providing technical standards.

From the high amount of different players working on the development of smart grids, it can be assumed that a certain amount of similar terminology already exists. Moreover, because of the high innovation degree of smart grid technology, a high amount of new concepts has and will be introduced by scientific organizations and businesses to standards. Therefore, terminology work of standardization bodies is not a mere collection of terminological elements but the creation of new ones. This implies thorough investigation of the discourse of a domain, its description, and probably even the introduction of completely new concepts into the domain. It is therefore to be expected that terminological problems like synonymy or homonymy will occur, since new developments take place and a lack of coordination leads to different terms for objects and phenomena that are in direct competition for a while until the harmonization of a terminology is conducted [2].

Actual integration of such technologies into ordinary society requires social acceptance, which then again is based on understandability of such technology by society [3]. The development of such understanding results through communication and use of specified vocabulary [4].

A terminological analysis presented in this article takes up the issue of terminological ambiguity and conceptual overlaps in smart-grid-related glossaries from domains like smart home, electric mobility, and smart meter. In conclusion, a suggestion for the improvement of terminological work in standardization groups will be made.

Basic concepts of terminology work

Terminology standardization is defined by DIN 2342:2011-08 as the

> *"standardization of concepts and their terms as well as of concept systems by authorized committees with appropriate domain-specific, linguistic and methodical qualifications aiming at terminological definitions in standards"*
>
> *(translated by the authors from the original quote: "Normung von Begriffen und ihren Benennungen sowie von Begriffssystemen durch autorisierte und dafür fachlich, sprachlich und methodisch qualifizierte Gremien mit dem Ziel, terminologische Festlegungen in Normen zu schaffen" (DIN 2342:2011-08))*

Concepts are mental entities that are achieved by abstraction of real world phenomena and objects based on their similarity according to characteristics. For innovative technologies, this abstraction process is a research task, which should in the end lead to an adequate account of the observed phenomena and objects. Experts of the respective domain should agree on that account when they are not able to falsify it. Only after conceptual standardization, terminological standardization can be achieved. This includes the structuring of concepts within their conceptual context, the concept system.

Conceptual standardization therefore is a prerequisite for the standardization of terms for those concepts. There are at least five aims:

1. Thorough but purpose-driven understanding of relevant concepts
2. Adequate and unambiguous definition of concepts
3. Identification of existing terms for concepts
4. Standardized use of terms
5. Use of understandable, precise, economic terms

The relation of real world phenomena and objects, concepts, terms, and definitions is demonstrated in Fig. 1.

Concepts as mental units of knowledge serve as a means of representation for real world objects. Since they are relatively tied to individual cognitive entities, their standardization must be driven by communication. The use of means of representation is therefore inevitable: concepts are referred to by terms, terms indirectly refer to objects. To standardize and synchronize concepts, they need to be explained in their contexts, which includes the identification of and an agreement on relevant contexts. Such explanations need to make use of further terms to specify the relations between concepts. In terminology standardization, these relations are usually given by definitions and short descriptive texts. The whole process of defining concept-term-relations follows

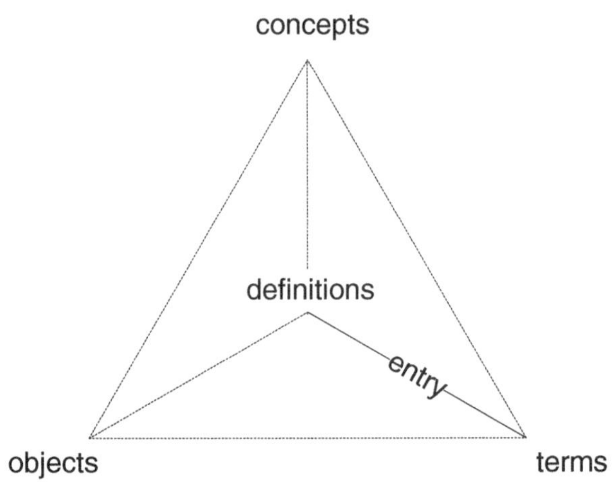

Fig. 1 Extended semiotic triangle based on Suonuuti's [35] focusing on concept representations forming terminological entries

the principle of concept orientation: the concept is the focus point for the structuring of terminological data.

A major problem for terminologies is that terms and concepts are not bijectively mapped onto each other:

> *"Concepts and terms develop differently in individual languages and language communities, depending on professional, technical, scientific, social, economic, linguistic, cultural or other factors. Harmonization is, therefore, desirable because*
>
> - *differences between concepts do not necessarily become apparent at the designation level,*
> - *similarity at the designation level does not necessarily mean that the concepts behind the designations are identical,*
> - *mistakes occur when a single concept is designated by two synonyms which by error are considered to designate two different concepts"* [5]

Mistakes can also occur when two very different concepts are designated by the same term. To avoid these mistakes, one needs to understand terminological standardization as "an integral part of standardization" [5]. A useful terminology should therefore make clear, which relations exist between its terms and concepts. Once these relations are known, concept systems can be made conscious. The uncovering of such relations maps sets of terms to their concepts. This implies the identification of concepts and their relations to each other. This allows for consistent and explicit labelling of designative ambiguities. Such relations between terms and concepts are illustrated in Fig. 2.

When one term represents different concepts, the concept-term-relation is called homonymy (1). Consider, for example, the term *virus* that receives distinct interpretations in medicine or computer sciences. When one term represents several very similar concepts, the concept-term-relation is called polysemy (2), for example, in *pull* as an act of inhaling as compared to *pull* as a deep draught of a drink, where the consumed medium differs. When several terms represent one concept, the concept-term-relation is called synonymy (3). For example, *car* and *automobile* can be considered synonyms.

Without terminological standardization, languages for special purposes cannot be effective tools for their specific purposes since they are likely to be vague and cause misunderstandings during communication. Nevertheless, even though terminological standardization is accepted as a necessity, there are certain boundaries.

The boundaries of terminology standardization

The definition of terminology standardization must be considered an ideal. Standardization bodies often cannot apply common state-of-the-art terminological knowledge and tools, thus they are restricted in their linguistic and methodical means needed for terminology-related tasks. It is, however, especially for motivational reasons, desirable not to bother domain experts with the acquisition of domain-external skills. Instead, trained terminologists should accompany standardization committees so that linguistic and methodic services are experienced as an added value.

Experts often work term-oriented, that is, experts often see their task as collecting a list of terms and defining each of them—disregarding related domains, alternative terms, and deviant usage of the term. Resulting glossaries are thus only marginally concerned with concepts and concept structures but rather with single terms. Conceptual structuring is followed only unsystematically, only when equivalents or acronyms play a role or when domain experts are already aware of ambiguities. The processing of terminological data is therefore often conducted with two-dimensional spreadsheets without measures for concept identification (see "Smart-grid-related glossaries" section).

Further complications are caused when standardization is concerned with converging technologies. As the standardization roadmap of German Commission for Electrical, Electronic and Information Technologies (DKE) and German Association for Electrical,

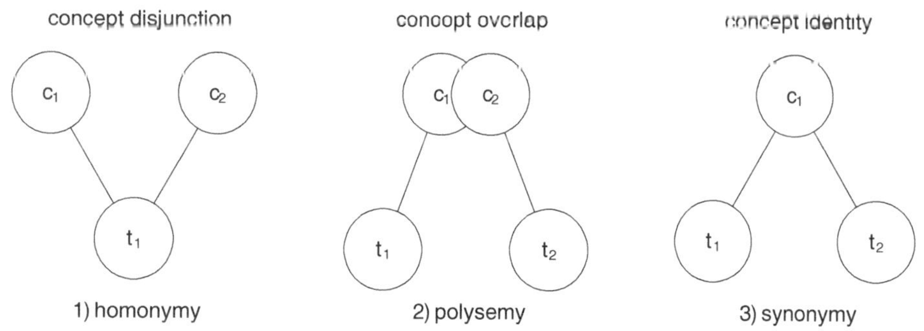

Fig. 2 Concept-term-relations: 1) homonymy, 2) polysemy and 3) synonymy (t = term, c = concept)

Electronic and Information Technologies (VDE) states, in such cases standardization is no longer "business as usual" [6]. Convergence affects standardization in different ways.

First, by making it difficult to determine which knowledge is the object of standardization. Converging technologies rely on innovative concepts and often research activities that try to explore the possibilities and feasibility of such concepts. Research activities, however, are initially characterized by diversity and exploration of different possibilities of bringing innovative concepts to practice. Especially for convergent technologies, results are not achieved by isolated projects so that the concept of R&D phase standardization might not go far enough to bring innovations to standards and preliminary results are introduced to standardization [6] (for the concept of R&D phase standardization see also [7]).

Second, by involving a multitude of different disciplines that make opting for interdisciplinary or even transdisciplinary approaches necessary. This requires experts of different domains to work on the same topic in a joint endeavor and to overcome boundaries of domain-specific knowledge, language, or methodology [6].

Third, by being under pressure from economical or societal parties that have an interest in the fast development of converging technologies, as is the case with smart grid technologies that ultimately serve the goal to prevent climate change and energy bottlenecks by integration of highly decentralized renewable energy sources. Further complication is brought by political efforts of legal regulation [6].

Fourth, by being highly relevant on an international scale, there are standardization activities all around the globe, which result in highly diversified conceptualizations of the smart grid. Furthermore, standardization efforts are not only segmentalized internationally but also locally. Committees are mostly organized hierarchically according to domains, and subjects are divided among them, as is demonstrated in Fig. 3.

Contrarily, it is necessary to consider thematic interrelations that blur the boundaries of committees and to understand the committee boundaries as more fuzzy. This has been done in the standardization roadmaps. The committees highlighted in Fig. 3, for example, have been identified as being commonly responsible for the operational safety of the charging infrastructure of electrical installations [8].

Strict subject segmentalizations, however, already caused problems in classic standardization processes, especially on the conceptual level. Due to the strict hierarchical organization of standardization bodies, interrelations between subjects may be neglected or overlooked and thus result in parallel activities on the conceptual level. Different groups happen to work on the same topic at least partially, since topics are related to each other in micro-aspects. The same referent when viewed by different experts may lead to different, inconsistent mental representations (concepts) to

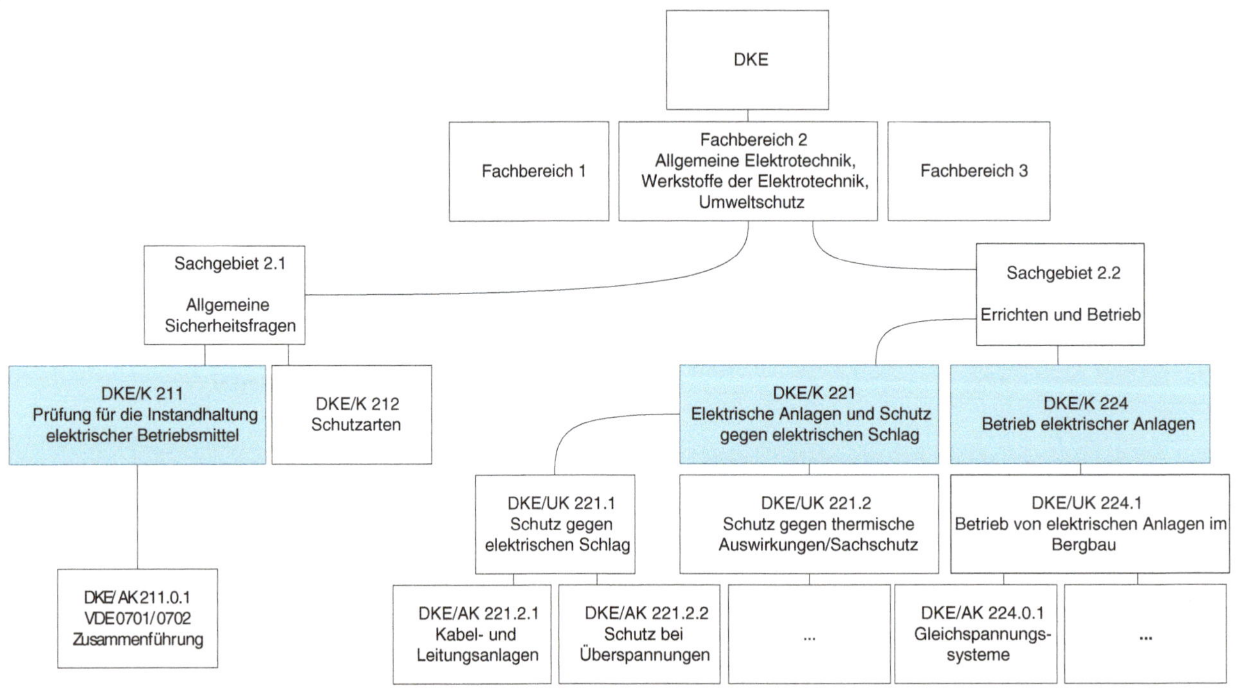

Fig. 3 Hierarchical organization of standardization committees exemplified for DKE (cf. www.iso.org, www.din.de, www.dke.de, www.cencenelec.eu)

emerge, which are differently represented by language, either in definition or terms.

It is thus very common for standardization bodies on all levels (regional, super-regional, international) often to provide an inconsistent, ambiguous set of concept, terms, and definitions. This is especially true for subjects that are relevant in a wide range of domains (for example, *safety* and *security* [9]). The full variety of definitions can be experienced by using the apposite databases, for example, the ISO Online Browsing Platform (OBP) [10], DIN-TERMinology Portal [11], or the International Electrotechnical Vocabulary (IEV, IEC 60050), as provided by [12].

Inconsistencies of concepts, concept systems, and their terms will ultimately effect standards and related standards as compared to each other. It is therefore vital to create consistent concept systems in accordance with existing standardization activities and existing standards (cf. "Consequences of conceptual inconsistencies" section).

What is needed are means of mediation between standardization bodies that ensure real-time synchronization for convergent technologies, innovative concepts, and R&D phase standardization. Synchronization tasks need to operate on a very fine-grained level while at the same time needing to remain an overview of the broader topics treated in the subject fields. As has been recognized for standardization of smart grids in general, steering groups for "inter-domain cooperation and coordination" are needed to avoid unnecessary efforts [6]. This should also be the case for efforts focusing on (or even only including) terminological issues. A respective coordination group could alleviate both problems by adding the relevant linguistic-methodical knowledge while at the same time having the overview over standardization efforts in different subordinate committees. Such groups could fulfill tasks of data governance, further disputes for settlement of concept- or term-related conflicts as well as assure formal or content-related data quality.

Consequences of conceptual inconsistencies

Terminology ambiguities are multidimensional and can bear serious consequences either for standardization organizations or for economic market and, in particular, for producing companies that develop and bring to the market their new technologies. For standardization bodies, it would cause a decreased product quality: ambiguous terminology will lead to ambiguous standards. In the area of economic market, it could have a negative impact on every step of the technology or innovation process. To present terminology-related problems with economic significance needs a rough division of this sector into its internal and external areas. Corporate internal area includes all activities needed for successful realization of a technology or innovation process: concept developing, product planning, system design, detail developing, product testing, production, and market launch. Activities of an external corporate sector include, by contrast, communication with business partners and customers, product marketing, distribution, logistic, delivering, and maintenance.

Internal corporate sector

In this sector, terminological ambiguities and duplicates can be caused by the division of labor between corporate departments. Due to company size, such differentiations lead to difficulties in managing currently ongoing company tasks resulting in an enormous lack of clarity. As a result, several different departments may simultaneously work on the same terminology without coordination with other departments [13]. Accordingly, many synonyms or homonyms will be introduced that, when added into central archives of internal technical documentation, will fossilize and the overall structure will be inconsistent. Such inconsistencies stay mostly unnoticed, until they cause significant loss events. Such inconsistencies can appear at every stage of a technology or innovation process and in every corporate department [14]. With regard to a development department, they could be the reason for undesirable developments. Cooperation of one department with production and manufacturing would cause terminological errors to spread which may be followed by manufacturing errors concerning the relevant products or technologies.

In such departments such as marketing or technical documentation that are responsible for development, creation, and subsequent dissemination of the technical documentation, terminological ambiguity could be the reason for inefficiency by management of documentation files: the ambiguities would be adopted into terminological databases, then be provided to technical editors for development of instructions for use, operating instructions or technical handbooks. It would reduce the efficiency of editor systems enormously and the quality of the text editing. According to the online survey of the Gesellschaft für Technische Dokumentation - tekom Deutschland e.V.—tekom e.V.—one to two thirds of all documentation errors are terminological errors [15].

Terminological ambiguity is also the most common cause of the reduction of work efficiency in sales departments. Managing of the synonymous or homonymous terminology could cause double warehousing or shortages as well as many wrong deliveries that dissatisfy customers and as a consequence impair corporate reputation [16].

The quality of internal corporate communication (which is conducted by use of company internal networks like intranet [17]) could also be influenced through inconsistent terminology. Search queries will be less resultful due to terminological variation that lessens the effectiveness of indexation. As a consequence, search queries deliver insufficient search results or wrongly select the requested information.

In the course of today's globalization, many corporations need to create multilingual intranet sites which provide contents from intranet sites in the language of the head office to corporate staff abroad. This should ensure that subject issues are communicated and understood professionally and in absolute equivalence in all corporate locations. As a consequence, the joint corporate internal knowledge network will be created which simplifies work-related cooperation.

The creation of terminologically well-managed corporate knowledge takes also place by further staff trainings in the form of specialized foreign language courses, because knowledge of foreign languages is one of the most important qualifications of every employee. Such education measures can bring poor results when learners use inconsistent specialized monolingual or bilingual dictionaries or glossaries which normally are the main terminological references for every language learner. In monolingual dictionaries/glossaries, such results are mostly caused by incorrectly determined meaning differences between synonymous, homonymous, or polysemous entries, which lead to difficulties of comprehension by dictionary users: by looking up a word, the user would not recognize the semantic side of the related entries and therefore would not be able to detect the meaning differences between them. This will result in the inability of the dictionary user to detect a correct usage environment (context) of the relevant synonyms and homonyms and misuse of terms in documentation and specifications. To solve this problem, other reference works like common word combination dictionaries need to be consulted. This, however, results in loss of efficiency and higher expenditure.

The fact that such monolingual dictionaries are mostly used by foreign language speakers makes important the detection of the exact semantic differences between synonyms and homonyms. As is known, many foreign language learners do not have the language intuition concerning semantic correctness about term combinatorics. Therefore, a wrong or inadequate information in such dictionaries could lead to incorrect term combinations, and memorizing could automatically lead to repeated misuse in the future.

Reaching terminological precision in bilingual dictionaries is essential for robust equivalence between terms of source and target languages. Every incorrect semantic relation between single synonyms and homonyms increases the probability to find incorrect equivalents in both languages. As a result of the fact that all terminological entries appear in such dictionaries without definitions, dictionary users cannot verify whether the semantic correctness of the foreign equivalence or the correctness of their usage environment is correct [18]. As a dictionary user, one thereby depends on term equivalence listed in the respective dictionary.

External corporate sector

The inconsistencies in source languages are the main causes of translation errors, like conceptual generalizations, specializations, alienation, or adaptation [19]. Therefore, one of the major communication problems caused by inconsistent terminology management occurs when translation services from a cooperative translation service provider must be requested [19].

False evaluation of the degree of synonymy in any source language can lead to limitations of term exchangeability and stylistic variability. Furthermore, partial synonyms may require different equivalent terms due to slight differences in meaning. Only when this fact is considered, valid equivalents can be found. Assuming absolute meaning equality between partial synonyms will lead to unreflecting use of equivalent terms and incorrect translations [20]. For example, a source text will describe some matter with a specific term while the target text uses a more general term. The semantic differentiations of the source text will be lost. Translations could also be made difficult by existence of homonyms in the source language. Allocation of many concepts to one term may prevent adequate translation because a translator may make the error of picking the wrong equivalent [15].

Regarding the translation process, inconsistent terminology will cause defects in the product as well as enormous time and cost losses. Increased needs of clarification and correction slow down decision-making and delay completion of translation assignments. Furthermore, it leads to defective terminology localization which at least makes necessary to repeat clarification situations permanently [21]. Localization is the process by which products and services are adapted to local peculiarities. This includes product documentation and its terminology. Localization of terminology encompasses cultural and technical assessment as well as linguistic and functional assurance. It should be presented in internal corporate glossaries and have ensured equivalence in foreign languages [22]. Using located terminologies by translation companies can therefore ensure a high-quality translation result.

Further defects can affect customer relationships established through sales or use of products or technologies. The main knowledge provider about operating instructions for products is technical documentation like operating instructions, manuals, etc. In such instructions, clear language and terminological consistency are the important feature for customer satisfaction: construction, use, and support for products heavily rely on clear communication. Product quality will be affected by defective corporate communication, and this again will affect the customers' disposition to identify with the brand.

In conformity with the German Civil Code and the German Product Liability Act, an instruction for use is an essential part of every product delivery and its creation is given the same importance and diligence like the remaining product components [23]. Based on this instruction liability, every instruction for use has to involve complete user instructions for every intended use, complete references to dangers, (residual) risks as well as foreseeable misuses of the relative products to ensure protection, and safety and health of the user [23]. "The right word at the right place at the right time" is, according to the European Commission, the condition for user safety [24]. The *right word* covers a clear and easy-to-understand language. Incorrect product use caused by terminological misunderstandings is classified as product or instruction defects which can lead a user to life-threatening situations. Fines may be imposed on manufacturers due to product liability, and prohibitions of marketing products may follow [13].

Methods

Smart-grid-related glossaries

Several glossaries currently under development in German standardization or taken from legal texts and authorities have been considered to assess whether these glossaries show violations of concept-orientation and whether they have conceptual overlap. The glossaries have mainly been chosen according to the criteria of relevance and availability. It must be taken into account that the presented terms and definitions are work in progress and may have changed during the course of the analysis. The glossaries taken into account have been provided by several sources relevant for the development of the smart grid but they are not exhaustive. Further glossaries are available but not considered here. The following glossaries have been considered:

- Glossaries of the German Bundestag as manifested by the German Renewable Energy Act [25] and the Energy Industry Act [26] [Bundestag][1]
- Glossaries from the Metering System 2020 of the German Association for Electrical, Electronic and Information Technologies (VDE) [Meter]
- The smart grid glossary being developed by the working group 111.0.5 of the German Commission for Electrical, Electronic and Information Technologies (DKE) of DIN and VDE [DKE GAK 111.0.5]
- Glossaries from the German Standardization Roadmap for Electromobility by DKE [Electromobility]
- The glossary of working group DKE/GAK 914.0.3 functional safety of electrical/ electronic/ programmable electronic safety-related systems (E, E, PES) for the protection of people and the environment [GK 914] [27]
- Glossary of the open metering system group [28] [OMS]
- Glossaries from the DKE activities focused on smart home and building [SmartHomes]

The glossaries have been gathered from online sources and by requests to DKE. Groups then provided their glossaries for further processing. As mentioned above, concept orientation is a rather marginal issue for standardization bodies, so that terminological data are often presented with only a minimum of term-related or concept-identifying information:

1. Term
2. Definition or other short describing texts
3. Source
4. Alternative terms (equivalen(s), acronyms, other short forms, orthographical and non-orthographical variants, synonyms)
   - Equivalent(s)
   - Acronyms
   - Other short forms
   - Orthographical and non-orthographical variants
   - Synonyms
5. Hyperonyms

Consider as an example the excerpts taken directly from the glossaries in Tables 1, 2, and 3.

Tables 1, 2, and 3 illustrate the diversity of the glossaries that are work products of standardization groups. Even though Table 1 shows a rather sophisticated approach in terminology data management, the spread sheet shows vagueness and inconsistencies. For example, it shows the vague field "alternative entries" which is used for acronyms, full forms, and translations. Table 2 shows only an English term that is not only accompanied by an English but also by a German definition with a German equivalent missing. Table 3 shows that brackets are used inconsistently to establish relationships between entries. The explanation in row 1 could imply a relationship of synonymy while the definition in line 2 suggests a relationship of hyperonymy. The same holds for cells 1 and 2 in row 1. Managing relationships between terms and concepts is a matter of ontology engineering or terminological ontology engineering (e.g., [29, 30, 31]).

The examples make apparent that there are violations as regards the ideal terminology standardization process as defined in [32]. Here, a strict procedure is stipulated that is outlined by three most basic tasks. A substantial

Table 1 Glossary entries taken from [SmartHomes]

Entry (ID)	Type	Language	Definition	Alternative entries
Plug-in electric vehicles	Term –	En	A vehicle with an electric drive (as only drive or in combination with a fuel engine) and a battery which can be charged at a charging station	PEV
Consumer	Term n	En	End user of electricity, gas, water, or heat. The consumer can also generate energy using a distributed energy resource	Letztverbraucher

methodological decision in this process is to work out concept systems separately for each language to be standardized. Only when all terms of all languages are structured conceptually, concept systems will be compared to find equivalents on the conceptual level. This means that "nach Möglichkeit die nationalen Systeme, die verschiedenen Organisationen, die verschiedenen Denkschulen usw. zu berücksichtigen sind"[2] [32]. Practical terminology standardization, however, is lagging behind this methodology. The most obvious reasons being lack of time, lack of familiarity and "practice", and a stronger focus on the whole standard to be worked out. Here, not even concepts of one language are structured systematically. The transfer to other languages is then not characterized by comparison of concept systems but by the translation of single terms without a closer look at their context.

Nevertheless, in our analysis, all terms defined by the glossaries have been processed in the way they have been provided by the committees. They have been managed in a terminology management system as term-oriented entries (short: *entries*) that are accompanied by their additional data (definitions, sources, relations, status). This structure has been chosen to keep the autonomy of all glossaries and to process them descriptively while at the same time using them to prepare concept identification. Where given or identifiable, entries with alternative terms have been assigned to the (alleged, not yet confirmed) primary entry and each other by concept-term-relations. Here, we make use of additional relations that imply that the related entries have the same meaning but different terms and are therefore synonyms. These

Table 2 Glossary entries taken from [OMS-Group] [28]

Term	Description, English	Description, German
Authentication	Ensures the integrity of the received data and proves the declared origin	Prüft Integrität (Richtigkeit) und Ursprung der Daten

Table 3 Glossary entries taken from [Meter]

Begriff	Erklärung	Quelle
Applikationsdaten (Lokale Daten)	Die TR 03109 verwendet in einigen Textpassagen den Begriff Applikationsdaten. Aus Sicht des FNN-Lastenhefts sind dies lokale Daten, die in einem SMGw zum Zwecke des Betriebes vorgehalten werden müssen	Lastenheft SMGw (Gateway, Funktionale Merkmale)
Tarifumschaltanweisung (Steuersignal)	Ein vom SMGW-Admin oder von einem autorisierten CLS übermittelte Anweisung für dieSteuerung von Tari fumschaltungen im SMGW	Technische Richtlinie BSI TR-0310943

relation types can accordingly be classified as synonymic but differ from general synonymy by giving additional information on the term or other parts of the related entries. These synonymic relations entail the following:

- Abbreviation: one entry is related to an entry with an abbreviated term for the concept (hasAbbreviation)
- Rejection: one entry is related to an entry with a synonymous but rejected term for the concept (hasRejected)
- Preference: one entry is related to an entry with a synonymous and preferred term for the concept (hasPreferred)
- Equivalence: one entry is an equivalent entry in another language (hasTranslation)
- Phrasal equivalence: one entry is related to an entry with an alternative but equivalent definition for the concept (hasEquivalentDefinition)

An example of such related entries is given in Fig. 4.

Figure 4 shows entries that have been related with each other according to the source glossaries. The relation types used are the synonymic relation types described above.

The population of these glossaries is as shown in Fig. 5.

Identification of violations of concept-orientation and conceptual overlaps

The investigated glossaries lack conceptual structuring so that the following procedure has been applied to identify violations and conceptual overlap:

Task 1: Term-duplicate analysis

1. Task 1.1: Glossary-internal term duplicate analysis
   (a) Identifies occurrences of term duplications
   (b) Gives simple criteria for quick classification of term duplications for the purpose of harmonization and merging of entries

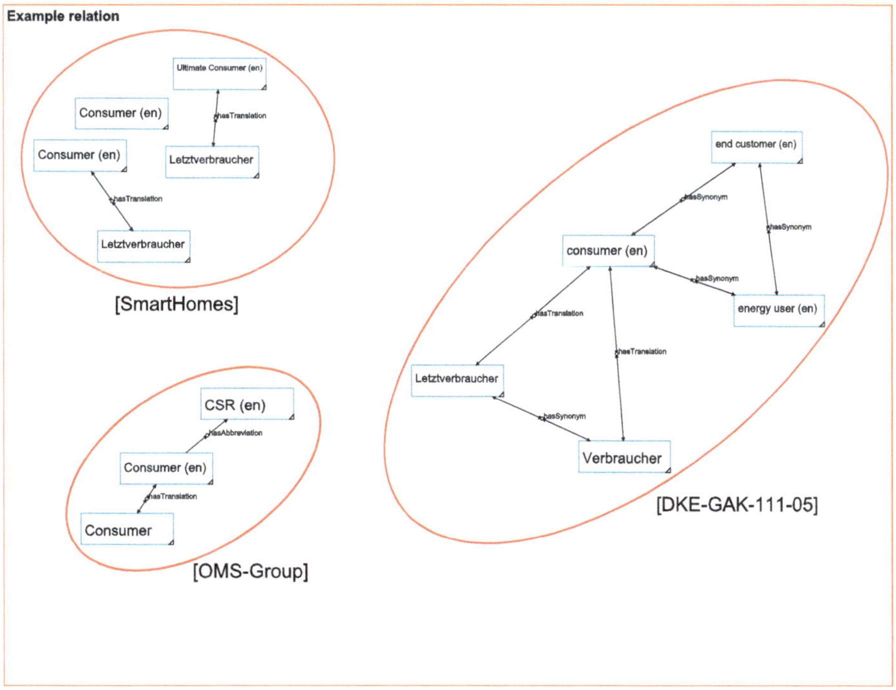

Fig. 4 Example relations

(c) Defines resulting actions for merging and harmonization
(d) Classifies term duplications and identifies resulting actions
(e) Makes recommendations for further actions to be performed by domain experts

2. Task 1.2: Glossary-extending term-duplicate analysis

Here, the same steps apply as in task 1.1.

Task 2: Synonym analysis

1. Task 2.1: Computer-aided synonym analysis
   (a) Definition-duplicate analysis: identify definition duplicates
   (b) Relation analysis: identify relevant relation types by analyzing metadata given in the source glossaries
   (c) Harmonization and merging of entries
2. Task 2.2: Manual synonym analysis
   (a) Pre-structuring of glossaries
   (b) Concept comparison

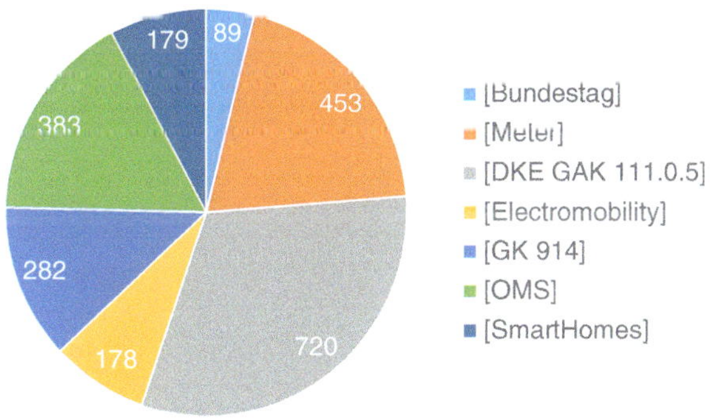

Fig. 5 Number of entries per glossary

Results and discussion

In the glossary analyses, the entries have been filtered and considered due to different criteria. The first entries to be analyzed were grouped by term (term-based entry groups in task 1). These entries varied in definition, source, or relations but had the same term. In the second analysis, entries were analyzed that were grouped by definition or by synonymic relations (task 2).

Task 1: term duplicate analysis

Task 1.1: summary of glossary-internal term duplicate analysis

Each glossary has been analyzed individually to identify glossary-internal occurrences of term duplication. This means that all entries labeled by the same term (= term-based entry groups) will be considered in the following descriptions. Yet undefined entries have been counted as well. The number of identified occurrences of term duplication for each glossary is listed in Fig. 6, thus representing an excerpt of the whole glossaries.

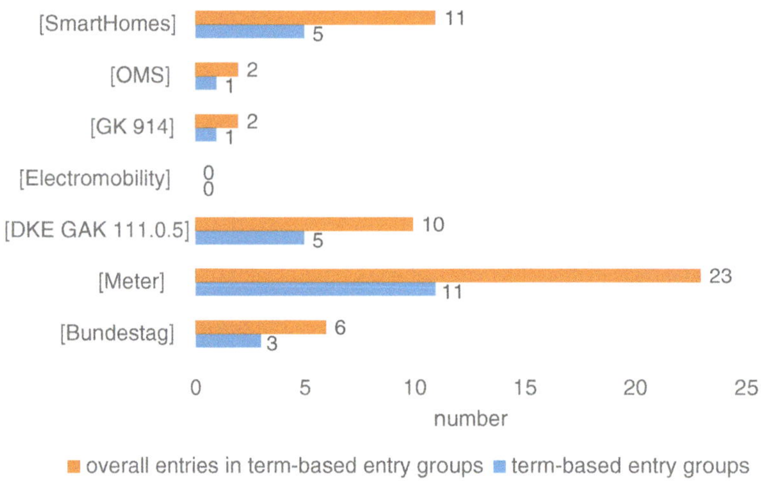

Fig. 6 Number of glossary-internal occurrences of term duplication

The total number of term-based entry groups is 26, with 54 entries involved. The terms under consideration here are 15 % acronyms and 85 % full forms.

The implications of the following analysis will be that the entries in term-based entry groups can be classified as either non-identical as regards their concepts (homonyms, polysemes) or as identical.

Figure 7 shows the number of entries in a term-based entry group for four examples with a full form term in the gray column. Next to it is the number of definitions in the orange column and the number of identical definitions in the blue column.

Table 4 gives several conditions for categorization of term duplications shown in Fig. 7 and some necessary actions that derive.

According to these criteria, every glossary entry can be classified for some necessary action that is a recommendation to domain experts. The plan of actions demonstrated in Table 5 results for each term-based entry group.

The identified occurrences of duplication for abbreviated terms, considered by the abovementioned glossaries and working groups, are shown by several examples in Fig. 8. The blue column here shows the number of entries that are represented by the abbreviation given as the descriptor, the orange column says how many definitions these entries carry, the gray column shows whether these definitions are identical. The yellow and green columns refer to a different set of entries, i.e., those that are represented by the full term the abbreviated term stands for. The yellow column shows the number of all entries related to the abbreviated

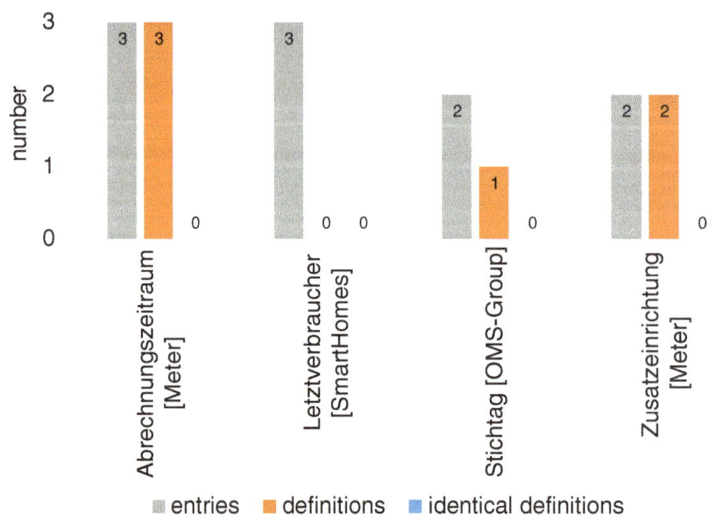

Fig. 7 Glossary-internal duplication of a selection of full form terms in smart-grid-related glossaries: opposition of number of entries and number of definitions of these entries

Table 4 Conditions for the identification of necessary further actions for glossary-internal terminology standardization (full form terms)

ID	Condition	Action
1	There is no definition within the respective glossary (e.g., *Letztverbraucher*)	An adequate definition for the glossary must be found, at best from an existing source (including other glossaries under development)
2	There are less definitions than entries in the respective glossary (e.g., *Stichtag*)	a) Duplicates need to be removed from the glossary
		b) Undefined duplicates represent different concepts (homonymy) and need to be defined accordingly
		c) Defined duplicates need to be treated according to action 3
3	There are as many definitions as entries in the glossary (e.g., *Zusatzeinrichtung*)	a) The definitions must be compared to see whether they are disjunct; then the entries are homonyms and distinct preferred terms should be determined where possible to help avoid misunderstandings
		b) The definitions must be compared to see whether they are formally identical; then the entries are conceptually identical and one of the duplicates can be removed
		c) The definitions must be compared to see whether they are semantically identical: then the entries are conceptually identical, the preferred definition must be determined and the duplicates can be removed
		d) The definitions must be compared to see whether they overlap semantically; then, the entries must be marked as polysemes
4	There are at least two definitions in the glossary and one definition refers to the other definition(s) (e.g., *Abrechnungszeitraum*)	The reference implies identity so that the entries can be considered conceptually identical and a merged entry could be created referencing both sources of appearance

entries, while the green column shows the number of full form terms that are used to represent these entries.

We assume here that actions 1 to 4 have been performed beforehand and that the glossary-internal conceptual structure of full form entries is clear. Then, the following conditions can be distinguished and two types of actions can be derived as is shown in Table 6.

The abbreviation entries can be characterized as follows and a plan of action derives in Table 7.

Which action needs to be taken may sometimes be a question of debate. Consider *LAN* and its related full form term *Local Area Network* and the two entries for that term. The related full form entries describe similar concepts that vary in their degree of abstraction:

- Definition of entry 1: *computer network located on a user's premises within a limited geographical area*
- Definition of entry 2: *Data communication network, connecting a limited number of communication devices (Meters and other devices) and covering a moderately sized geographical area within the premises of the consumer. In the context of this PP the term LAN is used as a hypernym for HAN and LMN*

While one of them applies the definition of LANs given by IEC 60050 (IEV 732-01-04), the other is a very specific interpretation for application in smart grids. What needs to be determined is whether both should be represented by the same term and whether they represent the same concept or have a subordination relation. In the latter case, the distinction between subordinate and superordinate concept should be given by the term so that it has the capability of evoking the appropriate context. The same holds for its abbreviation.

The next section will show the overlaps that transcend the boundaries set by one glossary. Therefore, an analysis for term duplicates has also been conducted for all glossaries taken together.

Table 5 Resulting actions for glossary-internal terminology standardization (full form terms)

Term-based entry group	Action #1	Action #2	Action #3	Action #4[a]
Abrechnungszeitraum [Meter]			x	x
Letztverbraucher [SmartHomes]	x			
Stichtag [OMS-Group]		x		
Zusatzeinrichtung [Meter]			x	x

[a]Whether condition 4 can apply to a term-based entry group can easily be detected automatically, whether condition 4 does apply needs to be analyzed manually.

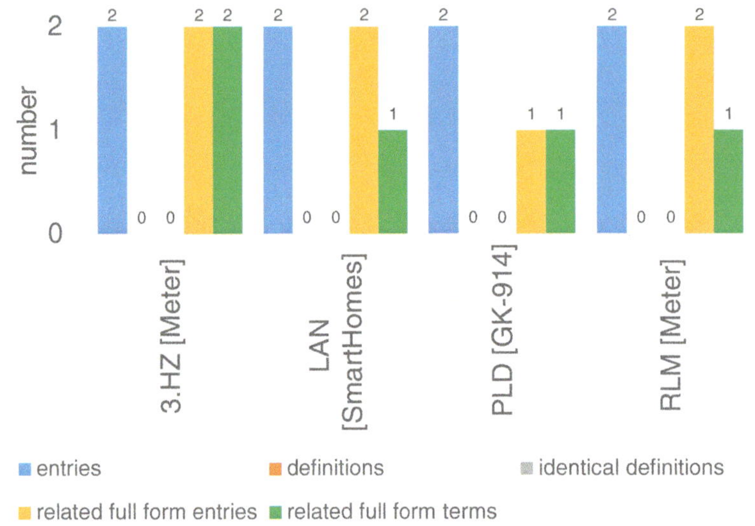

Fig. 8 The differentiation between full form entries and full form terms is made because different entries may be labeled with the same term so that there may be more related entries than terms.[3]

Task 1.2: summary of glossary-extending term duplicate analysis

In this analysis, 119 terms have been identified to occur more than once in several glossaries, representing 259 entries. The terms under consideration here are 36 % acronyms, 61 % full forms, and 3 % mixed forms.

Figure 9 shows an excerpt of the term-based entry groups which have been identified in all glossaries taken together. The number of full form entries is shown in the blue column, next to the number of definitions used in these entries in the orange column, as well as the number of identical definitions in these term-based entry groups in the gray column.

Here again, criteria 1–4 can be distinguished (see "Task 1.1: summary of glossary-internal term duplicate analysis" section) so that the plan of action, shown in Table 8, results by classification of each term-based entry group.

Term duplication has also been identified for abbreviated terms of which again only an excerpt will be presented in this paper. Figure 10 shows the number of entries represented by the same abbreviation in the blue column, the number of definitions involved in these term-based entry groups in the orange column, the number of identical definitions in the gray column, the number of related full form entries in the yellow column, as well as the number of related full form terms in the green column.

Here again, we assume that actions 1–4 have been performed beforehand and that the inter-glossary conceptual structure is clear. Then, the following cases with according types of actions can be derived as shown in Table 9.

The abbreviations listed here are interesting term duplications in several glossaries. The abbreviation entries themselves are already giving formal clues on their conceptual identity:

- Related full form terms are strongly hinting that different concepts are represented and homonymy applies (e.g., ERP for *enterprise resource planning* and *effective radiated power*)
- Related full form terms are orthographical variants so that the abbreviation entries are conceptually identical (e.g., *HES* for *Head End System* and *Head-End System*)

Table 6 Conditions for the identification of necessary further actions for glossary-internal terminology standardization (abbreviations)

ID	Condition	Action
5	The abbreviated forms represent the same concepts	One of the abbreviation entries can be removed from the glossary
6	The abbreviated forms do not represent the same concepts	Both abbreviation entries can be maintained but their relationship needs to be explicitly stated as homonymy

Table 7 Resulting actions for glossary-internal terminology standardization

Term-based entry group	Action #5	Action #6
3.HZ [Meter]		x
RLM [Meter]		x
PLD [GK-914]	x	
LAN [SmartHomes]	(x)	x

Abbreviations: *t.b.d.* to be determined; *(x) & x* one action is more likely than the other; *x & x* both actions apply

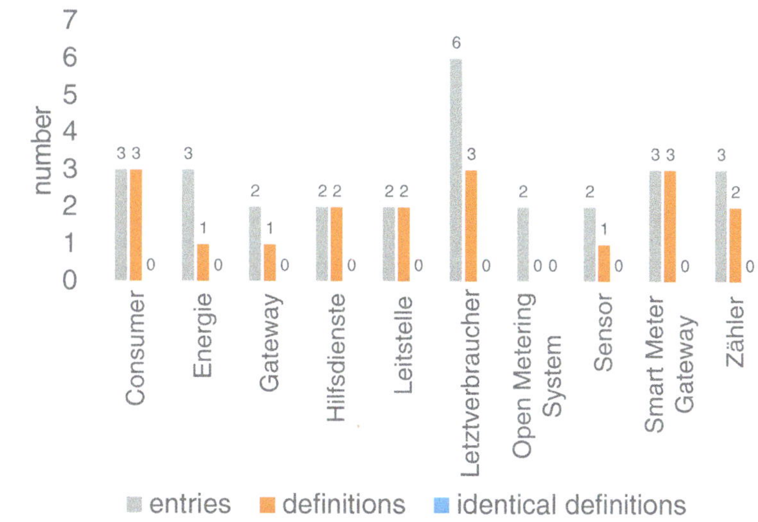

Fig. 9 Glossary-extending duplication of a selection of full form terms in smart-grid-related glossaries: opposition of number of entries and number of definition of these entries (full form terms)

- The full form terms suggest ontological differences that result in slightly different understandings while at the same time there is great formal similarity of the terms which leads to a high potential of misunderstanding (e.g., *DER* for *distributed energy resource* and *Distributed Energy Resources* where different conceptualizations may result from the singular/plural distinction)
- Their contexts or the contexts of their related full form entries may be different so that entries carry different semantic relations to other entries or show different viewpoints in their definitions and have different focuses; this makes it necessary to settle whether these are complementary views or whether they are conflicting (e.g., for *KWK-Anlage*)
- They or their related full form entries may reference different sources so that different conceptualizations are probable (e.g., *OMS* for *open metering system*)

Table 8 Resulting actions for glossary-extending terminology standardization (full form terms)

Term-based entry group	Action #1	Action #2	Action #3	Action #4
Consumer			x	x
Energie		x		
Gateway		x		
Hilfsdienste			x	x
Leitstelle			x	x
Letztverbraucher		x		x
Open metering system	x			
Sensor		x		
Smart meter gateway			x	x
Zähler		x		x

While the performance of task 1 primarily serves to identify homonymy of terms, it will also lead to the identification of conceptually identical entries. However, not all instances of conceptual identity can be identified by task 1. Therefore, task 2.1 and task 2.2 need to be performed additionally.

Task 2: synonymy identification in smart-grid-related glossaries: glossary-internal and glossary-extending analysis

Task 2.1: computer-aided synonymy identification

Synonyms have been identified by comparing two data categories of the glossaries:

1. The text of their definitions
2. Semantic relations of certain types (based on metadata analysis in the original glossaries, e.g., brackets, additional columns, and definition comparison)

In total 255 synonym entries have been identified (cf. Fig. 11).

For inter-glossary comparison based on relations, only those entries have been considered where the relation's subject and object are located in distinct glossaries. Since the glossaries so far have been tended separately, there are no inter-glossary relations. Furthermore, the glossaries are concerned with new terminology: although a recourse to existing terminologies could be possible to integrate known terms, it is not likely that these are the first documented terms in the different glossaries. Hence, identical definitions are not to be expected in the current stage of the glossaries. In summary, there are no inter-glossary synonyms.

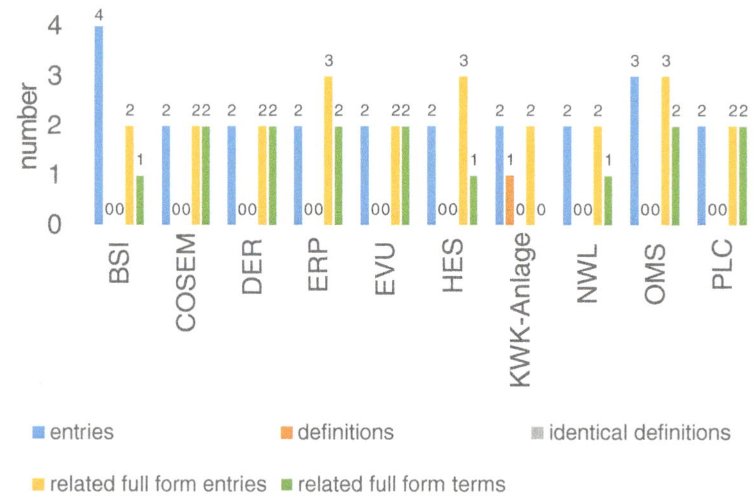

Fig. 10 Glossary-extending duplicates of abbreviations in smart-grid-related glossaries: opposition of number of full form entries, definitions, identical definitions, and related full form entries and related full form terms

For glossary-internal comparison, the occurrence of identical definitions and glossary-internal relations is much higher. Standardization groups are well aware of the non-bijective relationship of terms and concepts but do not explicitly manage these relations in their glossaries. The relations that helped to identify the synonyms are therefore based on spreadsheet data (see "Smart-grid-related glossaries" section). The explicit marking of synonymies should be conducted.

Table 10 shows a selection of synonym sets that have been identified by comparison of definition.

Table 11 shows a selection of synonym sets that have been identified by synonymic relations.

To be normative, synonyms need to be classified according to their permission for use: are they preferred, deprecated, or permitted? When several abbreviations are synonyms, then the abbreviation of the preferred related full form term should be chosen as preferred abbreviation. Analogically, abbreviations of deprecated full form terms should be deprecated as well.

Table 9 Resulting actions for glossary-internal terminology standardization

Term-based entry group	Action #5	Action #6
BSI		x
COSEM	x	(x)
DER	x	x
ERP	x	x
EVU		x
HES	x	(x)
KWK-Anlage	t.b.d.	t.b.d
NWL	t.b.d.	t.b.d.
OMS	x	x
PLC	x	(x)

Abbreviations: *t.b.d.* to be determined; *(x) & x* one action is more likely than the other; *x & x* both actions apply

Task 2.1 does not identify all cases of synonymy since it is only based on relations and formal identity of definitions. When definitions are not identical or entries are not related, there will be no findings, which is why task 2.2 needs to be applied as well. This, however, requires thorough study of sources.

Task 2.2: manual synonymy-identification

Since naturally growing terminologies will contain synonymy, a certain approach to terminology work needs to be followed to identify synonymic relations among entries with identical terms. This approach is for example described by [23, 32, 33]. Systematic terminology work is *corpus*-based: from relevant sources of a domain, term candidates (= primary information) and terms' meanings, uses, grammatical categories etc. (= secondary information) will be drawn. Potential terms are, in this process, supposed to be administered as term-based entries that include the secondary information. Alternative terms should only then be included to such term-based entries when they are abbreviations or orthographical variants. Alleged synonyms should be treated autonomously.

The information taken from the *corpus* is then instrumentalized by the terminologist in order to aid the reconstruction of the domain's conceptual system. Information on the terms' meanings (definition-like information on the characteristics of a concept) will be considered to identify the concept-term-relations. Accordingly, term-based entries will be allocated to their common concept and synonymic relations will be established. The comparison of concept systems of different

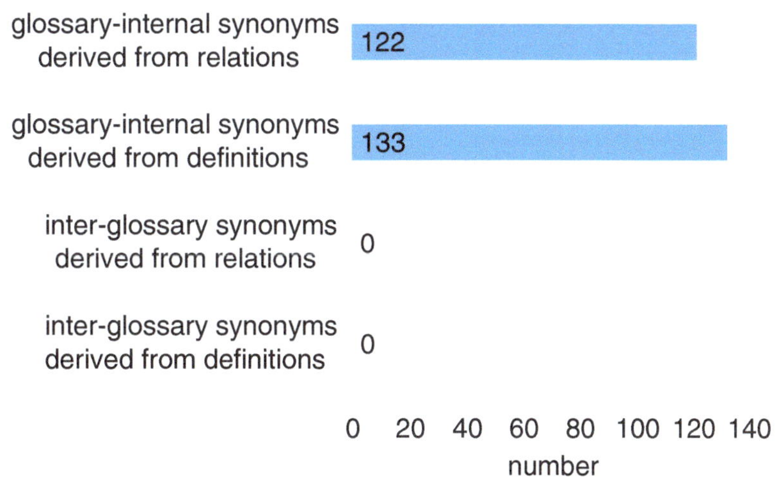

Fig. 11 Overview over synonymous entries in all glossaries

languages should proceed one concept at a time, so that a common structure can be uncovered.

The terminological literature makes very clear that the reconstruction of monolingual concept systems should be complete before the establishment of equivalence relations begun (especially emphasized by [2]). This also requires that each concept of a language must be fully defined. Picht et al. [2] recommends that the identification of synonyms needs to be confirmed by reliable sources, which include oral statements of domain experts. Going one step further, it will even become necessary to clear contradictory information with an expert who is sufficiently authorized to do this, before equivalence between systems is established. The comparison of concept systems with the aim of finding equivalents and probably even the adjustment of concept positions, and therefore systems, is then based on the comparison of definitions that describe characteristics of concepts. Equivalence relations should only be documented when conceptual identity is firmly identified.

Table 10 Selection of synonym entries identified by comparison of definitions given by source glossaries

Set ID	Glossary	Synonyms	Definitions
3	Meter	Steuersignal Tarifumschaltanweisung	Eine vom SMGW-Admin oder von einem autorisierten CLS übermittelte Anweisung für die Steuerung von Tarifumschaltungen im SMGW
4	Elektromobilität	Ladestation Stromversorgungseinrichtung für das Elektrofahrzeug	Eine Ladestation ist ein zum Laden von Elektrofahrzeugen vorgesehenes Betriebsmittel gemäß IEC 61851, das als wesentliche Elemente die Steckvorrichtung, einen Leitungsschutz, eine Fehlerstrom-Schutzeinrichtung (RCD), einen Leistungsschalter sowie eine Sicherheits-Kommunikationseinrichtung (PWM) enthält. Abhängig vom Einsatzort können ggf. noch weitere Funktionseinheiten wie Netzanschluss und Zählung hinzukommen. Die DIN EN 61851-1 (VDE 0122-1) definiert die Begriffe "Wechselstrom-Ladestation für das Elektrofahrzeug" und "Gleichstrom-Ladestation für das Elektrofahrzeug". In der Normungs-Roadmap wird auch der Begriff Ladestation daher als Oberbegriff verwendet und schließt AC-, DC- und das induktive Laden ein
5	GK-914	Application data Application software Configuration data	Part of the software of a programmable electronic system that specifies the functions that perform a task related to the EUC rather than the functioning of and services provided by the programmable device itself
7	OMS-Group	Eichwesen Gesetzliches Messwesen	Messung im geschäftlichen oder amtlichen Verkehr
24	DKE-GAK-111-0-5	Cumulating measuring unit Measuring device for supplied quantities Meter	Instrument intended to measure energy quantities by integrating power with respect to time

Table 11 Selection of synonym entries identified by synonymic relations based on source glossary meta-data

Set ID	Glossary	Synonyms	Definitions
71	DKE-GAK-111-0-5	Anforderungsseitige Führung Demand side management	Anwendungsfallszenario als Prozess, das den Versorger befähigt, in Notsituationen oder im geplanten Betrieb die Anforderungskurve zu managen und die Kunden-Lastkurve zu bestimmen, wobei die Funktionen die Laststeuerung und die Lasterhebung beinhalten Funktion, die den Versorger befähigt, in Notsituationen oder im geplanten Betrieb die Anforderungskurve zu managen und die Kunden-Lastkurve zu bestimmen, wobei die Funktionen die Laststeuerung und die Lasterhebung beinhalten.
84	DKE-GAK-111-0-5	Komponente Technischer Akteur Systemelement Aktor Systemkomponente	Erzeuger von elektrischer Energie oder Wärmeenergie über zentrale Erzeugungsanlagen. verteilte Erzeugunganlagen sowie Mikroerzeugung in den Objektnetzen der Prosumer
102	Meter	Abrechnungszeitraum Abrechnungsturnus	Der Abrechnungszeitraum legt die Start- und Endezeitpunkte des Zeitraums fest, über den eine Abrechnung erstellt wird. Ein SMGW kann diese, vertragliche Größen nicht kennen, da der Endezeitpunkt erst dann bekannt sein wird, wenn ein Letztverbraucher seinen Vertrag kündigt oder ändert. Hingegen kann ein SMGW den Abrechnungsturnus (jährlich, vierteljährlich, monatlich) kennen. Im Sinne der TR-Zielsetzung wird daher angenommen, dass der Abrechnungsturnus gemeint ist, wenn die TR den Begriff "Abrechnungszeitraum" benutzt. Der Abrechnungszeitraum legt die Start- und Endezeitpunkte des Zeitraums fest, über den eine Abrechnung erstellt wird. Ein SMGW kann diese, vertragliche Größen nicht kennen, da der Endezeitpunkt erst dann bekannt sein wird, wenn ein Letztverbraucher seinen Vertrag kündigt oder ändert. Hingegen kann ein SMGW den Abrechnungsturnus (jährlich, vierteljährlich, monatlich) kennen. Im Sinne der TR-Zielsetzung wird daher angenommen, dass der Abrechnungsturnus gemeint ist, wenn die TR den Begriff Abrechnungszeitraum benutzt.

The methodological consequence is that terminology work needs to include some sort of redundancy when it comes to the description of meaning: each term going into the concept comparison needs to contain information elements that serve to identify the concept and help decide whether two terms represent the same concept.

Summing up, the approach taken to reconstruct a concept system is based on information extracted from original domain sources, comparison of this information, descriptions of defining characteristics of concepts and fully formulated definitions. This is a very sound methodology for concept reconstruction that will take into account cultural, historical, and idiosyncratic features of a domain's concept system as well as contradictions and controversies of the domain.

There are, however, several drawbacks with this approach: first, the information extracted from the context may be irrelevant for the identification and description of the concept that is represented by the term. Second, comparability of concepts may be restricted due to several factors: a concept may not be reconstructed properly (underrepresentation), it may be represented from different perspectives, there may be a discrepancy on the amount of information in the term entries that represent the concept, and there may be different arrangements of style and information structure (which would be especially relevant for computer-aided comparison). Third, the *corpus* may not give any explicit hints on synonymic or homonymic relations between terms so that the ambiguities of the *corpus* texts may be transferred to the terminology. Fourth, the term-by-term-comparison is very time-consuming and expensive.

To conclude, the question whether term-based entities are to be related by conceptual identity is not easily decidable. For terminology standardization, efforts of pairwise comparison should be lessened, either methodically or automatically to make the whole procedure more economically feasible.

A feasible way to enhance comparability of (preliminary) definitions is the application of standards for information structuring. Information units needed to identify conceptual identity of terms can thus be organized according to common principles. A common standard for definitions is DIN 2342:2011-08, which gives recommendations on how to write definitions. A common template would enhance machine comparability of definitions. This, however, would require terminological rigor also in definitions. Other possibilities of information comparison could be provided by natural language processing techniques, identifying common semantic structures in definitions, preliminary definitions and secondary information extracted from documents. Methodologically, a pre-structuring of the terminologies may help reduce the number of comparison pairs for which formal or semantic criteria could be applied. The question as to what kind of criteria (e.g., compositional structure of terms) should be applied and how to identify them for a specific domain is, however, left to future work.

Task 2.1 and task 2.2 give proposals for mergeable entries. Experts need to agree or disagree with these proposals, and in case of disagreement, they need to start a process of clearing and settlement between working groups, not just within their own working group. When this has been done, common entries should be merged from the ones existing in working-group-specific glossaries. The end of the process would be a common terminological resource during the process of its development.

Conclusions

This article has shown how terminology standardization for smart grids produces overlapping glossaries that may contain different concept systems and contradictory definitions. A reason for this can be seen in the practical impossibility for domain experts to reach a high level of methodological practice as regards terminology management. Furthermore, the situation is complicated by the fact that standardization committees may even start out with preliminary terminology that is prone to changes and may consolidate outside the committees. Despite ongoing efforts to deal with convergence in standardization and coordination of standardization bodies by expertise centers and steering committees, the measures undertaken are not fully effective on the terminological level. There are overlaps between the glossaries that are in need of inter-committee harmonization.

To reach truly standardized terminology resources, it is necessary to include terminological experts into working groups and to reach an overall gain in efficiency for the task of identifying concept-term-relations of heterogeneous domain-specific sources and glossaries from standardization bodies. There should be better assistance for domain experts that participate in standardization and the task of glossary data management. Data governance mechanisms for standardized terminologies and terminologies currently being standardized could be helpful, probably in the form of a coordination group trained in linguistics and terminology methods.

The article shows a structured way of identifying common terminological problems like homonymy and synonymy and a fine-grained method for treating these phenomena on the concept and term level, where they appear. The method leads to proposals that can be put to vote among domain experts. This is exemplified by data from smart-grid-related glossaries currently under development. The identification of glossary elements that represent the same concept in several of the glossaries under development is necessary to overcome the boundaries of single working groups. When identified, those concepts should be treated as a common resource of the whole domain which makes it necessary to provide it as such to the interested parties. The single entries of the glossaries could be—after harmonization—merged to a single resource that is based on wider consensus among working groups during the process of development.

For the purpose of the analysis, all glossaries have been transferred to a common platform, a terminology management system prototype [34] of TU Braunschweig. This or a like common web-platform could be used for further development and synchronization of the glossaries, as well as for definitions of conceptual systems with explicit concept relations. Respective tools for published terminologies are already in use, e.g., ISO Online Browsing Platform (OBP) [10], DIN-TERMinology Portal [11], or the International Electrotechnical Vocabulary (IEV, IEC 60050), as provided by [12], which are the most valuable resources for terminology development. In our approach to bring together different glossaries under development, we furthermore started a process of establishing ontologically structured systems, which has most prominently been adopted in DKE/GAK 111.0.5. In such systems, inconsistencies can be more easily detected and common areas of definition activities can be detected. The resulting data can furthermore be brought into semantic applications, e.g., for information retrieval.

Endnotes

[1]In the following, all glossaries will be referenced by short reference given in square brackets.

[2]Translation by the authors: if possible, national systems, different organizations and different schools and practices etc. need to be considered.

Competing interests

The authors declare that they have no competing interests.

Authors' contributions

SA conducted the analysis of the glossary data and devised the structure of the paper and is responsible for the final draft. TS conducted the literature review, drafted the backgrounds section of the paper, and gave critical reviews on the analysis sections of the draft. CG conducted a review of the complete paper for English language, contents, and comprehensibility of the paper. All authors read and approved the final manuscript.

Authors' information

SA is a research associate at the Institute for Traffic Safety and Automation Engineering of TU Braunschweig. Her research topics are located in special language communication: (1) terminology in requirements engineering and standardization, (2) terminologies as ontological structures, and (3) term formation and intuition.

TS is a research associate at Germany's National Metrology Institute, Physikalisch-Technische Bundesanstalt in Braunschweig. Her research topics are language for special purposes: understandability of words and texts, language use in technical contexts.

CG is a research associate and postgraduate at the working group Networks and Distributed Systems at Faculty of Technology at Bielefeld University. His research topics are terminological requirements in standardization and system safety and security.

Acknowledgements
This paper has been made possible by funding from the Federal Republic of Germany for the SmartTerms project. The funding body is the Federal Ministry for Economic Affairs and Energy (BMWi). A critical review for the paper has been given by Dr.-Ing. Uwe Becker of TU Braunschweig, Institute for Traffic Safety and Automation Engineering. A part of the analyzed glossaries has been provided by working groups from the Deutsche Kommission Elektrotechnik Elektronik Informationstechnik in DIN und VDE, Verband der Elektrotechnik Elektronik Informationstechnik e.V. or have been taken from publicly available Federal Publications or publications from OMS-Group. We very much want to thank them for giving insights to preliminary working products.

Author details
[1]Institut für Verkehrssicherheit und Automatisierungstechnik, Technische Universität Braunschweig, Hermann-Blenk-Straße 42, 38108 Braunschweig, Germany. [2]Physikalisch-Technische Bundesanstalt, Bundesallee 100, 38116 Braunschweig, Germany. [3]AG Rechnernetze und verteilte Systeme, Universität Bielefeld, Universitätsstraße 25, 33615 Bielefeld, Germany.

References

1. Deutsche Kommission Elektrotechnik Elektronik Informationstechnik im DIN und VDE (DKE) (2013) Die Deutsche Normungs-Roadmap Smart Home + Building: Status, Trends und Perspektiven des Smart Home + Building-Normung., https://www.dke.de/de/std/informationssicherheit/documents/nr_smart%20home_de_version%201.0.pdf. Accessed 10 Feb 2015.
2. Picht H, Arntz R, Schmitz K-D (2014) Einführung in die Terminologiearbeit. Olms, Hildesheim
3. Sucharowski W (2009) Die Normierbarkeit der Kommunikation. In: Henn-Memmesheimer B, Franz J (eds) Die Ordnung des Standard und die Differenzierung der Diskurse: Akten des 41. Linguistischen Kolloquiums in Mannheim 2006, Teil 1. Peter Lang, Frankfurt am Main
4. Wendt S (1997) Terminus – Thesaurus – Text: Theorie und Praxis von Fachbegriffssystemen und ihrer Repräsentation in Fachtexten. Narr, Tübingen" 1204-1205 - change "Das neue produktsicherheitsgesetz (prodSG): leitfaden fur hersteller
5. International Standardization Organization (2007) ISO 860:2007: Terminology work – harmonization of concepts and terms. ISO, Geneva
6. Deutsche Kommission Elektrotechnik Elektronik Informationstechnik im DIN und VDE (DKE) (2013) The German roadmap E-energy/smart grid 2.0., German Version: https://www.dke.de/de/std/aal/documents/nr_energy%20smart%20grid_de_version%202.0.pdf; English Version: https://www.dke.de/de/std/excellenceclustersmartenergy/aktivit%C3%A4ten/documents/nr_eenergysmart%20grid_en_version_2.0.pdf. Accessed 16 Oct 2014
7. Deutsches Institut für Normung (n/a) DIN-leaflet for R&D phase standardization. http://www.ebn.din.de/sixcms_upload/media/2929/EBN-Broschuere.pdf. Accessed 01 Dec 2014
8. Deutsche Kommission Elektrotechnik Elektronik Informationstechnik im DIN und VDE (DKE) (2013) The German standardization roadmap for electromobility—version 2.0A., German Version: http://www.dke.de/de/std/aal/documents/nr_elektromobilit%C3%A4t_de_version%202.0a.pdf, English Version: http://www.dke.de/de/std/aal/documents/nr_elektromobilit%C3%A4t_en_version%202.0a.pdf. Accessed on 26 Oct 2014
9. Piètre-Cambacédès L, Chaudet C (2010) The SEMA referential framework: avoiding ambiguities in the terms "security" and "safety". Int J Crit Infrastruct Prot 3(2):55–66. doi:10.1016/j.ijcip.2010.06.003
10. Online Browsing Platform. [www.iso.org/obp]
11. DIN-TERMinology Portal. [www.din-term.din.de]
12. International Electrotechnical Vocabulary. [www.electropedia.org]
13. Loetzer M, Buck P, Schwabediessen A (2013) Rechtskonformes Inverkehrbringen von Produkten: In 10 Schritten zur Konformitätserklärung. Beuth, Berlin u. a
14. Kalogerakis K (2010) Innovative Analogien in der Praxis der Produktentwicklung. Gabler, Wiesbaden
15. Schmitz K-D, Straub D (2010) Erfolgreiches Terminologiemanagement im Unternehmen. Grundlagen, Umsetzung, Kosten-Nutzen-Analyse, Systemübersicht. TC and more GmbH, Stuttgart, Praxishilfe und Leitfaden
16. Budin G (1996) Wissensorganisation und Terminologie: Die Komplexität und Dynamik wissenschaftlicher Informations- und Kommunikationsprozesse. Narr, Tübingen
17. Bodrow W, Bergmann P (2003) Wissensbewertung in Unternehmen: Bilanzieren von intellektuellem Kapital. Erich Schmidt Verlag GmbH & Co KG, Berlin
18. Model BA (2010) Syntagmatik im zweisprachigen Wörterbuch. Walter de Gruyter, Berlin, New York
19. Stegemann J (1991) Übersetzung und Leser: Untersuchung zur Übersetzungsäquivalenz, dargestellt an der Rezeption von Multatulis "Max Hevelaar". de Gruyter, New York
20. Langenmayr A (1997) Sprachpsychologie. Ein Lehrbuch. Hogrefe, göttingen u. A
21. Körper D (2007) Terminologie in der Softwarelokalisierung – Probleme und Lösungen. Diplomatica Verlag, Hamburg
22. SDL (2014) Software localization: what is software localization?., http://www.sdl.com/technology/language-technology/what-is-software-localization.html. Accessed 07 Dec 2014
23. Schlagowski H (2013) Technische Dokumentation im Maschinen- und Anlagenbau: anforderungen. Beuth, Berlin
24. Wilrich T (2012) Das neue Produktsicherheitsgesetz (ProdSG): Leitfaden für Hersteller. Importeure und Händler, Beuth, Berlin
25. Erneuerbare-Energien-Gesetz vom 21. Juli 2014 (BGBl. I S. 1066), das durch Artikel 4 des Gesetzes vom 22. Juli 2014 (BGBl. I S. 1218) geändert worden ist.
26. Energiewirtschaftsgesetz vom 7. Juli 2005 (BGBl. I S. 1970, 3621), das zuletzt durch Artikel 6 des Gesetzes vom 21. Juli 2014 (BGBl. I S. 1066) geändert worden ist.
27. International Electrotechnical Commission (IEC) (2010) IEC 61508–4:2010 Functional safety of electrical/ electronic/ programmable electronic safety-related systems – part 4: definitions and abbreviations. IEC, Geneva. German version: DIN EN 61508–4:2011–02 funktionale sicherheit sicherheitsbezogener elektrischer/ elektronischer/ programmierbarer elektronischer systeme teil 4: begriffe und abkürzungen. Beuth, Berlin
28. OMS Group (2014) Open metering system specification. Glossary of terms used in or related to the OMS annex A to volume 1 general issue 2.0.0. FINAL DRAFT A (2014-01-27)., http://oms-group.org/download4all/. Accessed 01 Mar 2014
29. Stein C (2012) Terminologie-Ontologien im Einsatz: über Missverständnisse unter Experten und Methoden zu einem besseren Verstehen. In: Jumar U, Schnieder E, Diedrich C (eds) EKA 2012: Entwurf Komplexer Automatisierungssysteme 2012. ifak, Magedeburg, p 321
30. Stein C (2012) Von der Terminologie zum semantischen Netz: Wissensmanagement und Ontologien. In: tcworld GmbH (ed) tekom-Jahrestagung 2012 in Wiesbaden: Zusammenfassungen der Referate. tekom Jahrestagung, Wiesbaden, p 460
31. Uschold M, Gruninger M (1996) Ontologies: principles, methods and applications. Knowl Eng Rev 11(2):93–136. doi:10.1017/S0269888900007797
32. Deutsches Institut für Normung (2000) DIN 2344:2000-05 Ausarbeitung und Gestaltung von terminologischen Festlegungen in Normen. Beuth, Berlin
33. Deutscher Terminologie Tag e.V (2014) Terminologiearbeit Best Practices 2.0. Dt. Terminologie-Tag, Köln
34. *iglos* – the intelligent glossary [www.iglos.de/app]
35. Suonuuti H (1998) Terminologia gvidilo. Universala Esperanto-Asocio, Rotterdam

2

Localized vs. synchronized exports across a highly renewable pan-European transmission network

Rolando A. Rodriguez[1†], Magnus Dahl[2†], Sarah Becker[3†] and Martin Greiner[1,4*†]

Abstract

Background: A future, highly renewable electricity system will be largely based on fluctuating renewables. The integration of wind and solar photovoltaics presents a major challenge. Transmission can be used to lower the need for complementary generation, which we term backup in this article.

Methods: Generation data based on historical weather data, combined with real load data, determine hourly mismatch timeseries for all European countries, connected by physical power flows. Two localized export schemes determining the power flows are discussed, which export only renewable excess power, but no backup power, and are compared to a synchronized export scheme, which exports renewable excess power and also backup power.

Results: Compared to no or very limited power transmission, unconstrained power flows across a highly renewable pan-European electricity network significantly reduce the overall amount of required annual backup energy, but not necessarily the required backup capacities.

Conclusions: The reduction of the backup capacities turns out to be sensitive to the choice of export scheme. Results suggest that the synchronized export of local backup power to other countries is important to significantly save on installed backup capacities.

Keywords: Renewable energies, Energy system, Power transmission, Pan-European transmission grid

Background

Today's macro energy systems, mainly based on conventional fossil and nuclear resources, will in the future increasingly rely on renewable resources. At the moment, it is unclear what will be the best transitional pathway between the current and the future energy system. In this respect, it makes sense to think backwards, which means in a first step to get a good understanding of fully renewable energy systems [1–19] and then in a second step bridge to today's energy system [20, 21]. Since wind and solar power generation are expected to be dominant, the fluctuating spatio-temporal weather patterns will determine the design of highly renewable energy systems. Based on state-of-the-art high-resolution meteorological data [22], spatio-temporal modeling, and the physics of complex networks [23, 24], some fundamental properties of a highly renewable pan-European power system have already been derived. Amongst such characteristics are the optimal mix of wind and solar power generation [11, 12], the synergies between storage and balancing [13], the extension of the transmission network [14, 20] as well as the ramp down of conventional power generation [21] during the transitional phase. These results indicate that the pathways into future energy systems will be driven by an optimal systemic combination of technologies that reduce the system's reliance on backup power and energy.

The benefit of transmission in highly renewable energy networks has been discussed and explored in [14, 20]. A simplified model of the electricity system has been considered, consisting of intermittent wind and solar PV production and a perfectly flexible dispatchable backup system. In this context, the backup energy, which is the

*Correspondence: greiner@eng.au.dk
†Equal contributors
[1]Department of Mathematics, Aarhus University, Ny Munkegade 118, 8000 Aarhus C, Denmark
[4]Department of Engineering, Aarhus University, Inge Lehmanns Gade 10, 8000 Aarhus C, Denmark
Full list of author information is available at the end of the article

integral over the usage of backup units in order to balance temporal deficits of the renewable power generation, turns out to be a strongly convex function of the overall transmission line capacities. For a fully renewable pan-European power system with zero interconnector capacities, the required backup energy decreases from 24 % of the annual electricity consumption down to 15 % with sufficiently large transmission capacities which permit unconstrained power exchange. Most of this reduction is already achieved with transmission capacities three to four times larger than today's interconnector strengths. Although backup energy quickly decreases with increasing transmission capacities, it has been observed that the backup capacities are not similarly reduced [14].

The objective of this paper is to generalize the interplay between backup and transmission used in [14], and to discuss its potential for a combined reduction of backup energy and capacity in a highly renewable energy network. Figure 1 presents a toy model to clarify this objective. The first node in the triangular network (a) generates more renewable power than what it consumes; it has two units of renewable excess power. The other two nodes do not generate enough renewable power compared to what they consume, and have a deficit of three and two units. In (b), which is the export scheme used in [14, 20], the source node exports one excess unit to each of the two sink nodes, which then still have to balance two and one units. In (c), the source node also exports its two excess units, but now in such a way that after their imports, the two sink nodes have to balance the same amount of 1.5 units. In (d), the source node exports three units, which is more than its own excess power. Two units go to the second node and one unit to the third node. All three nodes are then left to balance one unit.

Behind schemes (b–d) is Thomson's principle of minimal squared flow. After the maximum export of the source node's excess power has been chosen, the resulting power flows through the network are determined by squared flow minimization, in accordance with the physical power flow on the network. Since the role of the three nodes might swap due to the fluctuating weather patterns, (b) leads to an average network-wide backup energy of three units, a backup capacity of two units per node and a transmission capacity of one unit per link. In (c), after the maximum export of the source node's excess power, the resulting flow leads to an equal usage of backup capacity for the sink nodes. The scheme (c) also leads to a network-wide backup energy of three units, but when compared to (b), the backup capacity per node is only 1.5 units, whereas the transmission capacity per link has increased to 1.17 units. The shared backup principle of (d) requires all nodes to balance the same amount after the imports and exports have taken place. It also keeps the network-wide backup energy to three units. The backup capacity at each node is reduced to one unit, and the transmission capacities of the links are increased to 1.67 units.

The three export schemes illustrated in Fig. 1b–d will be formalized in the "Methods" section. The "Results and discussion" section will use a simplified pan-European transmission network (see Fig. 2) to quantify the differences between the three export schemes with respect to backup energy, backup capacity, and transmission capacities. A conclusion will be given in the last section.

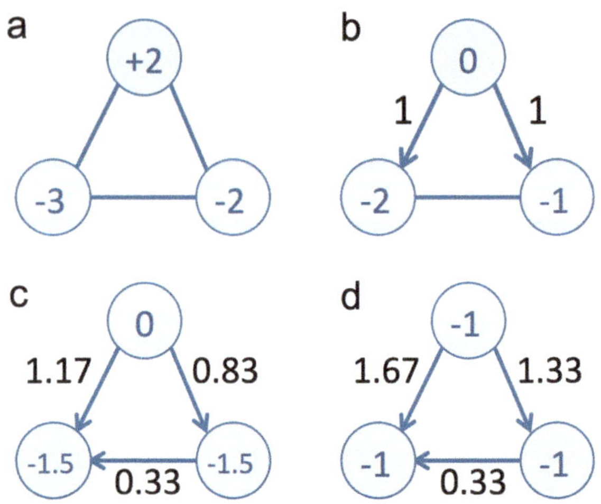

Fig. 1 Four export schemes. The network with one source and two sinks can be balanced via different export schemes: **a** zero exports, **b** maximum export of renewable excess power with minimum overall power flow, **c** maximum export of renewable excess power with maximum reduction of local backup, **d** combined export of renewable excess and backup power with maximum reduction of local backup

Methods

Renewable power generation

We define a country n to have a fully renewable power system once the average power generation from renewable sources $\langle G_n^{RES} \rangle = \langle L_n \rangle$ matches the average load. The renewable power generation

$$G_n^{\mathrm{RES}}(t) = G_n^W(t) + G_n^S(t) \tag{1}$$

is assumed to consist of wind $\left(G_n^W\right)$ and solar photovoltaics $\left(G_n^S\right)$. While large hydroelectric already offers significant amounts of renewable energy to the system, most growth is expected to take place in wind and solar. Due to the relative control possible in hydroelectric generation, we consider it as a part of dispatchable backup. Other renewable sources are either neglected (like run-of-river hydro) or considered as part of the backup system (like biomass). The ratio

$$\alpha_n = \langle G_n^W \rangle / \langle L_n \rangle = 1 - \langle G_n^S \rangle / \langle L_n \rangle \tag{2}$$

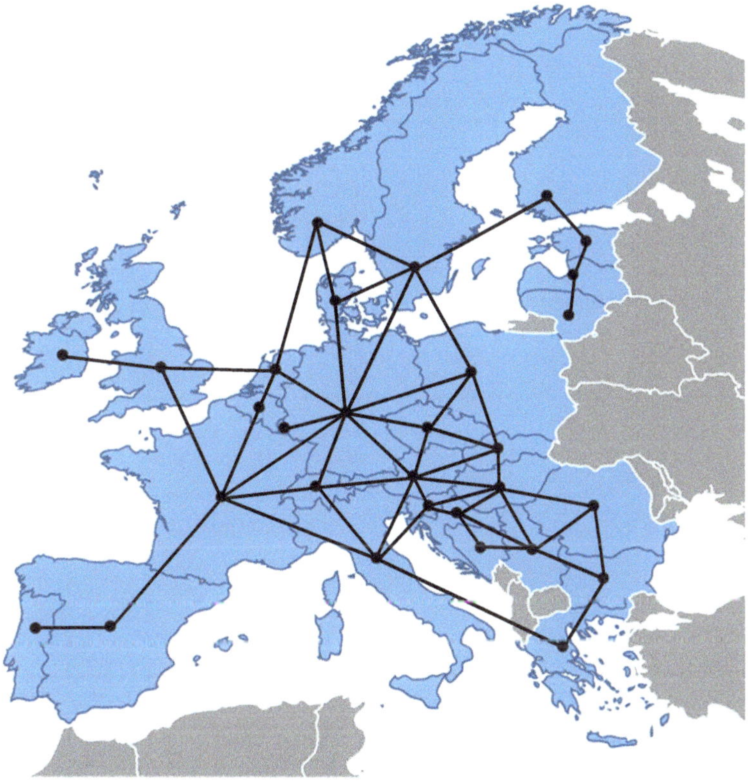

Fig. 2 Pan-European transmission network. A simplified pan-European transmission network, where countries are treated as ideal nodes and are linked by interconnectors

defines the mixing parameter between wind and solar power generation. Furthermore, we assume that intra-country transmission capacities are sufficient to allow for aggregation of load and generation on country level.

Realistic time series for the wind and solar PV power generation, including consistent spatio-temporal correlations, have been obtained from a recently developed renewable energy atlas [22]. The atlas is based on weather data from high-resolution forecast models. The data extends over several years with hourly time resolution and has a spatial grid cell resolution of $0.3125° \times 0.3125°$ (approximately 40×40 km^2 at European latitudes) covering all of Europe. The conversion from weather data to potential wind and solar PV power generation is done on a grid cell level and then aggregated to a country level. The load time series $L_n(t)$ have been taken from historical records with hourly resolution. For more details on the wind, solar, and load data, see [11, 12].

Backup energy and power

Although equal on average, the renewable power generation of a fully renewable country will be larger than the load for roughly half of the time, and smaller for the other half. The country then might want to export ($T_n > 0$) or import ($T_n < 0$) power. This defines the residual mismatch $\Delta_n(t)$ normalized to the mean load $\langle L_n \rangle$:

$$\Delta_n(t) = G_n^{\text{RES}}(t) - L_n(t) - T_n(t) \ . \quad (3)$$

If $\Delta_n(t)$ is negative, the country needs to balance $B_n(t) = -\min(\Delta_n(t), 0)$ with own backup power plants, and if positive, the country needs to curtail $C_n(t) = \max(\Delta_n(t), 0)$.

As mentioned above, dispatchable renewable generation, such as hydro power or biomass, is considered to be a part of backup power. We do not take into account any inertia of the backup system due to finite reaction times. Instead, we consider it to be ideally flexible. For a more detailed treatment of the backup system, distinguishing between a range of flexibility classes of a weather-driven backup power system, see [25, 26].

Due to fluctuations in the renewable power generation, the load and the export power, the backup power $B_n(t)$ changes from 1 h to the next. When multiplied with the number of hours $N_{\text{hours}} = 8760$ per year, the average backup power is equal to the annual backup energy $E_n^B = \langle B_n \rangle N_{\text{hours}}$. We take the 99 % quantile $\int_0^{\mathcal{K}^B} p(B)dB = 0.99$ of the backup power distribution as a measure for the backup capacity $\mathcal{K}_n^B$, as rare, extreme values define the 100 % quantile.

Splitting the residual mismatch from Equation (3) into backup and curtailment leads to the energy conservation law of power supply matching power demand

$$G_n^{\mathrm{RES}}(t) + B_n(t) = L_n(t) + T_n(t) + C_n(t) \,, \tag{4}$$

which holds locally at every country. Based on the assumption of lossless transmission, a second conservation law holds globally for the networked countries:

$$\sum_n T_n(t) = 0 \,. \tag{5}$$

Power transmission

In the DC approximation [27] for high-voltage systems, the AC power flows along the interconnectors between the countries are directly determined from the net exports:

$$\begin{aligned} T_n &= \sum_{m=1}^{N} \mathcal{B}_{nm}\delta_m, \\ f_{m\to n} &= b_{nm}(\delta_n - \delta_m) \,. \end{aligned} \tag{6}$$

Inversion of the first equation gives the voltage phase angles δ_n at all nodes. Their differences determine the power flows $f_{m\to n}$ along the links $m\to n$ between neighboring nodes (negative flows are used to denote flow in the direction $n \to m$). To reduce the complexity of a full solution to the AC problem, the DC approximation assumes zero link resistances and no reactance dependence on link lengths. The elements of the susceptance matrix $\mathcal{B}_{nm} = \left(\sum_k b_{nk}\right)\delta_{nm} - b_{nm}$ are therefore initially assumed to be unitary and equal to the Laplacian of the network graph; see [14] for more details.

To examine the validity of the distance-independent line susceptances, we calculated test cases. There, we assume non-unitary reactance values, using instead a typical value of 0.28 Ω/km [28]. As a metric for the line distance, we use the distance between capitals of the countries in question. This leads to $b_{nm} = 0.28 \cdot d_{nm}$ where d_{nm} is the line distance. Given that 0.28 is a linear factor, the actual number is irrelevant, and the only effect it brings is a preference for shorter paths over longer ones. In terms of total transmission capacities, this leads to a slight increase by about 9 %.

Since the exports $T_n(t)$ are functions of time, the power flows $f_l(t)$ will also fluctuate with time. The low and high quantiles $q = \int_{-\infty}^{F(q)} p(f)df$ of the unconstrained flow distributions $p(f)$ represent a measure for the required interconnector strengths, which we set to $\mathcal{K}_l^T = \max(-F_l(0.01), F_l(0.99))$. The total interconnector capacity $\mathcal{K}^T = \sum_l \mathcal{K}_l^T$ of the pan-European transmission network is given by the sum over all links.

Export schemes

Next, four different export schemes are outlined. They include zero export, two localized trade schemes of only renewable excess power, and synchronized backup operation with additional export of backup power.

Zero export

The zero-export scheme is expressed by

$$T_n = 0 \,, \tag{7}$$

and is illustrated in Fig. 1a. Each country then has to fully balance or curtail the mismatch between its renewable power generation and its load.

Localized export I

Rather than full curtailment, countries with surplus renewable power generation may want to export as much of it as possible to other countries with a current deficit:

$$\begin{aligned} \text{exporters}: &\quad 0 < T_n(t) \le \max(G_n^{\mathrm{RES}}(t) - L_n(t), 0) \,, \\ \text{importers}: &\quad 0 > T_n(t) \ge \min(G_n^{\mathrm{RES}}(t) - L_n(t), 0) \,, \\ \text{objective 1}: &\quad \min\left(\sum_n B_n(t)\right) \,, \\ \text{objective 2}: &\quad \min\left(\sum_l f_l^2(t)\right) \,. \end{aligned} \tag{8}$$

Two steps need to be taken to assign exporters and importers. The first objective is to use as much excess renewable power as possible to reduce the overall backup energy. In case that the overall export strengths are larger than the overall import needs, this objective makes sure that all importers will be fully served, but there is still the freedom to select different subsets of exporters. This freedom is fixed by the second objective, where the transmission system operator requires the power flows on the network to be as localized as possible. Those nodes which are at the end left unable to export everything have to locally curtail.

In case of larger overall import needs, all export opportunities will be used by the first objective, and the second objective fixes the subset of importers. Those nodes which are not able to import enough then have to locally balance for themselves. Compare again with Fig. 1b. Note also that from one time instance to the next, the role of exporters and importers can be reversed.

This export scheme has been employed in two previous publications [14, 20]. Given that generation from renewables has a lower marginal cost, it is not unrealistic that importers would be willing to import excess renewable power generation from other nodes. Excessive transmission capacities would increase the total system cost. A localization of exports—as forced by the square minimization of total flows—does result in smaller total transmission capacities, as shown in the results below.

Localized export II

The third export scheme uses the same constraints on exporters and importers as in (8), but replaces the first objective by

$$\text{objective 1}: \quad \min \left(\max_n \frac{B_n(t)}{\langle L_n \rangle} \right) . \tag{9}$$

This gives priority to importers with the largest relative deficit power, allowing reductions to the backup capacities. The second step is, again, identical to the one in (8). Compare with Fig. 1c.

Synchronized export

A very different export scheme forces all countries into a completely synchronized residual mismatch (3):

$$\Delta_n(t) = \beta(t)\langle L_n \rangle , \tag{10}$$

where due to (3) and (5)

$$\beta(t) = \frac{\sum_n \left(G_n^{\mathrm{RES}}(t) - L_n(t)\right)}{\sum_n \langle L_n \rangle} . \tag{11}$$

After exports and imports, every node is balancing ($\beta < 0$) or curtailing ($\beta > 0$) the same amount relative to its own average load as all the other nodes. This requires cooperation. For example in case of $\beta < 0$, nodes with excess renewable power generation not only export their excess to other countries but also balance power from their own backup capacities; see again Fig. 1d.

In the zero export case, the backup capacities are determined by the countries' weather patterns and demand. For the localized export schemes, the position of the country in the network also affects its backup capacity. An immediate result of the synchronized export scheme is a uniform distribution of backup capacities, relative to the countries' mean load.

All export schemes can be stated as a quadratic optimization problem and were computed numerically using the python interface to the commercial solver Gurobi [29].

Results and discussion

Residual mismatch distributions

The distribution $p(\Delta_n)$ of the residual mismatch (3) depends strongly on the mixing parameter α_n between wind and solar power generation, as previously discussed in [14]. Lowest backup energy needs can be obtained with $\alpha_n \approx 0.7$ in single countries. This value of the mixing parameter is therefore used in the subsequent analysis. Although zero on average, the distributions are very broad, cf. Fig. 3, showing the distribution of non-zero mismatches for the different export schemes.

The negative part of the mismatch distribution is of particular importance. Its average determines the average backup power, which is proportional to the annual backup energy. Its tail characterizes the most extreme backup

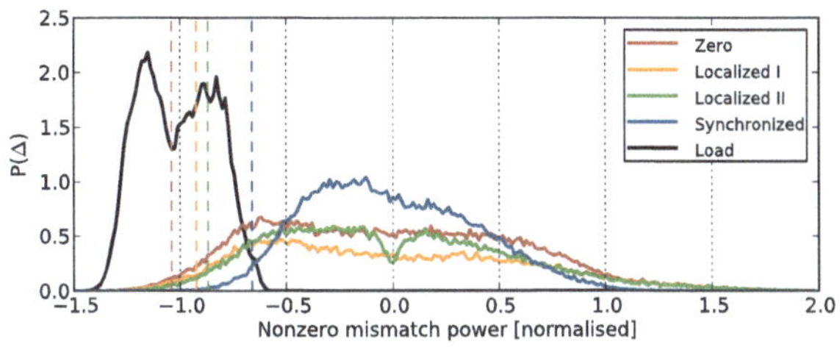

Fig. 3 Residual mismatch distributions. Distributions of residual mismatches (3) for Germany with $\alpha_{\mathrm{GER}} = 0.7$ resulting from the zero (*red*), the two localized (*yellow*, *green*) and the synchronized (*blue*) export schemes. For comparison, the negative load distribution (*black*) is also shown. Mismatches and load have been normalized to the average load $\langle L_{\mathrm{GER}} \rangle$. The residual mismatch distributions resulting from the two localized export schemes are shown without their singular peak at $\Delta_{\mathrm{GER}} = 0$. *Dashed vertical lines* indicate the 99 % quantile of the negative mismatch

power needs, which determine the required backup capacities.

Figure 3 shows the dependence of the residual mismatch distribution on the three non-zero export schemes (8)–(10), where each country is allowed to import and export from all 30 European countries. For comparison, the load and the zero export schemes are shown as well. Contrary to the synchronized export scheme, the two localized schemes produce a lot of $\Delta_n = 0$ events. The resulting singular peak has been removed from the plotted distributions to clarify the illustration. It can still be observed indirectly as the integrated area below those two normalized distributions is smaller than for the other two distributions resulting from the zero and the synchronized export scheme.

The total backup energy (or equivalently, the average backup power) is shifted to smaller values by transmission. In fact, as transmission line capacities are not constrained, all non-zero export schemes lead to the same total backup energy, equal to the backup needs of a European aggregation. This is because in the event of a Europe-wide deficit, all local excesses are exported to other countries in deficit and none are curtailed. Conversely, when there is Europe-wide excess, all local deficits are served before curtailment is activated. In the two localized schemes, this is achieved by the first objective of minimal backup energy needs. In the synchronized export scheme, it is a consequence of the Europe-wide aggregation of the mismatch, which is then distributed back to the single nodes as backup or curtailment, cf. (10). When compared to the zero export scheme, the two localized export schemes are able to lower the frequency of negative residual mismatches, causing a decrease of the average backup power. The residual mismatch distribution resulting from the synchronized export scheme changes in a different way, also leading to a reduction of the average backup power.

Contrary to the two localized export schemes, the synchronized export scheme also lowers the negative tail of the residual mismatch distribution and therefore significantly reduces the required backup capacity. As a proxy to the backup capacity, the 99 % quantile of the negative mismatches is indicated as dashed vertical lines in Fig. 3.

Backup energy and capacity

The two localized and the synchronized export schemes (8)–(10) have a different impact on the average backup power for the different countries. See the top panel of Fig. 4, which shows the average imports $\langle \max(-T_n, 0) \rangle$ for all countries. For example, Germany, the first localized export scheme leads to a slightly larger reduction of the average backup power when compared to the second localized scheme. This is because the first scheme allows Germany to import more on average than the second scheme. The synchronized export scheme leads to a significantly larger average import. However, this does not lead to a further reduction of the average backup power because each country's exports increase in parallel.

For some countries other than Germany, the two localized export schemes (8)–(9) lead to a smaller average backup power than from the synchronized export scheme (10). This is explained by the observation that the sum $\sum_n \langle B_n \rangle$ is identical for the three export schemes as discussed above. Consult also again Fig. 1b–d.

The lower panel of Fig. 4 shows the average net exports $\langle T_n \rangle = -\langle \Delta_n \rangle$ for the different countries. The average net export is the difference between average export and average import. Except for a minus sign, it is equal to the mean of the residual mismatch (3). For the synchronized export scheme (10), the net exports are zero for all countries. However, this is not the case for the two localized export schemes (8) and (9). For big countries like Germany, France, and Great Britain with $\alpha_n = 0.7$, it turns out to be of the order $\langle T_n \rangle / \langle L_n \rangle \approx -0.004$. Those countries become net importers. The biggest relative net importers are the two Nordic countries Norway and Sweden with $\langle T_n \rangle / \langle L_n \rangle \approx -0.023$. Spain and Italy are net exporters of the order $\langle T_n \rangle / \langle L_n \rangle \approx 0.014$. With $\langle T_n \rangle / \langle L_n \rangle \approx 0.023$, the strongest relative net exporters are the smaller countries in southeast Europe. Of course due to (5), the overall country sum over all net exports and net imports is again $\sum_n \langle T_n \rangle = 0$.

The backup capacity $\mathcal{K}_n^B$ also shows a strong dependence on the mixing parameter. See Fig. 5. For all four export schemes (7)–(10), it takes a minimum at $\alpha_n = 0.85$. Compared to the value $1.32\langle L_n \rangle$ of the 99 % quantile of the German load, the backup capacity $\mathcal{K}_n^B = 1.04\langle L_n \rangle$ at $\alpha_n = 0.85$ resulting from the zero export rule is not much smaller. Without transmission, a 100 % renewable country needs an enormous amount of backup capacity. The two localized and the synchronized export schemes have a bigger impact on the reduction of the backup capacities. For Germany with $\alpha = 0.85$, the first localized export scheme leads to $\mathcal{K}_n^B = 0.89\langle L_n \rangle$, the second localized scheme to $\mathcal{K}_n^B = 0.82\langle L_n \rangle$, and the synchronized scheme

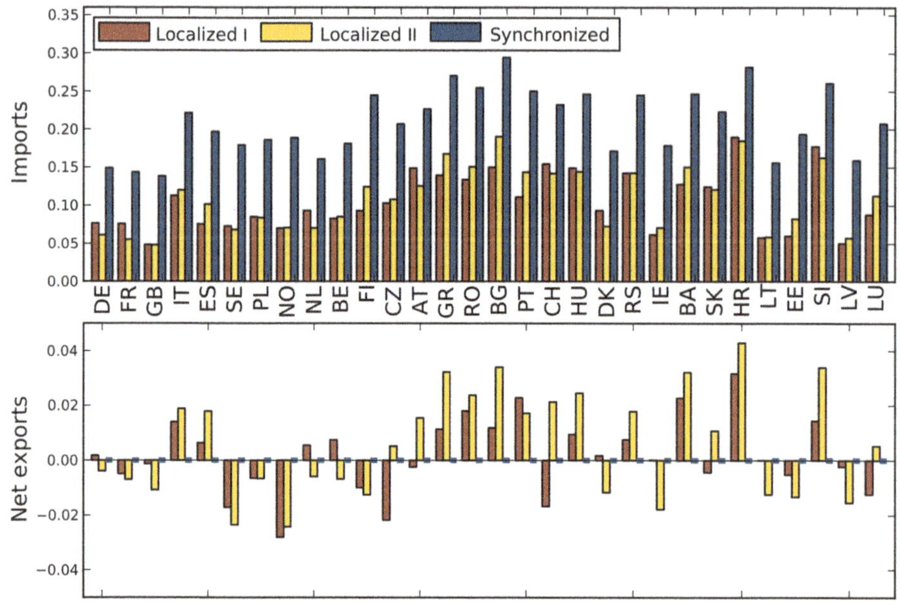

Fig. 4 Average imports and net exports. (*Top*) average imports $\langle \max(-T_n, 0) \rangle$ and (*bottom*) average net exports $\langle T_n \rangle$ for all countries following from the two localized and the synchronized export schemes (8)–(10). The renewable mixing parameter $\alpha_n = 0.7$ has been used, and the imports and net exports have been normalized to the average loads $\langle L_n \rangle$

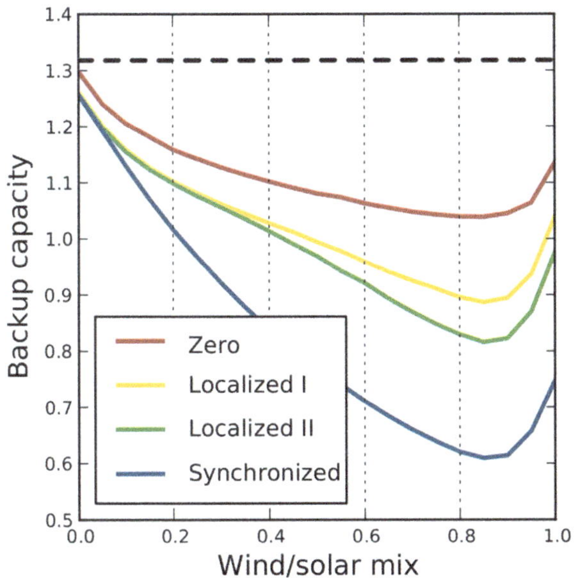

Fig. 5 Backup capacity. Backup capacity as a function of the renewable mixing parameter α_{GER} for Germany resulting from four different export schemes. The backup capacity has been normalized to the average load $\langle L_{GER} \rangle$. For reference, the 99 % quantile of the German load distribution is shown as the *black dashed line*

to $\mathcal{K}_n^B = 0.61 \langle L_n \rangle$. For the other countries, the relative backup capacities resulting from the two localized export schemes are slightly different than the German values (see Fig. 6). The synchronized export scheme leads to identical relative backup capacities for all countries. This is no surprise since the synchronized export scheme (10) with (11) enforces a synchronized balancing between all countries.

Since Equation (11) describes the all-European mismatch, we arrive at a strong conclusion: the synchronized export scheme leads to the largest possible reduction in overall backup capacities. More precisely, synchronized export takes advantage of the strongest possible mismatch smoothing by aggregation across Europe. In the absence of single countries with exceptionally stable resources that would make such sharing unattractive (see Fig. 6) and without countries taking advantage of each other (in game-theoretic terms, as long as free-riding and altruistic behavior are excluded), there is no scheme in which the mismatch quantiles of single countries can be decreased more. In particular, no other export scheme is able to reduce the backup capacities (measured as quantiles of the mismatch) further. For mixes between $0.70 \leq \alpha_n \leq 0.95$, the synchronized export scheme leads to backup capacities which are only half of the high load quantiles, the latter representing more or less the backup capacities of our current electricity system.

Transmission capacities

Reference [14] has shown that the unconstrained total interconnector capacities become relatively large. The same European country network as in Fig. 2 has been used with a mixing parameter around $\alpha_n \approx 0.71$, and only the first localized export scheme has been discussed. For unitary impedance across the whole network, the total interconnector capacities turned out to be $\mathcal{K}^T = 395$ GW, which is about a factor 5.4 larger than the present European layout; see Table two in [14].

Figure 7 extends the previous findings to the full range $0 \leq \alpha_n \leq 1$ of the mixing parameter, and to the second localized and the synchronized export schemes. The two localized export schemes produce very similar total interconnector capacities. The minimum capacities are reached at $\alpha_n = 0.35$. For the solar only $\alpha_n = 0$, the transmission capacities are larger because of the strong East–West power flows in the mornings and evenings, and because of the strong seasonal South–North flows. For the other extreme, the wind-only $\alpha_n = 1$, the transmission capacities become even larger. This is caused by strong power flows between decorrelated regions separated by distances larger than the observed correlation length of about 500 km for the wind power generation [15, 18].

The synchronized export scheme leads to a minimum at a slightly larger $\alpha_n = 0.42$. Over the whole range $0 \leq$

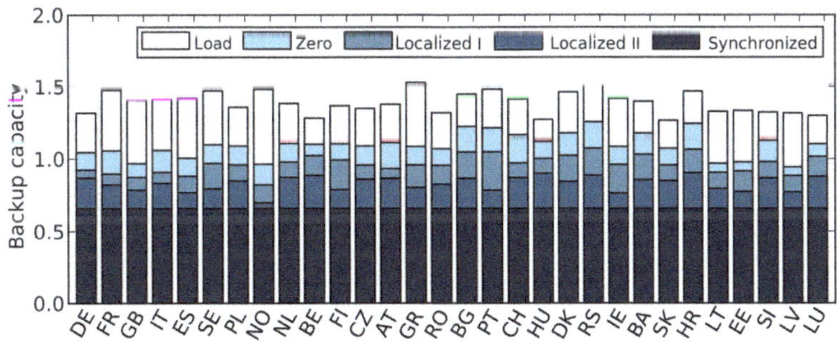

Fig. 6 European backup capacities. Backup capacities for all European countries resulting from four different export rules. The renewable mixing parameter $\alpha_n = 0.7$ has been used, and the backup capacities have been normalized to the average loads $\langle L_n \rangle$. For reference, the 99 % quantile of the load distributions is also shown

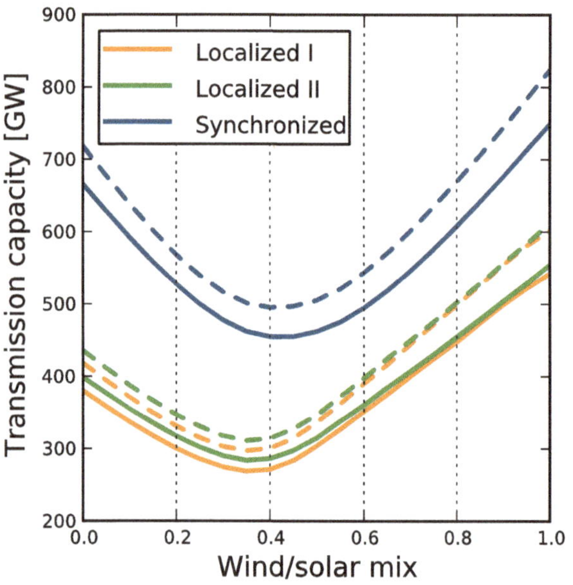

Fig. 7 Transmission capacity. Overall interconnector capacities $\mathcal{K}^T$ of the unconstrained European transmission grid as a function of the renewable mixing parameter α_n. *Dashed lines* indicate $\mathcal{K}^T$ for non-unitary, line-dependent impedances. The scale of $\mathcal{K}^T$ is set by the average European load $\sum_n \langle L_n \rangle = 345.5$ GW from the year 2007

$\alpha_n \leq 1$, the total interconnector capacities required from the synchronized scheme are a factor of 1.35–1.70 larger than for the two localized schemes. For $\alpha_n = 0.7$, the synchronized scheme produces $\mathcal{K}^T = 550$ GW, whereas it is only $\mathcal{K}^T = 400$ GW for the localized schemes. This difference is easily explained by the extra exports of backup power caused by the synchronized scheme (see again Fig. 4), which cause more power flows and result in higher required transmission capacities (see again Figs. 1b–d).

Reproducing the results considering non-unitary, length-dependent impedances for all lines brings only small changes to the results. Quantitatively, total backup energy and capacity are unchanged, as the objective that minimizes them has priority over the actual flow minimization. Since longer paths are more penalized in this calculation, the total amount of capacity needed is higher, but only by around 9 %, regardless of the export scheme used. The dashed lines in Fig. 7 show the transmission capacities resulting from calculating flows with impedance.

Conclusions

Fully renewable networked power systems have been discussed, for which the average renewable power generation is equal to the average load. Such systems still require backup generation units to cover the occurring temporal deficits of fluctuating renewable power generation to match the load. The amount of backup capacity depends on the export schemes applied to the transmission network. Compared to no exports at all, the export of local renewable excess power, which we have dubbed as localized, leads to some reduction of the overall backup capacities. Significantly more reduction is achieved when also local backup power is allowed to be exported. For a fully renewable pan-European transmission grid, such a synchronized export scheme leads to backup capacities which are only half of the maximum loads.

In its present form, the synchronized export scheme is based on a completely synchronized residual mismatch across the network. No other export scheme is able to reduce the backup capacities further. However, it requires larger transmission capacities when compared to the localized export schemes. From an economic perspective, this is not necessarily a drawback, since costs for backup capacities are usually one order of magnitude bigger than transmission costs [3]. It will be interesting to explore other, partly synchronized export schemes, including market-like schemes, which might be able to reduce also the transmission capacities. Such an exploration could be based on the minimization of an overall cost objective, which includes not only backup and transmission capacities but also renewable power generation capacities. These findings will have an impact on the future strategies of policymakers and energy-related companies.

Competing interests
The authors declare that they have no competing interests.

Authors' contributions
The article was jointly prepared by all authors. All authors read and approved the final manuscript.

Acknowledgements
SB gratefully acknowledges financial support from O. and H. Stöcker as well as M. and H. Puschmann.

Author details
[1]Department of Mathematics, Aarhus University, Ny Munkegade 118, 8000 Aarhus C, Denmark. [2]Department of Physics and Astronomy, Aarhus University, Ny Munkegade 120, 8000 Aarhus C, Denmark. [3]Frankfurt Institute for Advanced Studies, Frankfurt University, Ruth-Moufang-Str.1, 60438 Frankfurt am Main, Germany. [4]Department of Engineering, Aarhus University, Inge Lehmanns Gade 10, 8000 Aarhus C, Denmark.

References

1. Jacobson MZ, Delucchi MA (2009) A path to sustainable energy by 2030. Sci Am 301:58–65
2. Jacobson MZ, Delucchi MA (2011) Providing all global energy with wind, water, and solar power – Part I: technologies, energy resources, quantities and areas of infrastructure, and materials. Energy Policy 39:1154–1169
3. Delucchi MA, Jacobson MZ (2011) Providing all global energy with wind, water, and solar power – Part II: reliability, system and transmission costs, and policies. Energy Policy 39:1170–1190
4. Popp M (2010) Speicherbedarf Bei Einer Stromversorgung Mit Erneuerbaren Energien. Springer, Heidelberg

5. Lund H (2010) Renewable Energy Systems – The Choice and Modeling of 100 % Renewable Solutions. Elsevier, Amsterdam
6. Connolly D, Lund H, Mathiesen B, Leahy M (2011) The first step towards a 100 % renewable energy system for Ireland. Appl Energy 88:502–507
7. Mason IG, Page SC, Williamson AG (2010) A 100 % renewable electricity system for New Zealand utilising hydro, wind, geothermal and biomass resources. Energy Policy 38:3973–3984
8. Mason IG, Page SC, Williamson AG (2013) Security of supply, energy spillage control and peaking options within a 100 % renewable electricity system for New Zealand. Energy Policy 60:324–333
9. Elliston B, Diesendorf M, MacGill I (2012) Simulations of scenarios with 100 % renewable electricity in the Australian National Electricity Market. Energy Policy 45:606–613
10. Elliston B, Diesendorf M, MacGill I (2013) Least cost 100 % renewable electricity scenarios in the Australian National Electricity Market. Energy Policy 59:270–282
11. Heide D, Von Bremen L, Greiner M, Hoffmann C, Speckmann M, Bofinger S (2010) Seasonal optimal mix of wind and solar power in a future, highly renewable Europe. Renew Energy 35:2483–2489
12. Heide D, Greiner M, Von Bremen L, Hoffmann C (2011) Reduced storage and balancing needs in a fully renewable European power system with excess wind and solar power generation. Renew Energy 36:2515–2523
13. Rasmussen MG, Andresen GB, Greiner M (2012) Storage and balancing synergies in a fully or highly renewable pan-European power system. Energy Policy 51:642–651
14. Rodríguez RA, Becker S, Andresen GB, Heide D, Greiner M (2014) Transmission needs across a fully renewable European power system. Renew Energy 63:467–476
15. Widen J (2011) Correlations between large-scale solar and wind power in a future scenario for Sweden. IEEE Trans Sustainable Energy 2:177–184
16. Schaber K, Steinke F, Hamacher T (2012) Transmission grid extensions for the integration of variable renewable energies in Europe: Who benefits where? Energy Policy 43:123–135
17. Steinke F, Wolfrum P, Hoffmann C (2013) Grid vs. storage in a 100 % renewable Europe. Renew Energy 50:826–832
18. Kempton W, Pimenta FM, Veron DE, Colle BA (2010) Electric power from offshore wind via synoptic-scale interconnection. Proc Natl Acad Sci 107:7240–7245
19. Budischak C, Sewell D, Thomson H, Mach L, Veron DE, Kempton W (2013) Cost-minimized combinations of wind power, solar power and electrochemical storage, powering the grid up to 99.9 % of the time. J Power Sources 225:60–74
20. Becker S, Rodriguez RA, Andresen GB, Schramm S, Greiner M (2014) Transmission grid extensions during the build-up of a fully renewable pan-European electricity supply. Energy 64:404–418
21. Andresen GB, Rasmussen MG, Rodriguez RA, Becker S, Greiner M (2012) Fundamental properties of and transition to a fully renewable pan-European power system. EPJ Web of Conferences 33
22. Andresen G, Søndergaard A, Rodriquez RA, Becker S, Greiner M (2013) Weather-driven modeling of future renewable power systems. In: Proceedings of the 8th SDEWES Conference; Dubrovnik, Croatia
23. Schäfer M, Scholz J, Greiner M (2006) Proactive robustness control of heterogeneously loaded networks. Phys Rev Lett 96:108701
24. Heide D, Schäfer M, Greiner M (2008) Robustness of networks against fluctuation-induced cascading failures. Phys Rev E (Stat Nonlinear, Soft Matter Phys) 77:056103
25. Andresen G, Rasmussen MG, Rodriguez RA, Becker S, Greiner M (2012) Fundamental properties of and transition to a fully renewable pan-European power system. In: 2nd European Energy Conference; Maastricht, The Netherlands. Vol. 33. p 04001
26. Andresen G, Heide D, Rasmussen M, Greiner M (2011) Transition to a fully renewable power system in Europe. In: Proceedings of the 10th International Workshop on Large-Scale Integration of Wind Power Into Power Systems as Well as on Transmission Networks for Offshore Wind Power Plants. Aalborg University, Aalborg, Denmark. pp 322–325
27. Wood AJ, Wollenberg BF (1996) Power Generation, Operation and Control. John Wiley & Sons, New York
28. Zhou Q, Bialek JW (2005) Approximate model of european interconnected system as a benchmark system to study effects of cross-border trades. Power Syst IEEE Trans 20(2):782–788
29. Gurobi Optimization I (2014) Gurobi Optimizer Reference Manual. http://www.gurobi.com

3

Local renewable energy cooperatives: revolution in disguise?

J. A. M Hufen[1*] and J. F. M. Koppenjan[2]

Abstract

Background: The Eurostat statistics on the 28 EU member states' performance in relation to renewable energy show that the Netherlands comes 25th in the EU-28. Yet, 500 initiatives were started by citizens and social groups over the last years to produce and consume renewable energy. The central question is whether these local renewable energy initiatives can be seen as a radical innovation or whether—and to what extent—they can contribute to a transition towards renewable energy production or consumption in the Netherlands.

Methods: Content analysis and case studies were the methods used in this study.

Results: The establishment of local energy cooperatives is a niche innovation in an early stage of development, and their impact on the energy production system is small, whereas local wind energy cooperatives are qualified as in a mature stage of development. Two of the four case studies of front-running local energy cooperatives demonstrate an energy transition or radical innovation in their home or facility area. One of the energy cooperatives developed an innovative concept that has potential for the production of wind energy (Deltawind). The other three energy cooperatives use a concept (sell green energy and produce renewable electricity) that is probably not suitable for large groups of clients, and the possibilities of up-scaling this formula are limited. The success of one of these three energy cooperatives (Texel Energie) is driven by the characteristics of the facility area and the social and other conditions on the island of Texel.

Conclusions: Because of the small scale of these initiatives, the performance of the Netherlands in renewable energy statistics and the rise of local energy cooperatives are not contradictory. Local wind energy cooperatives have proven to be successful, and under specific circumstances, other local energy cooperatives can be successful as well. The potential of the concept to contribute to energy transition or radical innovation is better for the local wind cooperative than for the other energy cooperatives. However, the uniqueness of each of the case studies prevents an easy reproduction of successes. The idea of up-scaling successful concepts is too simplistic, but acquiring knowledge about the mechanisms underlying energy transitions or radical innovations seems to be valuable.

Keywords: Local renewable energy; Local renewable energy cooperatives; Bottom-up innovation; Radical innovation; Energy transition; Energy innovation systems

Background

Introduction

The Eurostat statistics on the 28 EU member states' performance in relation to renewable energy show that the Netherlands comes 25th in the EU-28 [1]. Around 4.5 % of gross Dutch final energy consumption comes from renewable energy sources. Although ambitious policy goals regarding renewable energy were formulated, so far, the Dutch government has not been able to realize its aspirations. This is surprising, because the Netherlands performs reasonable well in terms of innovation and competitiveness indexes [2].

The Eurostat statistics show a backlog in the production of renewable energy, but at the same time, many new initiatives to produce and use renewable energy have been started by citizens and social groups over the last couple of years. Recently, reports have identified 500 such initiatives in different communities [3]. There is great variety in the nature and scale of these initiatives,

* Correspondence: hh@qaplus.info
[1]Research and Consultancy in Leiden, Galileïweg 8, 2333BD, Leiden, The Netherlands
Full list of author information is available at the end of the article

but they share a strong commitment towards energy conservation and renewable energy. The magnitude of these initiatives came as a surprise [3]. It is interesting that these initiatives emanate from citizens and social groups, despite the fact that these groups are rarely addressed as target groups in governmental policies on renewable energy and conservation [4].

The combination of the Dutch backlog on renewable energy on the one hand, and the enthusiasm of civic and social groups in many recent local energy initiatives on the other, prompted the writing of this article. The central question is whether these local energy cooperatives can be seen as a radical innovation or whether—and to what extent—they contribute to a transition towards renewable energy production in the Netherlands.

The article first gives an overview of the Dutch energy sector. After the introduction to energy transition theory and innovation system theory, research questions are formulated, and the research method is described. The results of the research are presented. The analysis of the results is followed by the conclusions.

Context

The Dutch energy market is dominated by six private multinational energy companies (Electrabel Nederland, Eneco, E.On-Benelux, Essent/RWE, Nuon/Vattenvall, and EPZ) that produce energy mainly using fossil energy sources like gas and coal [5]. Consumer demand for green energy is covered by imports whose authenticity is claimed by a certificate. The renewable energy produced by the multinationals comes from large-scale biomass and wind energy. The share of renewable energy in the multinationals' energy production will probably not change or change only slightly, because they have invested in conventional production facilities [6]. Only one of the energy companies (Eneco) chose to prioritize investments in renewable energy [7]. Energy delivery to small-scale users is dominated by large multinational companies, whose market share of small-scale users is around 80 %, and this share is relatively stable [8].

In addition to the aforementioned multinational energy producers, four other energy producers operate on the Dutch energy market. These producers have chosen to distribute 100 % green energy and to produce renewable energy in the Netherlands. These companies have their own strategy in the production of renewable energy, that is, the combination of solar, wind, and biomass energy (Greenchoice), biomass energy (HVC energy), and wind energy (Raedhuys, Windunie). Their investments in production facilities are small in comparison to the large energy companies [9]. Apart from the energy production enterprises (the six multinationals and the four "nationals"), 18 energy delivery companies of some magnitude distribute energy to small-scale users. Some of these energy delivery companies use the license of another energy distribution company, whereas others have their own license [6, 7].

Since the liberalization of the energy market in 1995, small-scale users have become active on the energy market, as indicated by the growing number of users that switch energy provider. In 2013, 12.1 % of all small-scale users chose a new energy delivery company [10]. Despite the neglect of renewable energy in energy production, three million small-scale energy users (around 50 % of this market segment) currently choose energy that is labeled as "green energy" for their domestic needs; however, although the number of green energy consumers is relatively stable, it has decreased a little [8]. Consumer demand can only be met by the energy producers importing green energy [11]. The origin of this so-called green energy is, however, debated by the consumer organization, Consumentenbond, as around 65 % of this energy is not green [7, 12].

The consumer organization's criticism concerns the definition as well as the administration of green energy. The administration of green energy is criticized because the system is imperfect as it does not exclude double counting. The energy sector admits that the green energy delivered is not green in terms of the original meaning [12]. The consumer organization investigated the product characteristics of the different brands of green energy sold [6]. It proved impossible to draw clear conclusions about all the green energy labels on the market. Thus, the market for green energy is not transparent for small-scale users because false information prevents them from making a correctly informed decision about their preferred energy purchase option [7, 12].

The Dutch government has created laws and regulation of competition in the energy market and indirect incentives for consumers that have a strong impact on energy conservation and renewable energy. For example, under Dutch law, there is no obligation for energy companies to produce renewable energy. The discussion about supplier obligations to enhance energy conservation and renewable energy never resulted in a political decision to change the system [13, 14]. Furthermore, Dutch fiscal policy provides enormous tax exemptions for large-scale energy consumption. This system stimulates low prices as well as energy inefficiencies and impedes innovation such as the introduction of renewable energy [9]. At the same time, the Dutch government has been pursuing a stimulating energy policy for several decades. Although citizens were never addressed directly as a separate target group, many stimulating instruments were applied. Furthermore, the energy agreement that was signed in 2013 by government as well as private enterprises and societal groups includes the goal that one million small-scale users should be producing renewable energy for their own energy needs by 2020 [15]. The signatories agree to enhance the production of renewable energy by small-scale users [15].

Methods

Theory and method

Theory: energy transitions, innovation system theory

The establishment of many local energy cooperatives in a short period of time in the Netherlands raises the question of whether this trend can be interpreted as a transition in the energy production and delivery system. "Transitions" are radical societal changes in thinking, acting, and organizing that take several decades. These changes are not instantaneous but rather evolve gradually as a result of technological, social, economic, and institutional factors. The energy market is not an easy context in which to innovate production processes and products, but perhaps the energy cooperatives are a start—or even more—of an energy transition [11, 16].A successful energy transition is a bottom-up process in which local people and groups participate in new experiments and projects [16, 17]. The energy sector is inclined to reject change, but these processes are initiated outside the system. The risk of duplicating the problems and solutions of the existing energy sector is therefore mitigated [17]. The successes in small-scale applications (niches) can under certain circumstances be up-scaled and become the new standard in the energy market (new regime) [18–20]. This type of change is the result of the drive by citizens and other societal groups. The role of government in these processes in the Netherlands is modest [4, 21].

The spontaneous establishment of local energy initiatives can also be conceptualized as a radical innovation. A radical innovation is a successful introduction of a new concept, product, service, or production process on a market. Innovation system theory interprets innovations as a result of the interactions between dependent consumers, producers, research and education, the political system, the supply system, and the infrastructure [22, 23]. Unlike classical innovation theory, innovation system theory does not focus only on research- and technology-related factors. Since the essence of local energy cooperatives is non-technical innovation, an innovation system approach seems appropriate.

Innovations like local energy cooperatives can only be realized when buyers or consumers eagerly accept the new concept, service, or product. The protagonist role of consumers in an innovation system is widely recognized. The adaption curve of successful innovations reflects how more and more consumers accept the innovation. In Rogers' terminology, a small group of innovators (2.5 % of the market) is followed by a group of early adopters, then early, and late majority, and finally the laggards [24]. A radical innovation can only be realized when the innovations have specific characteristics. The most important conditions for successful innovations are 1) comparative advantages: an innovation can be successful if it offers consumers advantages in comparison to other products, concepts, and services; 2) lack of complexity: an innovation is less attractive when additional efforts to use the innovation are necessary; and 3) compatibility: an innovation that fits in the usual habits of consumers has a better chance of succeeding. The potential for innovation by local energy initiatives depends on these factors [25].

Innovation system theory points at the relevance of innovators' competitive environment. Local energy initiatives' innovations can only be successful when they can compete with the large multinational energy companies. An innovation in a local energy cooperative is the result of a bottom-up process in a horizontal network, whereas innovation in large multinationals is the result of a top-down process (pipe-line process). The development of innovations in horizontal networks can be successful in some markets even when competitors use a top-down model for innovation [26, 27].

The success of horizontal innovations has been described in many sectors but depends strongly on the characteristics of the innovation system. A necessary condition for the development of these bottom-up innovations is horizontal networks with 1) the incentive to innovate, 2) the incentive to reveal their innovation, and 3) low cost and competition advantage in relation to private enterprises [27]. If the third condition is not met, private enterprises on the market will pick up the bottom-up innovations and integrate them in their own products or services. These factors seem to be a useful conceptual framework to describe and assess the chances for success of the horizontal innovations that take place in local energy initiatives.

Main question and research questions

Main question:

Are the local energy cooperatives an energy transition or a radical innovation, or to what extent do they contribute to a change in renewable energy production and in small-scale users' energy consumption in the Netherlands?

Research questions:

1. What is the development stage of local energy cooperatives in terms of their contribution to a change in production and/or consumption of renewable energy in the Netherlands?
2. Have the frontrunners amongst the local energy cooperatives realized a radical innovation or energy transition in their facility area?
3. What are the dominant factors in the case of the frontrunners amongst local energy cooperatives

that stimulate(d) or inhibit(ed) the (further) development of these cooperatives?

The concept of local energy cooperatives can be defined by a number of characteristics [28]:

- The initiators live in a local community (citizens and small and medium enterprises (SMEs)) and aim to develop and/or distribute renewable energy and enhance energy conservation;
- The organization seeks collaboration with other local citizens, SMEs, and local government;
- The organization starts on the basis of the enthusiasm of volunteers but develops into a legal entity like a cooperative or another legal entity.

What are the development stages of an energy transition?

The development stages of an energy transition can be described as early, mature, or late. The response of consumers is "early" when small groups of consumers use the new product, "mature" when large groups do, and "late" when only a few do not use the service [24]. Furthermore, the development can be describe in terms of the successes in small-scale applications (niches) that are up-scaled and become the new standard in the energy market (new regime) [18–20]. Development is at an early stage when niche innovations do not, or almost do not, occur, and there are no changes in a dominant regime. The development stage is mature when niche innovations have proven to be very competitive, have changed the patterns in the energy market, and have become a new standard.

What is radical innovation in a local energy system?

An energy transition or radical innovation in a facility area occurs when the consumption and/or the production of renewable energy rises strongly as a result of the local renewable energy cooperative. Renewable energy can be renewable electricity (solar panels, wind energy), heat (CHP, geothermal energy, or solar collectors), or gas (green gas from biomaterials) [29]. Energy is "green" when it is produced by renewable sources in the Netherlands (as proven by a certificate of authenticity) or when it is produced elsewhere and is most likely produced by renewable sources [12, 30]. A strong rise in renewable energy in a facility area as a result of a local energy cooperative occurs in the following circumstances:

1. When the use of green energy in a facility area has become much more than before, as indicated by the number of clients who buy green energy from the local energy cooperative. Comparison with the national average of green energy users (36 %) can illustrate the strength of the rise [8, 12, 30].

2. When a lot of the energy needs of small-scale users in the facility area are covered by renewable energy production facilities produced by the local energy cooperative. The contribution of renewable energy production by a local energy cooperative can be indicated by the number of households whose energy needs are covered or by the size of the investment in relation to the size of normal investment [6].

3. When many more small-scale users (households, small, and middle-sized companies) start producing renewable energy for their own needs or re-sell it under the influence of a local energy cooperative. Statistics about the production of renewable energy by households in that facility area or in the Netherlands are indicated in the database of the production installation of energy users [31]. Another possible indicator is the percentage of average households whose energy needs are completely covered by households' renewable energy production.

The occurrence of an energy transition or radical innovation in a facility area is assessed on the basis of changes in renewable energy production and/or consumption, as illustrated in points 1–3. A strong change in renewable energy is not necessarily based on strong changes in both consumption and production of renewable energy. A change in either consumption or production can be convincing on its own.

What factors determine whether to develop or up-scale a local energy cooperative?

Innovation system theory is used in this study to identify and assess the relevance of the dominant factors that influence the development of the local energy cooperatives (see the section on theories). Firstly, the probability of an energy transition in a local energy cooperative is stronger when the following conditions occur in relation to local energy cooperatives and their clients: 1) comparative advantages, 2) lack of complexity, and 3) compatibility. Our four cases are thus analyzed in terms of these conditions. Secondly, energy transition has the potential to be successful in competition with other energy production or delivery organizations when the following crucial factors occur: 1) the incentive to innovate, 2) the incentive to reveal the innovation, and 3) low costs and competitive advantage in relation to private enterprises. Our four cases are thus analyzed in terms of these factors.

Method

The first research question was answered by content analysis of white papers, research reports, scientific or other articles, and websites. Information concerning local energy cooperatives in the Netherlands was gathered and analyzed. In addition, information was gathered in four interviews with frontrunners amongst the local energy cooperatives and two experts in the field.

The second and third research questions were answered by case studies of four frontrunners from their establishment and development until January 2014. White papers, (policy) reports, verbatim reports of meetings, and newspapers concerning the local cooperatives and their organizational environment were gathered and analyzed. The content analysis was facilitated by the abundance of reports by local governments and most of the local energy cooperatives. Liander [31] provided a database with statistics about the local production of different forms of renewable energy by small-scale users. Furthermore, the databases of Statistics Netherlands were used to describe the facility areas. The content analysis and the databases were used to answer the second research question, and some details were checked. The content analysis was also used to describe the background of the facility area, the services of the local energy cooperatives, and the network environment. In addition, data were collected from interviews with the local energy cooperative and the local government in the facility areas, the regional government, and the energy supplier that delivered energy before privatization and liberalization. In sum, 23 interviews were held with 17 persons connected to the case studies (some people were contacted twice). In addition, another four persons were interviewed. All the interviews were conducted using an interview protocol with relevant points, and 12 interviews were recorded in order to facilitate the analysis. Furthermore, for three of the cases, a conference about different aspects of the local cooperatives was attended. For the Goeree-Overflakkee case, the wind turbine sites were visited. The four case studies are described in terms of the relationship between local energy cooperatives and their clients, with special attention for the network in which this relationship is embedded (background, services provided, organization, and networks). In addition, research questions 2 and 3 are answered using the descriptive information about the case studies as indicated in the section on main question and research questions.

The reputation of local energy cooperatives as frontrunners was ascertained in interviews with Hier Opgewekt and an expert in this field. Hier Opgewekt acts as an independent information and communication manager in the field of Dutch local energy initiatives. The expert recently wrote a description of the Dutch local energy initiatives. Furthermore, specialist literature was used as well as information about prize-winning organizations amongst local energy cooperatives, participation on national committees of local energy cooperatives, and participation in the dissemination of knowledge about local energy cooperatives. Since several local energy cooperatives specialize in wind energy, it was appropriate to choose one such cooperative as a case study, whereas the others specialize in other renewable energy technologies. Since the number of members in a cooperative is probably part of the success of frontrunners, it is understandable that the frontrunners are larger than average, but we avoided choosing only the largest cooperatives. Most of the frontrunners are relatively old, but we also chose some cooperatives that were established not so long ago. Having taken into account the aforementioned background characteristics (technology, size, and age) of frontrunners, we selected Deltawind, Texel Energie, Grunneger Power, and Lochem Energie [21, 32].

Results and discussion

Development stage of local energy cooperatives in the Netherlands

Local energy cooperatives started to be established in the 1980s and specialized in wind energy. The first local wind energy initiatives involved researchers and technicians on university campuses. In a later stage, environmental groups and citizens started wind cooperatives that aimed to investigate the promises of new energy technologies [33–35]. To date, around 16 local wind cooperatives have been established [21, 32, 36]. Most of these wind cooperatives evolved into organizations that connected a stable number of members. The local wind energy cooperatives have contracts with energy producers in the energy markets and are bedded in as part of the energy system (regime). Local wind energy cooperatives are integrated in long-term energy production plans [37]. The wind cooperatives are an innovation with a relevant source of renewable energy that can be qualified as having left the early stage of development and having developed into a mature stage, but leaving room for further development.

Later, local energy cooperatives were established that did not specialize mostly in wind energy but stimulated all different sorts of renewable energy production and energy conservation. These cooperative experiments share a common idea, but each of them chooses its own form. We consider these local energy cooperatives as small-scale experiments or projects (niche innovations). The establishment of around 91 local energy cooperatives (not specialized in wind energy) in the last 7 years is a new reality in the Dutch energy market [21, 32, 36]. The large number of energy cooperatives means that these cooperatives cannot be seen as isolated experiments. Apart from these local energy cooperatives, the Association of Dutch Municipalities calculates that there are 400 local energy initiatives. A great variety of initiatives are included in these counts, such as technical feasibility studies, research on measures for social housing cooperatives, information evenings about insulation, and/or the installation of solar panels [3].

At this juncture, the establishment of local energy cooperatives (not specialized the wind energy cooperatives) is in an early stage of development [21, 38]. As 90 % of the local energy cooperatives have fewer than 250 green energy consumers, the volume of their demand for green energy is modest [21]. Their consumption of green energy represents a mere 1 % of the energy consumption in an average-sized community, but many local communities do not have a local energy cooperative. The renewable energy that is realized is often restricted to some renewable electricity (solar panels); there is no, or very little, renewable heat or renewable gas. The local cooperatives operate as a retail organization selling green energy under the license of one of the established energy suppliers in the energy market. Recently, the first new local energy cooperatives with their own license were founded to accommodate this new form of energy company. The gradual expansion of the number of clients or the production of renewable energy by these organizations in the years to come could take the local renewable energy cooperatives that are not specialized in wind energy out of the early stage of development.

Case studies

Background

Our frontrunner local energy cooperatives were established in different periods: Deltawind (1989), Texel Energie (2007), Lochem Energie (2010), and Grunneger Power (2012). The background characteristics of home/facility areas these cooperatives are different (see Table 1). Three of them are located in rural areas with many townships, whereas one is in an urban area. Two of the cooperatives are surrounded by water (Deltawind on a peninsula and Texel Energie on an island), and the other two are situated on land. The rural cooperatives are located in areas of natural beauty, with nice landscapes attractive to tourism.

The process to establish the cooperatives was broadly similar in all cases. Several people were enthusiastic about the production of local renewable energy and asked others to share this idea. A small group elaborated the idea and realized that success was possible. Because the concept of a local energy cooperative specializing in renewable energy is recognized as useful and inspiring, the first steps were relatively easy. Volunteers offered their services, and energy consumers joined the club. Each additional step was welcomed by others, and the organization started to grow.

Table 1 Background characteristics of the local cooperatives' home/facility areas

	Deltawind	Texel Energie	Grunneger Power	Lochem Energie
Surface area (land/water) in square km	262 (160)	170 (302)	84 (0)	213 (0)
Type of landscape	Rural	Rural	Urban	Rural/ urban
Type of area	Peninsula	Island	Land	Land
Number of townships	14	22	14	12
Local government	Goeree-Overflakkee	Texel	Groningen	Lochem
Regional government	South Holland	North Holland	Groningen	Gelderland

The initiators did not see how their ambition could be realized in the context of the existing energy sector. Their sense of that market's failure was strong, and consequently, a new organization was unavoidable. After extensive deliberations, the groups decided to create a new legal entity for the local energy organization, i.e., a non-profit local cooperative. The initiators and other early participants formed the hard core of each new cooperative. Local media reported on it, and the group of participants and members grew step-by-step. Local government welcomed the initiatives, communicated with the groups, and started collaboration.

Each of the local cooperatives engaged tens of volunteers to support service delivery of their output. The energy cooperatives had various ways to connect to their adherents. Membership was an easy way of connecting. Deltawind and Texel Energie required payment of a membership fee to join the cooperative. Texel Energie had 3200 members who bought shares in it for an amount of 350,000 Euro in 2013, and Deltawind had 1566 members who invested 3.9 million Euros through loans. Lochem Energie and Grunneger Power had 400 and 681 members, respectively (end 2013). These cooperatives do not sell shares or ask for loans as they become member of the cooperative.

Services of the energy cooperatives

The energy cooperatives offer more or less the same concept and a similar type of service to their adherents, although each has its own profile:

Energy services

Deltawind does not provide green energy, but between 2 and 7 % of the households in the 14 townships participate (end 2013: 1566 members) in the energy cooperative. In a few years, Texel Energie has become one of the largest energy suppliers to the island, with an estimated 3400 clients that consume its renewable electricity or gas (end 2013). Lochem Energie and Grunneger Power provide energy to 130 and 523 clients, respectively (end 2013). This is about 1 % of the households in Groningen and Lochem.

Production of renewable energy or installation of renewable energy by small-scale energy users themselves

The production of renewable energy by the local cooperatives concerns renewable electricity (solar panels, wind turbines). Renewable heat or renewable gas was not produced by the local cooperatives. However, Texel is interested in renewable gas and heat, and has developed projects for them that will be realized shortly. Deltawind invested in 22 wind turbines, whereas Texel Energie buys wind energy from one local producer. All the cooperatives are active with solar panels with one or more groups such as households, association of owners and tenants, and SMEs. In some cases, the cooperative is the owner of the solar panels and sells the electricity (Deltawind) or rents the solar energy to tenants or others (Lochem Energie). In other cases, the cooperatives support procurement by their members who buy the solar panels themselves. Deltawind's renewable energy production involves 22 wind turbines and a solar park with 2960 solar panels. The electricity needs of all households on Goeree-Overflakkee are completely covered by the wind turbines and the Deltawind solar park. The number of solar panels installed after consultation or with the help of Deltawind is not recorded. Statistics show that the number of solar panel systems of households on Goeree is twice as much as the national average. Solar panels provide around 3 % of the electricity needs of the households in the facility area. In Groningen and Lochem, an estimated 25 % of the clients of Grunneger Power and Lochem Energie have opted for solar panels. The local cooperatives reported that they contributed directly or indirectly to the installation of solar panels in Lochem (1103) and Groningen (1608). Since Grunneger Power's and Lochem Energie's share of clients is relatively small (around 1 %), so too is the contribution to electricity production in the facility area by the local cooperatives' clients.

Texel Energie reported an involvement in installation of solar panels on different buildings (in sum: 704 panels) after it had given advice and help. The number of solar panels for households to which they directly or indirectly contributed was not recorded. Statistical data from Liander show that on Texel, more than three times as many solar panel systems are installed than the national average, covering around 6 % of the electricity needs of households in the facility area. The supportive volunteers' enthusiasm for solar panels, and the education, and advice from Texel Energie and its partners have contributed to this result, according to the interviewees.

Advice, education

The volunteers in the local cooperatives offer personal advice about different aspects of energy conservation and renewable energy (for example, an energy conservation scan, a heat use scan). In some cases, information evenings are organized. The services of the cooperatives include technical and commercial advice, support of the procurement process, installation, and maintenance (using preferred suppliers), and financial arrangements (loans, subsidies, rent). Some of the local cooperatives are involved in education activities to the local community in general (Lochem) or give lessons to primary or secondary schools (Deltawind). Deltawind participates in local socio-cultural projects on Goeree.

Research and development

Local cooperatives are actively involved in research and innovation. The cooperatives are continuously looking for new services to provide. With the prospect of changes in policy or law in mind, the cooperatives are developing new plans and services to install solar panels. In addition, the feasibility of renewable energy projects is being investigated, often in collaboration with local governments. Two of the cooperatives are actively involved in smart grid research projects that aim to improve energy management at home. The research is undertaken by research institutes or partners, and the local cooperatives enroll participants for the experiments (Texel Energie: 300 participants, Lochem Energie: 30 participants). In addition, one cooperative (Lochem Energie) is participating in an experiment with smart energy displays.

Organization of the local cooperatives

The organization of the local energy cooperatives was a rather open process in which people became involved step-by-step. It was an improvised process led by the inspired initiators who wanted to produce and consume local renewable energy. The initiatives without exception chose the legal entity of a non-profit cooperative. This choice was induced by an aversion to private energy enterprises in the energy sector. The choice of cooperative as legal entity shows that these organizations were established by people and for people. In order to build up a professional organization, several employees were hired. In 2013, Deltawind had seven employees (4.4 fte); Texel Energie, six employees (5 fte); Grunneger Power, two employees (2 fte); and Lochem Energie, two employees (<1 fte). As already mentioned, the local cooperatives are supported by tens of volunteers that provide advice or other services. The cooperative option has become the model for the establishment of new local energy initiatives. In this legal entity, the members are decisive in all important matters such as the policy headlines and composition of the board.

Local cooperatives and their networks

The production and delivery of green energy requires collaboration with many other organizations. Table 2 outlines each of the operations' networks of private, semi-public, and public partners. As the freshly established local cooperatives did not produce energy, it was necessary to buy green energy and subsequently deliver it to their clients. The local cooperatives chose at the

Table 2 Networks of the local energy cooperatives

	Deltawind	Texel Energie	Grunneger Power	Lochem Energie
Energy distribution	▪ No energy distribution	▪ HVC	▪ Trianel	▪ Trianel
		▪ Greenchoice	▪ Sepa-Green/EABO	▪ Eneco
		▪ DGB	▪ NLD-Energy	
		▪ DE Unie		
Energy production	▪ E.on	▪ TEzon, TEgas, TEjoule	▪ Solar cooperative	▪ Not applicable
	▪ Three limited companies			
Preferred suppliers	▪ Greenchoice	▪ Installation firm	▪ Installation firms (5)	▪ Eneco
	▪ Eneco-Wind			▪ Installation firms (4)
Social housing organizations	▪ No	▪ Collaboration	▪ Collaboration	▪ Collaboration
		▪ Woontij	▪ Leeuwenborgh	▪ IJsseldal Wonen
Local government	▪ Independent	▪ Independent	▪ Independent	▪ Independent
	▪ Collaboration	▪ Involved	▪ Involved	▪ Involved

start not to buy green energy from the large energy cooperatives. The energy cooperatives not specialized in wind energy changed their preferred green energy provider for a variety of reasons. Some of the providers did not produce electricity as well as gas (HVC), were not flexible enough (Greenchoice), or went bankrupt (Trianel). After several years, two local cooperatives (Texel Energie, Grunneger Power) opted for a new organization with its own license to deliver green energy (DE Unie, NLD Energie). Independence was seen as an important advantage, and becoming independent seems to be their final choice after an elaborate search process. In the end, Lochem Energie chose one of the large energy cooperatives to deliver green energy (Eneco). The position of Deltawind is different from that of the other energy cooperatives. Consequent to mergers and acquisition processes, the energy production organization E.on has become a client for Deltawind's wind energy.

Some of the local energy cooperatives established new organizations for the production of renewable energy for fiscal and financial reasons. The autonomy of these private enterprises is a condition *sine qua non* for investors. One wind turbine park was developed in collaboration with a private company established by one of the Deltawind founders. In addition, Texel Energie decided to establish several private enterprises for their renewable energy production (TEzon, TEgas, TEJoule). One of the two other local cooperatives decided to establish an energy cooperative for collective solar panels in order to avail of tax exemptions.

The local cooperatives started partnerships with local government, and three of them collaborate on solar panel projects with social housing cooperatives, whose involvement was problematic for a long time because of the split incentive between investor and beneficiary. The energy cooperatives developed new services to solve these problems. Now, tenants can participate in solar panel projects operated by collectives in Groningen and Lochem. In addition, Lochem Energie invites tenants to rent solar panels for their own homes.

The collaboration with local governments is based on coinciding missions and goals. The local governments have a mission to achieve an energy neutral status in their territory, and the local energy cooperatives are established to produce renewable energy. Local governments respect the cooperatives' independent identity. The sympathy of local governments for the energy cooperatives is manifest in the way they make themselves accessible to the cooperatives, which have taken on some of the role that local governments played on the energy market for decades.

The partnership with local government has proven to be very important for the local energy cooperatives. Local governments facilitate the cooperatives in a variety of ways such as support for the establishment of technical feasibility studies on renewable energy. They give financial support to the cooperatives through loans, subsidies, and other support. The partnership supports new projects undertaken by the cooperatives because renewable energy requires a supportive local energy policy, and legal and regulatory support.

Mission or perspective of the local energy cooperatives

The missions of the energy cooperatives have many similarities when formulated in abstract statutory terms. The idea is to enhance energy conservation and to produce local renewable energy. However, the individual local cooperatives have differing perspectives.

- Deltawind has already succeeded in producing electricity for the inhabitants of Goeree-Overflakkee. When another 22 wind turbines are installed, the peninsula will be an exporter of wind energy. The Deltawind wind turbines are integrated in national plans for wind energy.

- In 7 years, Texel Energie has become one of the largest energy suppliers to the island. Its plans to produce renewable gas and renewable heat are a signification contribution towards the community's long-term objective: energy independence.
- Grunneger Power and Lochem Energie have a relatively small clientele, and their objective is to grow and to ensure a solid financial basis for their organizations. Grunneger Power chose to focus on solar energy, and this limits the potential for energy production.

Energy transition in local energy cooperatives (research question 2)

Deltawind At Goeree-Overflakkee, a radical energy transition took place because of Deltawind. The cooperative does not sell green energy but produces enough renewable electricity for the 15,000 households on the peninsula by wind turbines and a solar park. Furthermore, it has enhanced the production of renewable energy by solar panels, resulting in an additional contribution to renewable energy. Deltawind will produce more renewable energy by wind turbines in the near future and will contribute to the electricity needs of other parts of the Netherlands.

Texel Energie Texel Energie has caused a radical energy transition on the island. An estimated 55 % of all households on the island buy green energy from the new energy cooperative, and around 16 % of the other households buy green energy from the other energy suppliers. In sum, 71 % of the households on the island use green energy, thereby giving a market pull to this type of energy. Furthermore, a large proportion of the households on Texel produce renewable energy with solar panels; around 6 % of the electricity needs of households on the island is covered by these systems. So far, Texel Energie does not produce renewable gas and heat, but plans to do so have been developed.

Grunneger Power An energy transition was not realized in Groningen, Haren, and Ten Boer. The number of green energy consumers is insufficient, and not enough renewable energy electricity is produced. So far, no plans for renewable gas and heat have been developed.

Lochem Energie An energy transition was not realized in Lochem because the number of inhabitants using green energy has not changed enough. The same holds for the production of renewable energy electricity. So far, no plans for renewable gas and heat have been developed.

Energy producers and customers: comparative advantage, complexity, and compatibility (research question 3)

Deltawind

Comparative advantage, complexity, and compatibility

From the start in 1991, Deltawind found local customers who were interested in wind energy because they saw the product as having a comparative advantage over other energy sources. The local energy cooperative (EMGO) was only mildly interested in the new technology. However, the idea of local production by Goeree inhabitants was convincing enough for the local energy provider. In return for buying the wind energy, EMGO asked for new customers to consume the product. For Deltawind's later customer (E.on), the comparative advantage of wind energy was obvious because the popularity of "home-grown" green energy is growing. The ongoing societal discussion about the authenticity of green energy is an additional comparative advantage for Deltawind.

Neither is complexity an issue because wind energy is as useful as the energy from a coal-fired power station. The transportation process is not more difficult for wind energy because the same infrastructure is used. The use of wind energy was not obstructed by high complexity and compatibility, and the comparative advantages were clear and became more evident over the years.

Texel Energie, Grunneger Power, Lochem Energie

Comparative advantages

Concept From the start, the local energy cooperative idea offered an attractive new concept for energy conservation and renewable energy. The concept comprises several products (e.g., green energy, renewable energy) as well as the cooperative's democratic structure. The idea to further develop the cooperative is an intrinsic and intriguing part of the new concept. For participants, this concept has strong comparative advantages relative to large energy companies on the Dutch market. The comparative advantages of the local cooperatives are clear; however, they seem to attract a very motivated but relatively small number of customers. The popularity of the concept on Texel seems to be the exception to the rule. The cooperative responds to the island's idea of autonomy and cutting the cable for energy transport from the mainland. The beautiful nature and landscape are additional reasons to support renewable energy as the ultimate natural way of energy production.

Free and personal advice In comparison to other energy delivery companies, the local cooperatives offer free and accessible one-to-one personal advice about domestic energy conservation, heat leakages at home, the advantages of insulation, and sustainable lights. The free support for the procurement of solar panels is an additional advantage. The accessible advice of volunteers engaged by the cooperatives supports the idea of commitment. Although energy is no longer the low interest product that it used to be, many people are not interested in free advice about energy conservation or other issues.

Green energy, renewable energy A comparative advantage of the local cooperatives is that real green energy is provided. Since much of the marketed green energy is "grey" green energy, the added value of real green energy from the local energy cooperatives is clear enough. For most people, the energy price is by far the most important determining factor in the choice of an energy delivery organization [8]. In recent years, green energy organizations' market share has decreased nationwide, but, apparently, the clients of the local energy cooperatives are motivated to buy real green energy from the local energy cooperative. The perceived comparative advantage of green energy is so strong that even the bankruptcy of Groningen's and Lochem's licensed energy supplier, and the uncertainties about energy delivery that resulted from this, did not lead clients to choose another energy supplier. This shows that the clients appreciate the green energy of their local cooperative much more than the services of other suppliers. Furthermore, more than elsewhere, the local cooperative on Texel is seen as part of the community; it is an expression of the inhabitants' cultural identity, which values autonomy highly. The large number of clients on the island shows the perceived comparative advantage of the energy services of "their" Texel Energie.

Complexity

The concept of the local energy cooperatives (not specialized in wind energy) and some of the products that they provide are relatively complex. The idea of local cooperatives is to reduce energy needs and to produce as much renewable energy as possible for the remaining energy needs. The possibilities for producing renewable energy need careful consideration by clients and so do the measures for energy conservation. The high complexity is probably an obstacle to attracting large groups of customers to the local energy cooperatives. Participation in the three energy cooperatives is more demanding than using energy from a traditional supplier. Even though people are nowadays more interested in energy than before and switch to new providers relatively easily, participation in the local energy cooperatives is relatively complex. For Texel, complexity is not a very limiting factor, because local commitment to the island is strong. Choosing Texel Energie in this case means not only choosing green energy but also choosing the island and its local identity.

Compatibility

The energy behavior of the three cooperatives' customers is not easily compatible with much of the normal energy behavior of a standard household. In addition to the complexities in energy consumption and production, the cooperatives require active involvement. This involvement might include participation as a volunteer to provide support for the procurement of renewable energy installations. Membership of the cooperative requires attendance at the cooperative's annual meeting or other gatherings. The choice to produce renewable energy locally is the start of a strategy that demands an active search for different technical options and tests of their feasibility. The need to innovate is apparent in the energy cooperatives' participation in research projects that aim to find new ways to manage the consumption of renewable energy. Compatibility with normal energy use is low, but at the same time, this represents a comparative advantage to some consumers.

Potential to develop or up-scale innovations (research question 3)

Deltawind

Incentive to innovate and incentive to reveal innovation

Innovation is an important aspect of Deltawind's credentials as it was established to bring about change in the energy production sector. The idea of producing renewable energy and being self-sufficient in energy was a revolutionary idea and a strong impetus to innovate. From the start, the cooperative has added to and upgraded its assets, i.e., wind turbines, over the years. The innovation in the past years concerned, for example, the technology of the turbines and the size of the production location. Deltawind is supported by a large group of members that live on the peninsula. The installation of the wind turbines has therefore been accepted by the local community. Deltawind's open attitude and the visibility of the wind turbines exemplify the cooperative's strong urge to communicate its intentions, plans, and innovation to others. Obviously, in this case, there is no risk of the local cooperative's ideas remaining undiscovered in the ranks of the Deltawind volunteers.

Competitive advantages

Deltawind is in a good position to produce additional renewable energy through wind turbines on Goeree. The organization is strongly rooted in the local community on the peninsula. It seems to be no coincidence that the second large wind cooperative is established on an island. Furthermore, the price of the energy produced by the wind turbines is attractive in comparison to other forms of renewable energy and acceptable in relation to fossil fuels. The cooperative's solid financial position enables it to invest in wind turbines. As a non-profit organization, Deltawind is a more credible protagonist of renewable energy than private energy companies. Its plea for renewable energy is based on a societal assessment of the need for renewable energy. It is therefore understandable that one of the energy companies that want to build the wind turbines is another regional cooperative.

Deltawind is able to attract local inhabitants to become members of the cooperative and has a good position in relation to other energy producing organizations. The limiting factor for up-scaling wind energy production is not so much the competition from other energy organizations as societal acceptance in and outside Deltawind. The enormous size of an additional wind park is not acceptable for everyone. The recent initiative to install additional wind turbines, about which it consulted with its members, is probably an acceptable new goal for the cooperative. The original goal of self-sufficiency for Goeree has been realized, but the wind turbines are seen by many as a valuable technology for others outside the area as well.

Texel Energie, Grunneger Power, Lochem Energie

Impetus to innovate and incentive to reveal innovation Local cooperatives were established because of dissatisfaction with large energy companies. The cooperatives' members share the idea of producing local renewable energy and an economical use of energy. The new concept has been developed by and for the members, whose active participation in research and development projects exemplifies their commitment to innovative services and products.

Local cooperatives have a strong incentive to reveal their new concepts, services, and products. Their mission to change energy consumption and production demands an open attitude. The cases show proactive cooperatives that attract new members, energy consumers, investors, and collaboration partners to their organization. The cooperative is an appropriate legal entity to include as many different groups and partners as possible. The need to attract collaborating partners is partly induced by the small size of the organization. A small professional staff in the local cooperatives complemented by volunteers demands additional expertise and skills to fulfill the cooperatives' ambitions. The energy cooperatives prove very successful in connecting with partners and societal groups. The proactive attitude towards new partners is also necessary to secure the financial foundations of the private cooperatives. Public as well as semi-public and private partners are important investors in the local cooperatives.

Competitive advantages

The competiveness of any supplier of energy to small-scale users is continuously on the line because the market has been liberalized. This means that consumers are at liberty to choose another energy supplier. This offered the local energy cooperatives the chance to start their green energy delivery using the facility that enables consumers to switch. The price of green energy is higher than the price of energy from fossil fuel. The following competitive factors influence the considerations of the local energy cooperatives' clients:

Committed clients Local cooperatives are built on members' enthusiasm, clients' loyalty, and a strong commitment to renewable energy. The great number of energy cooperatives and the great number of participants demonstrate the societal support for this idea. Even the bankruptcy of a green energy supplier did not change clients' commitment. The social cohesion on Texel supports the commitment of its members, and it seems no coincidence that the second large energy in the Netherlands cooperative is also located on an island.

Promising coalitions Local cooperatives share the same goals concerning renewable energy as regional and local governments and social housing cooperatives. Collaboration in projects with local and regional partners is based on coinciding goals. Partnership with local government is promising because of the access to legal licenses, spatial reservations for renewable energy, and ownership of the energy network.

Social acceptance The collaboration between local energy cooperatives and local government provides energy projects a legitimacy that projects of other organizations lack. Projects like a solar park of solar panels (Lochem), solar panels on the town hall (Lochem) and tenants' houses (Groningen), and solar panels in the social housing sector (Lochem, Groningen) are realized in collaboration between energy cooperatives and local government.

Competitive disadvantages

Governance power The non-commercial spirit and the participation of volunteers is an important asset but also a weakness. Supplying an energy market demands technical knowledge, a professional organization, and entrepreneurship. The bankruptcy of the energy supplier in Lochem and Groningen shows cooperatives' vulnerability.

Financial constraints Local cooperatives employ a small staff and a relatively large workforce of volunteers. In order to be competitive on this market, the local organizations should invest in renewable energy projects, but large energy companies have much larger investment budgets [6]. Investment in (collective) renewable energy is difficult because of a lack of resources.

Energy networks Dutch energy networks have been built over the course of decades and are designed for the central production of energy [5, 17]. Local production of renewable energy demands a change in the local energy infrastructure. Two local energy cooperatives anticipate problems with local energy infrastructure and are participating in experiments to test new energy management systems that accommodate the supply of, and demand for, renewable energy.

Legislation The energy market is a liberalized but regulated market that provides rules for the production, transport, and delivery of energy. The relationship between local cooperatives and local government is potentially challenging or even problematic [38]. Local governments combine different roles such as shareholder in old energy suppliers, initiator and/or shareholder of new energy cooperatives, investor in cooperatives, and collaborating partner in many projects. Local governments run the risk of violating European and national rules concerning the energy market.

Conclusions

Energy transition or not?

The statistics on the performance of the Netherlands in renewable energy use, indicating a relatively small amount of final energy consumption from renewable energy sources on the one hand and the establishment of many local energy cooperatives specializing in renewable energy on the other hand, are not contradictory. The establishment of about 91 local energy cooperatives (not specialized in wind energy) in the last 7 years is a niche innovation in an early stage of development and not yet integrated in the energy sector. The local energy cooperatives have settled as an undercurrent in Dutch society and the energy market. The establishment and development of 16 wind energy cooperatives in the last three decades is an example of a niche innovation in a mature development stage that was successfully integrated in the energy production system (regime). Apparently, the contribution of the wind cooperatives was not sufficient to provide a better score for the Netherlands in energy statistics.

However, the contribution of renewable energy consumption and production appeared to be successful in two of the four case studies of local communities. The case studies demonstrated that on Goeree-Overflakkee and Texel, an energy transition or radical innovation took place, whereas none occurred in Groningen and Lochem. Deltawind already produces enough renewable electricity (wind turbines and solar panels) for all 15,000 households on the peninsula. The realization of another 22 wind turbines will further add to the production of renewable energy. Texel Energie can also be seen as the initiator of an energy transition because of its large number of green energy consumers and the substantial production of green electricity by households. The production of renewable gas and heat in the near future will give an extra boost to energy production on the island. The number of green energy consumers in Groningen and Lochem is but a small proportion of normal energy consumption, and the production of renewable energy is limited to renewable electricity (especially solar panels), and renewable gas and heat are not yet produced.

If and to what extent the concept of the local cooperatives can contribute to a transition towards production and/or consumption of renewable energy in the Netherlands is summarized in Table 3. The Deltawind concept is relatively attractive for clients or buyers in comparison to the local energy cooperatives not specialized in wind energy. Active participation in the energy cooperative appeals to a group of enthusiastic consumers, but, for many others, membership of this cooperative is too complex and demanding. The use of wind energy is very simple and compatible since the same energy infrastructure is used. It is not likely that the not-wind energy cooperatives will cause an energy transition or radical innovation in the near future.

The concept of the wind energy cooperative, Deltawind, has a better position on the energy market than that of the other local energy cooperatives. The organization is strongly rooted in the local community, the price of its wind energy is acceptable, and social acceptance of the organization is strong. Deltawind was able to build wind turbines without significant protest from citizens or environmental groups. The three energy cooperatives not specialized in wind energy sell relatively expensive green

Table 3 Potential for change of renewable energy production and consumption by local wind and energy cooperatives in the energy sector

Potential energy cooperatives		Deltawind	Texel Energie, Grunneger Power, Lochem Energie
Energy transition/ radical innovation in facility area		+	−/+
Success conditions radical innovation (relation producer-consumer)	Comparative advantage of the concept, services	+	+/−
	Complexity	+	−
	Compatibility	+	−
Success conditions bottom-up innovations	Incentive to innovate	+	+
	Incentive to reveal	+	+
	Competitive advantage in relation to other suppliers/producers	+	−

energy and produce some renewable energy by solar panels. Time will tell whether these local energy cooperatives will be able to maintain the commitment of their clients and use their social acceptance and their partnerships with public and semi-public organizations for further development. Competitors in the energy sectors have the advantage in terms of governance power, financial resources, and the existing energy infrastructures. The balance between competitive advantages and disadvantages will probably be influenced by the national government's choices about whether or not to stick with the privatized market regime.

The Texel Energie example demonstrates that, under specific circumstances, a local energy cooperative is able to cause an energy transition in its facility area despite using a difficult concept. The large number of small townships on the island and the abundance of space, wind, and sun are favorable circumstances for the production of renewable energy. The cooperative responds to the island's idea of autonomy and of no longer having to import energy from the mainland. The Deltawind example shows a powerful concept that caused a radical innovation in energy production on Goeree-Overflakkee. However, it is not certain that this concept can be easily reproduced in all other local communities. The active or even passive support of citizens and social groups on Goeree-Overflakkee seems to be rather the exception than the rule. In many other local communities, serious obstacles arise such the acceptance of wind turbines, spatial planning problems, and legal procedures and protest [35].

The successes of Deltawind and Texel Energie in terms of contribution to energy transition or radical innovation are obviously not a guarantee for the same results elsewhere. The uniqueness of the circumstances in the case studies prevents easy reproduction of the success mechanisms of an innovative concept. The idea of up-scaling the experiences of niche innovations to an existing regime is probable too simplistic [39, 40]. It is therefore too early to interpret these examples as a revolution in disguise in the Netherlands. The case studies in this article give an indication of the potential of several new concepts for energy transition, as well as obstacles that must be tackled. Because of the uniqueness of circumstances, additional case studies will probably not result in readymade solutions to contribute to energy transitions or radical innovations. However, knowledge about other energy transitions or radical innovations and about additional successful mechanisms in other local energy cooperatives is needed. A combination of success mechanisms in different case studies can probably facilitate the switch towards a zero energy society.

Competing interests

The authors declare that they have no competing interests.

Authors' contributions

JAMH contributed to the design and carried out the data collection and analysis. JFMK contributed to the theoretical background, the analysis, and conclusions. Both authors read and approved the final manuscript.

Acknowledgements

We would like to acknowledge Catherine O'Dea for her services as language editor as well as the participants in our workshop (Smart Energy System Transition in Cities and Regions) for their feedback on a concept version of this article during the ESEIA conference held on 24 and 25 April 2014 at the University of Twente.

Author details

[1]Research and Consultancy in Leiden, Galileïweg 8, 2333BD, Leiden, The Netherlands. [2]Department of Public Administration, Faculty of Social Sciences, Erasmus University Rotterdam, P.O. Box 17383000DR, Rotterdam, The Netherlands.

References

1. Eurostat Newsrelease (2014) Renewable energy in the EU28 (37/2014). Available via Eurostat Press Office, http://ec.europa.eu/eurostat/documents/2995521/5181358/8-10032014-AP-EN.PDF/91350d4a-4b57-4833-b9f0-32cfe0a6d360?version=1.0. Accessed 19 February 2015
2. European Commission (2013) Innovation scoreboard 2013. Brussels. doi:10.2769/72530. Accessed: 15 May 2015.
3. Buitelaar S (2013) AL 500 lokale energie-initiatieven. http://www.binnenlandsbestuur.nl/ruimte-en-milieu/nieuws/al-500-lokale-energie-initiatieven.9165637.lynkx. Accessed 19 February 2015
4. Planbureau voor de leefomgeving (2011) De Energieke Samenleving, Op zoek naar een sturingsfilosofie voor een schone economie. PBL, Den Haag
5. Afman MR, Bles M, Schepers BL, Wielders LML (2014) Electriciteitsleveranciers in kaart.. CE, Delft
6. Afman MR, Bles M, Schepers BL, Wielders LML (2014) Elektriciteitsleveranciers in kaart (referentierapport) CE, Delft
7. Consumentenbond, Greenpeace, Hivos, Natuur & Milieu, Vereniging Eigen Huis, WWF, Wise (2014) Onderzoek duurzaamheid elektriciteitsleveranciers. Utrecht. Available via Consumentenbond. http://www.consumentenbond.nl/tests/bestanden/woning-huishouden/onderzoek-duurzaamheid-elektriciteitsleveranciers2014.pdf. Accessed 19 February 2015
8. Marketresponse (2014) Stand van zaken op de energiemarkt. Available via autoriteit consument en markt, https://www.acm.nl/nl/download/publicatie/?id=12909. Accessed 12 February 2015
9. Krozer Y (2014) Smart grid and EU 27 countries policy. Paper presented at the ESEAI conference Twente University, Enschede, 24–25 April 2014 Krozer, Y (2014) Smart Grid and EU 27 countries policy. In: European Sustainable Energy Innovation Alliance conference; Enschede, 24–25 April 2014, 24-04-2014 - 25-04-2014, Enschede
10. Association of Energy Companies in the Netherlands and Association of Network Operators in the Netherlands (2012) Energy in the Netherlands 2011. Den Haag/Arnhem. Available via Windunie. http://www.windunie.nl/Documents/Energie-in-Nederland-2011rapport.pdf Accessed 19 february 2015
11. Ministerie van Economische Zaken, Landbouw & Innovatie (2011), Energierapport 2011, Den Haag. Available via Rijksoverheid. http://www.rijksoverheid.nl/bestanden/documenten-en-publicaties/rapporten/2011/06/10/energierapport-2011/energie.pdf. Accessed 19 february 2015
12. Schöne S (2014) Groene stroomconsumenten hebben recht op geloofwaardig product. Energiegids. Het vakblad over energie-efficiency 5:24–5
13. Ministerie van Economische Zaken (2005) Nu voor later. Staatsuitgeverij, Den Haag
14. Ministry of Economic Affairs, Agriculture and Innovation (2011), Available via Dutch parliament, http://www.tweedekamer.nl/zoeken?qry=2011Z07455%2F2011D19360&clusterName=Overige+Kamerstukken&Type=Kamerstukken&fld_prl_kamerstuk=Overige+Kamerstukken&fld_prl_soort=Brief+commissie+aan+bewindspersoon&srt=score%3Adesc%3Anum. Accessed 23 May 2015
15. Raad SE (2013) Energieakkoord voor duurzame groei. Sociaal Economische Raad, Den Haag
16. Ministerie van Volkshuisvesting, Ruimtelijke Ordening en Milieubeheer, Een wereld en een wil werken aan duurzaamheid (2001). Available via RIVM,

http://www.rivm.nl/bibliotheek/digitaaldepot/VROM2001NMP4.pdf. Accessed 19 February 2015

17. Rotmans J (2012) In het oog van de orkaan. Aeneus, Boxtel
18. Loorbach J (2010) Transition management for sustainable development: a prescriptive, complexity-based governance framework. J Policy Adm Inst 23:161–83
19. Koppenjan J, Frantzeskaki N, Loorbach D, Charles MB, Ryan N (2012) Introductory editorial. Int J Sustain Dev 15:1–18
20. Verbong G, Loorbach D (2012) Introduction. In: Verbong G, Loorbach D (eds) Governing the energy transition? Routledge, New York/London
21. Planbureau voor de leefomgeving (2014) Energiecoöperaties, Ambities handelingsperspectief en interactie met gemeenten. PBL, Den Haag
22. Lundvall BA (2010) National innovation systems of innovation, towards a theory of innovation and interactive learning. Anthem Press, London
23. Lintsen H (2005) Made in Holland, Een techniek geschiedenis van Holland. Walburg Press, Zutphen
24. Rogers EM (1983) Diffusion of innovations. New York Free Press, New York
25. Hekkert M, Ossebaard M (2010) De innovatiemotor, het versnellen van baanbrekende innovaties. Van Gorcum, Assen
26. Egyedi TM, Vrancken JLM, Ubacht J (2002) Inverse infrastructures: coordination in self-organizing systems. Available via Delft University of Technology, http://www.tbm.tudelft.nl/fileadmin/Faculteit/TBM/Over_de_Faculteit/Afdelingen/Afdeling_Infrastructure_Systems_and_Services/Sectie_Informatie_en_Communicatie_Technologie/medewerkers/tineke_egeydi/publications/doc/siit2007_Inverse_Infra.pdf. Accessed 19 February 2015
27. Von Hippel E (2009) Horizontal innovation networks—by and for users. Available via Massachusetts Institute of Technology, http://ssrn.com/abstract=1411719. Accessed 19 February 2015
28. Ministerie van Infrastructuur en Milieu, Rijkswaterstaat (2013) Handreiking, De rol van gemeenten bij lokale duurzame energie-initiatieven. The Hague. Available via Rijkswaterstaat. http://www.rwsleefomgeving.nl/onderwerpen/lokaal_klimaatbeleid/publicaties/downloads/handreiking-rol/ Accessed 20 february 2015
29. Rijksoverheid (2013) Visie lokale energie. Available via Rijksoverheid, http://www.rijksoverheid.nl/documenten-en-publicaties/rapporten/2013/11/08/visie-lokale-energie.html. Accessed 19 February 2015
30. Wielders LML, Cheriff S, Schepers BL (2014) Groenestroomklanten voor de consumentenmarkt. CE, Delft
31. The database of regional distribution systems operators is a register of production installations that are reported by energy users. Available via a website of regional distribution system operators. http://www.energieleveren.nl. Accessed 19 February 2015
32. Oskam HP (2011) Local sustainable energy companies and municipal involvement: designing a framework for municipal involvement. Available via Delft Technical University, http://www.tbm.tudelft.nl/fileadmin/Faculteit/TBM/Over_de_Faculteit/Afdelingen/Afdeling_Multi_Actor_Systems/Sectie_Beleid,_Organisatie,_Recht_en_Gaming_POLG/Onze_afgestudeerden/doc/HansPeterOskam_Thesis.pdf. Accessed 19 February 2015
33. Oteman M, Wiering M, Helderman JK (2014) The institutional space of community initiatives for renewable energy: a comparative case study of the Netherlands. Germany and Denmark, Energy, Sustainability and Society 4:11
34. Verbong G, Van Selm A, Knoppers R, Raven R (eds) (2001) Een kwestie van lange adem, De geschiedenis van duurzame energie in Nederland. Aeneas Uitgeverij, Boxtel
35. Agterbosch S (2006) Empowering windpower: on social and institutional conditions affecting the performance of entrepreneurs in the wind power supply market in the Netherlands. Dissertation, University of Utrecht
36. Oskam HP (2012) Rol gemeenten in lokale energiesector. Energie 6:28–30
37. Ministerie van Volkshuisvesting, Ruimtelijke Ordening en Milieubeheer, Nationaal ruimtelijk perspectief windenergie op land (2010). Available via Vereniging Nederlandse Gemeenten, https://www.vng.nl/files/vng/20100614_rpw20bij20bo20ruimte202420juni202010_rh.pdf. Accessed 19 February 2015
38. Arentsen M, Bellecom S (2014) Power to the people: local energy initiatives as seedbeds of innovation? Energy, Sustainability and Society 4:2
39. Van den Bosch S (2010) Transition experiments, exploring societal changes towards sustainability. Dissertation, Erasmus University Rotterdam
40. Koppenjan J (2012) The new public governance in public service delivery. Eleven international publishing, The Hague

4

Upscaling a district heating system based on biogas cogeneration and heat pumps

Richard P van Leeuwen[1,2*], Jirka Fink[2], Jan B de Wit[1] and Gerard JM Smit[2]

Abstract

Background: The energy supply of the Meppel district Nieuwveense landen is based on biogas cogeneration, district heating, and ground source heat pumps. A centrally located combined heat and power engine (CHP) converts biogas from the municipal wastewater treatment facility into electricity for heat pumps and heat for district heating purposes. Development of the urban district is influenced by the current economic and building decline. For the district heating energy concept, a migration strategy for the required infrastructure is required. The migration spans the district's small-scale starting phase involving 40 houses up to a scale of 176 houses.

Methods: An optimization model which maximizes profitability is developed which includes data from district heating and cooling demand patterns.

Results: With the optimization model, optimal CHP size, boiler size, and operational hours are determined for various scenarios.

Conclusions: From the scenario analysis, a migration strategy is developed which starts with a simple system concept supported by boilers to a larger system which includes a CHP. Sustainability in terms of CO_2 emission savings of the energy concept is compared with other possible energy concepts.

Keywords: Cogeneration; Biogas; District heating; Heat pumps; Renewable energy; Low-energy building; System migration

Background

In the Netherlands, at least 40% of the national carbon dioxide emissions are related to the heating demand of buildings. Due to the natural gas reserves, conversion by natural gas boilers dominates the heating supply for space heating and hot tap water. But natural gas reserves are gradually depleting and large-scale carbon dioxide emissions cause climate change. Hence, Dutch policies are aimed to gradually diminish the consumption of fossil fuels in general and increase the share of renewable energy. Recently, local governments take up their responsibility and create favorable policies to stimulate reduction of energy consumption and increase the share of renewable energy on a local scale. One of these initiatives is the smart grid demonstration project MeppelEnergie, which is funded by the Dutch program Switch2SmartGrids. For the project's energy system, we develop a smart grid control. The Meppel energy concept consists of a biogas combined heat and power engine (CHP), backup boilers, high temperature (HT) water storage, heat pumps, and aquifer underground thermal storage. The CHP generates electric and thermal energy. The thermal energy of the CHP is used for district heating, the electricity is used to supply heat pumps placed at houses with no connection to the district heating, or is sold through the external grid.

Cooling energy for the houses is provided by an underground aquifer consisting of a warm and cold well. During the heating season, the warm well provides low temperature (LT) heat for the heat pumps. Cooling energy of the houses also provides part of the required regeneration energy to maintain temperature balance within the underground thermal storage. Another part of the regeneration is provided either by a dry cooler or an effluent stream from the municipal wastewater treatment plant (MTP).

*Correspondence: r.p.vanleeuwen@saxion.nl
[1]Sustainable Energy Group, Saxion University of Applied Sciences, Maarten Harpertsz Tromplaan 28, P.O. Box 70.000, 7500 KB Enschede, the Netherlands
[2]Department of Computer Science, Mathematics and Electrical, Engineering, University of Twente, Drienerlolaan 5, P.O. Box 217, 7500 AE Enschede, the Netherlands

Due to the recent economic recession in the Dutch housing market, the planning of the district's expansion is shifted from ambitious to a more realistic planning. The district is now built in smaller steps, i.e., phase 1 for the period up to 2015 contains 40 houses. After that, phase 2 contains another 136 houses and is built up to 2017. Even after phase 2, the scale is not yet reached to justify all investments in the pictured energy concept. Hence, the energy concept is to be treated as a horizon relating to 300 to 400 houses in the district, which is not expected before 2020. We investigate in this paper a possible scenario for the evolution of the district heating system from a starting system with less investments for the first building phase as shown in Figure 1 up to the final system which includes aquifer thermal storage. The starting energy concept consists of heat generators which run on natural gas. It has to be determined which type of generators (boilers or a CHP with supporting boilers) are optimal in this case. Without biogas supply during the first years, integration of renewable energy is realized by purchasing certified green energy (green gas) from the grid. The starting concept also includes a refrigeration chiller to provide cooling for the houses.

The first goal of this paper is to determine the required heating and cooling generator capacity rates for a district heating and cooling system. The second goal is to determine the most economic heat converter (boiler or CHP) in relation to the number of houses. The third goal is to compare the sustainability performance of the starting energy concept with other options. Contributions of this paper are twofold, first we develop a simplified methodology to estimate heating and cooling demand data for a district heating network and a method to determine the operating time of a heating or cooling device based on such data. Second, we demonstrate the application of such methods for an integrative study on financial and sustainability aspects of a complex energy system.

The paper is structured as follows: the 'Background' section introduces the energy concept and problem statement. The 'Related work' subsection outlines work related to our investigation. The 'Methods' section defines the energy model, equations and case related parameters. The 'Results and discussion' section includes results, and finally, in the 'Conclusions' section, the conclusions are drawn.

Related work

Experience with district heating in the Netherlands is traditionally related to large-scale steam power plants (e.g., Amsterdam and Almere city heating network and many smaller city projects). More recently, decentral projects for new urban districts apply either biomass (wood) thermal conversion in boilers (e.g., muziekwijk Zwolle) or biogas cogeneration (e.g., Apeldoorn, Zeewolde, and Leeuwarden) [1]. An overview of Dutch district heating projects and profitability investigation is presented in [2].

Vallios et al. [3] develop a modeling approach for sizing biomass-fueled district heating systems. The district heating (DH) system only includes biomass boilers. Optimal size of a residential micro CHP and storage in Japan is investigated by Ren et al. [4] who apply a mixed integer nonlinear programming model. They determine optimal

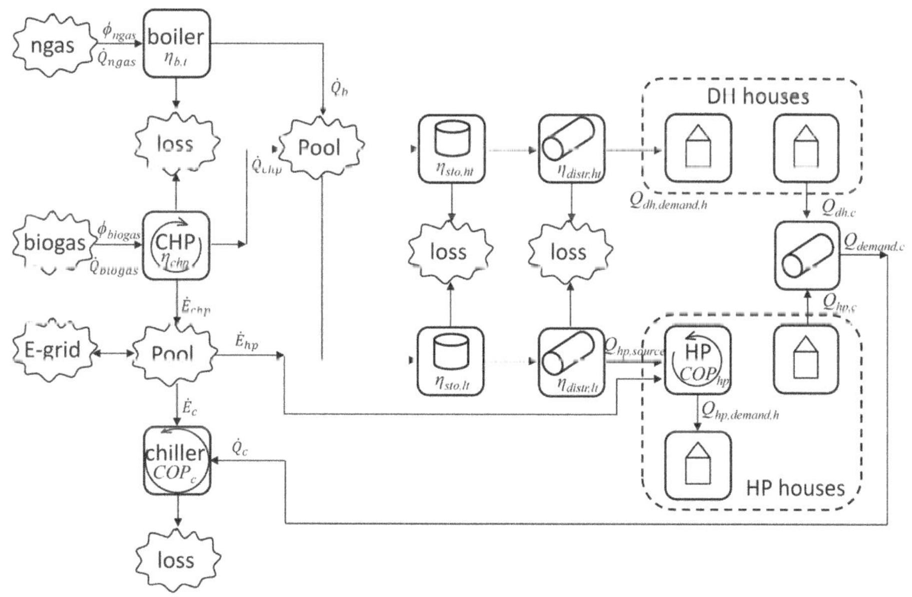

Figure 1 Meppel district heating system starting concept.

CHP size in relation to varying electricity prices. In a similar study, Cho et al. [5] determine with a linear programming model, for a micro CHP an optimal control scheme to reach economic optimal operation. Chinese et al. [6] develop two optimization models to investigate the design of biomass based district heating networks in Italy in relation to profit maximization and greenhouse gas minimization. The optimization is applied to the design of a new industrial district heating network. Curti et al. [7] investigate configurations for district heating heat pumps which draw low temperature heat from Swiss lakes, supported by cogeneration. They use genetic algorithms to find optimal system sizes. Hlebnikov et al. [8] investigate different scenarios to renovate an existing Estonian district heating network and they apply EnergyPLAN and RETScreen softwares. Lund et al. [9] investigate the role of district heating for the future energy system of Denmark, developing a model of the Danish energy system in EnergyPLAN. Pirouti [10] develops a nonlinear programming model to design district heating networks which are minimized on total energy and exergy consumption and operational costs. Finally, Østergaard [11] reports on the regional energy system of the Danish city Frederikshavn in which an EnergyPLAN model of the region's energy production and consumption is used to predict the outcome of geothermal energy application.

Most authors develop specific models for optimization of case specific energy systems, while others apply modeling capabilities of specific software like EnergyPLAN or RETScreen. Both approaches require specific input generation, e.g., home heating and cooling demand, energy price, and tax schemes. Besides that, constraints of the energy supply system are required, e.g., amount of available biogas, supplying order between the CHP and boilers, and system efficiencies. We consider using EnergyPLAN at a later stage of our research. Due to the complexity involved, we develop our own optimization model within this paper which at a later stage serves as validation tool for results generated by EnergyPLAN.

An important difference between the optimization carried out in, e.g., the papers of Vallios et al., Ren et al., Chinese et al., and Curti et al. and our optimization is that we have limited the amount of variables by developing an additional constraint which relates size of each generator (CHP, boiler) to its operational time. Another difference is that within the present paper, our focus is not limited to technical size but also to compare energy concepts on profit and sustainability performance.

Methods

District heating system energy model

The starting concept of the DH system is characterized by converter efficiencies, energy sources, energy demands, and energy flows, as indicated in Figure 1. In the following, we develop an energy balance model for which the nomenclature is given in Table 1.

Efficiency of the thermal storage is determined by the size, the wall insulation, and the average temperature difference between water within the storage and outside air. For the model, we assume that the size of the storages is sufficient to store and supply thermal energy within a daily cycle. As a rule of thumb, 125 L/household storage capacity is estimated for a DH system with maximum storage temperature of 90°C. With this, we determine an average storage efficiency. Hence, capacity or size of the storage is not a variable within the model.

The DH system is conceived as a utility which buys energy (biogas, natural gas) from the upstream market, converts it to electricity and heat, and sells heat to the

Table 1 Nomenclature energy concept

Term	Signification
$\eta_{chp,e}$, $\eta_{chp,t}$	CHP electric and thermal efficiency
$\eta_{b,t}$	Boiler thermal efficiency
$\eta_{sto,ht}$	High temperature (HT) thermal storage average efficiency
$\eta_{sto,lt}$	Low temperature (LT) thermal storage average efficiency
η_{distr}	DH distribution network average efficiency
COP_{hp}	Heat pump (HP) coefficient of performance (COP)
COP_c	Refrigeration chiller COP
$f_{netw,p}$	Fraction of required electric pump energy for network pumps related to transmitted thermal energy of the network
ϕ_{biogas}, ϕ_{ngas}	Biogas supply to CHP, natural gas supply to boilers (m^3/h)
LHV	Lower heating value (MJ/m^3)
$\dot{Q}_{biogas}$ or $\dot{Q}_{ngas}$	Fuel-related thermal energy input (kW)
$Q_{dh,demand,h}$	Total DH heating demand of (*n*) connected houses (GJ)
$Q_{hp,demand,h}$	Total heating energy demand of (*m*) houses with a heat pump (GJ)
$Q_{hp,source}$	Total low temperature source heat for the heat pumps (GJ)
Q_{dhw}, Q_{sh}	Domestic hot water demand and space heating demand (GJ)
$Q_{demand,c}$, $Q_{dh,c}$, $Q_{hp,c}$	Cooling energy demand, for DH network houses, for houses with heat pumps (GJ)
$\dot{E}_{chp}$	CHP electric rated power (kW)
$\dot{Q}_{chp}$	CHP thermal rated power (kW)
$\dot{Q}_b$	Boiler thermal rated power (kW)
$t_{op,chp}$ and $t_{op,b}$	Full load operational hours of the CHP and boiler (h/year)
$\dot{E}_c$, $\dot{Q}_c$	Refrigeration chiller electric rated power and cooling rate (kW)

downstream market, i.e., the households. The electricity is used by circulation pumps and by heat pumps. Surplus electricity produced by the CHP and not used by the local energy system is sold to the grid. Our first emphasis is on the primary choice between either a DH system with only natural gas boilers or with a biogas CHP and some supporting natural gas boilers.

Besides differences in capital and operational costs of the generators, we have to take cost differences of the downstream system (i.e., storage and network assets) into account caused by the type of converter being applied. The network assets are not influenced but we expect a larger HT storage in case of a CHP. When only boilers are applied, it is easy to adjust the generated thermal power instantly within a large range due to cascading and power modulation. With a CHP the thermal modulation range is not as large and part-load conditions lead to a loss of electrical generation efficiency. Besides that, for best efficiency and lifetime performance, a CHP should run more or less continuous when it runs and should not constantly follow the actual demand of the network. Therefore, increased HT storage capital costs are calculated for the CHP case.

Based on average climatic data, the model determines the base generator sizes for heating (CHP, boilers). How this is done is explained in the 'Relation between CHP size and operational time' subsection. In practice, an additional peak boiler is added to provide sufficient heating capacity during extremely cold days which occur infrequently. The peak boiler capacity is calculated from the coldest day during the last 5 years. For cooling, we determine the generator size directly from the warmest day during the last 5 years.

We assume energy tariffs (natural gas, electricity) are only related to total yearly amounts, although on the electricity spot market and day ahead market, prices vary hourly. But these variations are small and often fixed electricity prices are contracted for longer periods. In this way, the complexity and the number of required iterations for the optimization problem to find the number of houses for break even profit is significantly reduced.

The yearly heat balance for the converters is:

$$\left[\frac{Q_{\text{dh,demand},h}}{\eta_{\text{distr}} \cdot \eta_{\text{sto}}} + Q_{\text{hp,source}}\right] \cdot \frac{10^6}{3,600} = \dot{Q}_{\text{chp}} \cdot t_{\text{op,chp}} + \dot{Q}_h \cdot t_{\text{op},h} \tag{1}$$

When the CHP and the boilers together generate the required heat, the CHP has priority and the boilers have a supportive function. Hence, an additional expression is required which relates the operational hours of the CHP to the size of the CHP, which is developed in the 'Relation between CHP size and operational time' subsection.

The total yearly heat demand of the DH system ($Q_{\text{dh,demand},h}$) is related to the average household heat demand. For the average household demand, suitable expressions are developed in the 'Thermal demand specification' subsection. The following relation applies for n houses:

$$Q_{\text{dh,demand},h} = n \cdot Q_{\text{dh,hhdemand},h} \tag{2}$$

Equally, for m houses heated with heat pumps the following relation applies:

$$Q_{\text{hp,demand},h} = m \cdot Q_{\text{hp,hhdemand},h} \tag{3}$$

The source heat demand of the heat pumps is related to the household demand by:

$$Q_{\text{hp,source}} = Q_{\text{hp,demand},h} \cdot \left[1 - \frac{1}{\text{COP}_{\text{hp}}}\right] \tag{4}$$

CHP heat generation rate is related to CHP electric power rate as follows:

$$\dot{Q}_{\text{chp}} = \dot{E}_{\text{chp}} \cdot \frac{\eta_{\text{chp,th}}}{\eta_{\text{chp},e}} \tag{5}$$

The biogas and natural gas flows are related to the CHP electric and boiler thermal power rate, respectively, as follows:

$$\phi_{\text{biogas}} = \frac{\dot{E}_{\text{chp}} \cdot 3,600}{\eta_{\text{chp},e} \cdot \text{LHV}_{\text{biogas}} \cdot 1,000}$$
$$\phi_{\text{ngas}} = \frac{\dot{Q}_b \cdot 3,600}{\eta_{b,\text{th}} \cdot \text{LHV}_{\text{ngas}} \cdot 1,000} \tag{6}$$

The total yearly electric energy demand including generation of cooling energy is as follows:

$$\begin{aligned} E_{\text{demand}} \cdot \frac{3,600}{10^6} = {} & Q_{\text{hp,demand},h} \cdot \left[\frac{1}{\text{COP}_{\text{hp}}} + f_{\text{netw},p}\right] \\ & + f_{\text{netw},p} \cdot Q_{\text{dh,demand},h} \\ & + Q_{\text{demand},c} \cdot \left[f_{\text{netw},p} + \frac{1}{\text{COP}_c}\right] \end{aligned} \tag{7}$$

$f_{\text{netw},p}$ is an estimated average percentage which discounts the required electric pump energy for network pumps as a percentage of the thermal energy transmitted through the piping network. Real pump energies are to be estimated from network fluid friction calculations and frequency controlled pump characteristics. A value of 5% is derived from practical calculations.

Optimization model equations

Financial profit of the starting energy concept shown in Figure 1 is analyzed with an equation in which profit terms are considered related to selling energy to houses (heating and cooling) and to the network (surplus CHP electricity).

Cost terms that we take into account are as follows: the input energy costs (fuel and bought electricity) and the true capital and operational costs of specific energy generators (boilers, CHP, heat pumps, and related storages). In Table 2, the nomenclature of the profit function is given. We excluded infrastructural costs of the downstream network because in practice, all infrastructural investments are payed for by homeowners through connection fees, so these are not carried by the utility. However, there are also financial reservations to be made for re-investments in the infrastructure and in that case, there is a difference between the DH system and heat pumps as the infrastructure for the heat pumps is relatively cheaper. This difference is accounted for in the fixed part of the heat tariff and is translated into additional income in the case of heat pumps. We explain this in the 'Profit rate for selling heat' subsection.

The equation which expresses the net yearly profit as objective function for maximization is defined as follows:

$$P_{\text{net}} = P_{E,\text{grid,sell}} + P_{Q,\text{dh},h} + P_{Q,\text{hp},h} + P_{Q,\text{dh},c} + P_{Q,\text{hp},c} - C_{E,\text{grid,buy}} - C_{\text{biogas}} - C_{\text{ngas}} - \sum_{j} \{C_{\text{cap},j} + C_{\text{op},j}\}$$

$$j \in \left\{\text{chp}, b, \text{hp}, \text{sto}_{\text{chp}}, \text{sto}_{\text{hp}}\right\} \quad (8)$$

Electricity generated by the CHP is partly consumed by the electric equipment of the DH network (pumps, heat pumps, chiller). One of the objectives of the smart grid control is to balance power generation and consumption as much as possible. The objective is to match the daily CHP operating times with the operating times of electric and thermal consumers. The smaller the CHP, the longer it will run daily and the higher the chance of matching its generation power with consumption. Therefore, we have implemented a simple algorithm which calculates the fraction of CHP operating time matching operational times of electrical demand by the heat pumps, chiller, and network pumps. This is combined with an algorithm that determines the electricity costs and profit for buying energy from the grid and feeding into the grid.

Table 2 Nomenclature profit function

Term	Signification (unit: €/year)
$P_{E,\text{grid,sell}}$	Profit made on electricity sold to the grid
$P_{Q,\text{dh},h}$	Profit made on heat sold through the DH network
$P_{Q,\text{hp},h}$	Profit made on heat sold to houses fitted with a heat pump
$P_{Q,\text{dh},c}$	Profit made on cooling sold through the DH network
$P_{Q,\text{hp},c}$	Profit made on cooling sold to houses fitted with a heat pump
$C_{E,\text{grid,buy}}$	Costs of electricity bought from the grid
C_{biogas}	Costs of biogas supply to the DH system
C_{ngas}	Costs of natural gas supply to the DH system
$C_{\text{cap},j}$	Capital costs of investments and required future re-investments into specific equipment type j
$C_{\text{op},j}$	Operational costs (operator, maintenance, insurance, etc) of specific equipment type j
chp, b, hp, sto_{chp}, sto_{hp}	Specific equipment: CHP, boiler, heat pump, storage related to the CHP, storage related to the heat pumps

Investment capital rate I and investments (INV) (unit: €) for the CHP, boiler, heat pumps, and storage are related to size or capacity and are described with power functions. Suitable power function coefficients are obtained from cost engineering handbooks, field experts, and company quotations.

Capital costs are expressed as net present value and are determined by initial investments, the interest rate r which we assume the same for all equipment, and equipment life time (LT) are as follows:

$$C_{\text{cap},j} = \text{INV}_j \frac{r}{1 - \frac{1}{(1+r)^{\text{LT}_j}}} \quad j \in \{\text{chp}, b, \text{hp}, \text{sto}_{\text{hp}}, \text{sto}_{\text{chp}}\} \quad (9)$$

Operational costs for each equipment are assumed as a fixed percentage of the initial investment.

Thermal demand specification

Approach and assumptions

Thermal demand includes demand for domestic hot water and space heating. As the district does not yet exist, we develop a straightforward design calculation method in subsequent sections to generate an average demand profile from average daily ambient temperature data for the district.

Domestic hot water demand

For a district, the demand is determined by the amount of households and the number of persons per household. The following equation based on [12] expresses the variation of daily heat demand for domestic hot water due to holidays throughout the year:

$$Q_{\text{dhw},i} = nQ_{\text{dhw,av}} + n(1-p)Q_{\text{dhw,av}} \times \cos\left(\frac{2\pi}{365}[i - \text{day\_shift}]\right) \quad i \in \{1, \ldots, 365\} \quad (10)$$

In which n the number of households, p the fraction of households not on holiday during the holiday season, and day_shift the number of days that the minimum of the cosine function (peak day of the holiday season with the least domestic hot water demand) is shifted forward from the first of July. In reality, demand profiles are not smooth functions, hence a random daily variation on the profile is generated between 80% and 120%. This is shown in the case application in the subsequent subsection.

Equations for space heating demand

The energy demand for space heating can be predicted by taking a number of assumptions into account:

- Heat loss to the environment. In general, the method we apply to construct a daily demand profile is to relate space heating demand to the temperature difference between a constant base temperature (T_{base}) and the average daily ambient temperature ($\bar{T}_{a,i}$). The base temperature is defined as the average daily ambient temperature for which no space heating is required. Due to solar and internal thermal gains within the interior of a house, the base temperature is lower than the heating setpoint temperature and can be found from experience (i.e., observed number of heating days) or dynamic simulation. The following constraints signifying the space heating demand limit apply:

$$\Delta T_{heat,i} = \begin{cases} T_{base} - \bar{T}_{a,i} & \text{if } (T_{base} - \bar{T}_{a,i} > 0 \\ 0 & \text{if } (T_{base} - \bar{T}_{a,i} \leq 0 \end{cases} \quad (11)$$

- For the daily sum of the space heating demand, $Q_{sh,i}$ (unit MJ/day), a simple prediction profile for a single household is constructed from the following relation:

$$Q_{sh,i} = U \, \Delta T_{heat,i} \quad (12)$$

  In which parameter U (unit MJ/K) is a constant total heat transfer coefficient, signifying all heat loss to the ambient due to envelope conduction, air ventilation and infiltration, and gains by solar radiation, residents, and appliances. U is determined from a known yearly sum of space heating demand as follows:

$$U = \frac{\sum_i Q_{sh,i}}{\sum_i \Delta T_{heat,l}} \quad (13)$$

- The yearly sum Q_{sh} (unit MJ/y) indicated in Equation 13 is determined either by simulation, from literature or energy performance coefficient (EPC) calculation tools which are developed specifically for Dutch houses. We used the EPC verification tool EPG & Kosten [13] which is distributed by the Dutch government free of charge. Within this tool, various house types (AgentschapNL reference houses, [14]) with reference dimensions based on comparative studies are defined. Building and installation details can be selected and the program calculates the relevant yearly energy demands, based on EPC reference calculations, involving the reference Dutch climate year. The EPC value is the most important parameter for determining the yearly space heating demand. According to current building regulations, an EPC value of 0.4 will be effective from 1 January 2015 for all newly built houses within the Netherlands. Houses within a district heating system which is based on renewable energy receive an EPC reduction, according to [15]. An equivalent generation efficiency of the DH system is calculated with a calculation tool [16], also distributed by the Dutch government.

Case related thermal demand specification

Based on the methodology developed in the previous subsections, the following specific parameters are defined for the Meppel case:

- Domestic hot water demand. Blokker and Poorteman [17] investigate hot water patterns of an urban district built in 2002 which consists of 3,000 houses. The population composition (age, type of households) compares with the target population of the Meppel case. On average, 2.6 persons per household are calculated. According to [17], [18], and [19] the daily hot water demand per person is 60 L/day of 45°C. With an average cold water supply temperature of 12°C, this amounts to 8.3 MJ/day · person and for an average household of 2.6 persons, this yields $Q_{dhw,hh} = 21.6$ MJ/day and 7.88 GJ/year.
- Space heating and cooling demand. Ambient temperature data for Meppel is taken from weather station in Hoogeveen which is close to Meppel. We determine the average daily temperature for the preceding 5 years. By comparing with simulations, we calculate $T_{base} = 14$°C. The number and type of houses is estimated from the current progress of the Meppel house building project. The target EPC value is estimated at 0.4. A mixed DH system based on biogas CHP and natural gas boilers yields an equivalent generation efficiency of 1.5 and results in an EPC reduction of 0.15. Hence, we estimate that the real heat demand of houses is designed for an EPC value of 0.55. With this, the heating and cooling demand values presented in Table 3 are calculated. The resulting district heating demand profile per average household is shown in Figure 2. The profile is constructed by applying Equations 12 and 10 which are added together. The cooling demand profile is shown in Figure 3.

Relation between CHP size and operational time

When a CHP is operated, it is supported by boilers. A suitable size relation between CHP and boiler size is determined from the coldest day generation balance:

$$\dot{Q}_{max} = \dot{Q}_{chp} + \dot{Q}_b \quad (14)$$

Table 3 District household space heating and cooling parameters

House type	% of houses	$Q_{sh,year}$ (GJ)	U_{sh} (MJ/K)	$Q_{c,year}$ (GJ)	U_c (MJ/K)
Apartment	14	12.2	6.38	10.8	−41.4
Terraced house	57	15.3	8.00	4.6	−17.7
Corner house	29	20.3	10.62	9.2	−35.3
Average DH house		16.3		6.8	
Average HP house		21.1		6.0	

On the coldest day, both heat generators will run continuously for 24 h to generate the required thermal energy. Hence, the following relation between the heat demand and required peak thermal power applies for the coldest day:

$$\dot{Q}_{max} = \max\left[\frac{Q_{dh,demand,h}}{\eta_{distr}\cdot\eta_{sto}} + Q_{hp,source}\right]_1^{365} \cdot \frac{10^6}{3,600\cdot 24} \tag{15}$$

If the daily heat demand data is sorted to a data array Q_{demand} from days with the largest to the smallest heat demand, Equation 15 is equal to the first value in this array. In Figure 4, the sorted array is drawn. We have also drawn the shaded heat demand coverage of a certain CHP which has just enough capacity to generate the heat demand indicated with a dot. The remaining area above the dotted line is covered by natural gas boilers.

For the sorted array which we convert from GJ/day to kWh/day, let $\dot{Q}_i$ be a variable CHP generation capacity such that on day i, 24-h operating time with this capacity results in exactly the heat demand of that day, i.e., $Q_{demand,i}$. For the shaded area of Figure 4, it can be demonstrated that the following equations determine the yearly operating time t_{op} for CHP generation capacity $\dot{Q}_i$:

$$\begin{aligned}\dot{Q}_i &= \frac{Q_{demand,i}}{24} \quad ; \quad i \in \{1,\ldots,365\}\\ t_{op} &= \frac{\sum_{j\geq i} Q_{demand,i}}{\dot{Q}_i} + 24\cdot i\end{aligned} \tag{16}$$

In this relation, $\dot{Q}_i$ can be any CHP thermal capacity between 0 and $\dot{Q}_{max}$. Equation 16 then yields the operating time of that CHP. Figure 5 shows the resulting relation between CHP size and its full power total operation time for the whole year, $t_{op,chp}$, calculated with Equation 16.

Case related energy cost parameters

Profit and cost rate of electricity

In the Meppel case, the utility company Meppel Energie buys electric energy from the grid and in case of a CHP it will also feed in surplus electricity into the grid. Contractual buying and selling prices of energy are usually based

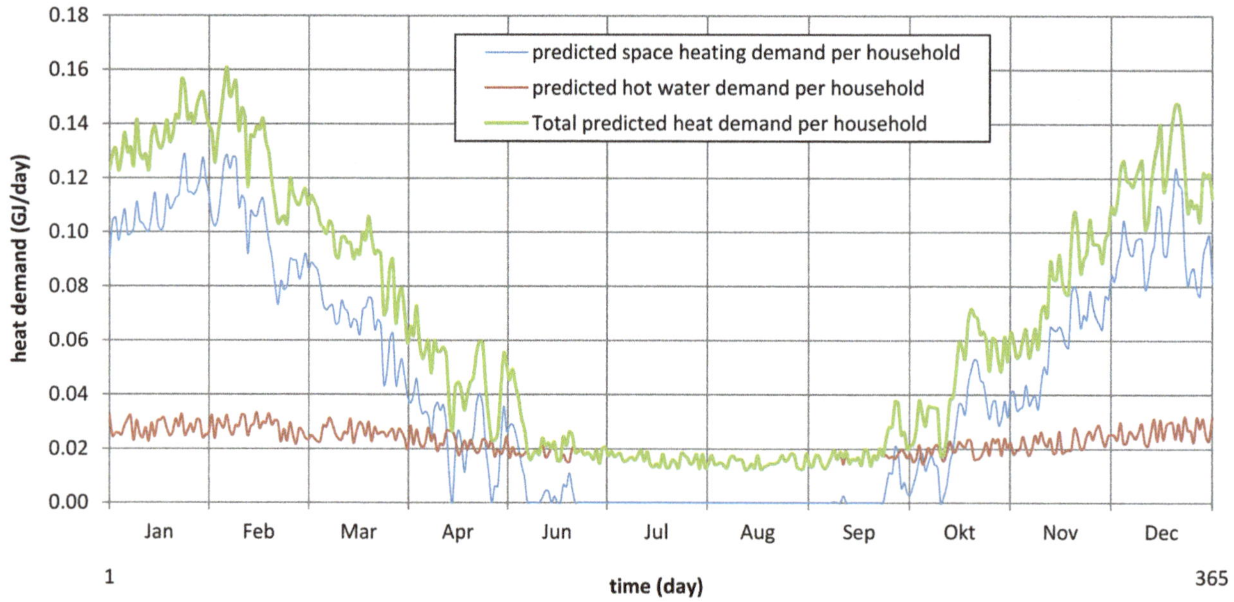

Figure 2 Heat demand profile for average household, Hoogeveen climate reference year.

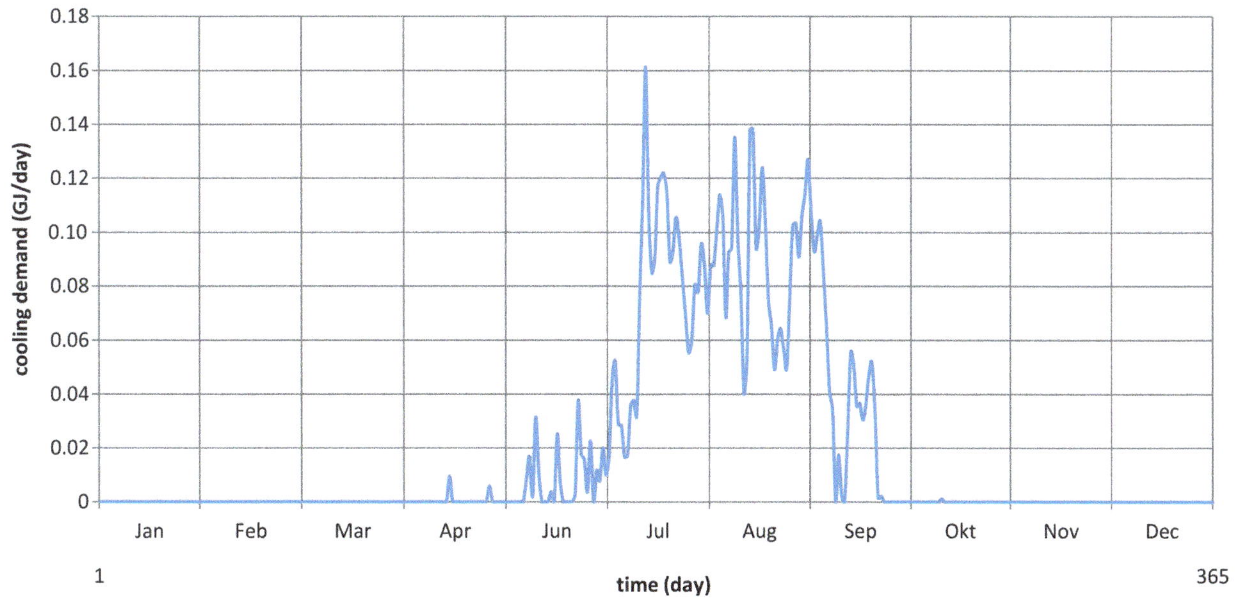

Figure 3 Cooling demand profile for average household.

on the expected amount of energy drawn from and fed into the grid within a certain time period. Following Dutch CHP practice, we assume the buying and selling prices are equal for equal amounts of energy transported to and from the grid over the period of 1 year.

In general, the price of energy drops when the amount of energy being bought increases. For very large amounts (i.e., above 500.000 kWh/year) electricity prices are close to the spot market or day ahead market price of electricity, which is currently around 5.5 €ct/kWh. In case of CHP generation with various electricity consumers, if the amount of sold (surplus) energy exceeds the amount of bought energy, network utilities are often reluctant to pay any compensation, or spot market prices at most. Specific energy price equations are developed for the Dutch situation and implemented into the model.

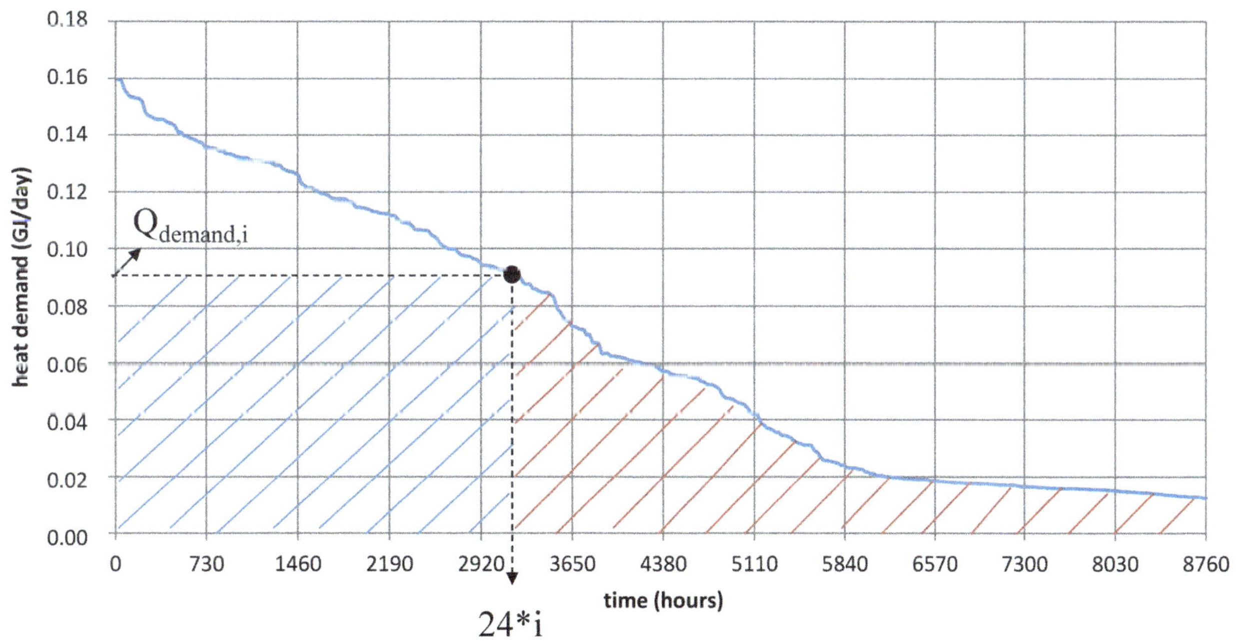

Figure 4 Sorted heat demand and indicated areas at a certain hour point *i*.

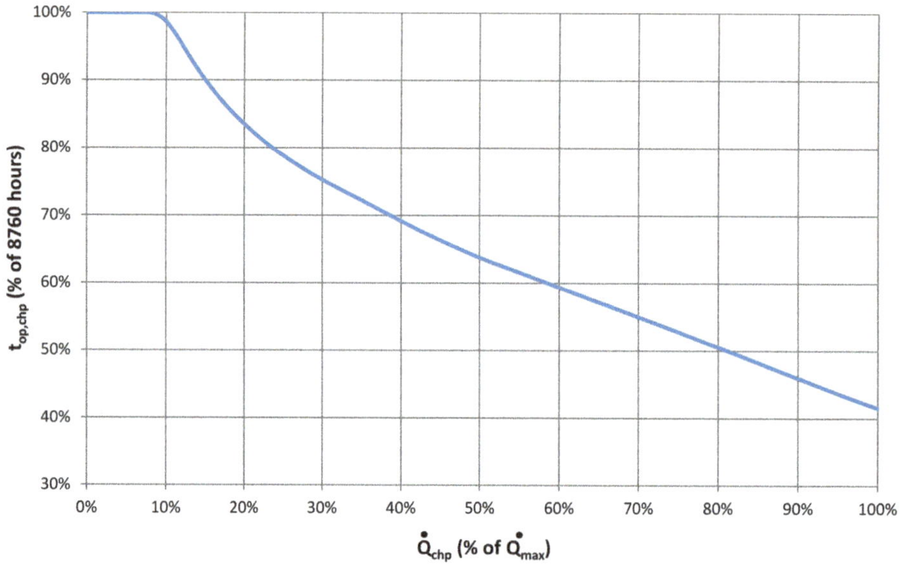

Figure 5 Relation between operational time and CHP size.

Profit rate for selling heat

As we explain in the 'Optimization model equations' subsection, infrastructural investments are payed off by the residents with a one-time connection fee. This fee is usually part of the building price of a house. Besides the connection fee, residents are charged for the consumed heat according to the following equation, i.e., the Dutch heat law, effective from January 2014 which defines the maximum possible tariff: $P_{Q,h} = 209,92+19.86 \cdot Q_{\text{demand},h}$ (excluding VAT).

Financial reservations have to be made for re-investments in the required infrastructure. For this, there is a difference between houses connected to the DH system and houses with heat pumps:

- In the case of houses connected to the DH system, we estimate the yearly costs of reservations for infrastructural re-investments at approximately: €146.00 /(year · household). This leaves as income for the profit model, refer to Equation 8: $P_{Q,\text{dh},h} = 63.00 + 19.86 \cdot Q_{\text{demand,dh},h}$
- In case of houses with a heat pump, we estimate lower reservations for infrastructural re-investments due to the use of non-insulated pipes and a generation building is not required for heat pumps, i.e., €65.00 /(year·household). This leaves as income: $P_{Q,\text{hp},h} = 145.00 + 19.86 \cdot Q_{\text{demand,hp},h}$.

Profit rate of cooling

Originally, the Meppel project was planned for a larger starting scale, including the use of the aquifer for cooling energy and source energy for the heat pumps. Cooling of the houses provides part of the required energy to balance temperature of the aquifers and is therefore offered for free to the residents. Meppel energy wants to keep its promise of free cooling to the residents and therefore the profit rate for cooling is 0, $P_{Q,c} = 0$.

Cost rate of biogas for CHP and natural gas for boilers

Cost relations for natural gas from the national grid, specific relations for biogas, and applicable energy taxes are developed and implemented as algorithms into the optimization model.

Results and discussion

Generator and storage capacities

For the two building phases (phase 1: 40 DH houses; phase 2: 160 DH houses, 16 HP houses) maximum required generation capacity results are shown in Table 4. The peak boiler capacity mentioned is only required for infrequent extremely cold days and also serves as a heat generation backup in case of CHP or boiler failure or during maintenance periods.

Profit generation, optimal CHP size, and operating times

The relation between profit and optimal size of the CHP in relation to boiler size is investigated by maximizing the

Table 4 Calculated generator capacities

	Heat generation $\dot{Q}_{max}$ (kW)	Peak boiler $\dot{Q}_{b,p}$ (kW)	Cooling $\dot{Q}_c$ (kW)
Phase 1	87	87	74
Phase 2	373	373	322

net profit equation (refer to Equation 8) with $\dot{Q}_{\text{chp}}$ as variable. For the two building phases with (n, m) houses, the optimal sizes in Table 5 are calculated.

Table 5 shows a negative profit for both building phases. Care should be taken in the interpretation of the profit values as they are influenced by assumptions explained in the 'Profit rate for selling heat' subsection. We may have exaggerated the re-investment costs, lacking detailed insight in the Meppel Energie business case. On the other hand, practical experience indicates problematic profitability of small-scale DH systems, which support our results. In the Meppel case, there is more pressure on profitability because the houses have a lower heating demand (and thus lower revenue per household) than usual, and there are generation costs involved for cooling while on the other hand, cooling is offered free of charge. If we calculate a fee for cooling based on break even operation, this yields €64/year and €70/year per household for phases 1 and 2, respectively. To place this in perspective of what households would normally pay if they generate the required cooling with home based air conditioning equipment, the average cooling demand of 6 GJ/year requires 555 kWh per household electric energy which would costs each household €127/year only for electricity. Hence, the calculated fees are reasonable propositions to the residents.

What is the influence of houses with heat pumps on the profit? By maximizing the net profit Equation 8 with $\dot{Q}_{\text{chp}}$ and m as variables, we find that for various numbers of n houses, $m = 0$. So in all cases, houses with heat pumps have a negative influence on profit. There are two reasons for this. First, in the starting energy concept, the source heat for the heat pumps is generated by the CHP and boilers, with fuel costs as a consequence. Second, individual heat pumps for each house are relatively expensive in comparison with DH delivery sets as HP's require more investments and maintenance and also have a shorter life time. Clearly, the compensation offered by a higher tariff ('Profit rate for selling heat' subsection) is not sufficient for an equally profitable operation of heat pumps compared to houses connected to the DH system.

If the number of houses connected to the DH system expands, when will operation become profitable? This depends mostly on (a) the applied interest rate r and (b) the household heating demand. If $r = 4\%$ then operation breaks even at $n = 490$ houses. The corresponding optimal generation sizes and operating times are as follows: $\dot{E}_{\text{chp}} = 426$ kW, $t_{\text{op,chp}} = 5,293$ h/year, $\dot{Q}_b = 448$ kW, and $t_{\text{op},b} = 1,340$ h/year. For $r = 2\%$ the break even point is $n = 277$ houses.

CHP migration steps

When the number of houses in the district increases, ideally the CHP and boiler sizes also have to increase according to the optimal profitability found in the previous subsection. It is relatively easy to install more boiler capacity, but replacing or adding a single CHP requires much higher investments and more accompanying equipment. CHP expansion is possible in two ways, either by leasing CHP units for a period of time or by investing in a starting CHP unit and investing into more CHP units when this is required. The latter option has the advantage that CHP capacity is gradually expanded and eventually the generation capacity exists of a number of smaller CHP units, which enables smoother power regulation and operation of each CHP under full load conditions, which is beneficial for fuel efficiency and technical reliability. However, this option requires more investments, although the Dutch tax scheme offers substantial tax reductions for CHP investments, so this is selected as the most interesting approach. Therefore, we investigate the relation between CHP size and profit in four steps from building phase 1 to phase 2, to estimate the proper CHP sizes. The result is shown in Figure 6.

For each line in the Figure, the value for (n, m) is indicated. Around the optimum of each line, the relation between profit and CHP size is rather flat. This gives flexibility to choose the CHP size according to logical migration steps for the district. An attractive migration route is to start with a 30-kWe CHP for a small district with 40 houses and to install a second, larger CHP, e.g., 140 kWe when 160 houses are connected.

Comparison on sustainability

Sustainability of the Meppel energy concept shown in Figure 1 is investigated by comparing fossil fuel requirements for building phase 2 ($n = 160$, $m = 16$ houses), for which we define four cases:

1. Case 1: starting energy concept with optimal sized CHP on biogas supported by a boiler on natural gas. The Sankey or energy flow diagram for this case is shown in Figure 7. Total natural gas input: 1,648

Table 5 Optimal CHP size for building phases 1 and 2

(n, m)	Profit without CHP (€/year)	Optimal CHP size $\dot{E}_{chp}$ (kW)	$t_{\text{op,chp}}$ (h/year)	Boiler size $\dot{Q}_b$ (kW)	$t_{\text{op},b}$ (h/year)	Profit with CHP (€/year)
(40, 0)	−6,911	6	8,659	78	3,077	−2,571
(160, 16)	−20,807	100	6,140	228	2,033	−12,380

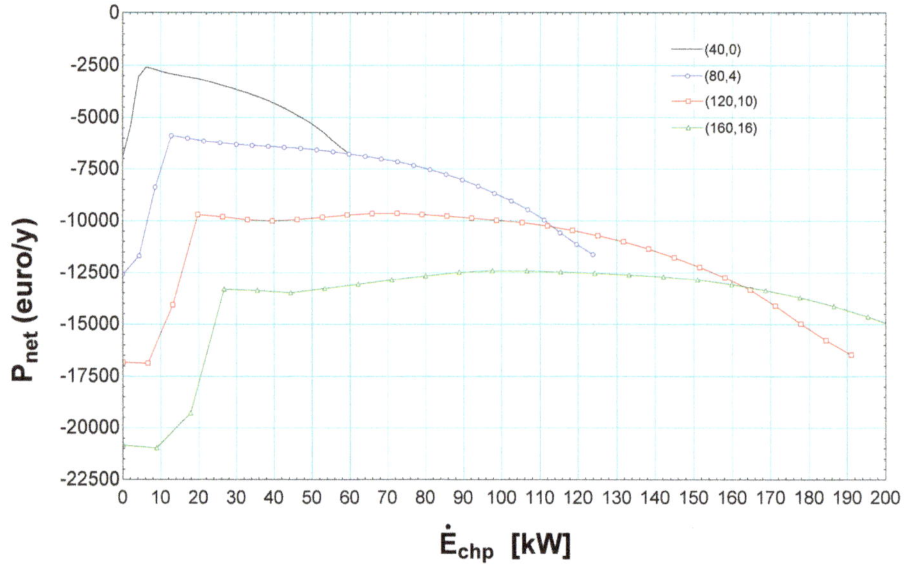

Figure 6 CHP optimum for possible migration steps.

GJ/year. Total biogas input: 6,177 GJ/year. If the CHP runs on natural gas, the total natural gas input is: 7,825 GJ/year.

2. Case 2: starting energy concept without a CHP. In this case, the boiler delivers 4,860 GJ/year with an equal natural gas input. The required electricity for heat pumps, cooling and network pumps, and energy delivered to the region total 2,222 GJ/year. This requires 4,938 GJ/year of natural gas input at a grid-based electrical power plant, taking Dutch national fuel efficiency of 45% into account. Hence, total natural gas input: 9,798 GJ/year.
3. Case 3: Dutch conventional house heating by individual home-fitted natural gas boilers. In this case for heating, 4,318 GJ/year natural gas is required. The required electricity for cooling and comparable electricity supplied to the region total 1,816 GJ/year. If a grid-based electrical power plant is used, 4,036 GJ/year natural gas input is required. In that case, the natural gas input totals: 8,354 GJ/year.

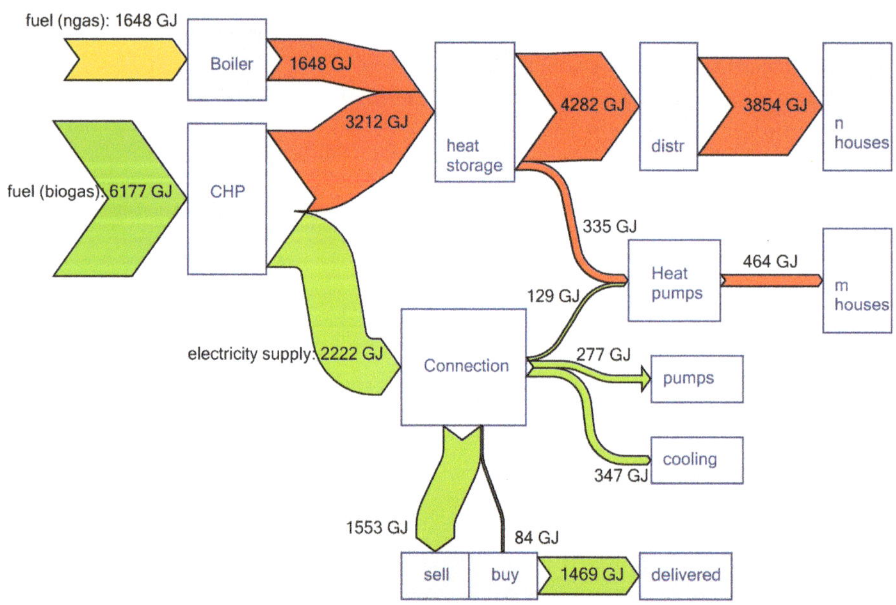

Figure 7 Sankey diagram starting energy concept, building phase 2.

4. Case 4: as an alternative for case 3, the 1,816 GJ/year electricity is generated by home installed solar PV. For 176 houses, this is 2,866 kWh/year per household. Under Dutch circumstances, this is feasible. Total natural gas input: 4,318 GJ/year only for heating is required.

The resulting natural gas input is shown in Figure 8. Case 1 shows the best sustainability results; however, if natural gas is used for the CHP, the concept is only 6% more sustainable than case 3, i.e., conventional house heating with natural gas boilers. Case 2 shows that the DH system without a CHP actually has the worst performance on sustainability.

Case 4 with home PV installations is interesting as improved sustainability is combined with less infrastructural investments than case 1 and is hence easier to implement in an unsure house building market. Case 4 also leads to another objective, i.e., to invest into more energy efficient houses, in order to further reduce natural gas consumption. We re calculated case 4 with passive house (PH) space heating requirements, and this results in the fifth column in Figure 8. This case can be taken further, replacing natural gas boilers with heat pumps. The district then contains passive houses, each with a solar PV roof and a heat pump connected to a ground source or aquifer. Natural gas input is then eliminated totally. The drawback of such a concept however is the imbalance between periods of major PV electricity generation and periods with the highest heating demand. Further comparisons of these concepts including financial aspects and loads on the electricity network are left for future work.

The use of biogas instead of natural gas reduces carbon dioxide (CO_2) emissions, while biogas is formed with CO_2 from the earth's atmosphere and the CO_2 is released when the biogas is burned. For natural gas, a conversion factor of 52 kg CO_2/GJ is derived. Case 1 with biogas results in 86 tons CO_2/year and case 2 in 509 tons CO_2/year. Hence, case 1 results in a CO_2 reduction of 423 tons/year.

The last question of interest for the Meppel case is how many houses and what size of the CHP are appropriate if all the available biogas of the municipality (i.e., 9,200 GJ/year) is applied? As optimum, we find $n = 232$ houses and $\dot{E}_{chp} = 159$ kW. The natural gas input required for boilers is then 1,775 GJ/year, resulting in net CO_2 emissions of 92 tons/year. The net profit (or loss in this case) is calculated at −€5,711/year. When a cooling fee of €50/year is asked from each household, the net profit is positive, i.e., €5,899/year.

Conclusions

In this paper, optimal heat and power generation sizes are determined for the new urban district Nieuwveenslanden situated in Meppel for two building phases starting at 40 houses and expanding towards 176 houses. Besides optimal CHP and boiler sizes, we determine corresponding profits and sustainability of the network on fossil fuel requirements. Profits appear to be negative, however more than 50% improvement is possible if a CHP is used for heat and power generation compared to operation with only boilers. With an interest rate of 4% on invested capital, profits are positive when more than 490 houses are connected, with 2% this is reduced to 277 houses. If households are charged a modest fee for cooling, the energy concept with a CHP is profitable from the start. Based on maximizing profit, a migration scenario for the CHP is developed starting with a single 30-kWe CHP for 40 houses connected to the DH system and adding a 140-kWe CHP when 160 houses are connected.

Sustainability of the energy concept is compared with the Dutch reference, i.e., houses heated by individual natural gas boilers. If a CHP on biogas is applied, the Meppel

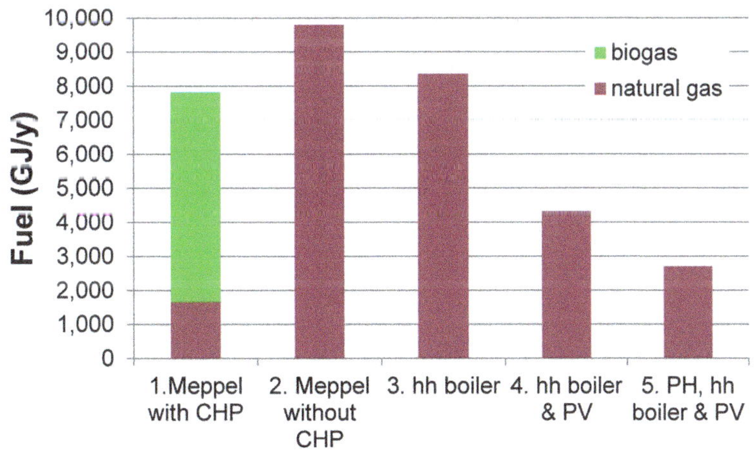

Figure 8 Case-related fuel input comparison.

energy concept reduces 351 tons CO_2/year or 80% reduction. However, if the CHP runs on natural gas, the CO_2 reduction is only 6%. It has to be considered that a scenario based on the Dutch reference combined with home solar PV is another possible route towards improved sustainability but involves less infrastructural investments. However, the Meppel DH system concept with a CHP on biogas still performs better on sustainability than this alternative route.

For the Meppel DH system, houses with a heat pump have a negative influence on the profitability. This is because individual household heat pumps are more expensive to operate than the DH system. It is interesting to compare this also for the final energy concept in which source heat and cooling are generated by an underground aquifer. In that case, we expect to reach better profitability. Investigating the final energy concept is left for future work. Also a more comprehensive comparison between the Meppel energy concept and more individual energy generation solutions requiring less infrastructure is left for future work.

Competing interests
The authors declare that they have no competing interests.

Authors' contributions
RVL carried out the research for this paper, developed the models, performed all simulations, and drafted the manuscript. JF participated in the model development and critically revised the paper. JDW defined alternative energy systems to compare on sustainability and supplied background case related data to compare with. GS critically revised the paper and gave final approval for the paper to be published. All authors read and approved the final manuscript.

Authors' information
JF is affiliated to the STW project i-Care and Dream.

Acknowledgements
The authors would like to thank the Dutch national program TKI-Switch2SmartGrids for supporting the project MeppelEnergie and the STW organization for supporting the projects i-Care and Dream that resulted in this work.

References
1. AgentschapNL (2014) Biogas-warmtekrachtkoppeling (WKK) voor Een Wijk. http://www.rvo.nl/onderwerpen/duurzaam-ondernemen/gebouwen/energieprestatie-nieuwbouw-epn/ontwerpen/referentiewoningen-nieuwbouw. visited January 2014
2. Schepers BL, Valkengoed MPJv (2009) Warmtenetten in nederland - overzicht van grootschalige en kleinschalige warmtenetten in nederland. Technical Report 09.3031.45, CE-Delft http://www.ce.nl/publicatie/warmtenetten_in_nederland/976, visited May 2015
3. Vallios I, Tsoutsos T, Papadakis G (2009) Design of biomass district heating systems. Biomass and bioenergy 33(4):659–678
4. Ren H, Gao W, Ruan Y (2008) Optimal sizing for residential CHP system. Appl Therm Eng 28(5):514–523
5. Cho H, Luck R, Eksioglu SD, Chamra LM (2009) Cost-optimized real-time operation of CHP systems. Energy and Buildings 41(4):445–451
6. Chinese D, Meneghetti A (2005) Optimisation models for decision support in the development of biomass-based industrial district-heating networks in Italy. Appl energy 82(3):228–254
7. Curti V, Favrat D, von Spakovsky MR (2000) An environomic approach for the modeling and optimization of a district heating network based on centralized and decentralized heat pumps, cogeneration and/or gas furnace. Part II: application. Int J Ther Sci 39(7):731–741
8. Hlebnikov A, Dementjeva N, Siirde A (2009) Optimization of Narva district heating network and analysis of competitiveness of oil shale CHP building in Narva. Oil Shale 26:269–282
9. Lund H, Möller B, Mathiesen BV, Dyrelund A (2010) The role of district heating in future renewable energy systems. Energy 35(3):1381–1390
10. Pirouti M (2013) Modelling and analysis of a district heating network
11. Østergaard PA, Lund H (2011) A renewable energy system in Frederikshavn using low-temperature geothermal energy for district heating. Appl Energy 88(2):479–487
12. Jordan U, Vajen K (2001) Realistic domestic hot-water profiles in different time scales. Report for IEA-SHC Task 26. http://sel.me.wisc.edu/trnsys/trnlib/iea-shc-task26/iea-shc-task26-load-profiles-description-jordan.pdf, visited May 2015
13. RVO (2014) EPG Rekentool. https://www.rvo.nl/onderwerpen/duurzaam-ondernemen/gebouwen/energieprestatie-nieuwbouw-epn/epc-berekenen/hulpmiddelen-epc-berekening?wss%l=1, visited January 2014
14. AgentschapNL (2013) Referentiewoningen Nieuwbouw 2013. Agentschap NL. http://www.rvo.nl/onderwerpen/duurzaam-ondernemen/gebouwen/energieprestatie-nieuwbouw-epn/ontwerpen/referentiewoningen-nieuwbouw visited January 2014
15. AgentschapNL (2014). http://www.rvo.nl/sites/default/files/Gebiedsmaatregelen%20voor%20het%20eerst%20gewaardeerd%20in%20de%20EPC_bepaling_0.pdf, visited January 2014
16. RVO (2014) EMG Rekentool. https://www.nen.nl/NEN-Shop/Norm/Update-rekentool-energieprestatie-gebied-beschikbaar.htm, visited January 2014
17. Blokker EJM, Poortema K (2007) Effecten levering warm tapwater door derden. Technical report, sKIWA
18. Geudens PJJG (2012) Drinkwaterstatistieken 2012. VEWIN 41(4):445–451
19. Blokker EJM (2006) Modelleren van afnamepatronen; beschrijving en validatie van het simulatiemodel simdeum. Technical report, KIWA. http://www.kwrwater.nl/WorkArea/DownloadAsset.aspx?id=8787, visited May 2015

5

The influence of parameters variability on biomass selection for energy use

María Elena Arce

Abstract

Background: Biomass selection integrates heterogeneous variables such as fuel composition, fuel morphology and operating conditions. One important challenge in fuel selection field is to determine the influence of each individual variable in the combustion process to obtain the most efficient energy.

Methods: This paper investigates an optimization design of different combustion parameters for biomass combustion in fixed-bed counter-current boiler. A new application of error propagation theory (EPT) and grey relational analysis (GRA) has been proposed to analyse the combined effect of the most influential variables in the operating conditions; i.e. the ignition mass flux and the maximum temperature. This methodology combines the GRA to rank the alternatives and EPT to compute input measurement uncertainty propagation through the subsequent computing operations.

Results: The behaviour of seven biomasses [i.e. brassica pellet, poplar pellet, refuse-derived fuel pellet, wood pellets, almond shell and olive stone] that are readily available in Spain, in southern Europe, was analysed. The results presented are based on 43 experimental tests with these fuels.

Conclusions: This paper proposes a new tool based on grey relational analysis and error propagation theory, for making quick and easy decisions regarding fuel performance. The main advantages of this new tool are its simplicity, versatility and reliability, which have been verified by comparing obtained results with other studies. The influence of operational parameters and fuel properties were analysed. The results showed that air supply injected on the bed basis and size particle are the most influenced variables on the combustion.

Keywords: Fuel behaviour; Fixed-bed combustion; Parameters variability; Multi-criteria analysis; Energy efficiency

Background

The widespread use of biomass for power generation has led to the expansion of the types of fuels used. These types of fuels are highly versatile when undergoing energy conversion treatments and differ significantly from each other in efficiency [1]. Although various thermochemical processes can obtain energy from biomass, combustion is the most widely used [2].

Research related to the combustion of biomass focuses on improving both the properties of the fuels and the characteristics of the boilers [3]. Thus, many projects have attempted to validate mathematical models of biomass combustion processes based on experimental data obtained from experimental plants [4]. Fixed-bed combustion technology is the most widely used because of its simplicity [5].

The main influences on the combustion process have been determined by research conducted in experimental plants and can be grouped into the following: fuel composition, fuel morphology and operating conditions [6]. No single criterion can be used to select the best fuel performance due to the multitude of influential parameters in the fixed-bed combustion process [7]. The ignition front velocity, the air mass flow rate and the maximum temperature reached are the most influential variables in the operating conditions [7,8]. Both variables depend on the air flow injected on the bed. In this context, determining the influence of each individual variable in the combustion process to select the most efficient fuel is a primary research focus [8]. The problem with determining combustion process of influential variables

Correspondence: elena.arce@cud.uvigo.es
Defense University Center at Spanish Naval Academy (Marin), Plaza de España 2, 36920 Marín, Spain

lies with the fact that parameters are interrelated. That is, it is a multi-criteria decision-making problem [9].

Thus, most papers published on this issue are based on statistical techniques such as multivariate data analysis [10] or mathematical models of dependence [11-13]. These multi-criteria decision analysis methods are used as tools for the classification and selection of the best alternative amongst a set of options. However, these analysis techniques have the disadvantage of requiring a minimum quantity of data for application, and for technical or economic reasons, this requirement cannot always be met. Therefore, grey relational analysis (GRA) was proposed [14] for accommodating inadequate data systems and solving decision-making problems in which there are multiple interrelated criteria [15]. This tool has been largely applied to multi-criteria decision-making problems in addition to energy systems, both the fuel selection and combustion process analysis [16-19].

In the biomass combustion process, several variables are optimised in opposite directions (i.e., minimise the maximum temperature and maximise the ignition mass flux), which leads to increased difficulty in multi-objective processes. In these cases, no single best alternative exists [20]. Grey relational analysis classifies different input sequences by determining the grey relational grade (GRG) [21]. Higher GRG values indicate better alternatives. Moreover, GRA may process a large number of variables independent of the measuring units [22].

This method facilitates the selection of the best alternative because it optimises a single GRG rather than multiple parameters. In addition, GRA identifies the factors that significantly influence the process [18]. However, the GRA does not consider the uncertainty in the measurements [23]. This omission can lead to confusion from similar GRGs measured at different positions or find that the GRG is uninteresting because of its high variability. In this study, the GRA is applied by selecting the best biomass fuel using the minimisation of the maximum temperature and the maximisation of the ignition mass flux as criteria. Experimental data are subject to measurement error; thus, the variable measurement error should be considered to determine how the error propagates through the operations [24]. Otherwise, factors that also influence the GRG may be not be considered, which could affect the GRG interpretation and selection [23].

Methods

Experimental plant

The experimental plant (Figures 1 and 2) and the subsequent discussion indicate that the results of this study are based on data presented by our research team [5]. The burner is a cylindrical steel tube with a diameter of 20 times the particle size to avoid edge effects resulting in the emission of product gas directly to the atmosphere [25].

Figure 1 Work module.

Along the tube, 12 type-K thermocouples are installed and distributed helically. The time required to reach a predetermined temperature between two adjacent thermocouples is used to calculate the velocity of the ignition front. The distances between the thermocouples are known with an associated uncertainty in both the temperature and position. Data from first and final thermocouples are rejected because in the first thermocouples the initial combustion phase is unstable and in the last thermocouples resonance phenomena appear [25]. The fuel is introduced in batches, forming the fixed bed. Each test is developed with a single fuel charge. The different air mass flow rates are supplied by a centrifugal fan and controlled by a flow meter. Figure 1 provides scheme of the tube. The plenum located on the tube is designed to ensure that the air flow in the output grate is homogeneous and completely developed.

Fuels

Seven different types of biomass found in southern Europe [26] were selected to include as many experimental

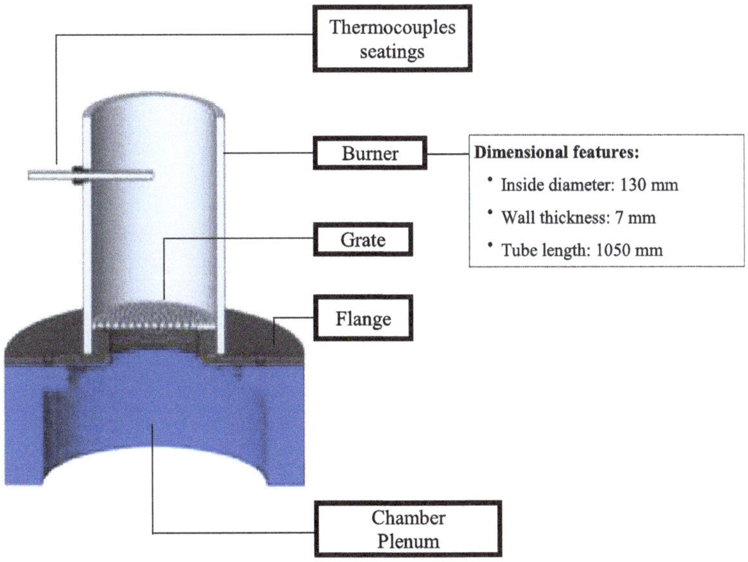

Figure 2 Detailed image of the joint between the plenum and the bottom tube base.

parameters as possible (for a more detailed description, see [5]) (Table 1). Five of them are pelletised fuels; two are mixed woods (wood pellet type 1, wp1, and wood pellet type 2, wp2). Two were derived from energy crops (brassica pellet, bp, and poplar pellet, pp), and one is made from municipal solid waste (rdfp). In addition, almond shell (as) reclaimed from the nut industry and olive stone (os), whose morphology is similar to a pellet but its composition and combustion behaviour are different, was used.

The immediate analysis (Figure 3) shows similar water content for the pellets (6% to 9.8%) except in the refuse-derived fuel pellet (rdfp) (17.9%), which is due to its origin and manufacturing process. The highest ash content levels were found in refuse-derived fuel pellet and the brassica pellet (13.4% and 7%, respectively), while the lowest ash contents were found in the almond shell and olive stone (0.6% in both cases). The minimal air requirement range varies from 4.39 (rdfp) to 6.96 (wp2). Wood pellet #2 also has the highest low calorific value $\left(18.3 \cdot \frac{\text{MJ}}{\text{kg}}\right)$ and the lowest concentration of oxygen in the equivalent molecule (0.55 mol oxygen per mol carbon). In terms of particle size (Er), lowest values were found in wood pellet #1.

Variables

Fixed-bed combustion process is influenced by multiple parameters [7]. This process complexity means that there is no single criterion for fuel selection. Thus, studies need to focus on a small number of variables. The main parameters influencing the process have been identified in the literature as the maximum temperature, air excess and ignition mass flux [6,7]. These parameters

Table 1 Properties of biomass fuels

Fuel type	Formula d.b.a.f.[a]	LHV (MJ kg^{-1})	Sa (–)[b]	*Er* (mm)[c]	*S*[d]
Wood pellet #1 (wp1)	$CH_{1.48}O_{0.71}$	16.3	5.79	3.8	0.87
Wood pellet #2 (wp2)	$CH_{1.49}O_{0.55}$	18.3	6.96	4.4	0.84
Brassica pellet (bp)	$CH_{1.41}O_{0.78}$	13.1	5.21	4.1	0.86
Poplar pellet (pp)	$CH_{1.55}O_{0.75}$	14.9	5.59	4.2	0.85
RDF pellet (rdfp)	$CH_{1.48}O_{0.57}$	14.6	4.39	7.4	0.87
Olive stone (os)	$CH_{1.50}O_{0.64}$	15.3	6.26	4.2	0.98
Almond shell (as)	$CH_{1.40}O_{0.68}$	15.6	5.89	6.5	0.54

[a]Dry basis ash free.
[b]Kg of dry air per kg of fuel burnt.
[c]Radius of equivalent sphere. $E_r = \left(\frac{3S_{vol}}{4\pi}\right)^{1/3}$; S_{vol} sphere volume (mm^3).
[d]Sphericity. $S = \frac{\pi^{1/3}(6V_p)^{2/3}}{A_p}$; V_p volume of the characteristic particle (m^3); A_p surface area of the characteristic particle (m^2).

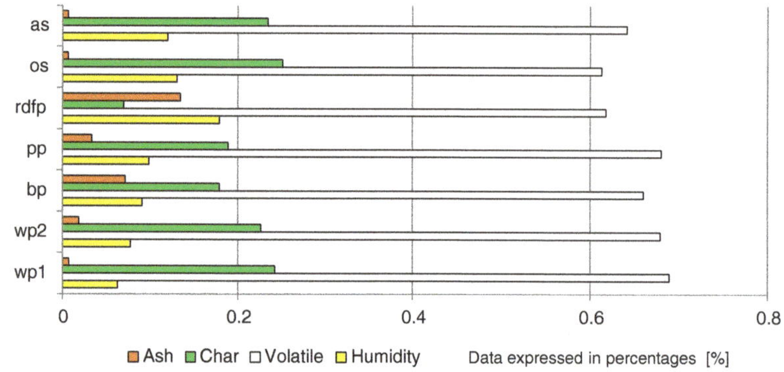

Figure 3 Properties of biomass fuels (immediate analysis, wet and ash basis).

provide information regarding fuel composition, morphology and the characteristics of the bed. These studies found that air excess is directly linked to the ignition mass flux and to the maximum temperature [13].

The fixed-bed combustion process is influenced by characteristic fuel-dependent parameters. The most important are as follows [11,27-29]:

- Particle size and shape
- Water, volatile and ash content
- Particle density, packing density and total porosity of the bed
- Calorific value
- Specific heat and thermal conductivity

The biomass parameters were considered constants for the purpose of this investigation, establishing that the packing density depends only on the particle size and shape [6].

The ignition mass flux (fuel mass burned) and the maximum temperature are dependent on the supplied air flow, which is controllable during the tests. Thus, the air mass flow rate is not used to calculate the GRG. In other words, the sequences analysed are combinations of the ignition mass flux and the maximum temperature. Furthermore, because the aim of this study is to choose the best-performing fuel regardless of its composition or calorific value, other parameters need not be considered. A good-performing fuel is generally one that achieves a high ignition mass flux with a low maximum temperature to avoid slag problems due to ash melting [30], which hampers the diffusion of the species and reduces the energy of the bed. However, the risk of ash melting is difficult to quantify because it is influenced by the temperature and other uncontrollable variables [31]. It is important to emphasize that ash melting temperature was not reached during fuel tests. On the basis of the above, the GRA may provide a useful tool for choosing a fuel based on its performance.

Experimental methodology

The experimental procedure may be summarised in the following steps [25] with the goal of avoiding the influence of other factors that could add additional uncertainty to the process. These eight procedures were rigorously followed during each test:

1. Move the combustor to the test area. The work module is connected to the power and control module through the air supply tube.
2. Place the grate at the bottom of the tube, and fill the tube with a fuel batch.
3. Introduce the thermocouples in their seatings, and ensure that they are straight and in the correct order.
4. Install the ignitor, and connect the temperature transducer power cables and the data bus.
5. Start the control panel and the data acquisition system.
6. Switch the ignitor feeding on, and start the test.
7. Remove the thermocouples from their seatings as the ignition front advances to protect them from the high temperatures. Once a thermocouple is removed, a hand-held probe is introduced in its seating to determine if the maximum ignition front temperature was reached.
8. Once the ignition front has run through the entire tube and the remains have been burnt over the grate, the test is complete.

Grey relational analysis and error propagation

Grey relational analysis

The GRA method may be summarised into four steps:

Step 1 Data normalisation

The first step in GRA is to normalise the measurements in the range between zero and one. To do this, reference sequence must be defined (the smaller-the-better or the larger-the better: Equations 1 and 2, respectively). The purpose of this pre-processing is to obtain dimensionless variables to compare the different sequences. It should be noted that variables are expressed in different units and scales.

A grey system with a set of I variables has y_i dependent factors; the J original measurement sequences may be normalised using the following expressions:

- If larger is the better:

$$x_{i,j} = \frac{y_{i,j} - \min_j y_{i,j}}{\max_j y_{i,j} - \min_j y_{i,j}} \in [0, 1];\ i = 1, \ldots, I;\ j = 1, \ldots, J \tag{1}$$

- If smaller is the better:

$$x_{i,j} = \frac{\max_j y_{i,j} - y_{i,j}}{\max_j y_{i,j} - \min_j y_{i,j}} \in [0, 1];\ i = 1, \ldots, I;\ j = 1, \ldots, J \tag{2}$$

For a given value of i, extreme values are achieved in the measurements $y_{i,m_i} = \min_j y_{i,j}$ and $y_{i,M_i} = \max_j y_{i,j}$ to yield $x_{i,m_i} \in \{0, 1\}$ and $x_{i,M_i} \in \{0, 1\}$

Step 2 Grey relational coefficient

After grey relational generation, the next step is to calculate the grey relational coefficient using Deng's formula as follows:

$$\xi_{i,j} = \frac{\min_{i,j}(x_{i,0} - x_{i,j}) + \zeta \max_{i,j}(x_{i,0} - x_{i,j})}{(x_{i,0} - x_{i,j}) + \zeta \max_{i,j}(x_{i,0} - x_{i,j})};\ i = 1, \ldots, I;\ j = 1, \ldots, J \tag{3}$$

in which ζ represents the distinguishing coefficient ($0 < \zeta < 1$). $x_{i,0}$ is the normalised ideal result of the ith variable and consequently, $x_{i,0} \geq x_{i,j}; j = 1, \ldots, J$.

The grey relational coefficient range will compress depending on the value of the distinguishing coefficient [20].

Step 3 Grey relational grade

The grey relational grade (GRG), γ, is calculated based on the average of the grey relational coefficients.

$$\gamma_j = \frac{1}{N} \sum_{i=1}^{I} \xi_{i,j};\ j = 1, \ldots, I \tag{4}$$

The GRG expresses the degree of correlation between the reference sequence (larger-the-better or smaller-the-better) and the comparison sequence [32]. The best alternative is selected based on the grey relational grade (GRG) value.

Step 4 Grey relational rank

Once the GRG is calculated, the different sequences are ranked according to their GRG values. Higher values indicate better options [33]. That is, the higher the GRG is, the more similarity exists between the reference sequence and the compared sequence.

Optimisation criterion

The grey theory is applied in this study to select the best fuel and air flow rate configuration. The best configuration is that which has the highest ignition mass flux with the lowest maximum temperature.

The fuel selection criteria depend on the combustion process, thus requiring multiple selection procedures. The analysed factors were the composition, the morphology and the supplied air mass flow rate. The systematic and statistical uncertainties associated with measuring processes are considered in calculating the GRG. The total uncertainty is the quadratic sum of the systematic (nominal) and statistical (associated with uncontrollable parameters) uncertainties: $\sigma_T = \sqrt{\sigma_N^2 + \sigma_S^2}$ [5]. Within the total uncertainty, the uncertainty associated with the uncontrollable factors is shown to have a greater weight than the nominal uncertainty, in terms of both time and space [34]. The straightness of the thermocouples is an important parameter for calculating both the ignition front velocity and the maximum temperature. A thermocouple tip deflection may result to variable distances between the thermocouples, and this could cause variations in the ignition mass flux value because temperature waves could reach the thermocouple at different points. Thermocouple tip deflection was analysed; the average variability in the positioning is 2.06 mm [25]. The uncertainty in the process is also affected by the time variable because data signals are sent every 4 s, while the thermocouple temperature time adaption requires less than a second. Thus, the limiting factor is the sampling time and the uncertainty may be expressed as $\Delta t = \Delta t \pm 4$ s [25]. Once the tests are performed and the uncertainties estimated, the uncertainty propagation can be calculated for each GRG to obtain the grey relational grade error (GRGe). The GRG value is not affected by the GRGe, but the GRGe provides additional information to facilitate the decision-making process.

Error propagation

To be thorough, measurement results should include information regarding uncertainty [35]. The uncertainty associated with a measurement represents the disparity between the measurement result and the actual measured value. Despite the importance of this parameter, it is not always quantified.

The EPT evaluates and quantifies how uncertainty in an input measurement is propagated to the final results.

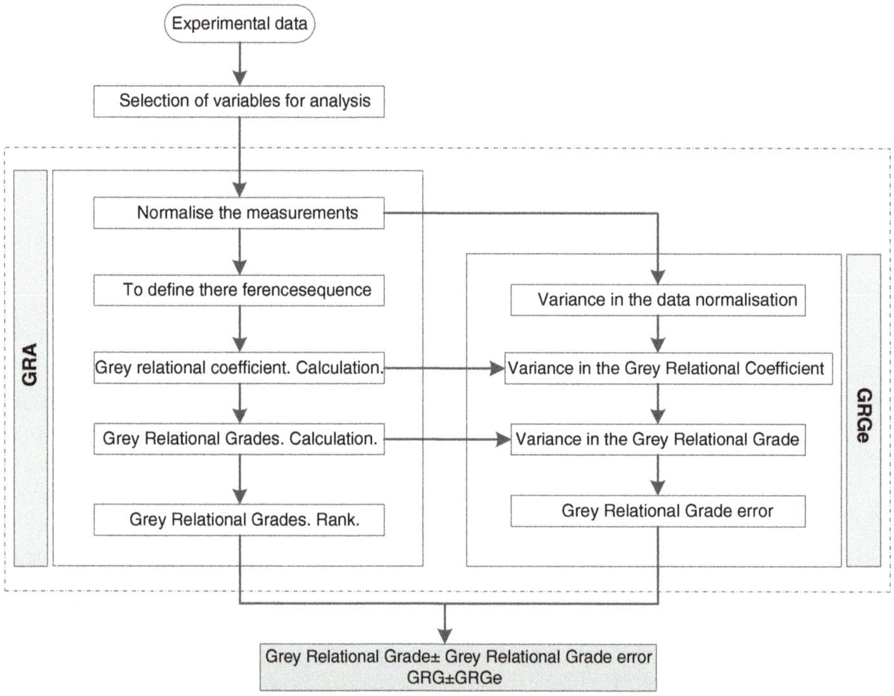

Figure 4 Calculation process scheme.

As the error associated with a variable combines other variable values, the error may not be calculated as the sum of the uncertainties in the constituent terms because the error would be overestimated. The error propagation theory allows one to know the error associated with the GRG, considering that the GRG is computed by averaging the grey relational coefficient corresponding to each process response. The error associated with the GRG calculation (GRGe) is determined by the following equation [22]:

$$\Delta\left(\gamma_j\right) \approx + \sqrt{\sum_{i=1}^{I} \frac{1}{I^2} \sigma^2\left(\xi_{i,j}\right) + 2 \sum_{i=1}^{I-1} \sum_{k=i+1}^{I} \frac{1}{I^2} \sigma\left(\xi_{i,j}\right) \sigma\left(\xi_{k,j}\right) \rho\left(\xi_{i,j}, \xi_{k,j}\right)} \quad (5)$$

in which $\sigma(\xi_{k,j})$ represents the grey relational coefficient error that depends on the error in initial uncertainty of the measurements and $\rho(\xi_{i,j}, \xi_{k,j})$ is the correlation coefficient. Grey relational coefficient error quantifies how the initial uncertainty is propagated through the

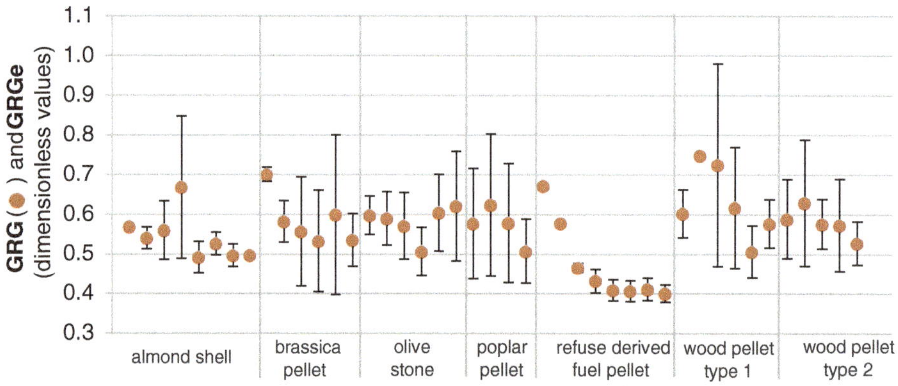

Figure 5 The GRG rank, GRG and GRGe values for seven fuels. Almond shell, brassica pellet, olive stone, poplar pellet, refuse-derived fuel pellet, wood pellet type 1, and wood pellet type 2. The GRGe is represented by the bars.

Table 2 Experimental results, grey relational grade (GRG) and grey relational grade error (GRGe)

Fuel	AMF	IMF	AE	MT	GRG	GRGe	GRGe(%)	Order (GRG)	Order (GRG + GRGe)
Almond shell (as)	0.08	0.02	0.65	800	0.57	4.51E-03	0.79	24	5
	0.13	0.04	0.52	910	0.54	2.72E-02	5.04	27	12
	0.15	0.05	0.57	1,000	0.56	7.31E-02	13.08	25	19
	0.23	0.06	0.60	1,010	0.67	1.80E-01	26.93	5	17
	0.35	0.04	1.26	1,030	0.49	3.96E-02	8.07	37	29
	0.38	0.04	1.53	940	0.53	2.91E-02	5.53	31	14
	0.43	0.04	1.87	970	0.50	2.86E-02	5.76	36	25
	0.48	0.03	2.80	910	0.50	8.25E-03	1.66	35	16
Brassica pellet (bp)	0.08	0.04	0.39	740	0.70	1.79E-02	2.56	3	2
	0.15	0.05	0.49	910	0.58	5.18E-02	8.91	16	8
	0.18	0.06	0.55	1,240	0.56	1.38E-01	24.76	26	36
	0.23	0.06	0.72	1,300	0.53	1.28E-01	24.10	29	37
	0.35	0.07	0.97	1,290	0.60	2.01E-01	33.66	12	39
	0.38	0.05	1.38	1,040	0.53	6.64E-02	12.42	28	24
Olive stone (os)	0.13	0.05	0.42	880	0.60	4.79E-02	8.03	13	6
	0.15	0.05	0.54	930	0.59	6.71E-02	11.40	14	9
	0.23	0.06	0.66	1,010	0.57	8.39E-02	14.71	23	18
	0.28	0.05	0.87	1,100	0.51	6.08E-02	12.03	33	30
	0.35	0.06	0.85	960	0.60	9.65E-02	15.99	10	13
Poplar pellet (pp)	0.18	0.06	0.52	1,030	0.62	1.38E-01	22.26	8	20
	0.23	0.06	0.66	1,160	0.58	1.39E-01	24.13	19	33
	0.28	0.06	0.77	1,130	0.62	1.79E-01	28.77	7	31
	0.33	0.06	0.94	1,200	0.58	1.50E-01	26.05	17	34
	0.43	0.05	1.33	1,170	0.51	8.06E-02	15.91	32	35
Refuse-derived fuel pellet (rdfp)	0.05	0.02	0.53	730	0.67	3.59E-03	0.54	4	3
	0.08	0.03	0.61	810	0.58	7.45E-03	1.29	18	4
	0.13	0.03	0.89	980	0.47	1.15E-02	2.46	38	28
	0.15	0.04	0.96	1,180	0.43	2.99E-02	6.92	39	38
	0.23	0.04	1.25	1,260	0.41	2.69E-02	6.59	41	41
	0.23	0.04	1.58	1,250	0.41	2.60E-02	6.40	42	42
	0.33	0.04	1.80	1,250	0.41	2.80E-02	6.82	40	40
	0.35	0.04	2.13	1,240	0.40	2.18E-02	5.46	43	43
Wood pellet #1 (wp1)	0.15	0.05	0.43	890	0.60	6.02E-02	10.01	11	7
	0.33	0.07	0.81	1,020	0.75	2.67E-03	0.36	1	1
	0.35	0.07	0.90	990	0.72	2.55E-01	35.20	2	23
	0.43	0.06	1.21	1,040	0.62	1.53E-01	24.76	9	26
	0.48	0.05	1.63	1,100	0.51	6.55E-02	12.95	34	32
Wood pellet #2 (wp2)	0.13	0.05	0.35	940	0.58	6.05E-02	10.50	20	10
	0.15	0.06	0.38	1,010	0.59	9.95E-02	16.92	15	15

Table 2 Experimental results, grey relational grade (GRG) and grey relational grade error (GRGe) *(Continued)*

	0.35	0.06	0.81	1,040	0.63	1.59E-01	25.28	6	22
	0.43	0.05	1.04	930	0.58	6.22E-02	10.81	21	11
	0.48	0.06	1.40	1,100	0.57	1.16E-01	20.31	22	27
	0.50	0.05	1.44	1,010	0.53	5.46E-02	10.35	30	21

AMF, air mass flow (kg m^{-2} s^{-1}); IMF, ignition mass flux (kg m^{-2} s^{-1}); AE, air excess; IMFU, ignition mass flux uncertainty (kg m^{-2} s^{-1}); MT, maximum temperature (°C).

calculations. It is important to remark that the GRG error is the result of the sequential calculation. That is, GRG error is determined from experimental data: input variable error propagation. Thus, the error propagation theory is an additional tool for GRA (Figure 4).

Results and discussion

The results presented herein are based on 43 experimental tests: eight almond shells (as), six brassica pellets (bp), five olive stones (os), five poplar pellets (pp), eight refuse-derived fuel pellets (rdfp), five wood pellet type 1 and six wood pellet type 2.

After all GRGs and GRGes have been calculated, the data are sorted according to the GRG value (Figure 5 and Table 2). The high GRGe values in the sequences are due to the uncertainty and values of the variables.

Figure 5 shows the GRG rank indicating that the wp1 is the best option as confirmed by the fact that GRGe associated with the first wp1 GRG is sufficiently small. Figure 5 indicates that best fuel performances come from those in pellet form. This result is consistent with the literature research on the influence of morphology on the ignition mass flow rate [27]. The references also found that a small particle size is associated with a rapid devolatilisation and that spherical particles are linked with tighter packing. Bed porosity is affected by size and shape of the particles. Maximum ignition front propagation registered by wood pellet type 1 is based more on the small size particle, rather than the shape (sphericity). Furthermore, pellet fuels have higher adiabatic temperatures [5].

In the specific case of the best alternative, wp1, the low water and ash content, together with and a small particle size and the characteristics of pellet fuels (high LHV and adiabatic temperature) caused the high GRG values that are consistent with those found in the literature [5,25,36]. This fact indicates that there is an inverse correlation between ash content and heating value, because the ash does not contribute significantly to the overall heat released [37,38].

Figure 5 also demonstrates that olive stone behaves similarly to the pellet-shaped fuels likely because the olive stone morphology is similar to that of the pellet-shaped fuels [5].

When data were analysed based on GRG + GRGe (Table 2), the order changes. This is because of the air excess coefficient influence. That is, the higher air excess coefficient used is, the worse combustion process efficiency is [39]. Previous works found that if oxygen supply increases, ignition mass flow increases until the reaction-limited mode is reached. This boundary line is defined by air excess. Moreover, air mass flow is a key factor to define and identify the combustion process regime. Three regimes were identified in fixed-bed combustion process: oxygen-limited mode, reaction-limited mode and extinction phase [40,41]. Best combinations attending GRG + GRGe criteria belong to reaction-limited mode, where maximum ignition flow values

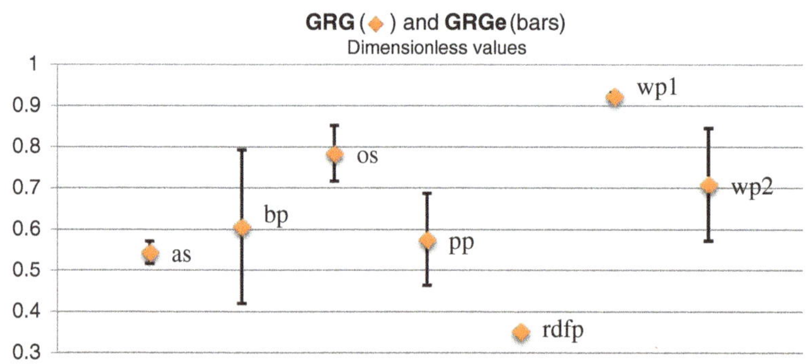

Figure 6 The GRG rankings for seven fuels at the same air mass flow $\left(0.352 \frac{kg}{m^2 s}\right)$: Almond shell (as), brassica pellet (bp), olive stone (os), poplar pellet (pp), refuse-derived fuel pellet (rdfp), wood pellet type 1 (wp1) and wood pellet type 2 (wp2).

were reached. This phase is characterised by increasing primary air supply. In this regime, the ignition mass flow is controlled by the oxygen supply to it [40]. Optimal combustion conditions occur in the sub-stoichiometric zone, because maximum rates are always reached under fuel-rich conditions [40]. As can be expected, optimal point conditions are a function of the fuel type.

It is important to point out the effect of maximum temperature uncertainty. Analysing changes in data order between the GRG rank and the GRG + GRGe rank, it is found that best alternatives are not related with points where highest maximum temperature values are obtained. This is due to the fact that maximum temperatures were registered with portable thermocouples which precision is reduced due to variability of the fit of thermocouples in their seatings.

Most of the studies on this subject indicate that the supplied air mass flow is the most important operative parameter [6].

The air mass flow range differs depending on the fuel because the ignition front is extinguished when certain air mass flows are applied. Thus, fewer sequences were examined in certain fuels such as olive stone or wood pellet 1. In order to compare fuel performance under the same conditions, a common air mass flow was selected $\left(0.352\,\frac{\text{kg}}{\text{m}^2\text{s}}\right)$.

Figure 6 indicates that wp1 remains the best option due to both the high GRG and to the small GRGe values. No significant changes in the rankings have been identified. The rdfp is clearly the worst option because of the tendencies exhibited in Figure 5. This variable rdfp behaviour is mainly due to its high water content that decreases the ignition mass flow velocity [25]. However, the water content does not affect the maximum temperature of the bed.

Thus, one can conclude that the fixed-bed combustion process is influenced by parameters that depend on the characteristics of the fuels. However, these parameters are interrelated, so selecting the best-performing fuel is difficult because their individual effects cannot be isolated and evaluated. Thus, the GRG + GRGe rank serves as a criterion for multi-objective decision-making.

Conclusions

In conclusion, this paper proposes and supports the effectiveness of the grey relational analysis and error propagation theory for making quick and easy decisions regarding fuel performance. The influence of operational parameters and fuel properties were analysed. The rank of the grey relational grade shows that morphology is a key factor on the combustion efficiency. The smaller size particle, the more bed porosity and heat transfer. In contrast, high water content in fuel composition is associated with low mass flow of ignition. However, the humidity does not affect the maximum temperature reached in the bed.

Fuel optimal conditions belong to the sub-stoichiometric regime. This fact reflects the importance of the adverse effect of convection counterflow fixed-bed combustion. Two value zones are distinguished in the GRG rank: sub-stoichiometric regime zone and excess air regime zone. This is due to the maximum combustion velocities that are reached in fuel-rich conditions. Naturally, the optimum point depends on the fuel analysed.

Besides, ignition mass flow, the other study parameter, is a function of air mass flow. Moreover, maximum temperature measurement uncertainty influence has to be considered on optimal fuel performance.

That is, air supply injected on the bed basis and size particle are the most influenced variables on the combustion. As can be expected, the ignition mass flux correlates greater with the size particle than with the air mass flow in order to select the best fuel combustion performance.

Abbreviations

AE: air excess; AMf: air mass flow; as: almond shell; bp: Brassica pellet; EPT: error propagation theory; GRA: grey relational analysis; GRG: grey relational grade; GRGe: grey relational grade error; GRGv: grey relational grade variation; IMF: ignition mass flux; IFMrc: ignition mass flux relational coefficient; IMFn: ignition mass flux normalised; IMFU: ignition mass flux uncertainty; IMFvn: ignition mass flux variation normalised; IMFvrc: ignition mass flux variation relational coefficient; LHV: low heating value; MT: maximum temperature; MTn: maximum temperature normalised; MTrc: maximum temperature relational coefficient; MTU: maximum temperature uncertainty; MTvn: maximum temperature variation normalised; MTvrc: maximum temperature variation relational coefficient; os: olive stone; pp: poplar pellet; rdfp: refuse-derived fuel pellet; wp1: wood pellet type 1; wp2: wood pellet type 2.

Competing interests

The authors declare that they have no competing interests.

Authors' contributions

The author conducted the analysis and drafted the manuscript.

References

1. Mehrabian R, Zahirovic S, Scharler R, Obernberger I, Kleditzsch S, Wirtz S, Scherer V, Lu H, Baxter LL (2012) A CFD model for thermal conversion of thermally thick biomass particles. Fuel Process Technol 95:96–108, doi:10.1016/j.fuproc.2011.11.021
2. European Environmental Agency (2013) EN16 Final energy consumption by sector, http://www.eea.europa.eu/data-and-maps/indicators/final-energy-consumption-by-sector-5/assessment. Accessed 21 Jul 2014
3. Chaney J, Liu H, Li J (2012) An overview of CFD modelling of small-scale fixed-bed biomass pellet boilers with preliminary results from a simplified approach. Energ Convers Manage 63:149–156, doi:10.1016/j.enconman.2012.01.036
4. Porteiro J, Patino D, Miguez JL, Granada E, Moran J, Collazo J (2012) Study of the reaction front thickness in a counter-current fixed-bed combustor of a pelletised biomass. Combust Flame 159:1296–1302, doi:10.1016/j.combustflame.2011.10.007

5. Porteiro J, Patino D, Collazo J, Granada E, Moran J, Miguez JL (2010) Experimental analysis of the ignition front propagation of several biomass fuels in a fixed-bed combustor. Fuel 89:26–35, doi:10.1016/j.fuel.2009.01.024
6. Porteiro J, Patino D, Moran J, Granada E (2010) Study of a fixed-bed biomass combustor: influential parameters on ignition front propagation using parametric analysis. Energy Fuel 24:3890–3897, doi:10.1021/ef100422y
7. Horttanainen M, Saastamoinen J, Sarkomaa P (2002) Operational limits of ignition front propagation against airflow in packed beds of different wood fuels. Energy Fuel 16:676–686, doi:10.1021/ef010209d
8. Miranda T, Arranz JI, Montero I, Roman S, Rojas CV, Nogales S (2012) Characterization and combustion of olive pomace and forest residue pellets. Fuel Process Technol 103:91–96, doi:10.1016/j.fuproc.2011.10.016
9. Arce ME, Saavedra A, Míguez JL, Granada E, Cacabelos A (2013) Biomass fuel and combustion conditions selection in a fixed bed combustor. Energies 6:5973–5989, doi:10.3390/en6115973
10. Arshadi M, Gref R, Geladi P, Dahlqvist S, Lestander T (2008) The influence of raw material characteristics on the industrial pelletizing process and pellet quality. Fuel Process Technol 89:1442–1447, doi:10.1016/j.fuproc.2008.07.001
11. Bidabadi M, Abedinejad MS, Fereidooni J (2011) Modeling of the propagation of a reaction front in fixed bed combustion of wood particles. J Mech 27:453–459, doi:10.1017/jmech.2011.48
12. Haseli Y, van-Oijen JA, de-Goey LPH (2011) A detailed one-dimensional model of combustion of a woody biomass particle. Bioresource Technol 102:9772–9782, doi:10.1016/j.biortech.2011.07.075
13. Sefidari H, Razmjoo N, Strand M (2014) An experimental study of combustion and emissions of two types of woody biomass in a 12-MW reciprocating-grate boiler. Fuel 135:120–129, doi:10.1016/j.fuel.2014.06.051
14. Deng J (1989) Introduction to grey system theory. J Grey Syst 1:1–24
15. Sakthivel G, Ilangkumaran M, Nagarajan G, Priyadharshini GV, Kumar SD, Kumar SS, Suresh KS, Selvan GT, Thilakavel T (2013) Multi-criteria decision modelling approach for biodiesel blend selection based on GRA-TOPSIS analysis. Int J Ambient Energy. doi:10.1080/01430750.2013.789984
16. Liu J, Jiang X, Huang X, Wu S (2009) Morphological characterization of superfine pulverized coal particles. 1. Fractal characteristics and economic fineness. Energy Fuel 24:844–855, doi:10.1021/ef900954m
17. Wang S, Jiang XM, Wang Q, Han XX, Ji HS (2013) Experiment and grey relational analysis of seaweed particle combustion in a fluidized bed. Energy Convers Manage 66:115–120, doi:10.1016/j.enconman.2012.10.006
18. Wang H, Jiang X, Liu J, Lin W (2007) Experiment and grey relational analysis of CWS spheres combustion in a fluidized bed. Energy Fuel 21:1924–1930, doi:10.1016/j.enconman.2012.10.006
19. Han X, Jiang X, Liu J, Wang H (2006) Grey relational analysis of N2O emission from oil shale-fired circulating fluidized bed. Oil Shale 23:99–109
20. Kuo Y, Yang T, Huang G (2008) The use of grey relational analysis in solving multiple attribute decision-making problems. Comput Ind Eng 55:80–93, doi:10.1016/j.cie.2007.12.002
21. Asian N, Shahrivar AA, Abdollahi H (2012) Multi-objective optimization of some process parameters of a lab-scale thickener using grey relational analysis. Sep Purif Technol 90:189–195, doi:10.1016/j.seppur.2012.02.033
22. Liu S, Forrest J (2007) The current developing status on grey system theory. J Grey Syst 19:111–123
23. Saavedra A, Arce E, Miguez JL, Granada E (2013) Potential effect of uncertainty on the GRG interpretation. Grey Syst Theory Appl 3:121–128, doi:10.1108/GS-11-2012-0043
24. Pazo JA, Granada E, Saavedra A, Patino D, Collazo J (2010) Heterogenic solid biofuel sampling methodology and uncertainty associated with prompt analysis. Int J Mol Sci 11:2118–2133, doi:10.3390/ijms11052118
25. Patino D (2009) Análisis experimental de combustión de biomasa en un quemador de lecho fijo. Dissertation, University of Vigo (Spain)
26. Tuck G, Glendining M, Smith P, House J, Wattenbach M (2006) The potential distribution of bioenergy crops in Europe under present and future climate. Biomass Bioenergy 30:183–197, doi:10.1016/j.biombioe.2005.11.019
27. Saastamoinen J, Taipale R, Horttanainen M, Sarkomaa P (2000) Propagation of the ignition front in beds of wood particles. Combust Flame 123:214–226, doi:10.1016/S0010-2180(00)00144-9
28. Ryu C, Yang Y, Khor A, Yates N, Sharifi V, Swithenbank J (2006) Effect of fuel properties on biomass combustion: Part I. experiments - fuel type, equivalence ratio and particle size. Fuel 85:1039–1046, doi:10.1016/j.fuel.2005.09.019
29. Mehrabian R, Scharler R, Obernberger I (2012) Effects of pyrolysis conditions on the heating rate in biomass particles and applicability of TGA kinetic parameters in particle thermal conversion modelling. Fuel 93:567–575, doi:10.1016/j.fuel.2011.09.054
30. James AK, Thring RW, Helle S, Ghuman HS (2012) Ash management review - applications of biomass bottom ash. Energies 5:3856–3873, doi:10.3390/en5103856
31. Anca-Couce A, Zobel N (2012) Numerical analysis of a biomass pyrolysis particle model: solution method optimized for the coupling to reactor models. Fuel 97:80–88, doi:10.1016/j.fuel.2012.02.033
32. Liao K, Yao Q, Wu X, Jia W (2012) A numerical corrosion rate prediction method for direct assessment of wet gas gathering pipelines internal corrosion. Energies 5:3892–3907, doi:10.3390/en5103892
33. Subbaya KM, Suresha B, Rajendra N, Varadarajan YS (2012) Grey-based taguchi approach for wear assessment of SiC filled carbon-epoxy composites. Mater Des 41:124–130, doi:10.1016/j.matdes.2012.04.051
34. Taylor RP, Hodge BK, James CA (1999) Estimating uncertainty in thermal systems analysis and design. Appl Therm Eng 19:51–73, doi:10.1016/S1359-4311(97)00093-8
35. Joint Committee for Guides in Metrology (2008) Evaluation of measurement data. Guide to the expression of uncertainty in measurement., Available online: http://www.bipm.org/utils/common/documents/jcgm/JCGM_100_2008_E.pdf. Accessed 21 Jul 2014
36. Perez JF, Melgar A, Benjumea P (2012) Effect of operating and design parameters on the gasification/combustion process of waste biomass in fixed bed downdraft reactors: an experimental study. Fuel 96:487–496, doi:10.1016/j.fuel.2012.01.064
37. Fagerström J, Näzelius IL, Gilbe C, Boström D, Öhman M, Boman C (2014) Influence of peat ash composition on particle emissions and slag formation in biomass grate co-combustion. Energ Fuel 28:3403–3411, doi:10.1021/ef4023543
38. Jenkins BM, Baxter LL, Miles TR Jr, Miles TR (1998) Combustion properties of biomass. Fuel Process Technol 54:17–46, doi:10.1016/S0378-3820(97)00059-3
39. González JF, González-García CM, Ramiro A, González J, Sabio E, Gañán J, Rodríguez MA (2004) Combustion optimisation of biomass residue pellets for domestic heating with a mural boiler. Biomass Bioenergy 27:145–154, doi:10.1016/j.biombioe.2004.01.004
40. Shin D, Choi S (2000) The combustion of simulated waste particles in a fixed bed. Combust Flame 12:167–180, doi:10.1016/S0010-2180(99)00124-8
41. Yin C, Rosendahl LA, Kær SK (2005) Grate-firing of biomass for heat and power production. Prog Energy Combust 34:725–754, doi:10.1016/j.pecs.2008.05.002

Market integration of renewable energies through direct marketing - lessons learned from the German market premium scheme

Alexandra Purkus[1*], Erik Gawel[1,2], Marc Deissenroth[3], Kristina Nienhaus[3] and Sandra Wassermann[4]

Abstract

Background: Increasing the market and system integration of renewable energy sources (RES) is regarded as key to reducing the costs of RES support and transforming the electricity system. In several EU countries, feed-in premium schemes have been implemented to better align RES production with electricity prices and to enhance the efficiency of RES marketing by increasing direct participation of producers in electricity markets. This paper examines the lessons learned from the German market premium scheme, which was introduced in the Renewable Energy Sources Act 2012 as an optional alternative to feed-in tariffs. The 2014 reform of that law has made direct marketing mandatory except for small RES plants, thus establishing the sliding market premium as the new primary instrument of RES support.

Methods: Combining a qualitative economic analysis with insights from sociological research on direct marketing and simulation results from agent-based modelling, we evaluate how well the optional market premium has performed in setting incentives for demand-oriented RES production and efficient marketing of RES electricity. Furthermore, we examine what efficiency gains can be realistically expected from the changes implemented in 2014, and discuss whether the direct marketing model adopted in Germany is a promising approach for improving the market integration of RES.

Results: We find that direct marketing has made a positive contribution to the marketing efficiency of dispatchable RES; for intermittent RES, it provides few structural changes compared to marketing through transmission system operators. The benefits of a greater demand-orientation of RES feed-in remain limited when considering the extent to which load shifting is incentivised. For intermittent RES in particular, incentives for demand-oriented feed-in are only effective in times of negative electricity prices when voluntary curtailment is encouraged.

Conclusions: For the further development of mechanisms to support greater market proximity, differentiating between the various RES according to their ability to respond to electricity price signals can prove advantageous. Also, objectives such as improving security of supply and reducing the costs of transforming the electricity system cannot be reached through RES market integration efforts alone but require an adaptation of electricity market conditions to the requirements of RES.

Keywords: Renewable energy policy; Market integration; Feed-in premium; Direct marketing

* Correspondence: alexandra.purkus@ufz.de
[1]Department of Economics, Helmholtz Centre for Environmental Research - UFZ, Permoser Str. 15, Leipzig 04318, Germany
Full list of author information is available at the end of the article

Background

RES market integration - objectives and instrumental implementation

As electricity generation from renewable energy sources (RES) continues to expand in Europe, EU member states are faced with the challenge of integrating the growing proportion of renewables, and especially intermittent renewables, into existing energy systems and electricity markets [1,2]. In this process, increasing the market integration of RES is viewed as an important step towards improving the alignment of RES feed-in to scarcity signals in electricity markets and systems, thus containing the costs of RES expansion; these relate not only to the costs of supporting RES but also to the wider costs of transforming the electricity system [2,3]. The new EU energy aid guidelines, for instance, stress that market-based mechanisms should promote the 'transition to a cost-effective delivery' (p. 23) and highlight the importance of direct market participation and the assumption of market responsibilities by RES producers [4]. In principle, it appears sensible to differentiate between two different dimensions of market integration (see Table 1) [5-7]. Firstly, market integration aims for a stronger alignment of RES feed-in to price signals in the electricity market as well as an efficient marketing of RES electricity [5,6]. Incentives for this dimension of market integration can be implemented in RES policy instruments where prices are administered by the state, by changing from a feed-in tariff which is paid independently from market prices to a feed-in premium model where electricity prices influence total remuneration. This approach is currently pursued by several EU member states using various different design options [8,9]. While fixed feed-in premiums offer RES producers a set bonus on top of the market price, sliding premiums balance out the difference between reference prices and average market prices [10]. Here, total remuneration remains based on state-determined prices, as is the case for fixed feed-in tariffs, so that the long-term market risks for producers of RES remain comparatively low. The second dimension of market integration, on the other hand, encompasses a competitive determination of the remuneration and a competitive steering of investment decisions [11,12]. This involves a transition to treating RES producers as regular market players with system responsibility, with competitive support mechanisms and electricity market revenues playing a significant role in their refinancing. At present, discussions about the instrumental implementation of this dimension of market integration focus primarily on competitive bidding processes, given that the new EU guidelines on state aid express a clear preference for this instrument type [4]. A further distinction is necessary between the market integration and system integration of RES - the latter aims at improving RES' contribution to grid stability [1,13] but is not the focus of this article.

One example of a sliding feed-in premium scheme is the German market premium model, which was initially introduced into the RES support regime in 2012 as an optional alternative to the fixed feed-in tariff in order to set incentives for demand-oriented RES feed-in and efficient marketing of electricity from RES. With the introduction of the amended Renewable Energy Sources Act (EEG) 2014, this approach was further extended by making

Table 1 An overview of the objectives of market integration of renewable energy sources

Market integration objectives	Dimensions of achievement of objectives	Contributions to overarching objectives
Demand-oriented generation of RES electricity and increased flexibility of RES plants	Prevention of surplus supply situations via voluntary curtailment in times of negative electricity prices	Contribution of RES to security of supply is increased;
	Shift of feed-in to times when demand is strong and prices are high:	Costs of RES promotion are reduced through an increase in RES market value;
	• Intermittent RES: Maintenance planning, installation design oriented towards market value and system requirements	Cost reductions in the overall system (e.g. lower balancing energy prices and system integration costs)
	• Dispatchable RES: targeted load shifting	
	Increased remote controllability of RES installations	
	Participation of RES installations in the balancing energy market	
Efficient marketing of RES electricity	Reduced transaction costs of marketing RES electricity	RES promotion costs are reduced
	Increased forecasting quality and reduction of costs of procuring balancing energy	
	Competition for efficient marketing forms	
Market-driven production and investment decisions	Competitive determination of RES remuneration	RES expansion costs are reduced
	RES producers become regular market players	

Based on BMU 2011 [7]; Klobasa et al. 2013 [6]; and Lehnert et al. 2014 [5-7].

direct marketing mandatory except for small plants, while elevating the market premium scheme to the standard remuneration model. In the literature, assessments of the market premium scheme come to divided conclusions. While some studies evaluate the promotion of market integration via the sliding market premium positively [6,14,15], others are critical of the instrument. On the one hand, market-determined incentives for RES producers are regarded as insufficient to promote market integration and reduce the costs of RES support [11,12,16,17]. On the other hand, however, critics question the usefulness of a direct marketing-driven integration of intermittent RES into the marginal cost-based electricity market, which may be structurally unsuitable to secure their long-term refinancing [18-21]. Furthermore, there are trade-offs between the objective of exposing RES producers to market price signals in order to increase the efficiency of production and investment decisions on the one hand, and the creation of a secure planning basis for RES investors, which is crucial to achieve a comprehensive transformation to an RES based electricity system, on the other [13]. If RES producers carry a high share of market risks, this would also tend to increase the costs of supporting the implementation of RES objectives, because higher risks have to be compensated by higher risk premiums in the remuneration [22-24].

The following article evaluates experiences gained in Germany from the introduction of the sliding market premium and discusses to what extent this form of government-supported direct marketing can serve as a promising instrument to promote the integration of RES into the electricity market. The article begins with an outline of the design of the sliding market premium instrument and supplementary regulations on the promotion of market integration in the EEG 2014, as well as an overview of the development of participation in direct marketing. This is followed by an evaluation of experiences gained with the market premium under the EEG 2012. Building on this, we examine to what extent the new regulations set forth in the EEG 2014 can be expected to make a greater contribution to the market integration objectives of realising a demand-oriented feed-in of RES and efficient marketing.

Methods

As a sliding feed-in premium with state-administered prices, the market premium scheme does not aim for a competitive control of RES investments [9,15]. Therefore, our assessment of the scheme's performance focusses on the market integration objectives of realising a demand-oriented generation of RES electricity and increased flexibility of RES plants, as well as an efficient marketing or RES electricity. Based on existing literature, measures were identified which contribute to these objectives; these are summarised in Table 1. Below, we discuss whether the market premium scheme sets incentives for the adoption of these measures and whether contributions to market integration objectives are effective and efficient. To conduct this assessment, we review literature findings and available data, and analyse the impact of incentives on RES producers' behaviour. This qualitative analysis is complemented by simulation results from the agent-based simulation model AMIRIS, which was developed at the Institute of Engineering Thermodynamics at the German Aerospace Center (DLR) in Stuttgart [25,26] The special nature of this article is an explicit interdisciplinary approach to the research topic which attempts to integrate existing studies from sociological research on markets [27], energy economics [9] and agent-based modelling [26].

Results and discussion

Promoting market integration via the German market premium model

In the market premium model under the Renewable Energy Sources Act (EEG), producers of electricity from RES receive the difference between technology-specific reference prices and the actual, likewise technology-specific, monthly average market price (EEG 2014 Annex 1). Under the EEG 2012, installation operators had the option of switching between direct marketing and feed-in tariffs, where the sale of electricity from RES is centrally managed by transmission system operators (TSOs). This is no longer possible under the EEG 2014. Only installations with a maximum installed capacity of 500 kW are now eligible to receive the feed-in tariff; and as of 2016, the obligation to participate in direct marketing will be extended to all installations with a capacity of over 100 kW (§ 37 (2) EEG 2014). In exceptional cases, operators will still be able to fall back on feed-in tariffs, with a 20% reduction in remuneration rates (§ 38 EEG 2014). The so-called 'green electricity privilege' provided under § 39 EEG 2012, whereby electricity suppliers benefitted from EEG surcharge reductions if a certain proportion of their electricity supply was sourced from directly marketed RES, has also been removed from the new EEG. A further amendment relates to the fact that under the EEG 2012, RES producers were eligible for a separately calculated management premium to compensate for the additional costs of trading and providing balancing energy. Under the EEG 2014, in contrast, the management premium is implicit in the reference prices. For dispatchable renewables, this is reflected by a reduction of 0.2 cents/kWh in the reference price on which the fixed feed-in tariff is based; while for intermittent renewables, the reduction is 0.4 cents/kWh (§ 37 (3) EEG 2014). At the same time, the remote control capability of RES installations, which was incentivised under the EEG 2012 by offering higher management premiums for installations fitted with appropriate systems, will now be a requirement for payment of the market premium (§ 35 sentence 1 no. 2 in conjunction with § 36 (1) EEG 2014).

As a measure aimed at strengthening the demand-orientation of all renewable energy sources, the EEG 2014 also introduces cuts in financial support in times of negative prices: In the case of installations commissioned after 1 January 2016, the reference prices on which the remuneration is based are to be reduced to zero when the value of the hourly contracts on the EPEX Spot exchange is negative in at least six consecutive hours (§24 EEG 2014). Given that installations with an installed capacity of less than 500 kW are exempt, this regulation applies only to larger installations for which direct marketing is obligatory (§ 24 (3) EEG 2014). As was the case under the EEG 2012, the market premium for biogas plants will be bolstered by a capacity-oriented flexibility premium (§§ 52 to 54 EEG 2014). In addition, for biogas installations with a capacity exceeding 100 kW, entitlement to feed-in tariff or premium payments has been limited to the share of the annual amount of electricity generated which corresponds to a power rating of 50% of the installed electric capacity (§ 47 (1) EEG 2014), a move intended to guarantee flexible plant operations.

Meanwhile, the transition to a more market-driven steering of investment decisions is supported in the EEG 2014 by the proposed switch to a competitive bidding process, which is geared towards establishing a competitive determination of remuneration by 2017 at the latest (§ 2 (5) EEG 2014). The details of its design are still unclear [28], but the model is to be tested on ground-mounted solar installations (§ 55 EEG 2014).

Experiences with the market premium under the EEG 2012: participation in direct marketing

Since its introduction in 2012, the market premium has successfully increased participation in direct marketing (see Figure 1): In April 2014, a total of 49% of the installed capacity eligible for support under the EEG was marketed under the market premium scheme [29,30]. At the same time, the green electricity privilege, as an alternative form of direct marketing, was largely replaced while the option 'other forms of direct marketing' without remuneration is practically irrelevant (see Figure 2). Initially, a high proportion of wind power producers in particular switched to the optional market premium, but in the meantime participation among other RES producers has also increased [9,31]: In April 2014, almost 60% of biomass installations were taking advantage of the model; and as much as 13% of solar installations were participating (see Figure 2). Nonetheless, the success of the sliding market premium can only really be measured by its contribution to the actual market integration objectives.

Experiences with the market premium under the EEG 2012: efficient marketing of RES electricity?

Under the fixed feed-in tariff system, quantities of RES electricity are traded on the day-ahead or intraday spot market in accordance with the provisions of the Equalisation Scheme Ordinance (§ 2 AusglMechV). By comparison, the sliding market premium would lead to more efficient marketing of RES if (a) direct marketing reduces the transaction costs of market participation, (b) balancing energy costs can be reduced as a result of better RES feed-in forecasts and better management of plant portfolios, e.g. by taking advantage of the ability to control plants remotely or (c) innovative forms of marketing can increase prices obtained for renewables by optimising the choice of markets (e.g. intraday and day-ahead spot market, balancing market, bilateral over-the-counter contracts) (see Table 1). Whether this applies to the market premium model will be examined in the following.

The majority of RES installation operators are not conventional players in the established energy sector and, as a result, most of them have neither infrastructure nor the knowledge required to sell the electricity they produce on the central electricity markets themselves. Introducing RES installation operators to the market was therefore only possible through electricity traders (hereafter referred to as 'direct marketers'). They act as service providers to RES installation operators and in that function offer a broad range of services such as preparing feed-in forecasts, communicating schedules to the grid operators as well as financially compensating for deviations or purchasing additional electricity to make up the scheduled amount. Based on their RES supply forecasts and expected electricity prices, these traders take the initial decision on behalf of the installation operator as to whether to remain within the feed-in tariff scheme or to market directly. In the latter case, they determine the time, the quantity and the market on which RES electricity will be offered for sale. Direct marketers also have access to the specific infrastructure needed to be able to trade on the energy exchange [25].

Regarding the reduction of transaction costs, compared with selling through TSOs, it can be said that initially additional costs are incurred because the TSOs still market the quantities of electricity in the feed-in tariff scheme. In addition, marketing costs are subject to economies of scale because fixed costs make up a large share of those costs [18,20,32]. Also, the larger the

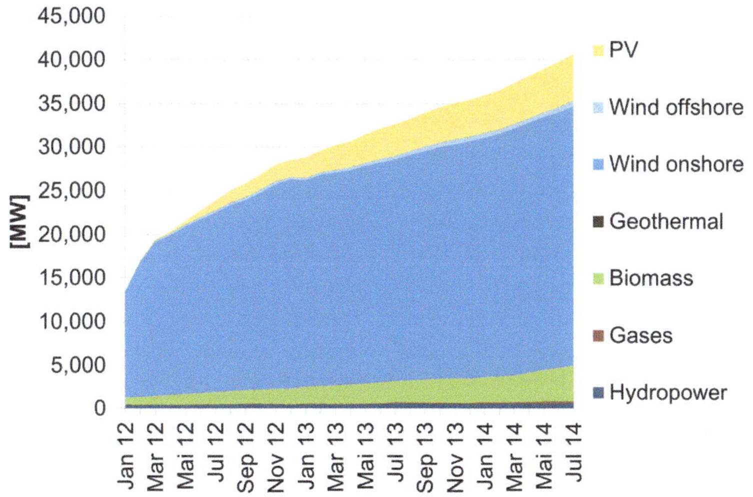

Figure 1 Development of installed capacity in the market premium model (January 2012 - July 2014). 'Gases' encompass landfill, sewage and mining gases. Authors' representation based on 50Hertz et al. 2014 [29].

management portfolio the lower the costs of balancing energy will be, because larger portfolios are associated with higher forecast accuracy [6]. This suggests that in terms of transaction costs, large direct marketers come closest to TSOs.

An approximate value for the additional costs of direct marketing under the market premium scheme is provided by the management premium: In 2012, this amounted to approximately 467 million euros, while in 2013 an estimated 354 to 400 million euros were paid out due to a reduction in management premium rates [33-35].[a] However, the management premium primarily acts as an indicator for the additional support costs associated with direct marketing, rather than for actual transaction costs. Especially in 2012, it was found that an overcompensation of the costs of direct marketing led to high windfall profits, subsequently giving rise to stronger cuts in the management premium rates for intermittent RES [36]. While the management premium pursuant to EEG 2012 forms the essential basis of direct marketing business models [18,37], its degressive design increases cost pressure on marketers over time.

When it comes to the marketing of wind and photovoltaics (PV), the marketing channels used are mainly

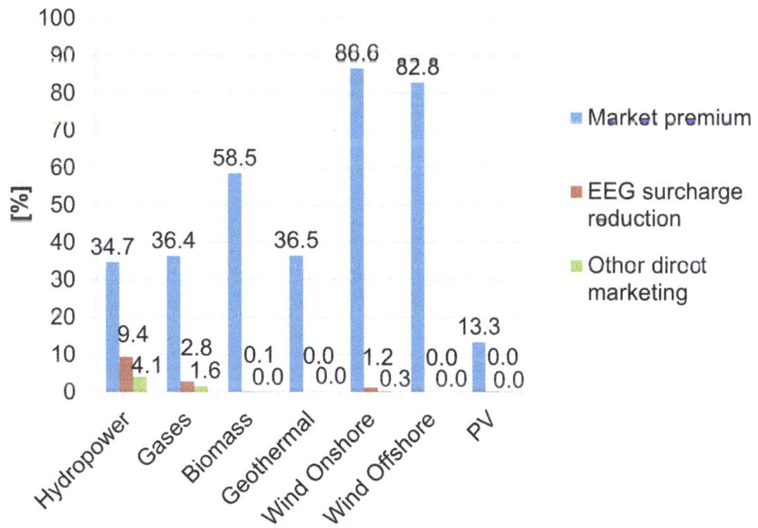

Figure 2 Direct marketing share of the total installed capacity eligible for EEG remuneration (April 2014). 'Gases' encompass landfill, sewage and mining gases. Authors' representation based on 50Hertz et al.; Bundesnetzagentur [29,30].

the same as those used by TSOs, i.e. the intraday and day-ahead spot market. This choice is structurally supported by the market premium because spot market prices are used to calculate the market value, and therefore selling on other markets would entail higher levels of revenue risk [6,38]. Biomass installations, on the other hand, show greater participation in balancing markets [6,39]. In principle, differences in flexibility options available to biomass installations in combination with heat utilisation concepts mean that they require more individualised marketing concepts than other RES, so that direct marketing is likely to increase efficiency. In the case of intermittent RES, marketing innovations can be found e.g. in the establishment of RES networks forming virtual power plants, which have spread rapidly [40]. However, targeted marketing of the 'green power' characteristic, as promoted by the green electricity privilege, is not possible under the market premium model (§ 56 (2) EEG 2012 and § 80 (2) EEG 2014).

Summing up, it can be said that the efficiency benefits that have been achieved with the marketing of wind and PV are mainly attributable to the fact that not only new installations but also existing ones were incentivised to provide for remote control capability and availability of real-time data [6]. However, the marketing channels used are largely similar to those employed by TSOs, with the exception of bioenergy and hydropower generators, for whom the supply of balancing energy is profitable. At present, direct marketing also leads to increases in the transaction costs of RES marketing. A lowering of these costs to a level that is comparable with marketing through TSOs alone could be achieved primarily by exploiting economies of scale. However, corresponding stronger price competition between direct marketers could promote processes of concentration [18,25,38].

Experiences with the market premium under the EEG 2012: incentives for demand-oriented RES production and flexibility?

A more demand-driven feed-in and greater flexibility in RES production can be achieved when the market premium sets incentives to encourage (a) voluntary curtailment in times of negative electricity prices, (b) in the case of intermittent RES, electricity price-oriented maintenance planning and plant designs that are oriented towards a maximisation of market value and aligned with system requirements, (c) targeted load shifting of dispatchable renewables, (d) increased remote control capability of RES installations (cf. also (b) in the section on efficient RES marketing above) and (e) participation of RES producers in the balancing energy market (see Table 1). Relevant factors for the analysis are not only whether these measures are taken but whether in this form they contribute to the overall objectives of security of supply and cost reduction.

For intermittent RES with near-zero marginal costs, incentives are set to make limited bids and/or to encourage curtailment in times of strongly negative electricity prices [6,41]. Voluntary curtailment becomes profitable when electricity traders participating in the spot market have to pay buyers a higher amount than that which is expected from the market premium payment including the management premium [9]. Currently, for example, wind energy plants in direct marketing curtail their production at negative prices of approximately – €65/MWh [42]. Their ability to respond depends upon the remote control capability of the installation, which has been growing since 2013 due, among other things, to corresponding incentives in the management premium [6]. Apart from optimising maintenance planning, intermittent RES have few other incentives for load shifting [9,20]. Price signals are either too low or not consistent enough to make it worthwhile to base decisions about the location and the design of installations on market value rather than on energy yield, or to stimulate investment in weak-wind installations or east-west facing PV systems [43,44]. What is more, the advantages of the market value of alternative installation concepts are being eroded by the increasing number of relevant suppliers [18]. In the case of balancing markets, prequalification of intermittent RES is only possible to a limited extent at present [5,38].

So far, there are no studies available on actual changes in feed-in behaviour for dispatchable renewables. Basically, the market premium sets incentives for voluntary curtailment when electricity prices fall below the difference between the marginal costs of production and the expected market premium including the management premium. For biogas plants, the uptake of the flexibility premium provides insight into whether the conditions for a demand-based, flexible operation are at least in existence [45,46]: In April 2014, 370 installations with approximately 216 MW_{el} claimed the flexibility premium, which corresponds to roughly 11% of the installed biogas and biomethane capacities participating in direct marketing [39]. After a slow start in 2012, the use of the flexibility premium is meanwhile increasing rapidly; the delay in uptake is explained by the fact that power plant flexibilisation is associated with financial investments and high information requirements [46].

Next to hydropower installations, biomass installations are the most active RES producers on the electricity balancing market. In April 2014, 24% of the directly marketed biomass power capacity and 43% of the directly marketed hydropower capacity were correspondingly pre-qualified. So far, mostly negative balancing energy is supplied [39,46]. Figure 3 gives an example of how biomass installation operators could generate additional revenues through participation in direct marketing including the balancing market for negative balancing energy. The figure shows the results of a simulation of the agent-based simulation model AMIRIS, which is being developed at the Institute of Engineering Thermodynamics of the German Aerospace Center (DLR) in Stuttgart [25]. Especially for solid fuel-based thermal power plants[b] and large biogas installations, the potentially achievable additional revenues of 6% and 11%, respectively, are worthy of note.

The difference in the revenues of the two installations mentioned above can be explained by different shares of directly marketed capacity, which is higher for large biogas installations. Although in this model small biogas installations with a maximum installed capacity of 350 kW cannot participate in the balancing market, they still profit indirectly from the other installations' sales, as can be seen from Figure 3. This results from the model assumption that, due to the additional revenues achieved on the balancing market, the direct marketers will also increase bonus payments to their other customers which is a possibility to improve their competitive position.

Finally, by exploiting balancing markets, the flexibilisation of installations can indeed have a positive effect on their economic efficiency while also supporting the system as a whole. Thus, it appears to make sense to incentivise a corresponding adaptation of installations and to enable their participation in the balancing market, where besides storage plant operators more and more other new participants are competing.

Viewed on the whole, however, the market premium currently has the character of a 'curtailment premium': for intermittent renewables, curtailment signals become effective when electricity prices are highly negative; for dispatchable renewables, this occurs earlier due to positive marginal costs, especially since the supply of negative balancing energy is profitable. For bioenergy installations, in particular, the response to electricity price signals sets important incentives for optimising the use of biomass resources.

However, so far, the model sets few incentives for dispatchable renewables to engage in positive load shifting, which might be desirable in order to increase the contribution of RES to the energy supply when feed-in from intermittent RES is low, and could contribute to the stability of the system. At present, this is also determined by relatively low peak/off-peak spreads on the spot market [47]. Furthermore, there are still few incentives for dispatchable renewables to supply positive balancing energy or for intermittent renewables to make market value-oriented decisions about the design of their installations.

By reducing the quantity of electricity that has to be remunerated in times of low and negative prices, voluntary curtailment eases the EEG surcharge burden, which is determined on the basis of the difference between the feed-in tariff's or market premium's reference prices and the market value for RES electricity [6]. However, the ongoing expansion of intermittent RES with near-zero marginal costs and the high temporal coincidence of their feed-in causes the market value of intermittent renewables to fall [25,48,49]. Even if the full costs of RES generation were decreasing, this would in turn lead to an increasing burden on the EEG surcharge and represents a fundamental problem for the integration of intermittent RES in the 'energy only' market. Moreover, analyses of the causes of negative electricity prices show that in times of negative prices, in addition to nuclear power plants, particularly emission-intensive lignite coal-fired power plants continue to operate at high load [42,50]. Here, RES curtailment reduces incentives for conventional power plant operators to curtail their operations or invest in flexibilisation measures. From the climate

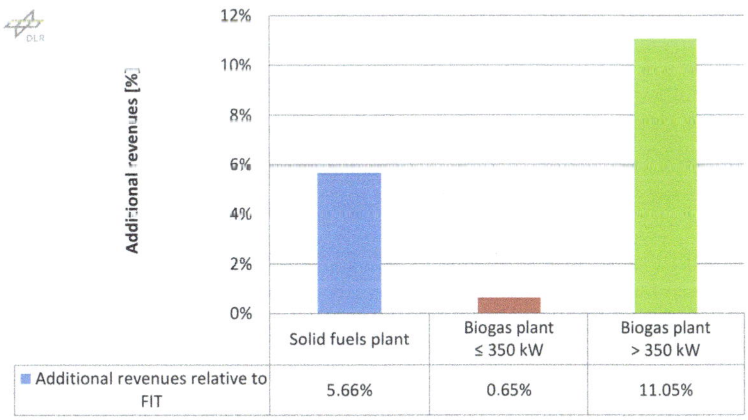

Figure 3 Additional revenues generated by biomass installation operators via participation in the balancing market. In comparison with their EEG feed-in tariffs.

policy and system transformation perspective, this situation is highly problematic [18,20,51].

The market premium under the EEG 2014 - a contribution to a more objective-based market integration?

The main changes in the EEG 2014 concerning RES market integration are the following: the mandatory character of direct marketing for larger, new installations; the pricing-in of the management premium in the reference prices used to calculate the market premium; cuts in payments when prices are negative for prolonged periods; and the introduction of remote control capability as a requirement for payment of the market premium. Further implications arise from the general decrease of remuneration rates, which is particularly pronounced for bioenergy [52]. So, the question is, compared with the EEG 2012, what changes can be expected regarding the contribution of the market premium to marketing efficiency and the promotion of demand-driven and flexible electricity feed-in?

In terms of marketing efficiency, the exploitation of the marketing channels used under the market premium model in the EEG 2012 seems likely to continue since there has been no change in the structure of the incentives created. For biogas installations, flexibility of operators to design site-specific marketing concepts is restricted by the newly introduced limit on the remunerated quantity of electricity to 50% of the installation's power rating [53]. In that respect, it lessens the advantage associated with direct marketing for these dispatchable RES of being able to optimise the marketing of electricity and heat to suit the specific installation. In particular, however, mandatory direct marketing in conjunction with the reduced management premium is expected to have a tendency to strengthen larger direct marketers who have lower marketing costs due to economies of scale and so can remain competitive [38]. The mentioned tendency and economic impact of the direct marketing costs in combination with a simultaneous abolition of the management premium were examined using the DLR simulation model described above (here, the obligation implies that the management premium paid out on top of the market premium under the EEG 2012 is no longer provided and is also not integrated in the reference prices, as is the case under the EEG 2014). The model accounts for fixed and variable costs of direct marketers, i.e. fixed costs represent office rents, IT and staffing costs and annual charges for trading, whereas variable costs refer to the amount of traded electricity and expenditures for electricity generation forecasts and balancing power needed in case of deviation from generation schedule. As depicted by Figure 4, by way of example, with the above-mentioned cost structure, the specific marketing costs of the direct marketers in AMIRIS assume values between €1 per MWh and €3.50 per MWh for the simulation years 2012 to 2019. The grey-shaded area illustrates the variation of the marketing costs as they are depending on the type of trader. The extent of the costs is strongly dependent on the composition of the trader's portfolio - in contrast to dispatchable renewables, high shares of intermittent renewables tend to result in higher costs to balance generation and demand. The decrease of the maximum specific costs until 2019 is due to an increasing size of portfolios of the direct marketers, i.e. they profit from economies of scale and improved forecast qualities over the years. The simulated costs correspond quite accurately with estimates given by direct marketers based on experience and show the eligible configuration of the model.

If marketing costs were no longer compensated by the management premium, direct marketers' revenues would be reduced. Figure 5 shows AMIRIS simulation results for the overall profit of the marketers under assumption of a decreasing management premium according to the EEG 2012 (light grey area) and a complete abolition of the management premium (dark grey area). The overall profit is defined as the difference between total income and costs, and the areas represent the variation of results of the different simulated direct marketers. As can be seen, with a decreasing management premium, the majority of the marketers still gain positive overall profit; the reduction of the premium is reflected in the diminution of the profit over the years, with explicit gradient in 2012 and 2013. In contrast, the abolition of the premium implies negative profits for all simulated types of direct marketers, as the dark grey area depicts. The decrease of the minimum profit until 2019 reflects the increasing expenditures for payments

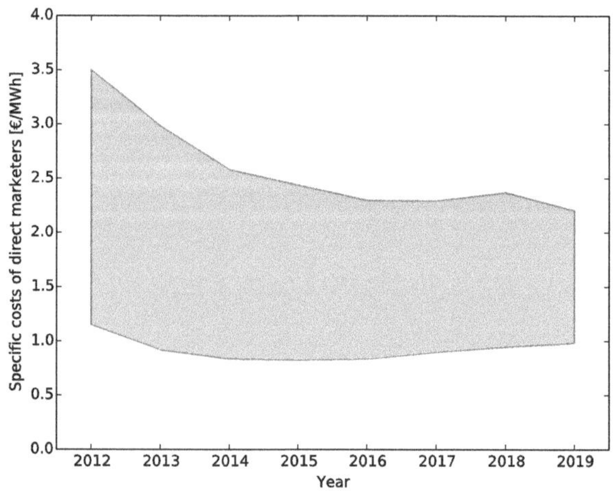

Figure 4 Specific direct marketing costs of different direct marketers.

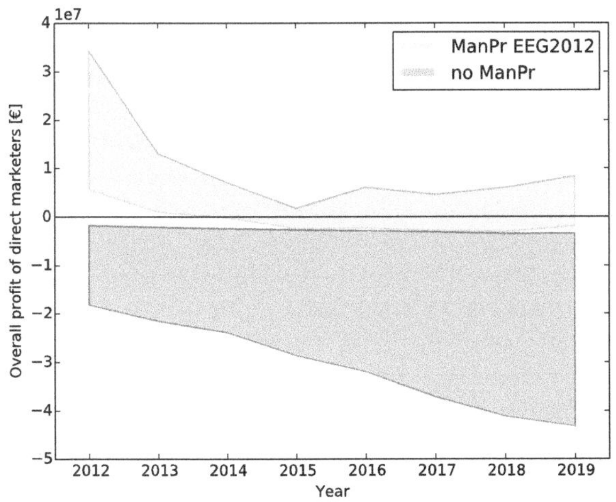

Figure 5 Overall profits of different direct marketers. Overall profits of different direct marketers in case of payment of a decreasing management premium according to EEG 2012 (light grey area) and in case of abolition of the management premium (dark grey area). Overall profits of direct marketers in tens of millions of euros.

to installation operators, as the portfolio of the direct marketers still grows in the simulation.

Profit losses are accompanied by an increased pressure to reduce costs that may lower the transaction costs of marketing and thus improve efficiency. But as not all direct marketers may have the possibility to reduce costs as needed, an oligopolisation in the direct marketers' market would weaken the bargaining power of RES producers [18]. Furthermore, direct marketers would have to 'retrieve' their costs from the installation operators in order to achieve an at least comparatively balanced account. This would mean that the specific direct marketing costs could also be interpreted by installation operators as a corresponding reduction of the remuneration rates.

Hence, these simulations point to the following possible risks:

- Market concentration: Direct marketers with large portfolios have comparatively lower marketing costs because economies of scale in power trading and the broad spatial distribution of RES feed-in installations reduce costs. This can lead to concentration effects and create dominant market positions for traders, whose market power can make conditions more difficult for further RES investments.
- Rising costs: If policy is aimed at preserving today's diversity among direct marketers and giving players with smaller portfolios a chance to compete in the market, it must allow higher specific marketing costs to be compensated, e.g. via a premium. The compensation premiums required to promote a heterogeneous market could subsequently result in rising costs of support for electricity from renewable energy sources.
- Lower remuneration: For installation operators, mandatory direct marketing without payment of compensation via the management premium implies an indirect reduction in remuneration in the amount of the specific marketing costs and a reasonable margin of profit for intermediaries. This effect could weaken investment incentives, especially for intermittent RES.

In contrast to the effects on intermittent renewable energy sources, the introduction of mandatory direct marketing for dispatchable renewables such as biomass installations presents a comparatively lower risk. In addition, bringing dispatchable renewables onto the market would increase their flexibilisation potential, provided the financing of a corresponding technological design is guaranteed.

The temporally unlimited fall-back option provided by § 38 EEG 2014, meanwhile, is essentially to be welcomed, as capital investors would otherwise impose significant risk premiums to account for cases where RES producers lose their direct marketers [18,38]. However, the 20% cut in remuneration results in substantial income losses so that the exploitation of the feed-in tariff in exceptional cases only provides a short-term option, which in turn strengthens the bargaining position of direct marketers. Producers are therefore all the more reliant on strong competition between direct marketers - but precisely this competition may be diminished if economies of scale reduce the number of direct marketing companies. Meanwhile, the remuneration of RES generation increasingly depends on the conclusion of individual contracts and so predictability and transparency suffer. How this will be reflected in risk calculations and investment inclination still has to be examined.

Regarding the promotion of demand-oriented feed-in, it again makes sense to differentiate between dispatchable and intermittent renewables. In combination with revenue prospects associated with a participation in balancing markets, the market premium scheme under the EEG 2012 has set meaningful flexibilisation incentives for bioenergy plants, particularly for biogas installations eligible for the flexibility premium. In the EEG 2014, there are interactions between market integration incentives and the significant general cuts in remuneration for bioenergy. To what extent new biomass installations are economically feasible is questionable [52], so that the role of new installations as a flexibility option in the electricity system is likely to be limited. Although, in principle, the limit placed on the amount of electricity eligible for remuneration in the case of biogas installations provides an incentive to sell electricity at the highest possible price (see above), it is likely to deteriorate further the profitability of new installations. Furthermore, in terms of climate impacts, the one-sided focus of the eligibility requirements on flexibility should be evaluated critically because concepts involving high levels of heat use - which lead to higher reductions in greenhouse gas emissions compared to pure electricity production [54] - are neglected. Hence, as a flexibilisation option, mainly existing plants which can claim the market premium based on the remuneration under earlier versions of the EEG are likely to play a role. Here, a flexibility premium that includes solid fuel-powered biomass installations might have been appropriate in order to encourage the relevant changes in feed-in behaviour and make flexibilisation investments worthwhile [55].

On the side of intermittent renewables, the introduction of remote control capability as an eligibility requirement for new installations represents a sensible amendment [32]. Apart from this, the lack of responsiveness of intermittent renewables to electricity price signals beyond curtailments continues to be a problem, and it will only be aggravated by cutting remuneration in times of negative electricity prices. While the market premium under the EEG 2012 primarily served to prevent negative price spikes, the new regulation is aimed at curtailing production once prices fall below zero for six or more consecutive hours. This encourages intermittent RES producers with near-zero marginal costs to curtail production before continuously operating conventional power plants with positive marginal costs do so - a situation which, from efficiency and climate policy perspectives, does not appear advantageous. At the same time, the existence of 'moderate negative prices' [41], p. 104, is considered to constitute an economically reasonable means of sending flexibilisation signals to the operators of conventional power plants [5,18,41]. However, the new regulation largely relieves the latter of responsibility for implementing the system transformation; RES operators tend to bear growing electricity price risks [42,56], while the less frequent occurrence of negative prices benefits conventional power plant operators, who can avoid the costs of investing in flexibilisation. While the new regulation might ease the EEG surcharge burden in the short term, the long-term costs of transforming the electricity system can be expected to rise, since flexibility incentives for the system as whole are reduced.

Conclusions

In summary, it can be stated that the sliding premium, in the form in which it has been introduced in Germany, has made a positive contribution to the marketing efficiency of dispatchable RES in particular, because it allows for an individual optimisation of marketing concepts including participation in balancing markets. Meanwhile, the benefits of a greater demand orientation of RES feed-in remain limited when considering the extent to which load shifting is incentivised. As a result of the sharp cuts in the reference prices for bioenergy in the EEG 2014, the recently introduced extended flexibilisation requirements for biogas plants are likely to be relevant for only a small number of new installations. Whether incentives for existing installations will be sufficient in the long term to encourage load shifts beyond the supply of negative balancing energy remains to be investigated. For intermittent RES, incentives to encourage a change in feed-in behaviour are only effective in times of negative electricity prices when voluntary curtailment is encouraged. However, this simultaneously leads to a reduction in urgently needed flexibilisation signals to conventional power plants and other electricity market players, not least to the demand side or storage concepts. This problem is exacerbated by the EEG 2014's cuts in remuneration during prolonged negative electricity prices. Meanwhile the fundamental problem of the market integration of intermittent RES with near-zero marginal costs - the fact that an increase in feed-in amounts is associated with a decrease in market value - remains unsolved.

Development trends in the market structure of direct marketing service providers towards few large players who can benefit from exploiting economies of scale and tendering broad RES portfolios on the spot market also show that direct marketing provides few structural changes compared to marketing through TSOs, at least in the case of intermittent RES. While the exploitation of economies of scale is linked to reductions in transaction costs and so can improve marketing efficiency, an oligopolisation in the direct marketers' market would represent a problematic development: insufficient competition between direct marketers would in particular weaken the bargaining position of intermittent RES operators, whose marketing is associated with higher costs compared to dispatchable renewables. This could have a negative impact on investment incentives.

So far, Germany's experiences with the market premium and direct marketing suggest that, for the further development of mechanisms to support greater market proximity, differentiating between the various RES according to their ability to respond to electricity price signals can prove advantageous [1,19,20]. Given the experience gained to date, it seems doubtful whether an increase in short-term electricity price risks can increase the efficiency of supply of intermittent RES, rather than just leading to higher risk premiums or reluctance to invest. Under these conditions, it has been suggested that rather than making direct marketing obligatory, it may be advantageous to allow intermittent RES producers a (non-reversible) choice between central marketing in combination with a low-risk but also low-profit margin feed-in tariff and direct marketing in combination with an instrument that promotes market integration more comprehensively than an administered sliding feed-in premium but compensates price risks with a chance of higher returns [19]. Candidates for such an instrument could be competitively determined feed-in or capacity premiums, the latter having the advantage of not distorting spot market price signals [12,19,57]. The continued marketing of intermittent RES by TSOs, on the other hand, would require improved access to production data and incentives to optimise central marketing of RES electricity. For dispatchable RES such as bioenergy, meanwhile, refinancing models which have a stronger competitive element in the determination of remuneration than the market premium could provide a promising option to encourage operators to design site optimised installation concepts, which also exploit heat marketing opportunities [19]. When designing support for new installations, it should, however, be taken into account that growing competition for cheap feedstocks could endanger the supply of raw materials and the profitability of existing plants [58].

Regarding the transition to competitive bidding processes as promoted in the EU energy aid guidelines, it is necessary to critically discuss the question of how a stronger integration of different RES in existing electricity markets can actually contribute to improving security of supply, reducing the costs of RES expansion and transforming the electricity system. The case study of the German market premium clearly points out that these targets cannot be reached through market integration efforts on the part of RES alone. Rather, electricity market conditions must also be adapted to the requirements of RES, and flexibilisation incentives are needed for other participants in the electricity market [41,59]. Considering the uncertainties and country-specific conditions that have to be taken into account in a combined reform of the design of the electricity market and RES support mechanisms, competition between systems with different approaches in the different member states could have advantages over a one-size-fits-all approach as formulated in the EU guidelines on state aid [60].

Endnotes

[a]Calculated according to [33,34] in conjunction with EEG 2012 Annex 4 and the Management Premium Ordinance (MaPrV) 2012 [35]. The exact value is determined by the share of intermittent RES installations that were eligible to claim the higher management premium rate for installations with remote control capability.

[b]Wood gasifiers, which are still in an infant stage and do not play a significant role in the stock of biogas installations, are excluded from the assessment since they do not participate in direct marketing.

Competing interests
The authors declare that they have no competing interests.

Authors' contributions
Several authors drafted the different parts of the first version of this manuscript. All authors contributed to revising and finalising this article. All authors read and approved the final manuscript.

Acknowledgements
Joint research on this paper has been undertaken in the context of the Integrated Key Topic 'Regulation, Instruments, Governance' of the Alliance ENERGY-TRANS established by the Helmholtz Association of German Research Centres. Further information can be found at www.energy-trans.de. Furthermore, we would like to thank two anonymous reviewers for helpful comments.

Author details
[1]Department of Economics, Helmholtz Centre for Environmental Research - UFZ, Permoser Str. 15, Leipzig 04318, Germany. [2]Institute for Infrastructure and Resources Management, University of Leipzig, Grimmaische Str. 12, Leipzig 04109, Germany. [3]Department of Systems Analysis and Technology Assessment, Institute of Engineering Thermodynamics, German Aerospace Center (DLR), Wankelstr. 5, Stuttgart 70563, Germany. [4]Stuttgart Research Center for Interdisciplinary Risk and Innovation Studies (ZIRIUS), University of Stuttgart, Seidenstr. 36, Stuttgart 70174, Germany.

References
1. Eclareon, Öko-Institut (2012) Integration of electricity from renewables to the electricity grid and to the electricity market - RES-INTEGRATION. Eclareon, Öko-Institut, Berlin, Freiburg
2. Imperial College London, Nera Economic Consulting, DNV GL (2014) Integration of renewable energy in Europe. Final report for the European Commission. Directorate-General Energy. DNV GL-Energy, Bonn
3. AEE (2014) Studienvergleich: Finanzierungsinstrumente für Strom aus erneuerbaren Energien. Metaanalyse von Vorschlägen für die künftige Finanzierung von Strom aus erneuerbaren Energien. Agentur für Erneuerbare Energien, Berlin
4. European Commission (2014) Guidelines on state aid for environmental protection and energy 2014–2020. Communication from the Commission (2014/C 200/01). Off J Eur Union C 200(28.6.2014):1–55
5. Lehnert W, Altrock M, Fouquet D, Vollprecht J, Nysten J, Dalibor M (2014) Vorbereitung und Begleitung der Erstellung des Erfahrungsberichts 2014 gemäß § 65 EEG im Auftrag des Bundesministeriums für Wirtschaft und Energie. Vorhaben III. Rechtliche und instrumentelle Weiterentwicklung des EEG. Wissenschaftlicher Bericht. Becker Büttner Held, Fraunhofer-ISI, Öko-Institut, Stiftung Umweltenergierecht, IKEM, Berlin
6. Klobasa M, Ragwitz M, Sensfuß F, Rostankowski A, Gerhardt N, Holzhammer U, Richts C, Lehnert W (2013) Nutzenwirkung der Marktprämie. Working paper sustainability and innovation no. S 1/2013. Fraunhofer-ISI, Karlsruhe
7. BMU (2011) Entwurf Erfahrungsbericht 2011 zum Erneuerbare-Energien-Gesetz (Stand 3.5.2011). BMU, Berlin
8. RES LEGAL (2014) Legal sources on renewable energy: renewable energy policy database. An initiative of the European Commission., http://www.res-legal.eu/. Accessed 10 Dec 2014
9. Gawel E, Purkus A (2013) Promoting the market and system integration of renewable energies through premium schemes - a case study of the German market premium. Energy Policy 61:599–609, doi:10.1016/j.enpol.2013.06.117

10. Kitzing L, Mitchell C, Morthorst PE (2012) Renewable energy policies in Europe: converging or diverging? Energy Policy 51:192–201, doi:10.1016/j.enpol.2012.08.064
11. Monopolkommission (2013) Sondergutachten 65: Energie 2013: Wettbewerb in Zeiten der Energiewende. Sondergutachten der Monopolkommission gemäss § 62 EnWG. Monopolkommission, Bonn
12. Kopp O, Engelhorn T, Onischka M, Bode S, Groscurth H (2013) Wege in ein wettbewerbliches Strommarktdesign für erneuerbare Energien. MW Energie AG, arrhenius Institut für Energie- und Klimapolitik, Takon, Ecofys, Mannheim
13. Gawel E, Purkus A, Korte K, Lehmann P (2013) Förderung der Markt- und Systemintegration erneuerbarer Energien - Perspektiven einer instrumentellen Weiterentwicklung. Vierteljahreshefte zur Wirtschaftsforschung 82(03/2013):123–136
14. Klobasa M, Winkler J, Sensfuß F, Ragwitz M (2013) Market integration of renewable electricity generation - the German market premium model. Energy Environ 24(1–2):127–146
15. Sensfuss F, Ragwitz M (2011) Weiterentwickeltes Fördersystem für die Vermarktung von erneuerbarer Stromerzeugung. Fraunhofer-ISI, Karlsruhe
16. EEX, EPEX Spot (2014) Positionspapier zur Weiterentwicklung der Fördermechanismen für erneuerbare Energien in Deutschland. European Energy Exchange, EPEX Spot, Leipzig, Paris
17. Bardt H (2014) EEG 2.0. Ein zweiter Schritt muss folgen. IW policy paper 5/2014. IW Institut der deutschen Wirtschaft Köln, Köln
18. IZES (2014) Herausforderungen durch die Direktvermarktung von Strom aus Wind Onshore und Photovoltaik. Greenpeace, Hamburg
19. IZES, Bofinger P, BET (2013) Stromsystem-Design: Das EEG 2.0 und Eckpfeiler eines zukünftigen Regenerativwirtschaftsgesetzes. IZES, Prof. Dr. Peter Bofinger, BET Büro für Energiewirtschaft, Saarbrücken, Würzburg, Aachen
20. Jacobs D, Schäuble D, Bayer B, Sperk C, Töpfer K (2014) Eckpunkte für die Gestaltung der Energiewende. IASS policy brief 2/2013. Institute for Advanced Sustainability Studies (IASS), Potsdam
21. Reeg M (2014) Entwurf eines ganzheitlichen Strommarktdesigns für hohe Anteile erneuerbarer Energien - Flexibilität statt reiner Kapazität. Diskussionspapier 01/2014 der Abteilung Systemanalyse und Technikbewertung. DLR, Stuttgart
22. Pahle M, Tietjen O, Joas F, Knopf B (2014) EE Förderinstrumente & Risiken: Eine ökonomische Aufarbeitung der Debatte zur EEG Reform. Diskussionspapier. Potsdam-Institut für Klimafolgenforschung PIK, Potsdam
23. Diekmann J, Kemfert C, Neuhoff K, Schill W-P, Traber T (2012) Erneuerbare Energien: Quotenmodell keine Alternative zum EEG. DIW Wochenbericht 45/2012. DIW, Berlin
24. Reeg M, Elkadragy MM (2014) Changed risk premiums and equity debt requirements due to different RES-E policy instruments for market integration of renewable energies in Germany. Paper presented at the 14th IAEE European energy conference - sustainable energy policy and strategies for Europe, Rome, Italy, 28–31 Oct 2014
25. Reeg M, Nienhaus K, Roloff N, Pfenning U, Deissenroth M, Wassermann S, Hauser W, Weimer-Jehle W, Kast T, Klann U (2013) Weiterentwicklung eines agentenbasierten Simulationsmodells (AMIRIS) zur Untersuchung des Akteursverhaltens bei der Marktintegration von Strom aus erneuerbaren Energien unter verschiedenen Fördermechanismen. Projektbericht. DLR, Stuttgart
26. Nienhaus K, Deissenroth M, Reeg M (2014) Policy instruments and market integration of renewables - an agent-based model analysis. Paper presented at the 14th IAEE European energy conference - sustainable energy policy and strategies for Europe, Rome, Italy, 28–31 Oct 2014
27. Wassermann S, Reeg M, Nienhaus K (2015) Current challenges of Germany's energy transition project and competing strategies of challengers and incumbents: the case of direct marketing of electricity from renewable energy sources. Energy Policy 76:66–75, doi:10.1016/j.enpol.2014.10.013
28. Kahl H, Kahles M, Merkel K (2014) Rechtlicher Klärungsbedarf zu Ausschreibungsmodellen bei der Förderung erneuerbarer Energien. Würzburger Berichte zum Umweltenergierecht Nr. 7, Würzburg
29. 50Hertz, Amprion, TenneT, Transnet BW (2014) Aktuelle Informationen zu den EEG-Strommengen: Monatsprognosen. Direktvermarktung nach § 33b EEG. http://www.netztransparenz.de/de/Monatsprognosen.htm. Accessed 10 Dec 2014
30. Bundesnetzagentur (2014) Kraftwerksliste BNetzA Stand 02.04.2014. http://www.bundesnetzagentur.de/DE/Sachgebiete/ElektrizitaetundGas/Unternehmen_Institutionen/Versorgungssicherheit/Erzeugungskapazitaeten/Kraftwerksliste/kraftwerksliste-node.html. Accessed 10 Dec 2014
31. Lüdemann V, Ortmann M (2012) Hält die Marktprämie, was sie verspricht? Eine Analyse anhand aktueller Zahlen. Zeitschrift für Neues Energierecht (ZNER) (4/20124):325–334
32. Beckers T, Hoffrichter A (2014) Grundsätzliche und aktuelle Fragen des institutionellen Stromsektordesigns - eine institutionenökonomische Analyse zur Bereitstellung und Refinanzierung von Erzeugungsanlagen mit Fokus auf FEE. Arbeitspapier. TU Berlin, Fachgebiet Wirtschafts- und Infrastrukturpolitik (WIP), Berlin
33. 50Hertz, Amprion, Tennet, Transnet BW (2013) EEG-Jahresabrechnungen 2012 (Stand: 26.07.2013). http://www.netztransparenz.de/de/file/2012_EEG-Jahresabrechnung_.pdf. Accessed 10 Dec 2014
34. 50Hertz, Amprion, Tennet, Transnet BW (2014) EEG-Jahresabrechnungen 2013 (Stand: 25.07.2014). http://www.netztransparenz.de/de/file/EEG_HoBA_2013_20140725.pdf. Accessed 10 Dec 2014
35. MaPrV (2012) Verordnung über die Höhe der Managementprämie für Strom aus Windenergie und solarer Strahlungsenergie (Managementprämienverordnung - MaPrV) vom 2. November 2012 (BGBl. I S. 2278)
36. Fraunhofer-ISI, Fraunhofer-IWES, BBH, IKEM (2012) Anpassungsbedarf bei den Parametern des gleitenden Marktprämienmodells im Hinblick auf aktuelle energiewirtschaftliche Entwicklungen. Fraunhofer-ISI, Karlsruhe
37. Munz I, Paltauf A, Walter G (2013) Subventionen für erneuerbare Energien in Deutschland. Das Marktprämienmodell und weitere Entwicklungen. Branche STROMMARKT Bulletin (5/2013):24–27
38. Energy Brainpool (2013) Auswirkungen der verpflichtenden Direktvermarktung für EEG-Neuanlagen. Kurzstudie im Auftrag des Bundesverbands Erneuerbare Energien e. V. Energy Brainpool, Berlin
39. Holzhammer U, Stelzer M (2014) Die Direktvermarktung: Aktueller Stand, Möglichkeiten und Ausblick. Paper presented at the Fachtagung Dezentrale flexible Strombereitstellung aus Biogas - Entwicklungen, Möglichkeiten und Perspektiven, Berlin, 03 April 2014
40. Wassermann S, Reeg M, Hauser W (2012) Auswirkungen der Marktprämie auf die Akteure der Direktvermarktung - ein agentenbasierter Modellansatz. Energiewirtschaftliche Tagesfragen 62(9):60–63
41. Connect Energy Economics (2014) Leitstudie Strommarkt. Arbeitspaket Optimierung des Strommarktdesigns. Studie im Auftrag des Bundesministeriums für Wirtschaft und Energie. Connect Energy Economics, Berlin
42. Götz P, Henkel J, Lenck T, Lenz K (2014) Negative Strompreise: Ursachen und Wirkungen. Eine Analyse der aktuellen Entwicklungen - und ein Vorschlag für ein Flexibilitätsgesetz. Energy Brainpool, Agora Energiewende, Berlin
43. Waldmann P, Bhandari R (2014) Vergleichende Analyse der technisch-wirtschaftlichen Bedingungen von PV-Anlagen mit Süd- und Ost-West-Ausrichtung. Zeitschrift für Energiewirtschaft 38(1):27–36, doi:10.1007/s12398-013-0123-7
44. Grothe O, Müsgens F (2013) The influence of spatial effects on wind power revenues under direct marketing rules. Energy Policy 58:237–247, doi:10.1016/j.enpol.2013.03.004
45. Häseler S (2014) Procuring flexibility to support Germany's renewables: policy options. Zeitschrift für Energiewirtschaft 38(3):151–162, doi:10.1007/s12398-014-0128-x
46. Krautz A (2013) Anlagen flexibel betreiben? DLG Mitteilungen (5/2013):46–49
47. Fraunhofer-IWES, Becker Büttner Held, Deutsche Windguard, Deutsches Institut für Wirtschaftsforschung (DIW Berlin), Leibniz Universität Hannover (2011) Flexible Stromproduktion aus Biogas und Biomethan. Die Einführung einer Kapazitätskomponente als Förderinstrument. Bericht zum Projekt "Weiterentwicklung und wissenschaftliche Begleitung der Umsetzung des Integrations-Bonus nach § 64 Abs. 1.6 EEG" im Auftrag des Bundesministeriums für Umwelt Naturschutz und Reaktorsicherheit (BMU). Fraunhofer Institut für Windenergie und Energiesystemtechnik (IWES), Kassel/Bremerhaven
48. Kopp O, Eßer-Frey A, Engelhorn T (2012) Können sich erneuerbare Energien langfristig auf wettbewerblich organisierten Strommärkten finanzieren? Zeitschrift für Energiewirtschaft 36(4):243–255, doi:10.1007/s12398-012-0088-y
49. Hirth L (2013) The market value of variable renewables: the effect of solar wind power variability on their relative price. Energy Econ 38:218–236, doi:10.1016/j.eneco.2013.02.004

50. Fraunhofer-ISE (2013) Kohleverstromung zu Zeiten niedriger Börsenstrompreise. Fraunhofer-ISE, Freiburg
51. Carstensen UT, Schulze L, von Geldern W (2013) Deutschland braucht ein Marktmodell Energiewende - ein Masterplan zur Energiewende. Zeitschrift für Neues Energierecht (ZNER) (6/2013):581–591
52. Thrän D, Krautz A, Scheftelowitz M, Lenz V, Liebetrau J, Daniel-Gromke J, Zeymer M, Nelles M (2014) Auswirkungen der gegenwärtig diskutierten Novellierungsvorschläge für das EEG-2014. Deutsches Biomasseforschungszentrum, Leipzig
53. Krautz A (2014) Bedarfsgerechte Anlagen passen sich den Marktbedingungen an - Zwischenbericht des Forschungsprojektes OptFlex Biogas. Paper presented at the Fachtagung Dezentrale flexible Strombereitstellung aus Biogas: Entwicklungen, Möglichkeiten und Perspektiven, Berlin, 03 April 2014
54. Lansche J, Müller J (2012) Life cycle assessment of energy generation of biogas fed combined heat and power plants: environmental impact of different agricultural substrates. Eng Life Sci 12(3):313–320, doi:10.1002/elsc.201100061
55. DBFZ, Fraunhofer IWES, INL, Bosch & Partner, UFZ (2014) Vorbereitung und Begleitung der Erstellung des Erfahrungsberichts 2014 gemäß § 65 EEG. Vorhaben IIa: Stromerzeugung aus Biomasse. Wissenschaftlicher Bericht. Deutsches Biomasseforschungszentrum, Leipzig
56. Energy Brainpool (2014) Zukünftige Auswirkungen der Sechs-Stunden-Regelung gemäß § 24 EEG 2014. Kurzstudie im Auftrag des Bundesverbands WindEnergie e. V. Energy Brainpool, Berlin
57. Öko-Institut (2014) Erneuerbare-Energien-Gesetz 3.0. Konzept einer strukturellen EEG-Reform auf dem Weg zu einem neuen Strommarktdesign. Agora Energiewende, Berlin
58. Purkus A, Röder M, Gawel E, Thrän D, Thornley P (2015) Handling uncertainty in bioenergy policy design - a case study analysis of UK and German bioelectricity policy instruments. Biomass Bioenergy, doi:10.1016/j.biombioe.2015.03.029
59. Leprich U, Hauser E, Grashof K, Grote L, Luxenburger M, Sabatier M, Zipp A (2012) Kompassstudie Marktdesign. Leitideen für ein Design eines Stromsystems mit hohem Anteil fluktuierender erneuerbarer Energien. BEE-Plattform System Transformation. Ponte Press, Bochum
60. Gawel E, Strunz S (2014) State aid dispute on Germany's support for renewables: is the commission on the right course? J Eur Environ Plan Law 11(2):137–150

7

Regional economic effects of renewable energies - comparing Germany and Japan

Jörg Raupach-Sumiya[1*], Hironao Matsubara[2], Andreas Prahl[3], Astrid Aretz[4] and Steven Salecki[4]

Abstract

Background: Renewable energy (RE) creates opportunities for sustainable, socioeconomic development of local communities as fossil fuel-based energy imports are substituted with local RE sources. Established methods such as input–output (I/O) analysis face limitations when assessing these regional effects due to insufficient model integration of RE sectors. For Germany, regional value-added modeling based on RE value chain analysis is recognized as a viable alternative approach. The international transferability of this model is tested for the case of Japan.

Methods: The model of regional value-added developed by the Institute of Ecological Economy Research (IÖW model) is adapted to Japan for five representative RE technologies. The estimate of regional value-added for Japan is based on standardized, technology-specific value chain structures with Japan-specific cost and revenue data for individual RE technologies, and applies Japanese statistical sources and tax schemes.

Results: The transferability of the IÖW model to other countries requires modifications to reflect country-specific industry structures, tax codes, and policy frameworks. The comparative analysis between Germany and Japan reveals substantial differences in the economics of individual RE technologies and the distribution of regional value-added. RE has a comparable high potential for regional economic development in Japan but depends largely on local ownership of RE sources.

Conclusions: The IÖW model offers a powerful tool for regional planners and policymakers to analyze and capture the potential of RE for regional economies. Its application for a comparative analysis between countries provides indications to enhance the effectiveness of RE-related policies for regional economic development.

Keywords: Renewable energy; Value chain analysis; Regional value-added; Sustainable energy supply; Japan

Background

Renewable energy (RE) such as solar, wind, hydro, biomass, or geo-thermal power is globally growing into an increasingly important power source for electricity, heating, and transportation, already accounting for 19% of total global energy consumption in 2012 [1]. This share is expected to rise to 30% to 45% in 2050, backed by fast growing global investment into RE amounting to $400 to $500 billion annually [2]. The rapid global advance of RE may well lead to a fundamental, structural transformation of national energy systems with far reaching economic, social and ecological consequences, as predominately centralized, rather rigid supply systems are transformed into largely decentralized, flexible ones [3,4]. The decentralized and distributed nature of a RE-based energy system leads to questions about the potential of RE deployment for regional economic development. In Germany, widely considered a pioneer in RE, the promise of local job and business creation has been a key motivation for the rapid surge of investment and rapid growth of RE which in 2013 accounted for 25.4% of Germany's gross electricity consumption [5-8]. Also in Japan, where the Japanese government is promoting RE in response to the disaster at Fukushima Daiichi nuclear power plant, researchers, policy makers, and local communities are turning their attention to the potential of RE for regional economic development [9-13].

This leads to questions about suitable methods to measure and evaluate possible regional economic effects as a result of the expansion of RE. Breitschopf/Nathani/Resch [14] provide a systematic review of various methodological

* Correspondence: raupach@fc.ritsumei.ac.jp
[1]College of Business Administration, Ritsumeikan University, Kyoto, Japan
Full list of author information is available at the end of the article

approaches that analyze the impact of RE on employment and other economic variables on a regional, national, and international level. The review covers more than 30 studies worldwide from 2003 to 2011 and compares them along a set of criteria such as the scope of impacts (e.g., direct, indirect, induced effects), the type of impulses (e.g., investment, trade, household consumption), the required type of input data, or the extent of dynamic effects that are considered (e.g., price and productivity effects, structural changes, scenarios). These modeling approaches are classified into gross models and net models. Net models assess the positive and negative, dynamic impacts of growing RE deployment on the overall economy as well as on non-RE industries. They employ comprehensive macroeconomic approaches such as econometric, general equilibrium, or system dynamic models. Gross models, on the other hand, only capture the positive economic effects within RE-related industries. They typically apply less complex analytical models such as employment factor analysis and focus on individual RE technologies or single regions and countries.

Input–output (I/O) modeling and related methods such as social accounting matrix or integrated econometric and I/O modeling are well-established, multi-sector approaches to analyze exogenous or policy-induced economic effects across industries on a national and regional level [15,16]. As its main strengths, I/O modeling builds on an integrated and consistent methodological framework that breaks down the estimated impacts into finely disaggregated industry sectors and then comprehensively measures direct and induced impacts on output or employment . Due to these advantages, I/O analysis has been widely applied also for the evaluation of the gross as well as net economic effects of advanced RE deployment. Examples for I/O studies on a national level are Lehr et al. [17] and O'Sullivan et al. [18] for employment effects of RE in Germany, APPA [19] for Spain, or NISTEP [20] for Japan. On a regional level, examples are Coon [21] for North Dakota, and Allan/McGregor/Swales [22] for Shetland Islands/Scotland.

However, the adoption of I/O modeling for RE on a regional level faces a number of serious obstacles which may compromise the reliability and accuracy of their results [23]. The required quality and sector-specific disaggregation of input data for I/O tables is often not available on a regional level such as municipalities or regional districts, and their generation is time-consuming and costly. Alternatively, the derivation of regional I/O tables from national statistics tends to compromise data quality such as the applied multipliers and reduces the level of sector disaggregation [24]. Regional I/O tables and the underlying data are often likely to be outdated. Furthermore, the RE industry is usually not yet reflected as an independent sector in the sector-specific structures of I/O models. I/O modeling, therefore, assumes general industry averages for sector-specific input structures, import relations, and employment coefficients that not accurately reflect the structural characteristics of the RE sector [14]. A more accurate I/O model would require the integration of the RE industry as an independent sector based on specific cost and revenue structures of RE-related enterprises.

In order to cope with these limitations, a number of research institutions in Germany such as the Institute for Ecological Economy Research (IÖW), the Institute of Economics at the University of Kassel, the Institute for Applied Material Flow Management (IfaS) at Trier University, or the Institute for decentralized Energy Technologies (IdE) have developed an alternative approach that applies value-added modeling in combination with technology-specific value chain analysis to measure and evaluate the direct regional economic gross effects of specific RE technologies [25-29]. While differing in respect to the covered range of technologies and scope of the value chains, these models capture the complexity of activities along the multi-tiered stages of technology-specific value chains in the RE industry and are based on a comprehensive set of technology-specific cost, revenue and investment data obtained through extensive enterprise and RE-related project surveys. Regional characteristics of the RE-specific value chains are then integrated into the region-specific analysis of the generated value-added and employment. Thus a high level of reliability of results can be expected from this sophisticated, bottom-up modeling approach. Furthermore, this modeling approach allows for a detailed analysis of the distribution of the generated value-added along the various stages in terms of income, business profit, and regional taxes [23].

A comparative analysis of the results of these studies reveals a high level of consistency and compatibility. They all apply the regional value-adding approach to several municipalities or regional districts and consistently highlight the substantial positive impact of RE for the economic development of these local economies [30]. Also the comparison with the results from I/O-based studies of national employments effects of RE for Germany provides evidence, though limited in scope, about the validity and reliability of the regional value-added modeling approach. The I/O-based study of O'Sullivan et al. [18], which made substantial efforts to integrate RE-specific indicators into their I/O model, estimates that RE provided 378,000 jobs in Germany in 2012, while the study of the IÖW (Hirschl B, Heinbach K, Prahl A, Salecki S, Schröder A, Aretz A, Weiß : Wertschöpfung durch Erneuerbare Energien – Ermittlung der Effekte auf Länder- und Bundesebene, forthcoming), based on regional value-added modeling, results in a comparable figure of 344,000 RE-related jobs. While further validation of the regional value-added modeling approach by means of a rigorous and systematic methodological comparison is still wanted, the above elaborations lead to the conclusion that I/O analysis is a suitable and universally accepted approach for impact analyses of RE on a national level, while the approach of regional value-added modeling based on technology-specific value chain analysis can be expected to deliver robust and reliable results for the impacts of RE for regional economic development.

However, with the exception of Germany and the UK [31], the approach of regional value-added modeling for RE has not yet been widely applied internationally.[a] This paper, therefore, attempts to evaluate the possibility and methodological constraints when applying regional value-added modeling for RE to other countries. As a test case, it introduces and analyzes the results when applying the model of the Institute for Ecological Economy Research (hereafter referred to as the 'IÖW model') to Japan, thereby intending

- to assess and test the transferability of the IÖW model to other countries and to evaluate required modifications of the model
- to compare and assess the RE potential for regional value-added across countries, and to identify factors for country-specific, structural differences in regional value-added across various stages of the technology-specific value chain

To the authors' knowledge, the IÖW model or similar approaches based on regional value-added modeling have not yet been applied to Japan. Also the required data on technology-specific cost structures, revenues, and investment for the construction of technology-specific value chains have not been available for Japan but had to be compiled by the authors based on expert interviews, the analysis of individual business plans and model cases. The data, therefore, are original but have been cross-checked - to the extent possible - with published data by the Japanese government and other Japanese research institutions. Efforts to match results with findings of the few available, regional I/O studies by researchers in Japan, however, have faced various methodological and practical difficulties which impede a direct comparison [32-36]. Most of the studies are rather explorative in nature, attempting to demonstrate how to integrate the RE sector into existing regional I/O tables by using a single RE technology or model plant as a reference case. To do so, they rely on national I/O tables for Japan from the year 2005, while the IÖW model for Japan is based on data from the years 2011 and 2012. The IÖW model for Japan analyzes the economic effects of RE based on an estimate for the generated regional value-added and its distribution, while the Japanese I/O-based studies focus on the direct demand effects, output and employment. This paper offers an alternative, comprehensive approach, tested extensively for Germany, to directly assess the gross economic impact of RE for Japan on a regional level. It provides estimates for the technology-specific regional value-added and its sources in terms of long-term business profit, employee income, and regional tax income that can be expected from the deployment of selected RE technologies in Japan. It is based on a modified IÖW model that incorporates Japan-specific features such as the national and regional tax schemes and relies on original, primary research of five representative RE technologies in Japan.

At first, the applied model of regional value-added will be introduced, describing the specific modifications from the original IÖW model to adapt to the circumstances in Japan. Then follows a comparative analysis of the underlying economics of selected RE technologies in Germany and Japan. Finally, the regional value-added of these selected technologies and its composition in Germany and Japan are compared, and possible causes for differences between Germany and Japan are investigated. The paper concludes with a discussion on methodology and practical applications of regional value-added modeling for the promotion of RE.

Methods

The IÖW model

The underlying IÖW model is composed from two main elements (Figure 1) [23].

In a first step, the model configures technology- and capacity-specific cost and revenue structures that correspond to the individual steps along the value chain of selected RE technologies and reflect the directly generated turnover of the respective technology. For this purpose, the value chains are broken down uniformly into four value-added stages: systems manufacture, planning and installation, operation and maintenance, as well as system management. Each value-added stage is further broken down into several technology-specific value-added steps. For example, in case of onshore wind power, the value chain stage of system manufacture is subdivided into individual value chain steps for the production of key components such as the tower, generator, or rotor blade. The value chain stage of planning and installation can be broken down into specific value chain steps such as system planning and design, foundation construction, or grid connection. The value chain stage of operation and maintenance covers the main value chain steps of service and maintenance, insurance, land lease, or administration, while the value chain stage of system management includes the profits and associated taxes of the system operator. Further, upstream value-adding stages and activities for example the material input for system components manufacturing are not included in this analysis. The

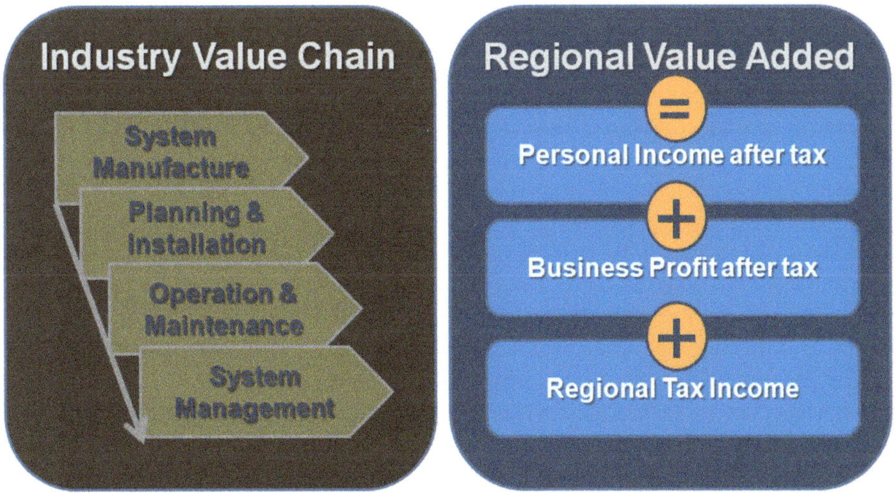

Figure 1 Elements of the IÖW model for regional value-added.

exception is the value chain for biomass which includes the provision of substrates required for system operation (e.g., wood waste).

The IÖW model assigns representative cost values to each individual value-added step of the disaggregated value chain. These reflect typical cost structures of individual RE technologies and serve as the basis for the calculation of the regional value-added generated by the respective technologies at each individual value-added step. The IÖW model follows the income approach to calculate value-added, which corresponds to the net value-added after depreciation obtained in the national accounts. The sum of net personal income, net business profit, and the regional tax revenue represents the technology-specific net regional value-added. Each component of the net value-added is capacity-specific, calculated for each value-added step, and can be extrapolated to a given regional RE capacity or to an average plant size. The model uses specific data sets for Germany, e.g., for wages and salaries, employment quotes, or tax rates.

Model building for Japan

The model for Japan applies the identical framework and structure of the one which was developed by the IÖW and applied for Germany. The model consists of two major parts:

- Construction of value chains for specific RE technologies based on detailed, technology-specific cost and revenue structures for individual steps along the technology's value chain
- Quantification of the technology-specific regional value-added based on the estimation and addition of step-specific employee net income, net business profit, and regional tax revenue

For the purpose of testing the transferability of the IÖW model to Japan, the base year for comparison was set for the year 2012 and five representative RE technologies were selected: 150 kW photovoltaic roof installation, 1 MW photovoltaic ground-mounted installation, 2 MW onshore wind power installation, 150 kW small-scale hydro power installation, 5 MW biomass installation fed by waste wood. In difference to the IÖW model, the model for Japan refrained from evaluating the regional economic effects at the stage of system manufacture, in order to simplify the comparative analysis. While the potential for value-added and employment at the stage of system manufacture is significant at the national level, RE system and components manufacturing companies have to be located within the respective region to generate value-added for a region. Because this is often not the case in many regions, it seems justified to exclude the stage of system manufacture from the comparative analysis and to focus on regional economic effects from RE at the stages of planning and installation, operation and maintenance, and system management. For each of these three stages, the model applied the same subdivided value chain steps as the IÖW model and allocated the same individual cost and revenue items for each individual step. Identical to the IÖW model, the cost and revenue data have been standardized at Japanese Yen per kilowatt.

The validity of the model rests on the availability of robust and reliable, technology-specific cost and revenue data. For Germany, the IÖW model has screened and incorporated an extensive set of detailed, technology-specific data that have been derived from a broad range of current literature such as evaluation reports, studies, or government statistics and have been updated over time. The analysis for Germany draws on the results of the

IÖW model for 2012 (Hirschl B, Heinbach K, Prahl A, Salecki S, Schröder A, Aretz A, Weiß : Wertschöpfung durch Erneuerbare Energien – Ermittlung der Effekte auf Länder- und Bundesebene, forthcoming). For the analysis of technology cost developments, additional data from 2009 have been used [25]. The data for Germany were computed into Japanese Yen at the exchange rate of 130 yen per euro which - incidentally - was roughly the average exchange rate in years 2009 and 2012, thereby simplifying the comparative analysis.

For Japan, a similar comprehensive and robust set of data is yet not available due to the comparatively sparse amount of current literature and published government data. The reports of the Feed-In-Tariff Evaluation Committee (*Chōtatsu Kakakutō Santeiiinkai*) of the Ministry of Economics, Trade and Industry (METI), which decides on the feed-in tariffs for individual RE technologies (first applied from July 2012), are supposedly the most comprehensive and reliable data source. However, compared to Germany, the official data for Japan on investment and operating costs proved to be rather fragmented, limited in scope and level of detail [37-39]. Further, comparison with data provided by various business associations in the RE industry and by Japanese research institutions reveal a high degree of variation of the provided data themselves as well as in regard to the cited reference base of installations, which mixes figures on actual operations, model plants, and estimates for ongoing projects [40-46]. Yet, no further details are provided on the breakdown of total system costs and possible reasons for the substantial deviations. The level of breakdown of technology-specific operating costs into individual cost items is also insufficient for the purpose of constructing technology-specific value chains for Japan.

Therefore, the Japan model draws on data from business plans and reports for a comparatively small number of reference cases for the selected RE technologies, derived from RE projects in the period of 2011 ~ 2012. In the case of solar power, two references cases for 150 kW roof installation and 1,000 kW ground-mounted installation were selected which were launched in 2011 as regional model cases with the support of the Japanese Ministry of Environment. Panels were supplied by Japanese manufacturers and planning and construction handled by regional firms. While for electrical installation and land rent, standard costs were assumed, and operational cost data were provided by the operator for 2012. In case of wind power, the Japan Wind Power Association provided data from a typical business plan for a wind farm with ten or more 2 MW installations. The system costs and operational cost structures for a 200 kW hydro power plant refer to a typical reference case provided by the National Small Hydro Power Promotion Association (*zenkoku shōsuiryokuriyō suishinkyougikai*) taking into account representative conditions for construction, water flow volume, and difference in altitude. For biomass, an installation of a 5 MW biomass facility based on waste wood in Fukushima Prefecture was taken as a reference applying the available business plan for the investment, operational, and substrate costs. Though limited in scope, these data proved to be consistent with the data ranges for average investment and operational costs provided by the Japanese government, related business associations, and Japanese research institutions. Nevertheless, the described variance and limited scope of the available data set for Japan generally inhibits the robustness of results.

The calculation of the regional, technology-specific value-added requires the estimation of net personal income, net business profit, and regional tax revenue on the individual steps of the respective value chains. Based on the rationale that individual steps of the value chain correspond with typical industry sectors, the IÖW model draws on industry-specific, average cost and profit ratios that are publically available for Germany from government institutions such as Germany's central bank, called *Deutsche Bundesbank*, and the German Federal Statistical Office. The application of industry-specific averages is justified by the notion that the companies involved along the various stages and steps of the RE value chains can be categorized within the existing classification of economic activities with corresponding cost and profit structures. In other words, the IÖW model assumes that companies would not engage in the RE business unless they can expect at least similar returns typical for the industries they associate with. For the short term, this assumption may underestimate the actual level of profitability of companies engaged in various stages along the RE-related value chains due to temporary supply–demand conditions or the use of subsidies or other incentives. Yet, it can also be expected that over the long term, these distortions level off and profits converge to a level typical for the respective industry. Furthermore, for the calculation of pre-tax profits at the stage of system management, the IÖW model applies the actual average return on equity for the individual RE technologies which were drawn from the Renewable Energy Sources Act (EEG) Progress Report [23].

Following the same logic, the model for Japan calculates the regional value-added based on industry-specific statistics by the Japanese government, namely the Financial Statements Statistics of Corporations by Industry (*Hōjin Kigyō Tōkei*) compiled by the Policy Research Institute of the Japanese Ministry of Finance [47]. The data are extrapolated from the results of the annual survey among approximately 28,000 representative non-financial and financial firms (with capital, contributions, or funds of 10 million yen or more) about their provisional profit-and-loss-accounts and balance-sheets. The surveyed companies are categorized into different classes of capital size and classified into different industries in accordance with the Japan Standard Industrial Classification which correspond in principle with the German classification. The results cover the account settlements of more than 2.8 million Japanese firms across all industries. Data include industry-specific sales revenues, gross margin, profits before and after tax, as well as personnel expenses (including salaries and bonuses) for full-time employees and management, depreciation, and interest payments.

For the purpose of this study, the key individual steps of the technology-specific value chain have been associated with the following industries, applying the respective ratios for business profit and personnel expenses, in order to derive the values for personal gross income and business profit before tax. For these industries average figures over the 3-year period from 2010 ~ 2012 were applied, in order to avoid one-time distorting effects (e.g., impact from Lehman shock)

- Planning and installation:
- Planning/project management: academic, professional and technical services (Code 83)
- Site preparation/construction/electric installation: construction (Code 15)
- Operation and maintenance:
- Service and maintenance: other technical services (Code 89)
- Insurance: insurance (Code 96)
- Banking: banking (Code 91)

With the exception of banking and insurance, large-size firms with capital of 1 billion yen or more were excluded from the database, in order to better reflect the structural characteristics of firms involved at the various stages of the RE business. This assumes that the planning, project management, construction/installation, as well as servicing and maintenance of RE projects are predominantly handled by smaller and medium-sized firms. This assumption is confirmed by analyzing the company profiles of 334 private companies in the field of RE that are registered in the Database of Renewable Energy Research Organizations & Companies in East Asia from the New Energy Foundation (NEF) [48]. The database includes a significant number of large, integrated manufacturing firms such as Mitsubishi Heavy Industries, Hitachi, or Kyocera who are major competitors in RE-related equipment. As these companies focus on manufacturing, which is excluded from this analysis, it can be safely assumed that the various RE-related service activities at the stage of planning, construction, operation, and system management are handled, if at all, by the companies' sales, service, or engineering divisions, or independent subsidiaries. The cost structures of these overhead divisions or subsidiaries are likely to resemble those of specialized service companies rather than those of integrated, manufacturing-oriented conglomerates. The same is true also in case of large, general construction firms like Taisei Corporation or Shimizu Corporation who are also active in the field of RE. The majority of non-manufacturing companies in the field of RE are specialized, medium-sized firms that provide services such as engineering/procurement/construction (EPC), project consulting, system integration, trading, service and maintenance, or the supply of substrates and input materials in the case of biomass and bio fuels. Even the largest EPC providers, system integrators, and operators like Softbank Energy (92 employees/746 million Yen equity), Eurus Energy (253 employees/1.8 billion Yen equity), Japan Wind Development (148 employees/1 billion equity), IHI Plant Construction (471 employees/471 million Yen equity), or NTT Facilities (5,000 employees/1.2 billion Yen equity) still fall more in the category of typical medium-sized companies. A certain exception are the operating firms of bio-fuel and biomass plants which often have been installed at the manufacturing sites of larger companies in the food, paper, and pulp industry to utilize their biomass waste. The NEF database has also been cross-checked with the member directories of the main Japanese business associations in the field of RE, such as the Japan Photovoltaic Energy Association (162 members), The Japan Wind Power Association (127 members), the Committee for the Promotion of the Use of Woody Biomass Energy (112 members), and the National Small-Scale Hydro Promotion Committee (99 members). Many of the member firms are also registered in the NEF database, and companies not registered are frequently smaller and medium-sized firms.

For the stage of system management, the IÖW model can draw on the official profitability data published in the Renewable Energy Sources Act Progress Report (*'EEG-Erfahrungsbericht'*) of the German government. No such data are yet made publically available by the Japanese government. Therefore, the model for Japan relies on the cash flow projection and internal-rate-of-returns (IRR) of selected RE reference projects. Revenues have been calculated based on the technology-specific feed-in tariffs for the year 2012.

Taxes

The tax system marks substantial differences between Germany and Japan, therefore requiring various adaptations of the tax model to reflect the characteristics of Japan's tax system. In regard to income-based taxes, the IÖW model takes into account whether a business is incorporated or self-employed as the tax base and rates differ. The statistical base for this distinction is provided by the enterprise statistics of the *Deutsche Bundesbank*. For Japan, the MoF statistics only refer to incorporated business; however, no major distorting effect is expected for the purpose of this study, as most businesses in Japan are incorporated as stock companies, limited liability partnerships, cooperatives, or public business associations (*kōjeki hōjin such as shadan hōjin*) who are subject to corporate income tax. It is assumed that businesses who deal with the considered RE technologies usually are likewise choosing an incorporated form of business association. An overview on the applied tax model is given in Table 1 which summarizes the main national and regional tax categories and the applied rates and formulas.

In order to calculate the net personal income and net business profit, the model for Japan takes into account the Japanese national and regional income-related, personal, and corporate taxes. The calculation of the national personal income tax (*shotokuzei*) assumes the average taxable personal income of 4.08 million Japanese Yen (in 2011), on which a 20% national income tax is levied [51]. For the calculation of the national corporate tax (*hōjinzei*), it is assumed - for the reasons stated above - that the majority of firms involved in the RE business are small- to medium-sized firms with capital less than 100 million Japanese Yen, applying a reduced national corporate tax rate of 18% valid in 2012. The threshold of 100 million Japanese Yen may be too restrictive, as also higher-capitalized firms engage in RE-related activities. However, the risk of overstating tax income is deemed to be significantly higher considering the fact that the break-even for system operators takes often several years and that tax breaks are rather common in the RE industry.

To these national income-related taxes, regional income taxes are added to derive the net personal income and net business profit. For Germany, the IÖW only refers to the municipal tax revenue as part of regional value-added. For Japan, however, regional value-added refers to both, municipal as well prefectural tax revenue as regional tax revenue. This distinction is important, as most of the 47 prefectures in Japan are far smaller in terms of population and geographic size than most German federal states (*Bundesländer*). Therefore, the regional economic impact of RE is deemed to be equally important on both, the municipal as well as the prefectural level. The distinction between effects on prefectural and municipal level is expected to provide further valuable insight for the evaluation of regional economic effects of RE in Japan.

The predominant source of municipality tax revenue in Germany is based on corporate and personal income, specifically the local trade tax (*Gewerbesteuer*) on business profits and the share of the national tax on salaries and wages. In Japan, there are three main bases for regional (municipal and prefectural) tax revenue: corporate and personal income, fixed assets, consumption [50].

Differing from the IÖW model, the regional tax model for Japan only considers those taxes that are directly levied by the regions but excludes the regional share of redistributed national taxes (*chihō kōfuzei*), because the criteria and bases for redistribution depend largely on the state of local public finances and allow for significant discretionary judgment by the Japanese national government.

Japanese regions levy two forms of corporate income taxes, the corporate residential tax (*hōjin juminzei*) and

Table 1 Overview on the Japanese tax system

National taxes		Regional taxes		
			Prefecture	**Municipality**
Personal income tax (percent of taxable income)	20%	Personal income tax		
Corporate tax (percent of taxable income)	18%	Residential tax (minimum rate in JPY)*	1,000	3,000
		Residential tax (percent of taxable income)	4%	6%
		Corporate income tax		
		Local business tax (percent of taxable income/not applied to system operator)		4%
		Electric utility tax (percent of revenue of system operator)	0.7%	
		Special regional corporate tax for electric utilities (percent of electric utility tax)	81%	
		Residential tax (minimum rate in JPY)**	20,000 ~ 50,000	50,000 ~ 150,000
		Residential tax (percent of taxable income)	5%	12.3%
		Asset-based tax		
		Operator fixed asset tax (percent of depreciated, registered asset value)		1.4%
		Consumption-based tax		
		Operator consumption tax (percent of value-added)	0.5%	0.5%

*Calculated as ratio of income from minimum residential tax to overall tax income from residential tax (2.37% in 2012).
**Calculated as ratio of income from minimum residential tax to overall tax income from residential tax (in 2012: prefecture = 22.9%/municipalities = 25.8%).
Source: MIAC (2014) [49], KPMG (2012) [50].

the corporate business tax (*hōjin jigyōzei*) . The overall tax rate for the corporate residential tax amounts to 17.3% of the taxable income, 5% on prefectural level and 12.3% on municipal level. In addition, a fixed corporate residential tax is charged to every business, the amount of which varies, however, depending on the size of equity and the number of employees of the firm. There are a total of 14 different tax classes, making it rather impractical for modeling. For the purpose of this analysis, the amount of the fixed corporate residential tax has been estimated by applying a mark-up on the income-related corporate residential tax. The mark-up has been calculated as the ratio of the regional tax income from the fixed corporate residential tax to the regional tax income from the income-related corporate residential tax.

In Japan the system operators of RE are in general not subject to the corporate business tax (*hōjin jigyōzei*) but are charged a special form of the corporate business tax called electricity utility business tax (*denkijigyōzei*). A tax rate of 1.3% is levied by the prefecture on the revenue from electricity sales by RE system operators. The tax rate is presently reduced to 0.7%. On top of this tax, a special regional corporate tax is charged that amounts to 81% of the special electricity utility business tax. However, companies who have also invested into RE next to their main business, but where revenues from electricity sales amount to less than 10% of their main business, are exempt from this tax and are even allowed to offset profits from their RE operation against losses in their main business. This exemption serves as an incentive for local businesses by regional government to invest into RE.

The corporate business tax is levied only on prefectural level, and tax rates vary according to the size of the company. For the reasons stated above, it is assumed that the majority of businesses in Japan that are involved in the RE-related business are small- to medium-sized firms, cooperatives, or public business associations with less than 100 million yen equity capital and less than 8 million yen in taxable income on average. The model assumes an average, uniform tax rate of 4% on profits of these businesses.

Prefectures and municipalities also levy a uniform personal residential tax (*kojin juminzei*) of 10% on personal income, salaries, and wages of individual citizen. In addition, there is a per capita personal residential tax of 1,500 yen on prefectural level and 6% or a minimum of 3,000 yen on municipal level. For the purpose including this rather insignificant tax into the model, a mark-up on the personal residential tax was applied, similar to the approach taken for the fixed corporate residential tax.

One of the main sources of tax revenue for municipalities in Japan is the fixed asset tax which contributed 42.7% of all municipal tax revenue in 2012. A uniform tax rate of 1.4% is annually levied on the registered, depreciated value of properties, real estate and other fixed assets, and investments into RE are also subject to the fixed asset tax. The modeling and estimate of the fixed asset tax for RE, however, poses substantial methodological difficulties, as the depreciation curve of the RE investment has to be taken into account. Furthermore, many prefectures and municipalities provide exemptions or reductions of the fixed asset tax as part of their promotional policies of RE. For the purpose of the comparative analysis, an average tax burden per kilowatt has been estimated based on the technology-specific cash flow plans and a depreciation period of 20 years. It has to be noted that region-specific circumstances have to be taken into consideration, when applying the model to a specific region.

A further source of regional tax revenue is their share of the Japanese consumption tax (*shōhizei*), consumption being defined as the purchase value minus the value of external purchases of input materials. In other words, it is a value-added tax. Purchases of RE equipment and related services, as well as the revenue generated from RE system operation are subject to a uniform 5% consumption tax, 1% of which is redistributed to regions (0.5% each on prefectural and municipality level). In 2012, the consumption tax already accounted for 17.1% of the regional tax revenue, and its significance is expected to grow further, as the Japanese government has raised the consumption tax to 8% from 1 April 2014 with further increase expected in the near future. For the comparative study, a uniform tax rate of 5% was applied only to the revenues on system operating stage, as the analysis does not take into account the regional economic effects of system manufacture.

Other kind of taxes that are charged on a regional level have been excluded as they are either not relevant for RE businesses or only applied in certain regions like Tokyo.

In order to calculate the amount and share of regional tax revenue for each RE technology, the income-based tax rates were applied to the gross personal income and gross business profit per kilowatt, thereby obtaining the net personal income and net business profit. The fixed asset tax and the special electricity utility tax are treated as cost items that reduce the system operator's pre-tax business profit. The consumption tax is levied on the system operator's revenue, deducting the costs for external material input, bank interest payments, and insurance premiums.

Just like the IÖW model, the sum of net personal income, net business profit, and the regional tax revenue leads to the technology-specific net regional value-added. The model results only reflect the regional value-added directly generated by RE technologies along the planning and installation, operation and maintenance, and system management stages of the technology-specific value chain. Further, indirect effects that are generated through these economic activities further upstream the RE value chains (e.g., production of raw materials) as well as induced effects such as multiplier income effects by the induced consumption or investment are not considered. Negative economic effects such as the replacement of economic activity in the field of conventional energy technologies and the subsequent loss of employment and business profit are also not considered.

The model results distinguish between one-time effects of regional value-added at the stage of planning and installation, and ongoing, annual value-added generation along the lifespan of RE at the stage of operation and maintenance as well as the stage of the RE system management. The model results reflect the overall *potential* for technology-specific regional value-added for all of Japan and Germany, assuming that all the value generating activities are located and performed within the two countries. The results are considered to be valid and robust, as the stage of system manufacture is excluded from the comparative analysis, thereby avoiding the cumbersome task of integrating differences in the structure of international trade in the field of RE technologies into the model.

Results

The comparative analysis between Germany and Japan is conducted in two steps. As a first step, the underlying economics of the five selected RE technologies are compared, applying return-on-equity (ROE) as well as the return-on-sales (ROS) as the key indicators for economic performance. The ROE indicates the overall investment performance of the respective RE technologies, while the ROS highlights the operational economic performance specifically at the system operating and management stage. The analysis allows to assess possible country-specific differences in the economic performance of RE businesses and to identify possible root causes for such differences at the stage of investment and system operation. Due to the substantial differences in the tax systems of Germany and Japan and the possible distorting influences of different tax bases and rates, the analysis is based on pre-tax business profit. Furthermore, as the legal depreciation period differs between Japan (17 years) and Germany (20 years), the data have been standardized for a depreciation period of 20 years to make them comparable.

In a second step, the technology-specific regional value-added per kilowatt for Germany and Japan are compared for each RE technology. Rather than merely comparing the absolute amount of net regional value-added by RE technology, the analysis concentrates on differences in the composition of the regional value-added, thereby highlighting differences in the sources of regional value-added and their relative weight. In addition, the comparative analysis distinguishes between the one-time value generation at the stage of investment or installation, and ongoing annual effects over the lifespan of the RE technology. The comparative analysis of RE technologies reveals significant differences between Germany and Japan in terms of total investment costs, economic performance, and the resulting composition of regional value-added.

Investment costs

As shown in Table 2, for the year 2012, the total investment costs - including direct investment for the equipment as well as the associated costs related to the planning and installation of the RE facilities - run 1.6 to 2 times higher in Japan than in Germany for all RE technologies under consideration; the only exception is biomass where total investment costs are similar in both countries. Further analysis of the 2012 data reveals that for solar and wind installations, the direct investment costs in Japan are 1.7 ~ 1.8 times higher than in Germany, while the direct investment costs for small-scale hydro power and biomass are lower in Japan. Especially in regard to wind power, the FIT system has not yet generated growth of wind power in Japan, and direct investment costs have remained at significant higher levels compared to Germany and other markets. Recent data for 2013 from the Feed-in-Tariff Evaluation Committee (*Chōtatsukakakutō Santeiiinkai*) of the Japanese METI, which reviews and decides on the annual feed-in tariffs for RE in Japan, also underline the significant higher investment costs in Japan, although equipment costs have dropped recently also in Japan, in particular for solar installations [52]. In case of solar power, one reason may be the dominance of Japanese suppliers, who command about 80% of the domestic solar panel market and tend to charge higher prices to Japanese consumers. Another reason may be the significant recent devaluation of the Japanese Yen, which has led foreign suppliers of equipment and components to raise prices on the Japanese market. This implies that the worldwide dynamic reduction of equipment costs for RE technologies have not yet fully found its way to the Japanese market. In general, investment costs in Japan in the year 2012 are at a similar level as those in Germany in the year 2009.

In addition, other investment costs related to planning, projecting, construction, and installation are also markedly higher in Japan as compared to Germany. While other investment costs for medium-scale solar roof installations record a comparatively modest mark-up of 30% in Japan, they are 2 ~ 3 times higher for wind, small-scale water power and biomass installations. The share of other investment costs as percentage of the total investment in Japan is between 10-percentage points (1 MW solar,

Table 2 Comparison of cost structures and performance of renewable energy by type of technology for Germany and Japan

Cost and revenue structures (JPY/kW) (for installations in the year 2012)*	Solar						Wind			Hydro			Biomass		
	150 kW (roof)			1,000 kW (ground)			2,000 kW (onshore)			200 kW			5,000 kW (wood)		
	Germany	Japan	*JP/GER*	Germany	Japan	*JP/GER*	Germany	Japan	*JP/GER*	Germany	Japan	*JP/GER*	Germany	Japan	*JP/GER*
At planning and installation stage															
Total investment	235,300	369,533	157%	208,000	372,000	179%	177,190	360,673	204%	968,108	1,587,500	164%	502,154	478,400	95%
Direct investment cost	149,240	260,000	174%	140,504	238,000	169%	130,000	230,000	177%	683,668	600,000	88%	432,464	287,000	66%
Indirect investment cost (e.g. planning/construction)	86,060	109,533	127%	67,496	134,000	199%	47,190	130,673	277%	284,440	987,500	347%	69,690	191,400	275%
At system operation stage															
Operating cost	20,246	34,005	168%	18,184	31,158	171%	19,444	33,269	171%	68,048	139,112	204%	114,984	184,957	161%
Operator revenue	26,430	42,048	159%	26,998	37,843	140%	25,966	38,544	148%	90,055	177,828	197%	136,577	201,830	148%
Profit before tax	6,184	8,043	130%	8,814	6,685	76%	6,522	5,275	81%	22,006	38,716	176%	21,593	16,873	78%
ROS from operation (before tax)*	23.4%	19.1%	−4.3%	32.6%	17.7%	−15.0%	25.1%	13.7%	−11.4%	24.4%	21.8%	−2.7%	15.8%	8.4%	−7.4%
ROE from operation (before tax)**	8.5%	7.3%	−1.2%	7.4%	6.0%	−1.4%	13.9%	4.9%	−9.0%	7.0%	8.1%	1.1%	11.6%	11.8%	0.2%
Investment-related ratios															
Operating cost/total investment	9%	9%	1%	9%	8%	0%	11%	9%	−2%	7.0%	9%	2%	23%	39%	16%
Other investment/total investment	37%	30%	−7%	32%	36%	4%	27%	36%	10%	29%	62%	33%	14%	40%	26%

*Own calculation based on primary research data on cost, investment, and equity for Japan and on IÖW data for Germany (Hirschl B, Heinbach K, Prahl A, Salecki S, Schröder A, Aretz A, Weiß : Wertschöpfung durch Erneuerbare Energien – Ermittlung der Effekte auf Länder- und Bundesebene, forthcoming); ROE figures for Germany are referring to the underlying assumptions of the German government and public research institutions on ROE industry averages for RE projects which are applied for the determination of technology-specific feed-in-tariffs; these assumptions are published in the review report on the Renewable Energy Law ('EEG Erfahrungsbericht').

**The underlying equity ratios differ slightly between Japan and Germany: for Japan 30%; for Germany: 25% (water/wind), 30% (150 kW solar/biomass), 50% (solar 1 MW)/exchange rate: 130JPY/Euro.

2 MW wind) and up to more than 30-percentage points (small-scale hydro, biomass) higher than in Germany. Significant differences in these so-called 'balance-of-system' costs are rather common worldwide, as international comparative studies have revealed; yet, the order of magnitude for Japan is remarkable [53,54]. A general reason is assumed to be higher construction and electrical installation costs, as well as overall project management costs. Higher construction and electrical installation costs may be caused by stricter construction standards to cope with the risks of earthquakes, typhoons or lightning, and land-related regulations, while lengthy approval processes and administrative standards (e.g., environmental assessment) may cause higher planning and project management costs. Furthermore, the reconstruction of the Northern Japan after the devastating earthquake and tsunami in March 2011 has led to a shortage of construction workers and a rise in construction wages. The recent review of METI provides further evidence by listing various administrative regulations related to grid access and connection, to the use of water or agricultural land rights (e.g., solar and water power installations), to forest administration (e.g., wind power, water power, biomass), and to building standards (e.g., wind power) as examples that lead to time-consuming planning and approval procedures which then drive up costs [55]. A recent report by the EU-Japan Centre for Industrial Cooperation also cites various regulations and administrative procedures that inhibit competition and tend to drive up costs [56].

Economic performance

Higher total investment costs are surely one reason for the Japanese government's decision to grant rather generous feed-in tariffs as compared to Germany. For 2012, the Japanese feed-in tariffs for RE technologies under consideration have been generally about twice as high as those that have been applied in Germany in the same year. It, therefore, could be assumed that RE installations in Japan would generate returns at least similar or even higher than similar installations in Germany. As summarized in Table 2, the comparative analysis of the performance in terms of ROS and ROE, both before tax, leads to different conclusions: medium-scale solar roof installations as well as small-scale hydro power installations achieved comparable returns in Germany and Japan in 2012 both in terms of ROS and ROE. But in case of larger-scale ground-mounted solar plant, wind power plant, and large-scale biomass facility projects in Germany delivered significantly higher operating returns. The differences in economic performance is particularly marked in the case of wind power (2 MW), where German installations on average generated a solid ROS of 25% and an ROE of almost 14%, as compared to a ROS of a 13.7% and a ROE of just 4.9% for Japan. Similarly, ground-mounted solar installations (1 MW) in Germany recorded an impressive ROS of 33.7% and a ROE of 8.5% as compared to an ROS of 13.4% and an ROE of 6% in Japan.[b] In case of a larger-scale biomass plant (5 W) fed on waste wood, the level of ROE is comparable, but the ROS in Japan of 8.4% is almost half of that for a similar installation in Germany in 2012 (15.8%).

A more detailed analysis of the underlying cost structures leads to interesting conclusions in regard to possible reasons for the significant divergence in the economic performance of RE facilities in both countries.

The ROS, on the one hand, depends on the revenue generated by the RE installation, and, on the other hand, on the level of operating costs to run the facility. As indicated above, rather generous feed-in tariffs in Japan lead to revenue per kilowatt that is between 1.4 times (solar ground-mounted installation) to 2.1 times higher (small-scale hydro power) than in Germany. Therefore, the comparatively weaker operational performance of RE installations in Japan is largely caused by operating costs in Japan that run higher by a factor of 1.5 to 2.0 than those in Germany. The imbalance of revenue to operating costs is most marked in the case of large-scale ground-mounted solar and wind power and to a lesser degree in case of biomass, which is reflected in the significantly lower ROS of these technologies as compared to Germany.

A main reason for these differences is the substantially higher depreciation burden due to the higher total investment costs in Japan. However, direct personnel costs as well as costs for service and maintenance run higher in Japan by a factor of 1.4 (solar) up to 3.8 (wind power). Further analysis into the root causes for higher share of direct personnel and service costs is required, but it may be assumed that administrative regulations which, for instance, require the employment of qualified engineering staff are pushing up operating costs. Furthermore, higher and further rising real estate prices are driving up the costs for land leases in the case of ground-mounted solar and wind power installations. In the case of biomass, the substrate costs exceed those in Germany by a factor of 3.3.

The differences in regard to ROE can only partially be explained by differences in the underlying equity ratio: for Japan, a unified equity ratio of 30% was applied, while for Germany the equity ratio varied between 25% (small-scale hydro power, wind), 30% (solar roof installation), and 50% (solar ground-mounted installation). The slightly lower equity ratio for small-scale hydro power and wind power does not sufficiently explain the significant divergence in ROE, and in case of large-scale ground-mounted solar installations, the ROE in Germany would even increase further when applying the lower equity ratio of Japan. The substantially lower ROE in Japan for mainstream RE technologies such as large-scale ground-mounted solar, wind power, and larger-scale biomass installations is mainly caused by the higher total investment costs, which reduce the asset turnover rate, and further depressed by the substantially higher operating costs which reduce the ROS.

Regional value-added

Total investment costs for RE at the stage of planning and installation, as well as the revenue from the sales of RE at the stage of system operation define the overall pool of value that a region can theoretically capture from the investment and operation of RE. As total investment costs for RE as well as the revenue for RE from feed-in tariffs differ significantly between Germany and Japan, a comparison of the absolute values for the generated regional value-added has only limited meaning. It, therefore, makes more sense to compare the relative shares of regional value-added of indirect investment costs at the stage of planning and investment, as well as of the annual sales revenue during operation. These ratios indicate to what extent the regional economy benefits from RE in the shape of personal income, business profits, and regional taxes.

As shown in Table 3, a significantly higher share of the indirect investment costs for planning and installation remain as regional value-added in Germany (22% ~ 44%) than in Japan (13% ~ 16%). The results show that personal income accounts for the predominant share of regional value-added in both countries at investment stage, while the share of business profit and regional tax income is comparatively smaller in both countries. However, in Japan employees receive a relatively larger portion from the available value pool than in Germany, while in Germany business profits appropriate a substantially higher share of regional value-added than in Japan.

While the share of regional value-added at planning and investment stage differs sharply between Germany and Japan, the share from annual revenues that remains as value-added in the region during system operation is rather comparable across all RE technologies (21% ~ 33%).

However, in respect to the composition of the regional value-added, there exist remarkable differences between Germany and Japan. In Germany, after-tax business profits account for 60% (hydro power, biomass) to more than 70% (solar, wind) of the regional value-added. In Japan, after-tax profits also contribute the highest share of regional value-added, but it ranges only between 28% (biomass) and 47% ~ 50% (solar, wind, hydro power). In Japan, a far higher share between 20% ~ 25% generally goes to municipal (and also prefectural) taxes (with the exception of biomass), while in Germany municipal taxes account for only between 11% ~ 15%. Similarly, the share of regional value that is distributed as personal income after tax is also about 1.5 to 2 times higher in Japan (28% ~ 32%/62% for biomass) as compared to Germany (12% ~ 23%/37% for biomass). These differences in the structural composition of the regional value-added in Germany and Japan can be explained, on the one hand, by the significant higher operating performance of RE facilities in Germany and, on the other hand, by distinct differences between the regional tax systems of both countries. Specifically, the fixed-asset tax that is levied on RE assets in Japan is a major source of municipal tax income accounting for about 50% of the municipal tax income from RE installations. In addition, Japanese regional tax authorities also benefit from the special electricity utility tax and receive comparatively higher share of the taxable personal income and business profit as their German counterparts.

Discussion

The overriding purpose of the comparative analysis of the regional value-added generated by RE in Germany and Japan has been to assess and test the transferability of the IÖW model to other countries.

The analysis has demonstrated that the IÖW model and its underlying combination of technology-specific value chain analysis with national industry statistics can be meaningfully applied to different countries and produces robust and reliable results for measuring and assessing the regional economic impact of RE technologies. However, it has also become evident that the transfer of the IÖW model depends on the availability of sufficient data and requires various modifications in order to reflect country-specific characteristics.

First, the reliability of the modeling results for regional value-added depends on the availability of a robust and broad set of data on technology-specific cost and revenue structures of RE along the various steps of the specific value chain. Compared to Germany, the available data in Japan are still rather limited in terms of scope, quality, level of detail, and historical development. The comparative analysis for Japan had to rely on a few reference cases in order to obtain the required details on cost structures and revenues. Although they generally proved to be compatible and consistent with other publicly available data from the Japanese government and think tanks, the published data sets are rather fragmented and reveal a significant variance, in particular in regard to total system costs. Such significant deviations in the total system costs have a fundamental impact on the economical viability of RE and impede the generalization of representative data. However, with the growing installation base of RE in Japan, a broader set of data should become available, in particular for solar power installations, thereby enhancing the reliability and robustness of the applied model for Japan.

Secondly, the estimation of the regional value-added relies on the availability of national, industry-specific statistics in order to estimate the profitability and labor

Table 3 Comparison of regional value-added (RVA) of renewable energy by type of technology for Germany and Japan

Regional value-added (RVA)*	Solar						Wind			Hydro			Biomass		
	150 kW (roof)			1,000 kW (ground)			2,000 kW (onshore)			200 kW			5,000 kW (wood)		
	Germany	Japan	*JP/GER*	Germany	Japan	*JP/GER*	Germany	Japan	*JP/GER*	Germany	Japan	*JP/GER*	Germany	Japan	*JP/GER*
At planning/construction stage (JPY/kW)	27,066	14,307		20,595	18,013		11,194	18,089		124,981	136,639		18,532	33,693	
RVA as percent of indirect investment cost	22.0%	13.1%	−8.9%	22.4.%	13.4%	−9.0%	23.7%	13.8%	−9.9%	43.9	13.8%	−30.1%	26.6%	17.6%	−9.0%
of which: business profits after tax	20.4%	6.6%	−13.8%	20.0%	6.7%	−13.3%	18.6%	6.9%	−11.7%	61.0%	6.9%	−54.1%	42.1%	7.1%	−35.0%
of which: personal income after tax	72.7%	87.7%	15.0%	73.1%	87.5%	14.4%	74.8%	87.4%	12.6%	36.1%	87.4%	51.3%	52.2%	80.3%	28.2%
of which: regional tax income	6.9%	5.7%	−1.2%	6.8%	5.8%	−1.0%	6.8%	5.8%	−1.0%	2.8%	5.8%	3.0%	5.8%	12.5%	6.8%
At system operation stage (JPY/kW)	9,926	12,614		11,917	10,616		6,963	9,396		24,873	56,808		33,639	42,004	
RVA as percent revenue from system operation	25.8%	30.0%	4.2%	33.4%	28.1%	−5.3%	26.8%	24.4%	−2.4%	27.6%	31.9%	4.3%	24.6%	20.8%	−3.8%
of which: business profits after tax	70.0%	47.1%	−22.9%	73.9%	46.7%	−27.2%	74.2%	46.2%	−28.0%	62.2%	50.3%	−11.9%	51.8%	27.6%	−24.2%
of which: personal income after tax	15.2%	32.3%	17.1%	12.3%	31.0%	18.7%	13.1%	28.5%	15.4%	22.5%	30.8%	8.3%	37.1%	61.8%	24.7%
of which: regional tax income	14.8%	20.5%	5.7%	13.8%	22.3%	8.5%	12.7%	25.2%	12.5%	15.3%	18.9%	3.6%	11.1%	10.6%	−0.5%

*Calculation of regional value-added (RVA) based on primary research data for Japan (in 2012) and from IÖW for Germany (in 2012; data refer to aggregated facility stock for solar installations) (Hirschl B, Heinbach K, Prahl A, Salecki S, Schröder A, Aretz A, Weiß : Wertschöpfung durch Erneuerbare Energien – Ermittlung der Effekte auf Länder- und Bundesebene, forthcoming)/exchange rate: 130JPY/Euro.

intensity of industries that are involved in activities along the value chain of specific RE technologies. These are obviously only approximations that do not necessarily reflect the RE-specific situation in regard to labor costs and returns. For long-term estimations, such an approach seems justified, because enterprises are unlikely to engage in RE-related activities unless at least industry-average returns are delivered. However, country-specific RE-related policies such as subsidies, quotas, or auction schemes may well impact the economics for specific RE technologies, thereby distorting the modeling results. As Germany and Japan have both adopted similar feed-in-tariff schemes, the transfer of the IÖW model can be expected to deliver reliable and comparable results. However, it is a subject for further research to test the transferability of the IÖW model to countries with different RE-related policy schemes. This may well require further modifications of the model that take into account these differences in policy frameworks.

Third, the comparative analysis of the regional value-added revealed a substantial difference in regard to the share of value-added of RE that is captured on the regional level at the stage of planning and construction. Also, German companies generally gain a significant higher share of business profit as compared to their Japanese counterparts at this stage of the RE value chains. A possible root cause for these differences may be that country-specific industry structures and business practices are not sufficiently reflected in the national, industry-specific statistics. For example, the industry statistics for the construction industry in both countries show a far lower labor ratio in Japan (17.9%) than in Germany (27.1%) [47,57]. Taken at face value, this would imply that Japan's construction industry in general employs substantially less direct labor than companies in Germany which seems rather unlikely. Instead, structural differences of the German and Japanese construction industry seem to be a more plausible explanation for these differences. Due to a peculiar licensing system administered by the Ministry of Land, Infrastructure and Transport (MILT), the Japanese construction industry extensively relies on a limited number of general contractors who then subcontract construction work to a multi-layered pyramid of first-, second-, and third-tier subcontractors [58-60]. Almost 60% of all construction work is subcontracted, and about half of all licensed contractors merely act as bidders without performing actual construction work [61]. This multi-layered, subcontracting-based structure of Japan's construction industry is not adequately reflected when applying national industry statistics, because business profits and labor income at the various levels of subcontractors are hidden in the material input costs of the subcontracting top layer firms. As a result, the estimated potential for regional value-added in Japan at the planning and investment stage is most likely understated.

For the case of Japan, it can be concluded that the IÖW model requires further adjustments to better reflect Japan's multi-layered subcontracting structures that are rather common in many Japanese industries. In general, therefore, a careful analysis of the prevailing industry structures is warranted when transferring the IÖW model to other countries.

Fourth, the case of Japan demonstrated that significant adaptations to the IÖW model are necessary in order to incorporate the specific characteristics of a nation's tax system. Differences in national and regional tax schemes have a profound impact on the level and distribution of regional value-added.

Finally, the present IÖW model allows only for an analysis of the direct regional economic effects of RE but does not capture indirect economic effects that are generated across related upstream industries or are induced by additional consumption and investment stimulated in the regional economy. However, it should be possible to couple this model with an I/O analysis to quantify further indirect economic effects generated at upstream value chain stages in the local economy. And by integrating income multipliers, the induced effects of additional consumption or investment can be captured and measured. At the same time, RE may also substitute other economic activities, in particular related to the incumbent, conventional fossil-fuel-based energy sector. In principle, it should be possible to incorporate such substitution effects of RE into the IÖW model by also integrating the value chains of conventional energy sources. In addition to the already noted need for further validation of the IÖW model by means of a systematic comparison with results from other methodological approaches, these final remarks point to possible extensions of the model and directions for further research.

Conclusions

This analysis has shown that the transfer of the IÖW model to other countries is principally feasible. It, therefore, can be concluded that the regional value-added modeling approach offers a powerful tool for regional planners and policymakers to analyze and capture the potential of RE for regional economies. By applying the approach for a comparative analysis between countries, it can also provide recommendations to enhance the effectiveness of RE-related policies for regional economic development.

Like in Germany, there is a similar promising potential for regional value-added by RE in Japan. However, as the model of regional value-added also allows for a comparison of the economic performance of RE technologies in different countries, it was possible to show that in case of Japan significantly higher total system costs in general, and the balance-of-system costs associated with the planning and installation of RE facilities in particular lead to a comparatively weaker economic performance than in Germany despite Japan's rather generous feed-in tariffs. The comparative analysis provided insights into possible structural causes for differences in economic performance, such as high construction and planning costs due to administrative regulations. Further growth of the RE sector in Japan, therefore, may well lead to significant improvements in the economic performance of RE technologies as RE system costs decrease in line with global market developments and as subsequent economies-of-scale raise productivity at the stage of planning and installation as well as at the stage of system operation. In addition, the relaxation of administrative procedures could further shorten the planning cycle in Japan and thereby lead to cost reductions.

The results of the IÖW model for Germany underline the importance of participation by regional players such as local business and farmers, energy cooperatives, or municipality-owned utilities for the generation of regional value-added. Similar to Germany, the results of the comparative regional value-added analysis for Japan show that business profit from RE system operation rather than income from local employment is the most important, long-term source of regional value-added. Therefore, local ownership is essential in order to generate and capture value-added for the region. Similarly, the availability of local capabilities along the various stages of the RE value chains is another important factor that determines the share of value-added captured by the region. While local manufacturing capabilities of RE technologies obviously benefit a local economy, capabilities at the planning and installation stage as well as during system operation including local services for service and finance are an equally important, long-term source of regional value-added. Therefore, when aiming for regional economic development by RE, regional planners and policy makers in Japan should consider policies to encourage or even mandate local investment and ownership as well as local capacity-building [62]. For example, as shown, Japanese municipal and prefectural taxes appropriate a comparatively higher share of regional value-added than in Germany. While higher tax income, especially from the fixed asset tax, may be tempting for local governments in Japan, the comparative analysis suggests that the relatively higher local tax burden in Japan may impede local RE investment. Instead, it seems to be more effective to adopt policies that effectively stimulate local investment into RE technologies, promote regional clusters and value chains in RE, foster local ownership and participation by local citizens, and raise the acceptance for RE by the local community. As a result, the expansion of RE based on local investment will then also create additional sources for regional tax income.

Endnotes

[a]The UK government has employed a similar methodological approach referred to as 'supply chain analysis' for Scotland, but the approach focuses on industry turnover and employment and provides little detail in respect to the underlying database.

[b]The ROE comparison for both countries in the case of large-scale solar installations on land is not fully compatible as the underlying equity ratios differ for both countries (Germany: 50%/Japan: 30%). When adjusting the equity ratios, the ROE for Germany increases significantly to about 13%, that is more than twice the level in Japan.

Abbreviations

kW: kilowatt; MW: megawatt; RE: renewable energy.

Competing interests

The authors declare that they have no competing interests.

Authors' contributions

All authors made a significant contribution to the design and adaptation of the IÖW model of regional value-added to the context of Japan, to the comparative analysis of RE economic performance and regional value-added between Japan and Germany. All authors read and approved the final manuscript.

Authors' information

At the time of our research, Mr. PRAHL was still working as a researcher at the IÖW and contributed to this article in this capacity.

Acknowledgements

The authors are deeply indebted to the members of the research team at the Institute for Sustainable Energy Policies (ISEP), namely Tetsunari IIDA, Noriaki YAMASHITA, Haruhiko DOHMAN, and Naoki WADA, who diligently compiled the database on technology-specific costs and revenue structures of renewable energy sources in Japan.

Author details

[1]College of Business Administration, Ritsumeikan University, Kyoto, Japan. [2]Institute for Sustainable Energy Policies (ISEP), Tokyo, Japan. [3]Ecologic Institute, Berlin, Germany. [4]Institute for Ecological Economy Research (IÖW), Berlin, Germany.

References

1. Renewable Energy Policy Network for the 21st century (REN21) (2014) Renewables 2014 – global status report, Paris. download under http://www.ren21.net/Portals/0/documents/Resources/GSR/2014/GSR2014_full%20report_low%20res.pdf.
2. Renewable Energy Policy Network for the 21st century (REN21) (2013) Renewables – global futures report, Paris. download under http://www.ren21.net/REN21Activities/GlobalFuturesReport.aspx.
3. Rifkin J (2011) The third industrial revolution - how lateral power is transforming energy, the economy, and the world. Palgrave Macmillan, New York
4. International Energy Agency (IEA) (2014) The transformation of power – wind, sun and the economics of flexible power systems. International Energy Agency Bookshop, Paris
5. Bundesministerium für Wirtschaft und Energie (BMWi) (2014) Erneuerbare Energien im Jahre 2013 – Erste vorläufige Daten zur Entwicklung der erneuerbaren Energien in Deutschland auf der Grundlage der Angaben der Arbeitsgruppe Erneuerbare Energien-Statistik (AGEE-Stat). download under
6. Stablo J, Ruppert-Winkel C (2012) The integration of energy conservation into the political goal of renewable energy self-sufficiency – a German case study based on a longitudinal reconstruction. Sustainability 5:888–916

7. Klagge B, Brocke T (2011) Decentralized electricity production from renewable sources as a chance for local economic development ? Qualitative study of two pioneer regions in Germany. In: Schäfer M, Kebir N, Philipp D (eds) Micro perspectives for decentralized energy supply. Proceedings of the International Conference Technische Universität Berlin, Universitätsverlag TU Berlin, Berlin, pp 21–26
8. Wuppertal Institut für Klima, Umwelt, Energie (2013) Stadtwerke-Neugründungen und Rekommunalisierungen. Energieversorgung in kommunaler Verantwortung, Wuppertal, download under http://wupperinst.org/uploads/tx_wupperinst/Stadtwerke_Sondierungsstudie.pdf
9. Ministry of Economy, Trade, and Industry (METI) (2014) Basic energy plan, Tokyo (*Keizai Sangyōshō, enerugi- kihon keikaku*). download under http://www.enecho.meti.go.jp/category/others/basic_plan/pdf/140411.pdf (*in Japanese*)
10. Institute for Sustainable Energy Policies (ISEP) (2014) Renewables Japan status report – executive summary, Tokyo: download under http://www.isep.or.jp/en/wp-content/uploads/2014/07/JSR2014_Summary_EN.pdf
11. Furuya S (2013) Community power plants. Ippanshoshinsho, Tokyo, *komyuniti- hatsudenshō*) (*in Japanese*).
12. Takahashi M (2012) Starting the renewable energy revolution – regional community-based energy. Ōtsuki Shōten, Tokyo, *Sheen energy kakumei-wo hajimeyo – chiiki-de tsukuru mina-no denryoku*)., (*in Japanese*)
13. Kurasaka H (2013) Local economic impacts of renewable energies, Proceedings at the Annual Meeting of the Society for Environmental Economics and Policy Studies (SEEPS) at Kobe, September 21/22, 2013 (*Saiseikanō enerugi- dōnyū-ni yoru chiiki keizai kōka-ni tsuite*). download under http://www.seeps.org/meeting/2013/submit/abst/2140_hi3Rsyfs.doc (*In Japanese*).
14. Breitschopf B, Nathani C, Resch G (2011) Review of approaches for employment impact assessment of renewable energy deployment, IEA-Renewable Energy Technology Deployment: download under http://publica.fraunhofer.de/eprints/urn:nbn:de:0011-n-1988024.pdf.
15. Miller R, Peter D (2009) Input–output analysis, foundations and extensions. Cambridge University Press, Cambridge
16. Loveridge S (2004) A typology and assessment of multi-sector regional economic impact models. In: Regional Studies, Vol. 38/3. Routledge, London, pp 305–317
17. Lehr U, Breitschopf B, Lutz C, Edler D, O'Sullivan M, Nienhaus K, Nitsch J, Bickel P, Ottmüller M (2011) Kurz- und langfristige Arbeitsplatzwirkungen des Ausbaus der erneuerbaren Energien auf den deutschen Arbeitsmarkt. Fraunhofer-Institut für Innovation- und Systemforschung, Osnabrück
18. O'Sullivan M, Edler D, Bickel P, Lehr U, Frank P, Sakowski F (2013) Bruttobeschäftigung durch erneuerbare Energien in Deutschland im Jahr 2012 - eine erste Abschätzung. Gesellschaft für Wirtschaftliche Strukturforschung mbH, Osnabrück
19. Spanish Renewable Energy Association (APPA) (2009) Study of the macroeconomic impact of renewable energies in Spain. http://www.appa.es/descargas/Informe_APPA_ENGLISH.pdf.
20. National Institute of Science and Technology Policy (NISTEP) (2013) Economic and environmental impact analysis for construction of renewable energy power plants by extended Input output table, Discussion Paper No. 96, Tokyo (*kakuchō sangyō kanrenhyō-ni yoru saiseikanō enerugi-hatsudennshisetsu no keizai kankyo-he no hakkyukoka bunseki)*. download under http://data.nistep.go.jp/dspace/bitstream/11035/2419/1/NISTEP-DP096-FullJ.pdf (*in Japanese*).
21. Coon R, Hodur N, Bangsund D (2012) Renewable energy industries' contribution to the North Dakota economy, In: Agribusiness and Applied Economics, 702. download under http://ageconsearch.umn.edu/bitstream/140122/2/AAE702.pdf.
22. Allan G, McGregor P, Swales K (2011) The importance of revenue sharing for the local impacts of a renewable energy project – a social accounting matrix approach. In: Regional Studies, vol. 45/9. Routledge, London, pp 1171–1186
23. Heinbach K, Aretz A, Hirschl B, Prahl A, Salecki S (2014) Renewable energies and their impact on local value-added and employment. Energy, Sustainability and Society 4(1):1–10
24. Flegg A, Thomo T (2010) Regional input–output tables and the FLQ formula: A case study of Finland. Regional Studies Association Annual International Conference 2010: Regional Responses and Global Shifts: Actors, Institutions and Organisations, Pecs, Southern Hungary, 24–26 May 2010, University of the West of England, Bristol. download under http://eprints.uwe.ac.uk/12338/.
25. Hirschl B, Aretz A, Prahl A, Böther T, Heinbach K, Pick D, Funcke S (2010) Kommunale Wertschöpfung durch Erneuerbare Energien, Schriftenreihe des Institut für Ökologische Wirtschaftsforschung (IÖW 196/10);Berlin.
26. Kosfeld R, Gückelhorn F (2012) Ökonomische Effekte erneuerbarer Energien auf regionaler Ebene. Raumforsch Raumordn 70:437–449
27. Institut für angewandtes Stoffmanagement (IfaS), Deutsche Umwelthilfe e.V. (DUH) (2013) Kommunale Investitionen in Erneuerbare Energien – Wirkungen und Perspektiven. download under http://www.stoffstrom.org/fileadmin/userdaten/dokumente/Veroeffentlichungen/2013-04-04_Endbericht.pdf.
28. Hoppenbrook C, Albrecht AK (2009) Diskussionspapier zur Erfassung der regionaler Wertschöpfung in 100%-EE-Regionen, In: DEENET (Hrsg.), Arbeitsmaterialien 100EE, Nr. 2: download under http://www.100-ee.de/index.php?id=schriftenreihe&no_cache=1.
29. Bundesministerium für Verkehr, Bau und Stadtentwicklung (BMVBS) (2011) Strategische Einbindung regenerativer Energien in regionale Energiekonzepte - Wertschöpfung auf regionaler Ebene. download under http://www.bbsr.bund.de/BBSR/DE/Veroeffentlichungen/BMVBS/Online/2011/ON232011.html?nn=423048.
30. Raupach-Sumiya J (2014) Measuring regional economic value-added of renewable energy - the case of Germany. In: *Shakai Shisutemu Kenkyū* (Social System Study), Vol. 29. Ritsumeikan University BKC Research Organization of Social Sciences Kyoto pp 1–31 download under http://www.rit-sumei.ac.jp/acd/re/ssrc/result/memoirs/kiyou29/29-01.pdf.
31. Department of Trade and Industry (2004) Renewable supply chain gap analysis. Department of Trade and Industry, London
32. Matsumoto N, Hondo H (2011) Analysis on employment effects of the introduction of renewable energy technologies using an extended input–output table. J Japan Institute of Energy (Nihon Energi- Gakkai) 90:258–267. *Kakuchōsangyōkanrenhyō-wo riyō shita saiseikanō enerugi- dōnyū-no koyōkōka bunseki*) (*in Japanese*)
33. Ishikawa Y, Nakamura R, Matsumoto A (2012) Economic effects of the introduction of renewable energy in the Tohoku region – analysis of the introduction of solar and wind power based on interregional input–output tables, RIETI Policy Discussion Series 12-P-014. Research Institute of Economy, Trade and Industry, Tokyo, *Tōhokuchiiki-ni okeru saiseikanō enerugi-dōnyū-no keizaikōka – chiikikansangyōkannrenhyōni yoru taiyōkōhatsuden fūryokuhatsudendōnyū-no bunseki*) (*in Japanese*)
34. Nakazawa R, Nakazawa J, Matsumoto A (2012) Economic and CO_2 reduction effects by wood-fed biomass – construction and new application of a regional I/O model. In: *Chiikigaku Kenkyū* (Studies in Regional Science), Vol. 42/4. The Japan Section of the Regional Science Association International, Tokyo, pp 799–817, *Kishitsu baiomasu-wo katsuyōshita* CO_2 *sakugen-to chiikikeizaikōka – Chiikisangyōkanren moderu-no kōchiku-to aratana tekiyō*) (*in Japanese*).
35. Hienuki S, Hondo H (2012) Employment life cycle analysis of geothermal power generation using an extended input–output table. In: Proceedings of the 28th Conference on Energy Systems, Economy and Environment by the Japan Society of Energy and Resources (*Enerugi- Shigen Gakkai), January 30th ~31st, 2012.* Japan Society of Energy and resources, Osaka, pp 209–212, *Kakuchōchōsangyoukanrenhyō-wo mochiita chinetsuhatsuden-no raifusaikuru koyōbunseki*) (*in Japanese*).
36. Shimomura S, Nakazawa J, Matsumoto A (2013) Regional economic effects of introducing solar power generation system applying extended regional I/O analysis - A comparative evaluation of different business forms in Kochi prefecture, paper presented at the 50th annual meeting of the Japan Section of the Regional Science Association International (JSRSAI), October 12 ~ 14th, 2013 at Tokushima University, (*Kankyōsangyōbunsekiyōchiikisangyōkanrenhyō-wo mochiita taiyōkōhatsuden-no chiikikeizaikōka – Kōchiken-ni okeru jigyōkeitai-ni yoru kōka-no hikakubunseki*) (*in Japanese*).
37. Cost Evaluation Committee of the Energy / Environment Conference (2011) Report of the Cost Evaluation Committee, December 19th, 2011, Tokyo (*Kosuto-nado hyōkaiinkai enerugi- kankyō kaigi, kosuto-nado hyōkaiinkaihyōkokushō*). download under http://www.cas.go.jp/jp/seisaku/npu/policy09/pdf/20111221/hokoku.pdf (*in Japanese*).
38. METI Evaluation Committee (Ministry of Economy, Trade and Industry Agency for Natural Resources and Energy, Evaluation Committee on Procurement Prices) (2012) Recommendations regarding tariffs and FIT period for 2012, April 27th, 2012 (*Keizai Sangyōshō Chōtatsu Kakakutō Santeiiinkai, Heisei 24 nendo chōtatsukakaku oyobi chōtatsukikan-ni kansuru iken)*. down load under http://www.meti.go.jp/committee/chotatsu_kakaku/report_001.html.
39. METI Evaluation Committee (Ministry of Economy, Trade and Industry Agency for Natural Resources and Energy, Evaluation Committee on Procurement Prices) (2013) Recommendations regarding tariffs and FIT period for 2013,

March 11th, 2013 (*Keizai Sangyōshō Chōtatsu Kakakutō Santeiinkai, Heisei 24 nendo chōtatsukakaku oyobi chōtatsukikan-ni kansuru iken)* download under http://www.meti.go.jp/committee/chotatsu_kakaku/pdf/report_002_01_00.pdf
40. Energy New Industrial Technology Development Organization (NEDO) (2014) NEDO renewable energy technology whitebook, 2nd edn. NEDO, Morikita Shuppan, Tokyo, *saiseikanō enerugi- gijutsu hakusho dai2ban* (*in Japanese*)
41. Japan Renewable Energy Foundation (JREF) (2013) The state of the solar power business and cost 2013, Tokyo (*Taiyōkōhatsudenjigyō to kosuto 2013*). download under http://jref.or.jp/images/pdf/20131220/reports_20131220.pdf (*in Japanese*).
42. Japan Renewable Energy Foundation (JREF) (2013) Present state and analysis of the cost of wind power generation in Japan, Discussion Paper, Tokyo (*Nihon no rikujōfūryokuhatsuden kosuto no genjō to bunseki*): download under http://jref.or.jp/activities/reports_20131023.php (*in Japanese*).
43. Public Electric Utility Business Association (2012) Requests related to the Feed-In-Tariff System for hydro power, April 3rd, 2012, Tokyo (*Kōeidenkijigyōkeieishakaigi, Suiryokuhatsuden-no zenryōkaitoriseido-ni kansuru yōbō*). download under http://www.meti.go.jp/committee/chotatsu_kakaku/pdf/004_03_00.pdf (*in Japanese*).
44. National Small-Scale Hydro Promotion Committee (2012) Requirements for the expansion of small-scale hydro power under the FIT system, April 3rd, 2012 (*Zenkoku shōsuiryokuriyōsokushinkyōgikai, FIT-ni yoru shōsuiryokuhatsudenkakudai-no jōken*). download under http://www.meti.go.jp/committee/chotatsu_kakaku/pdf/004_04_00.pdf (*in Japanese*).
45. Japan Wind Power Association (2012) Materials provided for the 3rd Meeting of the Evaluation Committee, March 19th, 2012, Tokyo (*Nihonfūryokuhatsudenkyōkai, Dai3kai chōtatsukakakutōsannteiinikai gosetsumeishiryō*). Download under http://www.meti.go.jp/committee/chotatsu_kakaku/003_05_00.pdf (*in Japanese*).
46. Japan Project Industry Council (2013) About the cost estimate for wood-fed biomass, Tokyo, (*JAPIC Shinrinsaiseijigyōkaiinkai, Shinrinbaiomasuhatsuden kosuto-ni tsuite*). download under http://www.japic.org/report/pdf/national_strategy_group30.pdf (*in Japanese*).
47. Homepage of the Japanese Ministry of Finance (MOF), access on 23.6.2014; http://www.mof.go.jp/english/pri/reference/ssc/index.htm.
48. Website of New Energy Foundation, Database of Renewable Energy Research Organizations & Companies in East Asia. http://jrdb.asiabiomass.jp/index.php?lang=en access on December 28th, 2014.
49. Homepage of the Japanese Ministry of Internal Affairs and Communications (MIAC), access on 23.6.2014. http://www.soumu.go.jp/main_sosiki/jichi_zeisei/czaisei/czaisei_seido/ichiran01.html.
50. KPMG Taxation Corporation (2012) Taxation in Japan 2012, Tokyo: download under.
51. National Tax Agency (NTA) (2011) The 137th National Tax Agency Annual Statistics Report FY2011, Tokyo, p. 12 (*Dai 137kai kokuzeichō tōkeinenpō heisei 23 nendōban*). download under http://www.nta.go.jp/kohyo/tokei/kokuzeicho/minkan2012/pdf/001.pdf (*in Japanese*).
52. METI Evaluation Committee (Ministry of Economy, Trade and Industry Agency for Natural Resources and Energy, Evaluation Committee on Procurement Prices) (2014) Presentation materials distributed in 15th committee meeting on March 7th, 2014 (*Keizai Sangyōshō Chōtatsu kakakutō Santeiinkai*). down load under http://www.meti.go.jp/committee/chotatsu_kakaku/pdf/015_s02_00.pdf, http://www.meti.go.jp/committee/chotatsu_kakaku/pdf/015_s03_00.pdf, (*in Japanese*).
53. International Renewable Energy Agency (IRENA) (2013) Renewable power generation costs in 2012. An overview: download under http://www.irena.org/DocumentDownloads/Publications/Overview_Renewable%20Power%20Generation%20Costs%20in%202012.pdf.
54. Seel J, Barbose GL, Wiser RH (2013) Why are residential PV prices in Germany so much lower than in the United States? A scoping analysis, presentation, Lawrence Berkeley National Laboratory. download under http://eetd.lbl.gov/node/50173.
55. METI Evaluation Committee (Ministry of Economy, Trade and Industry Agency for Natural Resources and Energy, Evaluation Committee on Procurement Prices) (2014) Presentation materials distributed in 15th committee meeting on March 7th, 2014 (*Keizai Sangyōshō Chōtatsu kakakutō Santeiinkai*). down load under http://www.meti.go.jp/committee/chotatsu_kakaku/pdf/015_s02_00.pdf.
56. EU-Japan Centre for Industrial Cooperation (2014) The clean energy sector in Japan – An analysis on investment and industrial cooperation opportunities for EU SMEs, Tokyo. download under http://www.eu-japan.eu/sites/eu-japan.eu/files/clean-energy-paper-27feb-finale.pdf#search='clean+technology+in+Japan+EU.
57. Homepage of the Deutsche Bundesbank, access on 23.6.2014 http://www.bundesbank.de/Navigation/DE/Statistiken/Unternehmen_und_private_Haushalte/Unternehmensabschluesse/Tabellen/tabellen.html.
58. Homepage of Ministry of Infrastructure, Land and Transport (MILT), access on 23.6.2014 http://www.mlit.go.jp/sogoseisaku/const/kengyo/kyoka-e.htm.
59. Japan Federation of Construction Contractors (JFCC) (2013) Construction industry handbook 2013 (*Kensetsugyō handobukku 2013*) download under http://www.nikkenren.com/publication/handbook.html.
60. Research Institute of Construction and Economy (RICE) (2007) The structure of the construction industry and its contribution to society, No. 49, October 2007, Tokyo, (*Kensetsukeizaikenkyūjo, Kensetsusangyō-no kōzō to shakai-he no kōken, kensetsukeizai ripōto*). download under http://www.rice.or.jp/regular_report/construction_economic_report.html (*in Japanese*).
61. The Economist (2001) Japan's construction industry – pass the scaffolding, print edition, July 12th. The Economist Newspaper Limited, London, download under: www.economist.com/node/693543
62. Wada T, Toyota Y, Taura K, Ito S (2014) Establishing energy cooperatives by citizens and communities (*shimin chiiki kyōdōhatsudensho no tsukurikata*). Shin Nippon Press. 168–170. in Japanese.

8

Approaches for modelling the energy flow in food chains

Baboo Lesh Gowreesunker* and Savvas A Tassou*

Abstract

Background: The heavy reliance of the global food chain on the use of fossil fuels and anticipated rise in global population threatens future global food security. Due to the complexity of the food and energy systems, the impact of adequate food, climate or energy policies should be carefully examined in a modelling framework which considers the interaction of the food and energy systems. However, due to the different modelling approaches available, it can be very difficult to identify which method best suits the required purpose.

Method: This paper presents the three main modelling approaches as 'top-down', 'bottom-up' and hybrids. It reviews different models under each category in terms of the practicality, benefits and limitations with reference to different past studies.

Results: Bottom-up approaches generally tend to provide high levels of details, but their specificity to particular products/processes detracts their application to holistic models. On the other hand, top-down approaches consider the holistic aspects of the food chain, but the limited level of disaggregation prevents the identification of energy and environmental hot-spots. As a result, hybrid models seek to reduce the limitations of the individual approaches.

Conclusions: This paper shows that the choice of one modelling approach over another depends on a variety of criteria including data requirements, uncertainty, available tools, time and labour intensity. Furthermore, future models and studies have to increasingly consider the inter-dependence of implementing social, demographic, economic and climate considerations in a holistic context to predict both short- and long-term impacts of the food chain.

Keywords: Food and energy chain; Modelling approaches; Top-down models; Bottom-up models; Hybrid models

Background

Overall perspective

The Food and Agriculture Organisation of the United Nations (FAO) has expressed concern over the high dependence of the global food sector on fossil fuels and the projected 70% increase in current food consumption by 2050 due to the rise in global population (FAO [1]). The food sector accounts for 30% of the global energy consumption and 20% of global greenhouse gas (GHG) emissions, with a major contribution from fossil fuels (FAO, [1,2]). Developed economies such as the UK used approximately 18% of the total energy consumption for the food sector, which produced approximately 32% of the country's GHG emissions in 2011 [3]. The disparity in energy consumption is, however, significant between developed and developing countries, whereby the former use the majority of energy in processing and distribution, whilst the latter use energy mainly for retail, preparation and cooking (FAO [1]).

Due to increased use of depleting fossil fuel resources, the energy-food-climate nexus (FAO, [1]) has been found to be a crucial and complicated challenge for the planet. Energy, food and climate change are intricately linked such that actions taken or policies imposed in one area are very likely to have consequences in the other areas. The nexus can be summarised as follows: high usage of fossil fuels impacts the climate due to GHG emissions - the food sector is heavily dependent on fossil fuels and becoming even more so due to rise in global population - but fossil fuel reserves are depleting and climate change is expected to lower average agricultural yields (Lobell *et al.* [4]; Roberts *et al.* [5]). The scenario is therefore complex, and the Food and Agriculture

* Correspondence: lesh_g@yahoo.com; Savvas.Tassou@brunel.ac.uk
RCUK National Centre for Sustainable Energy Use in Food Chains, Brunel University, UB8 3PH, Uxbridge, Middlesex, UK

Organization (FAO [1]), the United Nations Environment Programme (UNEP [6]), the Clinton Global Initiative (CGI [7]), Grace Communications Foundation [8] and Mistry and Misselbrook [9] suggest that a reduced dependency on fossil fuels and an increased use of renewable energy technologies are imperative in an attempt to tackle this nexus.

Current barriers to adopting renewable energy technologies are mainly the high capital costs [10] and relatively low-energy efficiencies of some common systems (e.g. ≈3 to 20% for photovoltaic systems [11]). Although various governmental incentives are available, subsidies for renewable energy technologies worldwide are low, hindering their rapid adoption relative to fossil fuels [12]. Nonetheless, renewable energy systems are proving to be cost effective in countries where there is a heavy reliance on diesel electricity-generators and which have high fuel costs (WFP [13]). A study by Maggio and Cacciola [14] conveyed that although there is considerable uncertainty in the lifetime of fossil fuel reserves, the global production of oil will start declining by 2015, whilst gas and coal production will peak in 2035 and 2052, respectively. Historically, the prices of oil and food commodities have been interlinked (FAO [1,15]; Heinberg and Bomford [16]), but some studies also suggest that in the long run, the price of agricultural commodities have a higher impact on food prices than oil commodities ([17]; Lambert and Miljkovic [18]). This ambiguity in projections can be associated with the assumptions made and modelling approach used in the respective studies [19]. It is generally agreed, however, that due to the current intensive use of fossil fuels in the food sector, the uncertainties in fossil fuel energy availability and prices may threaten food security and affect political stability in the future (FAO [1,2]). It is therefore imperative that the impact of increased use of renewable energy technologies and the adoption of new more efficient technologies and energy supply systems on food security and supply chain sustainability be investigated further (FAO [1,20,21]).

Scope of this review

This review mainly targets readers interested in furthering their understanding with respect to developing models which study parameters influencing the energy and GHG emissions flows in product-specific, national and international levels of the food chain. It aims at providing an appreciation of existing models and hence informs the reader of the potential benefits and drawbacks of employing different modelling methodology. This review focuses on the food and energy/GHG chain, where the growing field of sustainable consumption suggested that food, home energy and transportation together form a large share of most consumers' personal impacts [22]. Food represents a unique opportunity for consumers to lower their personal footprints due to the high impact of food, high degree of personal choice and a lack of long-term 'lock-in' effects which limit consumers' day-to-day choices [23]. In this regard, Garnett [24] summarised three perspectives on tackling the food security and sustainability issues as: efficiency oriented; demand restraint and food system transformation, of which efficiency oriented measures have been advocated by governments and food industry decision makers [23,24]. As such, it is imperative to understand the flow of energy in the food chain in order to identify optimum energy efficient and sustainable pathways.

The food chain refers to the successive collection of the farming, food processing, distribution, packaging, retail, catering, household operations, waste and disposal industries, for different types of food products. The emphasis of this review is on the direct and indirect energy consumption and the associated GHG emissions of the overall food chain (or specific food product chains), where the definition of food products abides by the EU foodstuff law (Regulation (EC) No. 178/2002) which refers to foodstuff as 'any substance or product whether processed, partially processed or unprocessed, intended to be, or reasonably expected to be ingested by humans'.

As alluded in the 'Overall perspective' section, the linkages between the food and energy sectors are complex and depend on a variety of factors. The energy systems are currently at a crossroad whereby policies need to determine a balance between sustainable development, competitiveness and supply security [20]. As a result, the interactions involved in the food and energy system should be addressed in a quantitative manner and a modelling framework, so as to aid effective policy design [25]. These models can be employed to evaluate energy effective pathways and the implementation of renewable energy technologies to deliver the energy/GHG emissions reduction targets, at present and in the future. However, in order to allow the proper selection of a modelling method, it is important for the user to understand the particularities of the model. The rationale for this paper therefore relates to the need for examining the benefits and drawbacks of current modelling approaches employed in relation to the food chain and its products and to understand the degree to which such approaches capture the complex and nonlinear interactive behaviour of the food-energy components [25,26]. The modelling approaches considered in this review are 'bottom-up', 'top-down' and hybrid, and their appreciation will aid in identifying and consolidating new developments in food-energy/GHG models and in elaborating on the performances and discrepancies of current models. In this regard, the paper is divided into the different modelling methods in the 'Methods' section -

which describes the models, as well as various studies that employed these models; the 'Results and discussion' section then compares the models and presents and analyses the benefits and limitations of each approach and the 'Conclusion' section then concludes on the general progression of adopted modelling approaches in the literature and suggests future modelling pathways.

Methods

Bottom-up approach

Bottom-up approaches adopt a view of assembling the local disaggregated influences in order to determine the global impacts associated with a particular product, process, service or industry. It is a detailed approach and therefore requires compiling inventories of energy, environmental, economic and material inputs for the various processes. The following sections do not attempt at describing the technicalities of the modelling methods, but rather to depict the applicability, practicality and benefits/limitations of these methods.

Life cycle assessment (LCA) LCA refers to a product- or process-based analysis of the GHG emissions and energy consumption, usually employing a 'cradle-to-grave' approach. In the food chain, it considers all stages from the farming/agricultural process through to consumption and waste disposal [27]. This method has been deemed important to examine the intricacies attached to food products or systems, where a current dearth of data exists and where future research is crucial [24,28,29]. As LCA and life cycle inventory (LCI) studies are extensive and apply to particular regions, this section will explore studies related to the food chain in the UK. AEA group and partners report on the comparative LCA of seven food commodities procured for UK consumption through different supply chains (DEFRA report no. FO 0103 [30]). The analysis was performed with reference to primary energy use and global warming potential, with assumptions relating to agricultural yields, transportation, and uncertainties due to imports increasing the overall uncertainty of the study. Fisher *et al.* [31] adopted a cradle-to-retail approach to study the GHG and secondary energy impacts of UK groceries, with special emphasis on high sales-volume products. The food products include alcoholic drinks, ambient products (breakfast cereals, canned food etc.), bakery, dairy, fruit and vegetables, meat, fish, poultry, eggs, non-alcoholic drinks and chilled and frozen products. Data were obtained from various sources including journals, industries, government reports and eco-labels. The study therefore assumes that the different sourced data can be combined together to form a whole chain analysis, but acknowledges the implications of this assumption in the final results of their study. Similar to the previous study, caution is suggested by the authors before using the quantitative results from this study due to the high level of uncertainty. Lillywhite *et al.* [32] studied the embedded energy associated with producing, processing and distributing a range of food products, with a view to addressing the food chain security issue in the UK. Data were also obtained from academic and grey literatures, where the lack of LCA data for multi-ingredient products such as pasta sauce, soup and pizza were derived by the authors' own LCA. This study further explored the price volatility and elasticity of food products with regard to food security and showed that the UK's food supply is almost completely dependent on fossil energy, raising concerns for future food security. The common trait of LCA studies to assume different sourced data can be combined, increases uncertainty in the analysis, where the authors generally caution the reader of the implications in the final results of their study.

In the UK, publicly available specification (PAS) 2050 was developed in 2008 to provide a consistent method for quantifying carbon footprints. The PAS 2050 standard was adopted from the ISO 14044 standard and has been refined with consultation of various research and user communities (PAS 2050 [33,34]). The stepwise procedures are as follows: (i) defining the 'system boundary' of the product life cycles, (ii) data collection, (iii) compilation and validation of emissions flows and (iv) identification of hot-spots and emissions reduction opportunities. The resource and energy use data are therefore crucial for this method, and primary data generally improve the accuracy (PAS 2050 [34]). This method requires that assumptions - generally relating to primary energy conversion factors, transportation energies, refrigerant leakages, waste disposal and agricultural emissions - are made clear and conservatively. The PAS 2050 method is valid if the assumptions are $< 5\%$ of the total footprint and the sample size is adequate (PAS 2050 [34]).

Holmes *et al.* [35] conducted a PAS 2050 life-cycle study for five food products in the UK, starting with agricultural production, up to the delivery of food to the catering site, employing past studies' data to evaluate the GHG emissions resulting from raw materials, energy use in manufacture, distribution, retail and wastes. The main assumptions related to GHG emission factors from fertilisers and pesticides, and transportation distances. The study provided emission reduction practices and showed that the level of assumptions, time and resources required pose potential barriers to LCA methods. Tassou *et al.* [36] employed the PAS 2050 method to explore the GHG impacts of food retailing. The report focused on emissions from energy consumption, refrigerant leakage and waste for products ranging from fresh meat to bread. The authors demonstrated that PAS 2050 can be used to quantify GHG emissions from food retail operations if stores are submetered to a sufficient level. They showed that appropriate functional units and boundaries must be selected when studying services with PAS 2050. Campden and Chorleywood Food Research Associates and Partners (DEFRA FO0409 [37]) employed PAS 2050 to study the GHG emissions from the preparation and consumption of complex meals such as cottage pies, bread and apple juice. The authors made assumptions relating to the energy use in cooking processes, storage temperatures and amount of water used. This study similarly conveyed that product-based studies should be preferred for PAS 2050, to minimise the uncertainties associated with technology models in process-based studies.

Although LCA methodologies can be applied to products, processes or even industry, it is observed that LCA approaches in the food chain relate mainly to the supply chain of specific food products. Its use tends to focus on GHG emissions rather than energy consumption, which is also an integral part of the assessment. The high level of details possible from LCA can indicate resource efficiency hot-spots and allow targeted actions for improvements [29]. However, the literature suggests that although there are vast LCI of various products, because research is done on a random basis, the compilation of data and hence the comparison of different food supply chains becomes complex (The Ecoinvent database is an example of consistency in LCIs (Ecoinvent [38])). Thus, the system boundaries, assumptions made and uncertainties of the different studies should be clearly specified to increase the potential and practicality of this approach.

MARKAL model

The MARKAL model is a multi-period linear programming and optimisation tool which allows the simultaneous assessment of several technologies for specific industries or the whole economy (Rath-Nagel and Stocks [20,39]). The competition between technologies is affected by energy/GHG emission policies, associated costs and technical constraints. MARKAL models can determine the trade-offs between various objective functions such as costs, environmental indicators, oil security, renewable primary energy etc. (Rath-Nagel and Stocks [39]).

MARKAL models are demand-driven models (Rath-Nagel and Stocks [39]), with exogenous energy demands specified by the modeller for a specific network of processes linked through their inputs and outputs, technical, economic and policy based parameters [20]; see Figure 1. As such, the model operates from energy technology databases which detail the current energy system as well as the technical and cost parameters of potential energy systems (Rath-Nagel and Stocks [39]). The model aims to find a partial equilibrium on the energy market which satisfies the maximisation of the net surplus of consumers and suppliers, via linear programming [40]. It should be noted that since the development of MARKAL models by the International Energy Agency in 1976, the MARKAL family has grown to include various improvements. The specifics of the extrapolated models can be found in Loulou *et al.* [40,41].

The TIMES model is an extension of the original MARKAL model and includes variable time periods, time-data decoupling, higher flexibility energy processes, age-dependent parameters and climate equations (Lolou *et al.* [41]). Seck *et al.* [20] employed the TIMES model to analyse the impact of heat pumps on the French food and drink (F&D) industry. The authors used this bottom-up approach as they argue that the French F&D industry ('a non-energy intensive' industry) requires a fine and disaggregated understanding of the existing and emergent technologies. The benefits of adopting such an approach was found to be as follows: (i) the detailed and explicit formulation of the technologies and processes, (ii) the ease of modelling the effect of different policies and (iii) the explicit modelling of the evolution of demand and energy prices. However, the authors state that the TIMES model does not incorporate feedback from other economic sectors and the large quantity of data required for the analysis make the application of the model difficult and perhaps uncertain. Concerning the former point, Loulou *et al.* [40] argue that the change in energy demand is itself the main economic feedback. Seck *et al.* [20] assumed an expected evolution of demand for different F&D products, future energy price scenarios and the technical performance of heat pump systems, and showed that the use of heat pumps is a promising technology, possible of 21% reduction in CO_2 emissions and 13.6% reduction in energy consumption in 2020, compared to a scenario without heat pumps.

Gerlagh and Gielen [43] developed a supplementary module 'MATTER 2.0' for the agriculture and food sector of Western European countries. The impetus for this module relates to the increasing importance of competing interest of land use for food, energy and material production, GHG emission reductions and food consumption lifestyle changes. The model assumes a constant mass flow of resources and various proportions of product/energy wastage for different processes, noting that most emissions are converted from the calorific energy content of food. The development of this module shows the versatility of MARKAL for providing a platform for further research and model development.

It should be noted that the MARKAL is now superseded by TIMES, which is supported by the International Energy Agency, and as mentioned before, TIMES has a similar underlying basis as MARKAL, but with several improvements in the techno-economic aspects of the model. MARKAL/TIMES is seen to depict a more aggregated bottom-up approach compared to LCA, therefore allowing a broader overview of the supply chain

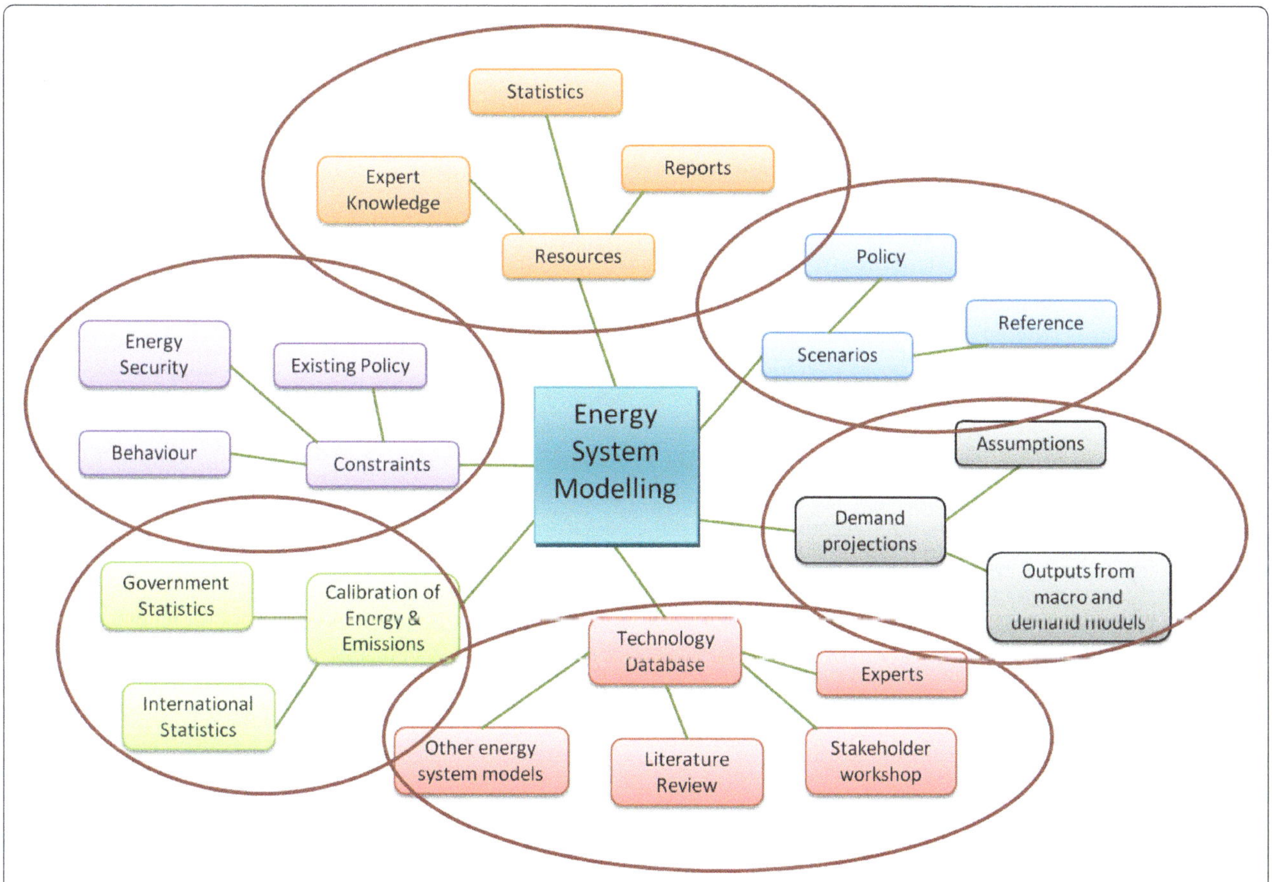

Figure 1 Parameters/database considered in the development of a MARKAL model [42].

of various food products or the food chain. The use of MARKAL/TIMES for the food chain is limited in the literatures; however, its use can be promoted especially as it provides a valid and IEA-supported linear optimisation platform for various user-defined scenarios [44]. A more in-depth evaluation is given in the 'Results and discussion' section.

Regression models

Regression models aim at understanding the causal relationship between two or more variables - where simply adopting correlations may not actually represent this causality - and to quantify how close and well determined the relationship is. It is therefore important in regression models to determine the influential dependent and independent variables, through statistical methods or empirical observations [45].

Spyrou *et al.* [46] studied the electricity and gas demand drivers for a UK food-retail organisation's buildings using a linear regression method. The authors identified a list of variables that affect the electricity and gas demands based on theoretical concepts and preliminary correlation tests, and observed that the level of detail is limited based on the availability of information. The dependent variables were electricity and gas intensity consumption, whilst the independent variables were divided into physical, operational and regional parameters. The main assumptions relate to the operation and efficiency of the electrical and gas systems, obtained from academic and grey literatures. The authors were able to develop three regression models for the electricity consumption ($R^2 = 0.75$, $p < 0.001$), gas consumption without CHP ($R^2 = 0.62$, $p < 0.001$), and gas consumption with CHP ($R^2 = 0.77$, $p < 0.001$). Employing such a simple model allows the food organisation to quickly identify retail buildings that are under-performing, but does not direct the modeller towards the factors causing the inefficiency.

Boyd [47] developed a performance-based energy efficiency indicator (EPI) for the US food processing sector using linear regression models. Due to lack of data and the diversity involved in food processing, the author adopted a more segmented approach using primary data for specific sectors in the food processing industry, as opposed to modelling the entire processing industry. The study is based on primary energy sources and calculates an EPI for each sector by comparing the actual energy consumption of each processing plant. The optimum scenario is obtained following a stochastic frontier regression analysis, which applies the ordinary least-square regression method to the standard linear regression model. The model predicted R^2 coefficients of 0.8 to 0.96 and variances varying between 0.06 and 0.6. The benefits of using the EPI are that it allows industry leaders to benchmark the performance of various plants and to simply assess the average performance of the sectors, crucial for effective policy designs. However, the large data requirements, the diversity in processing methods and the need for constant updating of the database (due to technical and business innovations) pose limitations in the implementation of such models for longer term.

Tassou *et al.* [48] employed a regression model to analyse the electrical energy intensity variation with respect to the sales area of UK retail buildings. The study predicted the efficacy of using simple models when the modeller has a firm idea of the influential variables of the system. The authors separated a sample of 2,570 UK food retail buildings based on the sales areas: convenience stores, supermarkets, superstores and hypermarkets. Power law models were employed for each type of store, observing that the electrical energy intensity reduces with increasing sales area, with the relative rate of energy intensity reduction decreasing as the store sales area increases above 2000 m^2. The standard deviations in the models were found to be higher for small stores (22 to 24%) compared to larger stores (15%). A general relationship was also developed, which showed that the potential for energy savings is higher for small stores, especially if the refrigeration energy consumption is reduced to the mean energy intensity for each category.

Amundson *et al.* [49] outlined the major steps for developing linear regression models for monitoring and reporting energy savings. The stepwise procedure is illustrated in Figure 2. The authors employed this procedure to develop a model characterising the effects of time resolution on the model coefficients for a large food processing plant and a 'high-tech' factory. The independent variables were the ambient temperatures, production periods etc., whilst total electrical energy consumption was the dependent variable. The authors assessed the fitness of their models using the R^2, coefficient of variation of the root mean square error (CV-RMSE) and model residuals, and quantified the validity of the models using fractional uncertainties. It was observed that the employment of daily resolution improves the uncertainty levels in the models, but the increase in

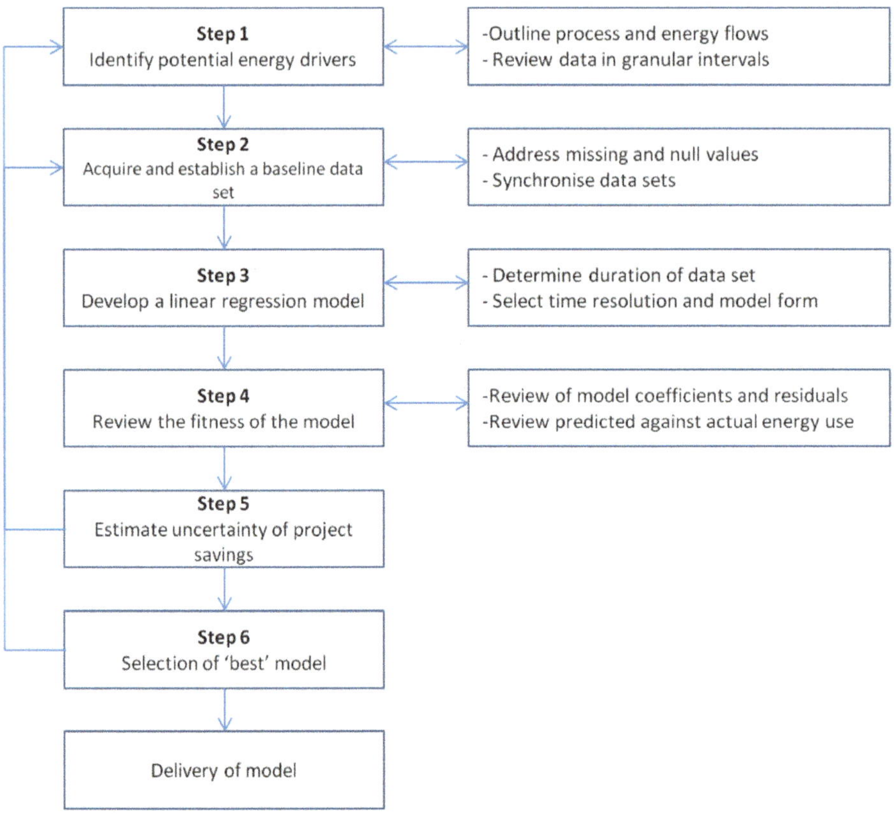

Figure 2 Development process of linear regression model for monitoring and reporting energy savings for industrial maintenance and operation [49].

hardware and software requirements and the additional time and complexity of smaller time resolutions should be justifiable. The selection of a suitable model was found to be subject to the opinions of the modeller and the user reaching a compromise.

Regression models are found to represent a simplistic approach to a complex system, by using a set of influential variables. It is important that these models do not only relate these variables but to also include the effects of causality in the representation. The use of regression models (mainly linear regression models) extends in various disciplines (energy, GHG emissions and costs analyses [50]) because of their relative ease of use and assessment of the model errors. The particularities of this approach are evaluated in the 'Results and discussion' section.

'Top-down' approach

Top-down approaches refer to the decomposition of a scenario into a set of equations where the 'required parameter' is obtained from a combination of the variables considered as 'underlying causes' [51]. The choice of factors depends both on a conceptual model and on data availability. There are two types of equations used in such macroeconometric models: stochastic (or behavioural) and identities. Stochastic equations are estimated from historical data, whilst identities are equations that hold by definition, i.e. they are always true [52]. The variables can be separated into endogenous variables (variables explained in the model) and exogenous (variables imposed on the model) [52], and the model generally applies to an annual basis/time-step. A top-down model can apply to a whole economy or a section of an economy [53]. This section does not attempt to detail the workings of such models, but rather to portray their practicality, benefits and limitations with respect to their use in the food and energy (and/or GHG emissions) chain.

Input-output model

Economic input-output analysis can be regarded as a collection of the aggregated (intermediate and final) value or amount of goods and services that flows in an economic system and/or as an analytical technique describing and predicting the behaviour of that economic system [54]. The data are usually presented in tabular form and can be obtained from national statistical offices on a yearly basis. Generally, governmental input-output (IO) tables are quantified in terms of monetary value and must therefore be adequately converted energy/emissions values [55,56]. This is accomplished by following the principle of conservation of embodied energy to create a hybrid monetary-energy table. This principle states that the energy burnt or dissipated by a sector of the economy is passed on, embodied in the product [55]. Since final demand is considered the output of an economy in economic IO analysis, conservation of embodied energy implies that all energy entering an economy is entirely embodied in the final market sales of goods and services [2]. The energetic IO process separates the economy into energy and non-energy sectors and tracks the flow of energy for each sector. It assumes linear homogeneous production technologies, where the output of a sector varies linearly with the production inputs. As such, the concept of energy intensity is often employed [55].

Canning *et al.* [2] employed the IO analysis of the national US food system to trace the energy flow of roughly 400 industries, using data obtained from two federal sources. The study aimed at understanding the factors influencing the US food-energy system over three time periods (1997, 2002, 2007) using a structural decomposition analysis (SDA) and a supply chain analysis (SCA). The SDA evaluated the effects of changes in US population, food-related budget, product mix and food system technologies, whilst the SCA provided a more in depth analysis of the changes, in terms of the contribution of agriculture, processing, packaging, transportation, retail, food service and household operations to the aggregate change in energy flows. The authors acknowledge that although the boundary of IO analysis is the domestic food system, imported and waste flows are crucial in evaluating the performance of the food chain. Thus, imported embedded energies were obtained by assuming the state of technologies in other countries to be similar US technologies, and waste energy flows were estimated from EPA statistics. The authors observed that energy-intensive technologies accounted for half of the food-related energy increase over the years, whilst the rest was due to increase in population, prepared food and eating out. Household operations accounted for the highest energy use, whilst food processing showed the largest increase from 2002 to 2007, as both households and foods service sectors outsourced manual food preparation and cleanup activities to the manufacturing sector, which relied on energy-intensive technologies.

Zhang *et al.* [57] employed multi-regional input-output model to track the embodied energy for various sectors in China in 2007. A wide range of data were obtained from the Chinese Academy of Science and the National Bureau of Statistics of China, where the linkage between each region and the economy was made by considering the direct primary energy use and imported embodied energies. Amongst the sectors considered, the food production and processing industry accounted for 4.5% of the total Chinese embodied energy use. The study not only showed the potential of IO analyses in terms of flexibility and the number of variables that can be incorporated into the model, but also depicted the statistical knowledge requirements to develop such complex models. Bekhet and Abdullah [58] studied the agricultural energy chain in Malaysia, in an attempt to reduce food imports, minimise energy consumption and increase the yield of the agricultural industry. The authors employed secondary data from National Statistical Databases, considering three energy industries. The study showed a more significant increase in the dependence of the agricultural industry on petrol and coal, compared to the other energy industries, although agriculture is a relatively weak energy consumer. The fisheries sector was found to be the largest consumer of energy, followed by forestry and logging and oil palm estates. The authors acknowledge that because IO data are normally obtained at 5-year intervals, the study fails to identify the changes in energy consumption using a time-series approach, but nonetheless suggest that electricity and gas should be promoted, instead of petrol.

Although the concept of IO analysis was developed for national economic systems, the principles have been extended to specific products. Essengun *et al.* [59,60] explored the energy flow of dry apricot and tomato production in Turkey, by collecting primary energy input, quantities and costs of inputs and outputs. The authors employed various multipliers to construct a relatively simple model that allows the modeller to track the energy and monetary flows in the agricultural production. Such model determined the energy efficiency and intensity of the production and suggested different energy improvement measures. Kuswardhani *et al.* [61] studied the energy and economic IO of greenhouse and open-field production in Indonesia. The authors obtained primary data from surveys and used appropriate conversion factors to obtain the energy values. The study identified the linkages between energy input and crop yield for greenhouse and open-field production, and depicted the energy efficiency ratio of different products. A number of other studies also employed simple adaptation of IO analysis for the agricultural sector [62-68]. The primary motivation for choosing the product-specific IO method was the flexibility in collecting primary data, determining energy efficiency ratios and extending the analysis to cost-benefit analyses.

This section has shown that food-energy IO analyses have not only been used at the national or regional level, but also at a more product-specific level. The approaches to IO energy analyses have been to convert national IO tables' monetary values to energy values (using hybrid tables) or to directly employ energy values (or closely related energy parameters). When using the national hybrid monetary-energy IO models, structural decomposition analysis (SDA) has been used to study the sources of changes obtained from IO analysis over different time periods [2,69-72], whilst supply chain analysis (SCA) explores the contributions of different stages of food production to the overall energy flow [2]. The analysis of embodied (i.e. direct and indirect) energy flows has been found to identify and optimise low energy-efficient processes and to propose improvements at both the energy and cost/profitability levels.

Index decomposition analysis models

An index decomposition analysis model refers to the definition of a governing function that relates the aggregate to a pre-defined number of decomposed factors, in order to measure the impact or weight of these factors on the aggregate, over specific time periods. The two most popular approaches for energy analyses are based on the Divisia index and the Laspeyres index where the choice is problem dependent [73]. The Laspeyres index measures changes in an aspect over time by letting the related variables change, but fixing all other variables at the base period values [74]; the Divisia index uses a weighted sum of logarithmic growth rates, where the weights are the components' share in the aggregated value [73].

Hammond and Norman [75] examined the causes and weights of the effects that resulted in the reduction in carbon emissions in the UK manufacturing sector, between 1990 and 2007. They adopted the log mean Divisia index (LMDI) approach, on the basis that LMDI has no residuals and the additive method is easier to interpret. Secondary data were obtained from various sources, and the changes in carbon emissions were decomposed in terms of changes in production volume, inter-sector structure, secondary energy intensity, fuel mix and carbon emission factor. Furthermore, in order to ensure the changes are associated with the correct effects, the manufacturing sector was divided into subsectors, which include the manufacture of food and beverages. The individual energy efficiency was taken to be inversely proportional to the energy-intensity effects, which does not imply that improvements in the energy-intensity effects are due to technological effects, unless the energy system is disaggregated to high enough level [76]. The results showed that improvements in energy intensity were the main cause of reduction in carbon emissions in UK manufacturing - whereby more efficient technology, better control and housekeeping, moving towards an increased use of electricity and natural gas, and inter-sector structural change are believed to have contributed to lower energy consumption during the period.

Hasanbeigi *et al.* [77] decomposed the Chinese manufacturing industry using the additive LMDI approach, for both past (1995 to 2010) and future years (2010 to 2020). The authors assumed a constant share of the value-added for each sector, and the future energy projections assumed to be cumulatively decreasing based on the government's reduction targets. The decomposition analysis investigated the effects of aggregate activity, sectoral structure and energy intensity. The additive LMDI approach was selected as it leaves no residuals, and the study showed that the food and beverage sector had the second largest sector rise in value-added from 1995 to 2010, whilst primary energy use remained relatively constant and the energy intensity decreased for the same period. The forecasted energy intensity for the food sector was found to decrease at a lower rate for the period of 2010 to 2020. The decomposition analysis showed that if China intends to meet its goal of re-structuring its economy and moving towards less energy-intensive and polluting sectors, then specific scenarios should be followed. The forecasted decomposition analysis therefore aided evaluation of the impact of the different scenarios on both primary energy use and value-added, at a more disaggregated level.

Generally, the log mean Divisia index model was found to be the most commonly used method [73] for economy, industry or sector level analyses, where the main effects are often separated into production volume, inter-sector structure, energy intensity, fuel mix and/or carbon emission factor. A finer disaggregation level of the dataset allows a more accurate evaluation of the relative impacts of the disaggregated effects, whereby the IEA identified the disaggregated energy indicator hierarchy as shown Figure 3. Furthermore, physical intensities (material, energy and emissions) are suitable, but not preferred for holistic industry decompositions, because of the large varieties in physical units involved. Instead, it is recommended that the energy and GHG intensities be based on monetary units [78].

In general, the Divisia index model was used in decomposition studies ranging from specific industries, such as the manufacturing industry - which includes food manufacturing [80-83], the agricultural industry [84], the service industry [85], to whole economy decompositions [86-90]. The Laspeyres index approach was less popular, employed mainly by the International Energy Agency (IEA), due of its ease of use and to ensure consistency in IEA publications (ETO/ESD/LTO, 2007; [79]). The reasons for the lower popularity of the Laspeyres approach were as follows: the model produces residuals when the aggregate values are calculated [74], the choice between the additive or multiplicative Laspeyres model affects the final results of the analysis [91] and the generally lower accuracy of the model compared to the Divisia index method [73].

Dynamic models

Dynamic models incorporate time into the breakdown of a structural framework using systems of difference or differential equations, in an attempt to provide future expectations/impacts of various policies [92]. These models are more complex and detailed than static models, as they involve assumptions on the rate of economic growth, time preferences, population growth rates, inflation rates, depreciation rates etc. [92,93]. Researchers to date have adopted dynamic models in different ways and for different applications, but application to the food-energy chain has been limited.

Irz *et al.* [17] studied the dynamics of price formation of food commodities with respect to agricultural, energy and labour commodity prices, for the case of Finland, by considering attributed relating to the demand and supply sides of the Finnish economy. The

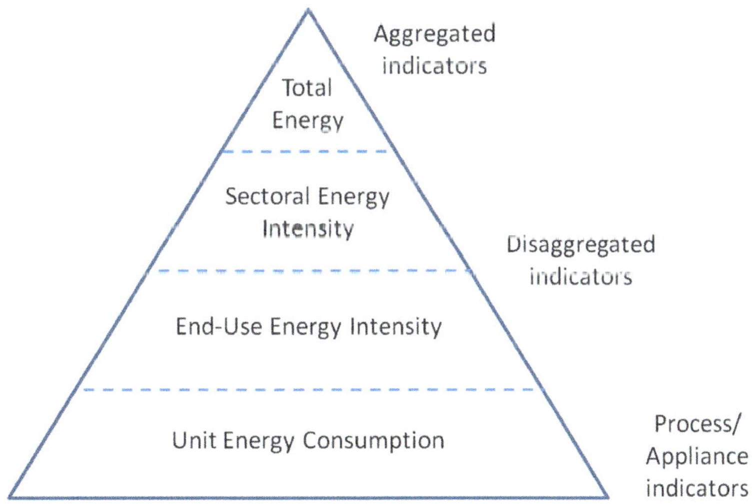

Figure 3 Disaggregated energy indicator hierarchy (adapted from [79]).

authors employed a co-integration analysis using both stochastic and identity sets of equations. The conceptual framework was based on the equilibrium of agricultural, labour and energy commodities for the supply side and disposable income and demographic distribution for the demand side. However, Finland was assumed to have a negligible demographic distribution and disposable income was ignored, due to the unavailability of data. Hence, the empirical model only explained the relationship between food prices and agricultural, energy and labour commodity. The authors studied the system dynamics using a time-series approach by testing for the presence of unit roots and using the Johansen approach to study long-term equilibrium. The causality aspect of the model was analysed using Granger causality tests. The outcome of the model was a long-run relationship of food prices with respect to the other commodity prices, assuming linear growth. The study showed that farm prices represent the main determinant of food prices, followed by wages in food retail and the price of energy. It should however be noted that technological change is implicitly proxied in the time-trend series. Other studies with similar conclusions can be found in Lambert and Miljkovic [18] and Baek and Koo [94].

MAgPIE is a nonlinear recursive dynamic optimisation model, developed by Potsdam Institute for Climate Impact Research Land-use group (PIK [95]). The model requires a set of exogenous demand parameters for each time-step and the yield growth from agriculture is obtained in relation to investment forecasts. The model predicts the impacts of agriculture on land use and GHG emissions. Applications of this model can be found in Dietrich *et al.* [96], Schmitz *et al.* [97] or Popp *et al.* [98].

Hence, dynamic models can be used to study the transient impacts of specific parameters on the economy, but require validation with historical data. The lack of application of dynamic top-down models related to the food-energy chain can be attributed to the fact that the dynamic behaviour and different timeframes (between production and consumption and waste) associated with food is difficult to be forecasted on an aggregated basis due to the large variance in types of food products and technological processes. Furthermore, the level of assumptions that may be required to decompose aggregated energy data for specific food sectors may themselves become a source of question, especially when the temporal aspect is considered. Hence, for such forecasting purposes, hybrid models (such as IO-based LCI or MARKAL-MACRO models) may be more suitable as they allow a higher disaggregation level in terms of technologies and/or food products. Thus, the following section explores the use of hybrid approaches.

Hybrid approach

A hybrid modelling approach seeks to combine both bottom-up and top-down approaches to allow the modeller to analyse specific details of processes and consider the entire supply chain simultaneously [99]. In the case of the food chain, hybrid IO and LCA were found to be most commonly used in assessing energy and carbon footprints. IO tables provide complete and aggregated data within national boundaries, whilst process-based LCI provide detailed and accurate process information. Suh and Huppes [100] identified three approaches to hybrid IO-LCI methods:

1. Tiered hybrid analysis employs process-based LCI data for consumption, waste and upstream activities, whilst the remaining information is obtained from the economic IO-based LCI. Although a simple and fast approach, limitations are (i) the demarcation between process-based and IO-based LCI should be carefully selected, (ii) double counting may occur and (iii) there is no feedback between the two approaches.
2. IO-based hybrid analysis requires the disaggregation of the industry sectors in the IO table into subsectors and employs the tiered hybrid method for product life cycles outside the IO table boundary. The interactive relationship between pre-consumer stages and the rest of the product life cycles is often difficult to model.
3. Integrated hybrid analysis: is a matrix representation of the physical product system, whereby the IO table is connected upstream and downstream of the matrix. The linkages between the product system and the economy can be obtained in terms of purchases and sales records from LCI. This approach puts LCI at the core of the hybrid model and allows full interactions between the individual processes and industries in a consistent framework.

Elgimez *et al.* [101] employed an IO-based LCI hybrid approach to study the sustainability impact of the US food manufacturing sector. The study consisted of disaggregating the industry into 33 sub-sectors (or product systems), which allowed the consideration of both the direct and indirect effects at a more detailed level than using IO only. In addition to the IO-LCI model, this study also included a data envelopment analysis (DEA) model which evaluated the impact of the sectors through a sustainability performance index (SPI), to allow comparative study. The authors therefore adopted a two-stage hierarchical process where the IO-LCI model provided the environmental outputs, which are then fed to the DEA model to evaluate the SPI of each sub-sector. The authors obtained the imported footprints assuming the same level of technology as the US food manufacturing industry and ignored the regional variations associated with carbon emissions due to the regional homogeneity obtained in IO tables. The study concluded that 19 of the 33 food manufacturing sub-sectors are inefficient and that fisheries and animal slaughtering, rendering and processing are the dominant 'carbon footprint' sectors in the US food manufacturing industry.

Virtanen *et al.* [102] studied the GHG emissions of the Finnish food-chain using process-based LCA and an IO-based LCA hybrid model. The process-based LCA was obtained from 30 typical lunch plates, used to augment the national IO tables. The hybrid IO-LCA model was derived from a combination of economic input-output tables associated with environmental emissions, LCI of agricultural sub-sectors and related emissions and publicly available LCI and conversion factors for imports (ignoring waste and disposal). Particular emphasis was placed on the agricultural industry because of its relatively high emissions. As a result, this industry was further disaggregated to 44 sub-sectors, with emissions obtained from the ENVIMAT model. The authors observed from the IO-LCI model that the Finnish food chain accounts for 14% of the Finnish GHG emissions, with agriculture accounting for 69% and the energy industry 12%.

Weber and Matthews [23] studied the effects of food miles on the US environmental impact of food. An IO-based hybrid model was employed, where the disaggregation of the transportation sector was done with life cycle information from secondary sources. A commodity-based functional unit of ton-km was employed, where the model assumed that all users of a commodity require the same amount of ton-km per dollar purchase of a commodity, where energy use and carbon emissions were obtained by assuming standard fuel conversion factors. The authors identified limitations of grouping different goods and time lag of data in the IO-LCA model which increase uncertainties in the analysis. Nonetheless, the study depicted that holistic transportation contributes an average of 11% for life cycle GHG emissions and that red meat is more energy-intensive than other food products. The study concluded that dietary shift is a more energy efficient means than 'buying local' in the US.

Wood *et al.* [103] employed an IO-based LCA hybrid approach to study the comparative energy, water, land and GHG emissions impacts of organic farming and conventional farming in the Australian food chain. The authors conducted a survey of organic farmers to obtain primary data for a process-based LCA for organic farming, whilst conventional farming data were deduced from IO tables. Farm operations were obtained from LCAs and the remaining indirect effects from IO tables. The authors used the SCA decomposition of the hybrid IO-LCA table to obtain the various energy and emissions contributions of the agricultural industry, assuming homogeneous organic farm operations across the industry. The results showed that total embodied energy is generally lower for organic produce, compared to conventional produce, and synthetic chemicals and fertilisers are a major source of energy use, where organic agriculture would reduce these impacts.

In the context of the food and energy/emission chain, the use of hybrid IO tables and LCA models has been found to focus on emissions, rather than energy, because of the relatively higher importance placed by environmental policies on emissions. As such, most hybrid IO-LCA models generally include the details from LCI mainly to the agricultural and manufacturing sections of the food chain, thereby providing more details for these sectors where a majority of the emissions take place in the food chain [101-103]. These models have been used as they allow further disaggregation of the conventional economic IO models (as well as including up- and down-stream analyses, beyond the IO model boundary) and incorporate the economy-wide effects to the segregated LCA models. The level of analysis is therefore enhanced, allowing for the detection of environmental hot-spots and better implementation of environmental policies.

Results and discussions

This paper describes different modelling approaches used in the food chain. These approaches are differentiated as bottom-up, top-down and hybrids, which are qualitatively evaluated as follows, in relation to their applicability to modelling the energy and emissions flow in the food chain.

LCA models provide a process or product based analysis of emissions and energy use in the food-chain using a cradle-to-grave approach. Owing to the complexity of the food chain and the high level of detail possible from LCA, efficiency hot-spots can be identified, therefore targeting actions for improvements [29]. Whilst LCA is good at identifying the intricacies and complex nature of the food chain, this very complexity presents an obstacle to the development of specific recommendations for the future [24]. The level of details required limits the accuracy of this approach and increases the associated uncertainties when such data are not available. Furthermore, it is important to have a standard procedure through which life cycle impacts are measured, with the system boundaries, assumptions, uncertainties and the definition of the functional units of the different studies clearly specified [31,104]. LCAs need to include ways of measuring outputs that are not only multiple but also intangible - such as social aspects of the chain [24] - and to consider the integration of individualistic models as components to holistic models. Nonetheless, current academic and grey literatures, although performed on a relatively random basis, can provide qualitative assessments of energy and GHG emissions, as well as opportunities for improvements in specific cases [27].

As opposed to LCA models, MARKAL and its sub-models are mainly employed at the national and regional levels, to analyse the implications of different national technology mix ([105]; Teri [106]). The national models and the disaggregated impacts of a particular sector are dependent on the level of detail obtained from the country's MARKAL databases. There are uncertainties when analysing future energy scenarios with regard to the discount rates. The two theoretical concepts of social time preference and social opportunity cost tend to show a divergence in the choice of discount rates between the values adopted by the private sector and the government (Rath-Nagel and Stocks [39]). Hence, different discount rates are often used for different situations and countries (3% for Switzerland, 5 to 10% for the USA, 10% for China or 10% for the UK) [105,107]. A high discount rate value shows high uncertainty in the energy impacts of a new technology [105].

In most cases, except when the main variables are definite, the development of regression models is an iterative process, requiring a large amount of data and an understanding of the engineering and statistical processes involved. The underlying assumption of linear regression models is that the residuals (difference between predicted and actual data) follow a normal distribution from the mean. This helps to identify outliers (which distort the regression model), model weaknesses and process changes [49]. As such, the judgement of the modeller is crucial in determining the 'best' representative model; e.g. Spyrou *et al.* [46] considered outliers to be at three standard deviations. Further to the residuals, the fitness and model coefficients are crucial. The use of various indicators such as R^2, variations of root mean square errors, p-value, auto-correlation coefficients, standardised (β) coefficient and fractional uncertainties have been found to be common. Although there are limitations associated with the 'error indicators' [49,108], the final adoption of a regression model will depend on a compromise between the judgement of the modeller and the satisfaction of the user.

IO models are static linear equilibrium models, where the national IO system boundary for the food-energy chain starts from energy production up to final consumption of food, hence requiring simplifying assumptions to consider the impacts of imports and wastes in the chain. Economic IO tables are usually several years old and in some cases at large time intervals and are mainly used to analyse past energy flows as opposed to predicting these flows. They are therefore not convenient for time-series-type analyses. The results obtained from economic IO analyses are usually exhaustive and applicable to large-scale questions, which are less useful for micro questions [55]. Nonetheless, such an aggregated method usually performed using data from single sources, reduce ambiguities related to acquiring data from different sources. It can also account for the aggregated inter-regional and socio-economic aspects of the chain [109]. When used in the simpler product-specific manner, the efficiency of IO models can be improved when primary data are collected. Furthermore in such studies, the conversion of monetary values to energy values is often not required, eliminating the assumption of sector homogeneity, as energy values are not determined from homogeneous energy prices.

The unique feature of index decomposition analysis (IDA) to provide macro results based on myriad detailed energy indicators gives policymakers quick access to findings from technical data [77]. In IDA analyses, there are issues of data quality, level of sector disaggregation, measurement of output/activity levels and the choice of indicators which would affect the quality and validity of the decomposition results [73] - these are however only dependent on the datasets being used and independent of the actual methodology. Only direct effects can be evaluated with IDA models [110], as opposed to structural decomposition analysis (SDA) that can evaluate embodied effects. An advantage of IDA over SDA is the lower data requirement. However, this is also a disadvantage, since IDA is capable of less detailed decompositions of the economic structure [75,110].

The comparative aspects of the aforementioned models are presented in Table 1. This comparison has been done in a qualitative manner with respect to the ease of use, benefits, limitations and the assumptions of the models to provide information in the choice of a specific modelling method. Table 1 generally shows that the benefits of one approach are the limitations of the other approach, for instance; bottom-up approaches provide the benefits of high level of detail for a specific product/ sector, whilst top-down approaches have limitations in the disaggregation of the economy to a sector/ product level. This therefore led to the development of the commonly used hybrid IO-LCI models.

Such hybrids allow the following: further disaggregation of the economic IO model, incorporate the up- and downstream analyses and include economy-wide effects to segregated LCA models. The level of analysis is therefore enhanced, allowing for the detection of environmental hot spots and better implementation of environmental policies. Hybrid IO-LCA analyses can pose problems with regard to temporal discrepancies. IO tables are usually published at typically long time intervals (1 to 5 years), making the information in the energy/emission IO tables usually older than a process-based LCA [100] and creating complexities in the clear demarcation of IO and process-

Table 1 Qualitative evaluation of 'bottom-up' and 'top-down' modelling approaches referred in this paper

Models	Brief description	Common benefits	Common limitations	Common assumptions
Bottom-up approaches				
LCA models	A process or product based evaluation method of the energy use and GHG emissions, typically using a 'cradle to grave' approach	- Measure high quality of energy and GHG emissions data	Level of detail required increases the complexity of data collection	- Secondary to primary energy conversion factors
		- Capture the intricacies and complex nature of the food-energy chain	- Analysis is usually very specific to country and product/process	- Transportation distances and fuels
		- Allow the identification of energy and GHG emission hot-spots	- Information from different sources cannot be combined, unless uncertainties and assumptions are clearly specified	- Imported energies and GHG emissions
			- Choice of functional units and demarcations between system add complexity to the method	- Agricultural sector energies and GHG emissions
			- Ignores the holistic industry impact on the product/process, i.e. static model	- Waste disposal and storage emissions
			- LCA needs constant update	
MARKAL and sub-models	- Demand-driven multi-period linear programming and cost optimisation tools	- Technologies and processes can be explicitly modelled in detail	- Results depend on the accuracy of demand-inputs and description of technological processes	- Description of technological processes
	- Simultaneously assess the impact of several technologies through the partial equilibrium of demand and supply of energy	- Currently being employed in various energy research studies, and various extensions/innovations of the model are being developed	- Discount rates of technologies impact the partial equilibrium when forecasting energy demands	- Discount rates and future energy demands
		- Allows the explicit modelling of the evolution of demand and energy prices	- Does not have information feedback to the wider economy	
			- MARKAL databases have to be regularly updated	
Regression models	Find the causal relationship between the dependent and independent variables in the food-energy chain	- Relatively easy to use and construct	- Require large amount of data to ensure proper correlations	- Determination of errors, and validity of models are subjective to modeller
		- Provide a simplistic description of problem, and allows quick approximations of different policies	- Provide only a quantitative evaluations, and does not show the factors causing the inefficiencies (depends on the level of disaggregation of the model)	- General assumptions relate to the energy consumption and GHG emissions in the data collection phase of the study
			- Assumptions are implicit in the model, hence constant updates required	- Data points outside three standard deviations require further investigation

Table 1 Qualitative evaluation of 'bottom-up' and 'top-down' modelling approaches referred in this paper *(Continued)*

Top-down approaches				
Economic Input-output (IO) models	Provide the aggregated monetary/ energy flow through an economy	- Analyse the impact of the entire economy on each industry, and inter-industry relationships	- Stop at the point of purchase, and ignore waste and imports	- Employ the principle of embodied energy to convert monetary to energetic values
		- Data are usually obtained from same sources, which provides consistency in analyses	- Aggregated data analysis prevents detection of specific energy/ environmental hot-spots	- Economic IO analysis requires energy/GHG assumptions for wastes and imports
		- Quantify the impact/weight of each sector, and therefore allows identification of low-performing sector	Frequency of national IO tables is low	- Linear production technologies [2]
				- No capacity constraints [111]
				- Sector homogeneity [111]
				Usually assume a constant level of technologies for future analyses [2,111]
				- Import emissions usually based on domestic production technologies [2]
Index decomposition analysis models (IDA)	Decompose aggregate energy and GHG emissions data into pre-defined factors to measure the relative impacts over specific time periods has been used for: (i) energy demand and supply, (ii) energy-related gas emissions, (iii) material flows and dematerialisation, (iv) national energy efficiency trend monitoring and (v) cross-country comparisons [70]	- Method is relatively quick and simple to implement	- Require an adequate level of disaggregation, else actual effects are not clearly identified	Energy and GHG intensities are usually based on monetary outputs, as opposed to physical outputs
		- LMDI approach has no residuals in decomposition process	- Laspeyres index is simple to implement, but calculates with residuals	
		Provide quick access to assessing the overall impact of policy measures on the economy		
Dynamic models	Aim at predicting future energy and GHG expectations of the food-energy chain	Can provide indication of future energy and policy expectations	- Technological effects are often implicitly accounted in models	- Economic growth rates
			- Can require significant level of assumptions, which questions the validity of such models	- Time preferences
			- Studies related to the food chain are scarce	- Population growth rates
				Inflation and depreciation rates

based LCA models. From the various literatures reviewed in this paper, it is apparent that the IO-based hybrid approach has been most popular for the food chain, due to the easy availability of national IO tables and extensive amount of LCA studies. However, it is not that one hybrid model outperforms another. The choice of a particular approach is dependent on a variety of criteria such as the following: data requirements, uncertainty of source data, upstream system boundary, technological system boundary, geographical system boundary, available analytical tools, time and labour intensity, simplicity of application, required computational tools and goal and scope of the model [100]. Generally, the use of hybrid IO-LCI models has been found to focus on emissions due to the high importance placed by environmental policies. As such, most hybrid IO-LCA models focus on the agricultural and manufacturing sectors where most emissions take place [101-103]. Other hybrid models such as the MARKAL/TIMES-MACRO model from the IEA exist and are worth examining in the context of the food chain.

Conclusion

This paper presents a review of modelling approaches for the energy and GHG emissions in the food chain. These methods can be classified as follows: bottom-up, top-down and hybrid approaches. The impetus for this study stems from the need to accurately model the holistic food-energy chain, in order to effectively design and implement policies to tackle the food-energy-climate nexus and food security issues. Top-down approaches have been found to consider the impact of the economy on the food-chain and be used to develop national policy measures. However, the limited level of disaggregation due to unavailability of data and the homogenisation of the economy when using top-down models are drawbacks, which do not help in the identification of energy/environmental hot-spots. On the other hand, bottom-up approaches generally provide a high level of detail and capture the intricacies of the food-energy system. However, their specificity to products/processes limits their application to holistic systems if the individual bottom-up models do not follow a standardised procedure. Hence, the predominant trend has been towards hybrid models, which seek to combine the advantages of both bottom-up and top-down approaches. Furthermore, although three hybrid modelling approaches (tiered hybrid, IO-based hybrid, integrated hybrid) have been identified, the IO-based hybrid has been more commonly used in the food-chain.

The choice of one modelling approach over another depends on a variety of criteria including data requirements, uncertainty, technological systems to be modelled, available analytical tools, time and labour intensity amongst others. It should also be noted that the same modelling approach may lead to different results depending on the assumptions made. As a result, the method (i.e. the mathematics and economics), assumptions and limitations of a particular model should be clearly stated in every study. A simple and quick model (such as regression models) may be useful in obtaining a rough indication of policy impacts for specific cases, but when used for the holistic food-energy chain may increase errors and uncertainties due to the complexity of the food chain.

The modelling of the agriculture and waste parts of the food chain was found to involve relatively more assumptions than the other sectors. This is particularly the case for modelling the GHG emissions because of the need to account for biological processes in agriculture and waste management, based on various conversion factors which increase the probability of inaccuracies in the model. Generally, these inaccuracies can be quantified in various forms, such as R^2 in regression models or relative-percentages in other models.

Most modelling approaches and studies to date consider both the energy and GHG emission aspects of the food-chain. However, in the majority of cases, the emphasis has been on the estimation of GHG emissions and their impact on climate change. Although GHG emissions and energy impacts are complementary in some sections of the food chain, their holistic interdependence is not uniform. Hence, the energy part of the nexus is equally important, especially as the food chain becomes more complex, food security becomes more prominent, and food and fossil fuels decouple as more renewable energy systems are implemented in the chain. This study serves as a background to the application of holistic and integrated approaches to modelling the energy and GHG emissions of the food chain. Such approaches have the potential to better represent energy technologies, can integrate different modelling methodologies and can incorporate social, demographic, economic and climate considerations in a holistic context to predict both short- and long-term impacts. Although the application of dynamic models to the food chain was found to be scare, the importance of considering the temporal impact of policy is crucial and requires further research.

Abbreviations

CHP: combined heat and power systems; CLEWS: climate, energy, water and land use strategies; DECC: UK Department of Energy and Climate Change; DEFRA: UK Department of Environment, Food and Rural Affairs; EC: European Commission; EPA: United States Environmental Protection Agency; EPI: energy performance indicator; F&D: food and drink industry; FAO: Food and Agriculture Organisation of the United Nations; GHG: greenhouse gas; IDA: index decomposition analysis; IEA: International Energy Agency; IO: input-output; ISO: International Organization for Standardization; LCA: life cycle assessment; LCIs: life cycle inventories; LMDI: log mean Divisia index; MAgPIE: model of agricultural production and its impact on the environment; MARKAL: market allocation model; PAS: publicly available specification; SCA: supply chain analysis; SDA: structural decomposition analysis; SIC: standard industry classification; UNEP: United Nations Environment Programme; WFP: World Food Programme.

Competing interests

The authors declare that they have no competing interests.

Authors' contributions

Dr. BLG carried out the review of different modelling techniques presented in this paper, structured and drafted the manuscript. Prof. SAT was involved in finalising the format and content of the manuscript, as well as putting the paper into the overall context of the food chain. Both authors read and approved the final manuscript.

Acknowledgements

This study is a result of funding from the Research Councils UK to set up the RCUK National Centre for Sustainable Energy Use in Food Chains (CSEF), grant no. EP/K011820/1. We would like to acknowledge the contributions of the EPSRC, ESRC and 'Manufacturing the Future' and those of the industry and academic partners in CSEF.

References

1. Food and Agriculture Organisation (FAO) (2011) Energy-smart food for people and climate. Issue paper, © FAO 20112
2. Canning P, Charles A, Huang S, Polenske KR, Waters A (2010) Energy use in the US food system. United States Department of Agriculture Economic research report No. 94
3. DEFRA (2013) Food Statistics Pocketbook, 2013, 71 pgs. https://www.gov.uk/government/statistics/food-statistics-pocketbook-2013. Accessed on 12 June 2014
4. Lobell DB, Schlenker W, Costa-Roberts J (2011) Climate trends and global crop production since 1980. Science 333(6042):616–620
5. Roberts MJ, Schlenker W, Eyer J (2011) Agronomic weather measures in econometric models of crop yield with implications for climate change. Am J Agric Econ 95(2):236–436, United Nations Environment Programme (UNEP) (2011) Towards a green economy: pathways to sustainable development and poverty eradication - a synthesis for policy makers. United Nations Environment Programme www.unep.org/greeneconomy
6. United Nations Environment Programme (UNEP) (2011) Towards a green economy: pathways to sustainable development and poverty eradication - a synthesis for policy makers. United Nations Environment Programme www.unep.org/greeneconomy
7. Clinton Global Initiative (CGI) (2012) http://www.clintonglobalinitiative.org/aboutus/tracks.asp, Accessed on 02 January 2014
8. Grace Communications Foundation (2013) http://gracelinks.org/1309/renewable-energy-at-the-nexus. Accessed on 29 Nov 2013
9. Mistry P, Misselbrook T (2005) Assessment of methane management and recovery options for livestock manures and slurries. DEFRA report: AC0402
10. Krozer Y (2013) Cost and benefit of renewable energy yin the European Union. Renew Energy 50:68–73
11. Parida B, Iniyan S, Goic R (2011) A review of photovoltaic technologies. Renew Sust Energ Rev 15(3):1625–1636
12. International Energy Agency - IEA (2012) World Energy Outlook 2012: Renewable energy outlook. http://www.worldenergyoutlook.org/media/weowebsite/2012/WEO2012_Renewables.pdf. Accessed on 29 Nov 2013
13. World Food Programme (WFP) (2013) http://www.greeningtheblue.org/what-the-un-is-doing/world-food-programme-wfp, Accessed on 29 November 2013
14. Maggio G, Cacciola G (2012) When will oil, natural gas and coal peak? Fuel 98:111–123
15. Kim GR (2010) Analysis of global food market and food-energy price links - based on systems dynamics approach, Hankuk Academy of Foreign Studies, South Korea, Scribd, 18 pages. http://www.systemdynamics.org/conferences/2009/proceed/papers/P1332.pdf, Accessed on 5 December 2013
16. Heinberg R, Bomford M (2009) The food and farming transition - towards a post-carbon food system, Post Carbon Institute, Sebastopol, California. 39 pages. http://www.postcarbon.org/publications/food-and-farming-transition/, Accessed on 21 December 2013
17. Irz X, Niemi J, Liu X (2013) Determinants of food price inflation in Finland - the role of energy. Energy Policy 63:656–663
18. Lambert DK, Miljkovic D (2010) The sources of variability in US food prices. J Policy Model 32:210–222
19. Ciaian P, Kancs D'A (2011) Interdependencies in the energy-bioenergy-food price systems: a cointegration analysis. Resour Energy Econ 33(1):326–348
20. Seck GS, Guerassimoff G, Maizi N (2013) Heat recovery with heat pumps in non-energy intensive industry: a detailed bottom-up model analysis in the French food and drink industry. Appl Energy 111:489–504
21. Welsch M, Hermann S, Howells M, Rogner HH, Young C, Ramma I, Bazilian M, Fischer G, Alfstad T, Gielen D, Le Blanc D, Röhrl A, Steduto P, Müller A (2014) Adding value with CLEWS - modelling the energy system and its interdependencies for Mauritius. Appl Energy 113:1434–1445
22. Tukker A, Jansen B (2006) Environment impacts of products - a detailed review of studies. J Ind Ecol 10(3):159–182
23. Weber CL, Matthews HS (2008) Food-miles and the relative climate impacts of food choices in the United States. Environ Sci Technol 42(10):3508–3513
24. Garnett T (2013) Three perspectives on sustainable food security: efficiency, demand restraint, food system transformation. What role for LCA? Journal of Cleaner Production. In Press, doi:10.1016/j.jclepro.2013.07.045
25. Bazilian M, Rogner H, Howells M, Hermann S, Arent D, Gielen D, Steduto P, Mueller A, Komor P, Tol RSJ, Yumkella KK (2011) Considering the energy, water and food nexus: towards an integrated modelling approach. Energy Policy 39:7896–7906
26. Ingram JSI, Wright HL, Foster L, Aldred T, Barling D, Benton TG, Berryman PM, Bestwick CS, Bows-Larkin A, Brocklehurst TF, Buttriss J, Casey J, Collins H, Crossley DS, Dolan CS, Dowler E, Edwards R, Finney KJ, Fitzpatrick JL, Fowler M, Garrett DA, Godfrey JE, Godley A, Griffiths W, Houlston EJ, Kaiser MJ, Kennard R, Know JW, Kuyk A, Linter BR et al (2013) Priority research questions for the UK food system. Food Sec 5:617–636
27. Garnett T (2011) Where are the best opportunities for reducing greenhouse gas emissions in the food system (including the food chain)? Food Policy 36(Sup 1):S23–S32
28. Dron D (2012) Sustainable food, a component of the green economy. In: Corson MS, van der Werf HMG (eds) Proceedings of the 8th International Conference on Life Cycle Assessment in the Agri-Food Sector (LCA Food 2012), 1–4 October 2012, Saint Malo. France. INRA, Rennes, pp 1–7
29. Mistry P, Cadman J, Miller S, Ogilvie S, Pugh M (2007) Resource use efficiency in food chains: priorities for water, energy and waste opportunities., An AEA Energy and Environment Report to Defra No. AEAT/ENV/R/2457 (ED05226)
30. DEFRA report No. FO 0103 (2008) Comparative Life cycle assessment of food commodities procured for UK consumption through a diversity of supply chains., Report by AEA, Cranfield University, Ed Moorhouse, Paul Watkiss Associates, AHDBMS, Marintek, submitted to DEFRA
31. Fisher K, James K, Sheane R, Nippress J, Allen S, Cherruault JY, Fishwick M, Lillywhite R, Sarrouy C (2013) An initial assessment of the environmental impact of grocery products. Report submitted to the Product Sustainability Forum (PSF), Code: RPD002-004
32. Lillywhite R, Sarrouy C, Davidson J, May D, Plackett C (2013) Energy dependence and food chain security. DEFRA report FO0415
33. PAS 2050 (2008) Guide to PAS 2050: how to assess the carbon footprint of goods and services., © Crown 2008 and Carbon Trust 2008, ISBN 978-0-580-64636-2
34. PAS 2050:2011 (2011) The guide to PAS 2050:2011, how to carbon footprint your products, identify hotspots and reduce emissions in your supply chain, © British Standards Institution 2011, ISBN 978-0-580-77432-4
35. Holmes M, Wiltshire J, Wynn S, Lancaster D (2010) PAS 2050 informing Low carbon procurement: pilot study - food, ADAS report to DEFRA
36. Tassou SA, Hadawey A, Marriott D (2008) Greenhouse gas impacts of food retailing. DEFRA report No. FO405
37. DEFRA report No. FO 0409 (2008) Understanding the GHG impacts of food preparation and consumption in the home., A report by Campden & Chorleywood Food Research Association, Food Process Innovation Unit (Cardiff Business School), Food Refrigeration & Process Engineering Research Center (University of Bristol)
38. Ecoinvent (2013), The ecoinvent database, Available at http://www.ecoinvent.org/database/, Accessed on the 22 January 2014
39. Rath-Nagel S, Stocks K (1982) Energy modelling for technology assessment: the MARKAL approach. Omega Int J Manag Sci 10(5):493–505

40. Loulou R, Goldstein G, Noble K (2004) Documentation for the MARKAL Family of Models, Energy Technology Systems Analysis Programme. http://www.etsap.org/tools.htm
41. Loulou R, Remne U, Kanudia A, Lehtila A, Goldstein G (2005) Documentation for the TIMES Model PART 1, Energy Technology Systems Analysis Programme. http://www.etsap.org/tools.htm
42. Remme U (2012) Capacity building through energy modelling and systems analysis. Presentation slides from the IEA Experts' Group on R&D Priority-Setting and Evaluation Developments in Energy Education: Reducing Boundaries. Copenhagen 9-10 May 2012. Accessible from http://www.iea.org/media/workshops/2012/egrd/Remme.pdf. Accessed on 01 February 2015
43. Gerlagh T and Gielen DJ (1999) MATTER 2.0: A module characterisation for the agriculture and food sector. ECN-C-99-048, http://www.iea-etsap.org/web/reports/c99048s.html, Accessed on 5 December 2013
44. International Energy Agency - IEA (2011) http://www.iea-etsap.org/web/Times.asp. Accessed on 19 October 2014
45. Ramcharan R (2006) Regressions: why are economists obsessed with them? Finance and Development: a quarterly magazine of the IMF, Vol. 43, No.
46. Spyrou MS, Shanks K, Cook MJ, Pitcher J, Lee R (2014) An empirical study of electricity and gas demand drivers in large food retail buildings of a national organisation. Energy Buil 68:172–182
47. Boyd GA (2011) Development of performance-based industrial energy efficiency indicators for food processing plants. Report to the US Environmental Protection Agency ENERGY STAR Programme, obtained from http://www.energystar.gov/buildings/tools-and-resources/read-about-food-processing-plant-epis, 05 December 2013
48. Tassou SA, Ge Y, Hadawey A, Marriott D (2011) Energy consumption and conservation in food retailing. Appl Therm Eng 31(2–3):147–156
49. Amundson T, Brooks S, Eskil J, Martin S, Mulqueen S (2013). Elements of defensible regression-based energy models for monitoring and reporting energy savings in industrial energy efficiency operation and maintenance projects. In proceedings of the 2013 ACEEE Summer Study on Energy Efficiency in Industry, 25 July 2013, Paper no. 192
50. Esmaeili A, Shokoohi Z (2011) Assessing the effect of oil price on world food prices: application of principal component analysis. Energy Policy 39 (2):1022–1025
51. Seibel (2003) Decomposition analysis of carbon dioxide-emission changes in Germany - conceptual framework and empirical results, Federal statistical office of Germany: Environmental economic accounting division. © European Communities, ISBN 92-894-5167-X, 32 pp
52. Fair RC (2004) Estimating how the macroeconomy works. ISBN: 0-674-01546-0, The President and Fellows of Harvard College, Cambridge, Massachusetts, London, England
53. Dagoumas A, Barker T, Scrieciu S, Stretton S (2009) Top-down technological modelling of stabilisation pathways: UKERC scenarios for the UK to 2050 using E3MG-UK. IOP Conference Series. Earth Environ Sci 6:272003, doi:10.1088/1755-1307/6/7/272003
54. Christ CF (1955) A review of input-output analysis. In Input-output analysis: an appraisal. Princeton University Press (ISBN: 0-870-14173-2), pp. 137–182. Available at www.nber.org/chapters/c2866. Accessed on 21 December 2013
55. Bullard CW, Herendeen RA (1975) The energy cost of goods and services. Energy Policy 3(4):268–278
56. Casler S, Wilbur S (1984) Energy input output analysis: a simple guide. Resour Energy 6(2):187–201
57. Zhang B, Chen ZM, Zia XH, Xu XY, Chen YB (2013) The impact of domestic trade on China's regional energy uses: a multi-regional input-output modelling. Energy Policy 63:1169–1181
58. Bekhet HA, Abdullah A (2010) Energy use in agriculture sector: input-output analysis. Int Bus Res 3:3
59. Essengun K, Erdal G, Gunduz O, Erdal H (2007) An economic analysis and energy use in stake-tomato production in Tokat province of Turkey. Renew Energy 32(11):1873–1881
60. Essengun K, Gunduz O, Erdal G (2007) Input-output energy analysis in dry apricot production of Turkey. Energy Convers Manag 48(2):592–598
61. Kuswardhani N, Soni P, Shivakoti GP (2013) Comparative energy input-output and financial analyses of greenhouse and open field vegetables production in West Java, Indonesia. Energy 53:83–92
62. Franzluebbers AJ, Francis CA (1995) Energy output: input ratio of maize and sorghum management systems in eastern Nebraska. Agric Ecosyst Environ 53(3):271–278
63. Pahlavan R, Omid M, Akram A (2012) Energy input-output analysis and application of artificial neural networks for predicting greenhouse basil production. Energy 37(1):171–176
64. Ren LT, Liu ZX, Wei TY, Xie GH (2012) Evaluation of energy input and output of sweet sorghum grown as a bioenergy crop on coastal saline-alkali land. Energy 47(1):166–173
65. Ozkan B, Ceylan RF, Kizilay H (2011) Energy inputs and crop yield relationships in greenhouse winter crop tomato production. Renew Energy 36(11):3217–3221
66. Bojacá CR, Casilimas HA, Gil R, Schrevens E (2012) Extending the input-output energy balance methodology in agriculture through cluster analysis. Energy 47(1):465–470
67. Michos MC, Mamolos AP, Menexes GC, Tsatsarelis CA, Tsirakoglou VM, Kalburtji KL (2012) Energy inputs, outputs and greenhouse gas emissions in organic, integrated and conventional peach orchards. Ecol Indic 13(1):22–28
68. Salehi M, Ebrahimi R, Maleki A, Mobtaker HG (2014) An assessment of energy modelling and input costs for greenhouse button mushroom production in Iran. J Clean Prod 64:377–383
69. Rormose P, Olsen T (2005) Structural decomposition analysis of air emissions in Denmark 1980-2002, 15th International Conference on Input-output Techniques Beijing, China June 27 to July 1, 2005., Accessible from http://iioa.org/conferences/15th/pdf/roermose_olsen.pdf. Accessed on 01 February 2015
70. Wachsmann U, Wood R, Lenzen M, Schaeffer R (2009) Structural decomposition of energy use in Brazil from 1970 to 1996. Appl Energy 86 (4):578–587
71. Nie H, Kemp R (2013) Why did energy intensity fluctuate during 2000-2009? A combination of index decomposition analysis and structural decomposition analysis. Energy Sustain Dev 17:482–488
72. Cao S, Xie G, Zhen L (2010) Total embodied energy requirements and its decomposition in China's agricultural sector. Ecol Econ 69:1396–1404
73. Ang BW (2004) Decomposition analysis for policymaking in energy: which is the preferred method? Energy Policy 32:1131–1139
74. Cahill CJ, Gallachoir BPO (2012) Combining physical and economic output data to analyse energy and CO2 emissions trends in industry. Energy Policy 49:422–429
75. Hammond GP, Norman JB (2012) Decomposition analysis of energy-related carbon emissions from UK manufacturing. Energy 41:220–227
76. Jenne CA, Cattell RK (1983) Structural change and energy efficiency in industry. Energy Econ 5(2):114–123
77. Hasanbeigi A, Price L, Fino-Chen C, Lu H, Ke J (2013) Retrospective and prospective decomposition analysis of Chinese manufacturing energy use and policy implications. Energy Policy 63:562–574
78. Reddy BS, Ray BK (2010) Decomposition of energy consumption and energy intensity in Indian manufacturing industries. Energy Sustain Dev 14(1):35–47
79. Taylor PG, D'Ortigue OL, Francoeur M, Trudeau N (2010) Final energy use in IEA countries: the role of energy efficiency. Energy Policy 38:6463–6474
80. Jeong K, Kim S (2013) LMDI decomposition analysis of greenhouse gas emissions in the Korean manufacturing sector. Energy Policy 62:1245–1253
81. Hasanbeigi A, Rue du Can S, Sathaye J (2012) Analysis and decomposition of the energy intensity of California industries. Energy Policy 46:234–245
82. Salta M, Polatidis H, Haralambopoulos D (2009) Energy use in the Greek manufacturing sector: a methodological framework based on physical indicators with aggregation and decomposition analysis. Energy 34:90–111
83. Bhattacharya SC, Ussanarassamee A (2005) Changes in energy intensities of Thai industry between 1981 and 2000: a decomposition analysis. Energy Policy 33:995–1002
84. Robaina-Alves M, Moutinho V (2014) Decomposition of energy-related GHG emissions in agriculture over 1995-2000 for European countries. Appl Energy 114:949–957
85. Mairet N, Decellas F (2009) Determinants of energy demand in the French service sector: a decomposition analysis. Energy Policy 37(7):2734–2744
86. Mulder P, de Groot HLF (2013) Dutch sectoral energy intensity developments in international perspective, 1987–2005. Energy Policy 52:501–512
87. Md S, Alam K (2013) Changes in energy efficiency in Australia: a decomposition of aggregate energy intensity using logarithmic mean Divisia approach. Energy Policy 56:341–351
88. Inglesi-Lotz R, Pouris A (2012) Energy efficiency in South Africa: a decomposition exercise. Energy 42(1):113–120

89. Balezentis A, Balezentis T, Streimikiene D (2011) The energy intensity in Lithuania during 1995-2009: A LMDI approach. Energy Policy 39:7322–7334
90. Ediger VS, Huvaz O (2006) Examining the sectoral energy use in Turkish economy (1980-2000) with the help of decomposition analysis. Energy Convers Manag 47(6):732–745
91. Ang BW, Huang HC, Mu AR (2009) Properties and linkages of some index decomposition analysis methods. Energy Policy 37:4624–4632
92. Paltsev S (2004) Moving from static to dynamic general equilibrium economic models (notes for a beginner in MPSGE), MIT joint program on the science and policy of global change, technical note 4 (June 2004)
93. Rout UK, Voβ A, Singh A, Fahl U, Blesl M, Gallachoir BPO (2011) Energy and emissions forecast of China over a long-time horizon. Energy 36:1–11
94. Baek J, Koo WW (2010) Analyzing factors affecting U.S. food price inflation. Can J Agric Econ 58(3):303–320
95. PIK Landuse Group (2011), MAgPIE Mathematical description, obtained from http://www.pik-potsdam.de/research/sustainable-solutions/research/landuse-group/magpie-mathematical-description. Accessed on 14 December 2013
96. Dietrich JP, Schmitz C, Lotze-Campen H, Popp A, Müller C (2014) Forecasting technological change in agriculture—an endogenous implementation in a global land use model. Technol Forecast Soc Chang 81:236–249
97. Schmitz C, Biewald A, Lotze-Campen H, Popp A, Dietrich JP, Bodirsky B, Krause M, Weindl I (2012) Trading more food: implications for land use, greenhouse gas emissions, and the food system. Glob Environ Chang 22 (1):189–209
98. Popp A, Lotze-Campen H, Bodirsky B (2010) Food consumption, diet shifts and associated non-CO2 greenhouse gases from agricultural production. Glob Environ Chang 20(3):451–462
99. Acquaye AA, Wiedmann T, Feng K, Crawford RH, Barrett J, Kuylenstierna J, Duffy AP, Lenny Koh SC, McQueen-Mason S (2011) Identification of "carbon hot-spots" and quantification of GHG intensities in the biodiesel supply chain using hybrid LCA and structural path analysis. Environ Sci Technol 45 (6):2471–2478
100. Suh S, Huppes G (2005) Methods for life cycle inventory of a product. J Clean Prod 13(7):687–697
101. Elgimez G, Kucukvar M, Tatari O, Bhutta MKS (2014) Supply chain sustainability assessment of the US food manufacturing sectors: a life cycle based frontier approach. Resour Conserv Recycl 82:8–20
102. Virtanen Y, Kurppa S, Saarinen M, Katajajuuri JM, Usva K, Mäenpää I, Mäkelä J, Grönroos J, Nissinen A (2011) Carbon footprint of food - approaches from national input-output statistics and a LCA of a food portion. J Clean Prod 19 (16):1849–1856
103. Wood R, Lenzen M, Dey C, Lundie A (2006) A comparative study of some environmental impacts of conventional and organic farming in Australia. Agric Syst 89(2–3):324–348
104. Garnett T (2006) Fruit and vegetables and UK greenhouse gas emissions: exploring the relationship., Produced as part of the work of the food climate research network, FCRN working paper 06-01 Rev. A
105. Goldstein G and Tosato GC (2008) Global energy system and common analyses: Final report of Annex X (2005-2008). International Energy Agency (IEA), http://www.etsap.org/FinReport/ETSAP_AnnexX_FinalReport-080915.PDF. Accessed on 5 December 2013
106. The Energy and Research Institute (Teri), Government of India (2006) National energy map for India: technology vision 2030. Press, TERI. ISBN 81-7993-099-8
107. Strachan N, Anandarajah G, Pye S, Usher W (2010) UK Energy-economic modelling. An overview of the UK MARKAL/TIMES family., Presentation at the UCL Energy Institute, University College London, July 2010
108. Reddy TA, Claridge DE (2000) Uncertainty of "measured" energy savings from statistical baseline models. HVAC&R Res 6(1):3–20
109. Martinez SH, Jv E, Cunha MP, Guilhoto JJM, Walter A, Faaij A (2013) Analysis of socio-economic impacts of sustainable sugarcane-ethanol production by means of inter-regional input-output analysis: demonstrated for Northeast Brazil. Renew Sust Energ Rev 28:290–316
110. Hoekstra R, van der Bergh JJCJM (2003) Comparing structural and index decomposition analysis. Energy Economics 25(1):39–64
111. Karkacier O, Goktolga ZG (2005) Input-output analysis of energy use in agriculture. Energy Convers Manag 46:1513–1521

Decentralised combined heat and power in the German Ruhr Valley; assessment of factors blocking uptake and integration

Birte Viétor[1], Thomas Hoppe[2*] and Joy Clancy[2]

Abstract

Background: In Germany, the energy system is undergoing reorganisation from a centralised system based on fossil fuels and nuclear power to a sustainable system based on decentralised production and consumption of energy, the so-called *Energiewende*. Recently, there has been more attention to improving energy efficiency in those regions where conventional energy production activities and energy-intensive industries are located, such as the Ruhr area. Although the potential for decentralised combined heat and power (CHP) units is high in this region, local action plans show only modest developments for this technology. In this paper, we address this issue by answering the following research question: Which factors block the uptake and integration of decentralised CHP in the German Ruhr area's energy system?

Methods: The multilevel perspective (MLP) was used to analyse the state of system innovation in relation to the uptake and integration of decentralised CHP technology. Prior to the MLP analysis, a stakeholders' analysis was conducted to identify stakeholders' views, positions and experienced barriers regarding the uptake and integration of decentralised CHP technology. Data collection included review of text documents and conducting 11 interviews.

Results: Due to many regime barriers blocking niche development, the uptake of decentralised CHP technology is limited. Identified barriers relate to lack of market services and mismatches with user preferences, (sectoral) policies and industrial interests.

Conclusions: Observed barriers relate to (i) lack of market services such as financial means for making investments; (ii) user awareness such as unawareness and information deficit regarding the benefits of decentralised CHP to potential users, (iii) the presence of centralised district heating systems, (iv) policy issues such as lack of sufficient policies supporting diffusion of decentralised CHP units, legal stipulations from social housing policies that prevent housing cooperatives from becoming energy producers and district heating systems owned by public and private owners (via concessions contracts); (v) sector issues such as lack of skilled service-providing companies, and (vi) industrial interested such as the vested interests of the coal and gas industry. Moreover, many of the mentioned barriers seem interrelated, especially those concerning policy and finance available for making upfront investments.

Keywords: Energy transition; Decentralised CHP; Energy efficiency; Multilevel perspective

* Correspondence: t.hoppe@utwente.nl
[2]Department of Governance and Technology for Sustainability (CSTM), Institute for Governance and Innovation Studies (IGS), University of Twente, Drienerlolaan 5, 7522, NB Enschede, The Netherlands
Full list of author information is available at the end of the article

Background

Following the 2011 Fukushima nuclear disaster in Japan, the political decision was made in Germany to phase out nuclear power production by 2022 and to initiate a countrywide energy transition, the *Energiewende*. In short, it stands for a transition of the entire energy system in Germany which 'involves replacing, or supplementing, established technologies with new ones' [1] and thereby performing the 'inevitable shift away from cheap, centralized, largely fossil-based energy systems towards decentralised energy systems to a large extent based on renewable energy sources' [2] (translation by the authors). Next to increasing the share of renewable energy within the total primary energy supply to 50% by 2050, the program mentions the urgent need for action particularly in those regions where conventional energy production activities and energy-intensive industries are located. In these areas, there is large potential for using energy more efficiently. One way to foster energy efficiency is by supporting adoption of decentralised combined heat and power (CHP) units. The Federal Government of Germany strives to raise the share of combined heat and power generation to 25% by 2020 [3]. Since the opportunities for centralised district heating systems have already been utilised to a large extent, and are difficult to change over the short term for infrastructural and contractual reasons, raising the share of CHP would mean increasing the share of decentralised CHP units.

The Ruhr Valley in the Federal State of North Rhine-Westphalia (NRW) is a very striking example of a region for increasing energy efficiency levels since this is where many conventional energy production activities and energy-intensive industries are located. It is home to the majority of the coal power plants in Germany, and around one third of all German GHG emissions (313 million CO_2eq in 2010) are registered in this area. Furthermore, almost 30% of the national electricity demand is produced in this area (180 million kWh in 2011 in NRW; 608 million kWh in 2011 in Germany [4]), and around 25% of German final energy is used here. Moreover, 40% of the German energy demand for industrial processes originates in the Ruhr Valley [5].

The report 'Deutschlands Energiewende. Ein Gemeinschaftswerk für die Zukunft' ('Germany's energy transition. Collaborative work for the future'; translation by the authors) that was published by the Ethical Commission on behalf of the Federal Government in May 2011 underlines the important role decentralised governments have in implementing *Energiewende* policies. For instance, the report points out that it is the municipalities, rather than the central or state government, that have the planning authority when it comes to locating sustainable energy production utilities. Furthermore, municipalities are responsible for public buildings and public transport; via their municipal energy suppliers and housing societies, they are delivering electricity and heat [6,7], and potentially, they can have policies in place to influence energy consumption by local stakeholders. Moreover, it is at the local level that environmental matters (including energy transition) manifest and citizens engage with government [8]. Hence, cities and regions can become powerful promoters of sustainable transitions [9]. Against this background, it is reasonable to discuss the energy transition approaches from a situational perspective as we deem the situational context and hence specific regional conditions, of great importance to strategies to effectuate energy transitions in cities and regions. In particular, situational conditions might favour adoption and rollout of one particular energy transition pathway, whereas it might disfavour alternatives pathways.

It is for these reasons that we explore the potential uptake and integration of decentralised CHP in the Ruhr Valley. Although the *Energiewende* is widely known to support diffusion of wind and solar systems for power production; in particular, it focuses on electricity generation from renewable sources; it also highlights the need for action towards using energy more efficiently in those regions where conventional energy production activities and energy-intensive industries are located, notably the Ruhr Valley. Given its very nature, the industrial, highly populated urban region of the Ruhr Valley offers high potential for further take up, integration and upscaling of decentralised CHP. In NRW, the share of CHP total electricity production, however, amounted to only 10% [10].[a] Hence, it is far removed from the national target of 25%.

Analysis of the local climate change protection action plans addressing the uptake and integration of decentralised CHP units provides an overview on the importance that is given to the uptake of this technology by nine Ruhr Valley cities (see Table 1).

Eight out of the nine cities have formulated measures that aim at increasing the share of heat and power produced in decentralised CHP units.[b] The actions planned by these eight municipalities differ in various ways. Whereas some municipalities mention in their local action plans that decentralised CHP should be used in municipal buildings, others target local stakeholders, like households and commercial firms. Different policy actions are mentioned such as constructing apartment block-heating installations, integration of decentralised CHP units in urban development planning and establishing contracts with private home owners for the provision of heat. In some municipalities, decentralised CHP is seen as a means to feed in decentralised-produced heat into the existing district heating grids. In this way, decentralised CHP can be seen as a means to 'extend' existing (district)

Table 1 Overview is policy action plans Ruhr city governments have prepared regarding CHP

City	Year of origin of policy action plan	Inhabitants	Actions regarding decentralised CHP
Bochum	2009	373,000	Feasibility study: integration of small CHP (buildings) in the regional supply concept ('virtual power plant').
Bottrop	2011	116,000	Local heating grid Kirchhellen: construction of a CHP plant of 15,000 kW with local heating grid on the basis of renewable energies (biogas, wood).
Dortmund	2011	580,000	Potential study of utilising CHP units in private households and the commercial sector.
			Block-heating station-'Push' in municipal properties.
Duisburg	2009	488,000	None.
Essen	2009	573,000	Integration of urban development contracts into town plans to regulate use of CHP.
Gelsenkirchen	2011	256,000	Block-heating station-'Push' in municipal properties.
			Pilot project: 'Extension of CHP'.
			Local heating grid in interconnection (potential analysis for block heating stations).
			Combination of geothermal energy and micro-CHP.
Oberhausen	2012	212,000	Heat supply in the form of heat contracting for private house owners.
			Decentralised CHP for the local heating grid in Barmingholten.
Recklinghausen	2012	118,000	Extension of (renewable energy) CHP in the city area.
Witten	2013	98,000	'Push' for combined heat and power.

heating grids. Not surprisingly, multiple Ruhr Valley local authorities deploy 'extension of CHP' actions in their respective programs due to the presence of centralised district heating grids. Studies regarding the technical and economic potential or feasibility of decentralised CHP have been prepared in four municipalities, and plans for constructing decentralised CHP units are mentioned by at least four (other) municipalities. The ways the nine city governments address the uptake and integration of decentralised CHP shows little coherence: between the nine municipal plans, there is little alignment in the formulation of goals and policy actions. In summary, the policy actions to spur the uptake and integration of decentralised CHP by the nine Ruhr Valley municipalities give the impression that uptake of decentralised CHP is at a rather early stage.

These findings stimulated us to find out why the uptake and integration of decentralised CHP is currently not at a more developed stage. *A priori,* we would expect that there would be a large-scale uptake and integration of CHP in the Ruhr Valley, but we do not see it in reality. Therefore, we aim to identify the factors that might explain why the potential for decentralised CHP is not being realised. The main research question of this paper is: *Which factors block the uptake and integration of decentralised CHP in the German Ruhr area's energy system?*

In order to answer the question, we use stakeholders' and systemic analytical perspectives for sustainable transitions. In line with Truffer and Coenen [9], we view sustainability transitions as political projects where stakeholder interrelationships play a crucial role in the transformation process. Therefore, we conduct a stakeholder analysis to identify relevant stakeholders' roles, positions, interrelationships, views and experienced barriers concerning the uptake and integration of decentralised CHP. Next, we analyse the systemic status quo regarding the uptake and integration of decentralised CHP into the energy system of the Ruhr Valley by using the multilevel perspective (MLP) framework for transitional change [11,12].

Background: the Ruhr Valley region context

The Ruhr Valley is one of the biggest urban agglomerations in Europe with 5.2 million inhabitants. The region can be divided into 11 urban districts (Kreisfreie Städte; translation by authors) (Duisburg, Mülheim a. d. Ruhr, Oberhausen, Essen, Bottrop, Gelsenkirchen, Herne, Bochum, Dortmund, Hagen und Hamm) and four counties (Kreise) (Wesel, Unna, Recklinghausen und Ennepe-Ruhr). These in turn are composed of 43 independent cities and municipalities within the county. Figure 1 presents the siting of the region in Germany and the siting of administrative divisions within the Ruhr Valley [13].

The coal deposit of the Ruhr area provided the natural precondition for the development of the biggest coal and steel industry in Europe. Thus, historically, the Ruhr area was the hotspot for steel and chemistry industries - economic sectors that are strongly linked to greenhouse gas emissions. During 150 years of industrialization, the region underwent strong spatial changes as settlements expanded and economic activity accelerated. Industrial and socio-economic institutions were essentially designed

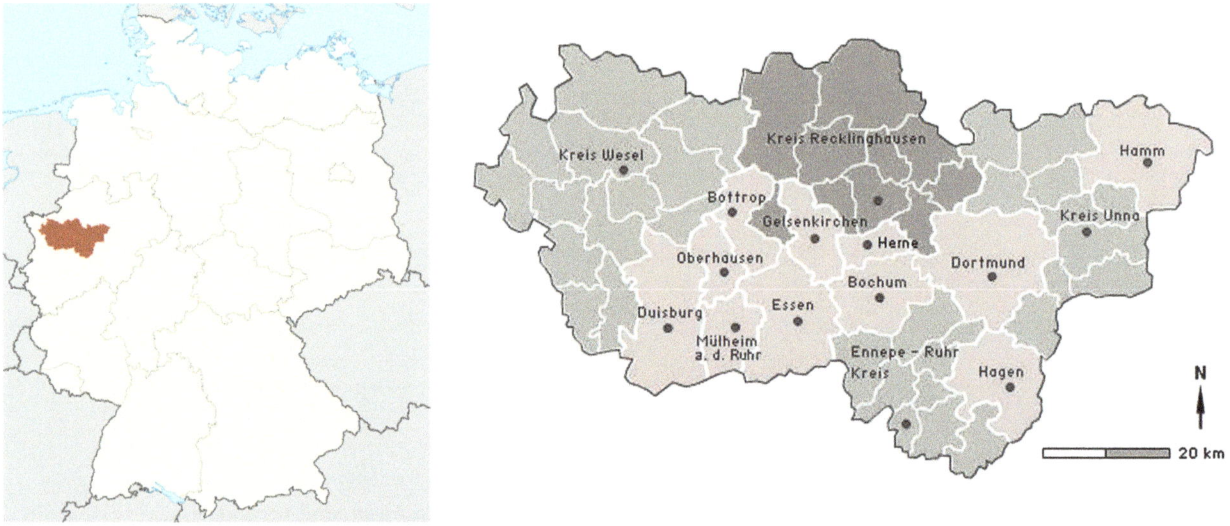

Figure 1 Siting of the Ruhr Valley region in Germany, and the Ruhr Valley's administrative divisions.

to serve the mining industry. While in the second half of the 19th century 280 coal mines were active in the region, this number diminished dramatically. In 2009, only four mines were left [14]. Nowadays, more than 70% of the population in the area is employed in the tertiary sector. This impressive development from an industrial to a knowledge society could not solve all problems that are linked to the 1950s crisis in the coal industry (e.g. high unemployment, financial debts) [15]. Some cities in the Ruhr area are nowadays among those in Germany with the highest unemployment rates (e.g. Duisburg, Dortmund). The unemployment rate contributes to the weak financial situation of the municipalities[c] which limits the scope for action.

Background: decentralised CHP technology

Low efficiency during the electricity generation processes results in energy being lost in the form of waste heat [16]. Collecting waste heat for reuse combined with electricity generation is a way to use primary energy more efficiently [17]. It is the combination of the electrical generation and heat production process that forms the basis of the CHP concept [18]. The overall efficiency can reach up to 90%, whereas single-electricity generation only reaches efficiency levels of 30% to 60% [19]. Waste heat from the exhaust gases, or used steam, is recovered for instance to produce hot water for the use in district heating schemes (in large-scale CHP systems) or to heat spaces in buildings (using smaller decentralised CHP plants; [16]). Three categories for use of CHP have been defined by the German Federal Ministry of the Environment, Nature Protection and Nuclear Safety, one of them being decentralised CHP. The latter is typically used to supply heat and power for consumption in residential or commercial buildings [19]. In this paper, 'decentralised CHP' relates to CHP units used at the district level, street level or at the level of (multistorey) housing complexes.[d]

Decentralised CHP offers considerable benefits as compared to conventional ways of heat and power production. First, it offers improved energy efficiency and preserves non-renewable energy reserves. Second, due to more efficient conversion of primary energy, reductions in ground-level particle and gas emissions and pollution can be realised, hence improving urban air quality [16]. Third, both fossil and renewable sources can be used in decentralised CHP plants; hence, technology can make an important contribution to reducing greenhouse gas emissions and the primary fossil fuel energy consumption in the future energy system [19].

Background: decentralised CHP technology uptake in the Ruhr Valley energy system

The integration of decentralised CHP into the German energy system forms an important cornerstone of the energy policy of the Federal Government of Germany. The Integrated Energy and Climate Protection Program (Integriertes Energie- und Klimaschutzprogramm) (2007) of the Federal Government of Germany includes the political target to raise the share of combined heat and power generation in Germany from the current level of about 13% to 25% by 2020 [3]. The important role CHP plays in the framework of the energy system transition was affirmed in a renewal of the respective law (Act on CHP) and the creation of different funding policies (e.g. the Act on Renewable Energies). Since 2003, power and heat generation via CHP has increased by 20% and 8%, respectively. In 2005, 18.2 TWh power was produced by CHP in NRW. Related to the entire power production in NRW (ca. 180 TWh/year), the share of CHP amounted to only 10% [10].

In January 2013, the Federal State NRW adopted the first Climate Protection Act in Germany which contains legally binding climate protection objectives. The act aims at contributing to the achievement of the national climate change mitigation goals and acknowledges the potential that decentralised CHP has in helping reach these objectives. It further acknowledges the significant responsibility the region has in this respect, notably when compared to other German States. According to the Climate Protection Act, by 2020, GHG emissions are to be reduced by 25% relative to the 1990 level. By 2050, a reduction of 80% relative to the 1990 level is targeted. NRW also adopted the national CHP objective based on which the state government developed a program, the 'CHP Impulse Program', to increase the share of decentralised CHP in the area. The CHP Impulse Program, together with the national political conditions, makes the current political framework conditions seem supportive towards the integration of decentralised CHP into the energy system in NRW.

Recent studies on the potential of CHP in NRW (May 2011) and on district heating in the Ruhr area (May 2013) have stressed the significant potential for this technology in meeting policy goals. Both studies argue that it is economically viable to extend the district heating system, in particular to supply heat to more residential areas in the Ruhr Valley. However, the large-scale, centralised CHP plants that supply heat to the district heating grid are emphasised. At the same time, the studies mention some barriers that could have strong impacts on the operating efficiency of the district heating system. Namely, due to improvements in energy efficiency and insulation of buildings, it is predicted that there will be a decrease in demand for heat and electricity. Furthermore, demographic changes will have an impact on the number of customers demanding heat and electricity. At the same time, many coal power plants are expected to close down in the near future, because the profit margins are decreasing rapidly. Consequently, less heat will be generated and supplied to the grid. Interestingly, these barriers do not seem to have a significant impact on the formulation of recommendations for the future energy system.

The study 'Investigation of the potential of CHP in Northrhine-Westphalia' (in German: *Potenzialerhebung von Kraft-Wärme-Kopplung in Nordrhein-Westfalen*; translation by the authors) [20] states that an increase in the number of decentralised CHP plants will lead to decreased connection rates to the district heating grid which in turn will have a negative impact on the cost effectiveness of the latter. Although this sounds logical, it did not lead to any critical reflection by policy makers on how to overcome this potential conflict between centralised and decentralised supply of energy. The study 'Perspectives of district heating in the Ruhr Area until 2050' (in German: *Perspektiven der Fernwärme im Ruhrgebiet bis 2050*; translation by the authors) [21] mentions briefly that there is a need for more flexibility and more heat sources to be able to maintain the economic viability of the district heating network. Responding to this point is very important as decentralised CHP units could theoretically ensure this flexibility and at the same time provide more heat sources.

In summary, there are five reasons why the potential for large-scale adoption of decentralised CHP in the Ruhr Valley should be considered worthwhile exploring. First, the two studies mentioned above underline the high potential regarding the further integration of CHP into the energy system as well as connecting to the existing district heating grid [21,22]. Second, the high-geographical proximity of cities in the Ruhr Valley combined with the high-population density in the area offer good preconditions for the efficient use of CHP - both centralised and decentralised. Third, the decentralisation of energy production and consumption is a stated political objective (in the *Energiewende* program) and as such could provide a window of opportunity for policymaking to support decentralised CHP uptake and integration in the future energy system. Moreover, promoting energy efficiency is viewed as an urgent priority *vis-à-vis* the future of energy transition in Germany. Although supply of heat and power from cogeneration and energy efficiency in buildings have taken a backseat in the German energy transition agenda, these sectors have recently been identified by researchers as having significant potential [22]. Fourth, increased uptake of decentralised CHP allows for greater flexibility in responding to fluctuations in energy demand and supply. Fifth, decentralised CHP also allows for local use of renewable energy sources. For this reason, CHP can play an important role in the envisaged energy system transition in Germany and contribute to attaining climate change policy goals. In conclusion, decentralised CHP could potentially offer substantial benefits to energy producers, end users, policymakers and governmental actors.

Theoretical framework: using the multilevel perspective to analyse regional energy transitions

The challenge for the uptake and integration can be viewed as fostering a socio-technical and sustainable transition. A *socio-technical transition* is a set of processes that leads to a fundamental shift in socio-technical systems (e.g. [23,24]). A socio-technical system (such as the energy system) consists of '(networks of) actors (individuals, firms, and other organisations, collective actors) and institutions (societal and technical norms, regulations, standards of good practice), as well as material artefacts and knowledge. The different elements of the system interact, and together they provide specific services for society' [25]. The systems concept highlights the fact that a broad variety of elements is tightly interrelated and depends on each other [26].

Sustainability transitions, for instance those intended to attain the status of a low-carbon society, are also considered by researchers to be socio-technical transitions. Energy transitions, for example the *Energiewende*, can also be viewed as *sustainability transitions*. These are long-term, multidimensional and fundamental transformation processes through which established socio-technical systems shift to more sustainable modes of production and consumption. A key feature of sustainability transitions is that guidance and governance often play a significant role in the transition [27], when decisive interventions from state and non-state actors are needed to overcome the inertia and lock-in typically found in the prevailing socio-technical systems [26,28,29].

Multilevel perspective on system innovation and transitional change

Currently, the theoretical framework most used to understand and systematically analyse socio-technical transitions, in particular the energy transition (e.g. [30]), is the MLP [11,12]. The framework has been developed based on findings from evolutionary theory and systems analysis. According to Geels 'the stability of established socio-technical configurations results from the linkages between heterogeneous elements' [12], p. 1259. For MLP, these are linkages between three conceptual levels: macro, meso and micro.

On the *macro level* 'landscape events' occur. Landscape is associated with the material context of society. It is made up of various macro factors such as oil prices, interstate geopolitical relationships and forthcoming events (like treaties and wars), political and governmental coalitions, cultural values and major environmental problems. The socio-technical landscape forms the external context for action of, and interaction between, actors.

The *meso level* is referred to as the 'socio-technical regime'. These regimes encompass social and institutional rules that enable and constrain activities between actors [12]. These rules are related to several institutionalised factors, such as markets, user preferences, (sectoral) policies, industries, science, culture and technology. As a rule of thumb, socio-technical regimes only change incrementally and contain defence mechanisms to fend off attempts to replace them with alternatives, typically radical innovations developed at the micro level in 'socio-technical niches'.

The *micro level* is referred to as the analytical level in which 'socio-technical niches' develop. The niche is one of the central concepts in transition research. Niches form protective spaces in which radical innovation can develop, while being protected from regime defence mechanisms. An important question regarding the role of the niche is upscaling (e.g. addressing the question how to increase the take up and integration of the niche within society). Smith and Raven [31] differentiate three functions for niche protection: shielding, nurturing and empowering. Shielding refers to processes that hold at bay selection procedures from mainstream selection environments. Nurturing refers to processes such as learning, networking and expectation formation. Finally, empowering refers to the process that makes niche innovations competitive within unchained selection environments ('fit-and-conform') and processes that restructure mainstream selection environments in ways favourable to the niche ('stretch-and-transform'). Political, administrative, managerial and academic interest in how to encourage (sustainable) transitional change have led to the development of managerial transitional change frameworks, notably the Strategic Niche Management (SNM) [32,33] and Transition Management (e.g. [34,35]).

The interlinked character of the macro, meso and micro levels means that regimes are embedded within landscapes and niches within regimes. Innovations (and hence attempts to bring about transitional change) take off in niches in the context of existing regimes with their specific problems, rules and capabilities. Thus, in a transition process, interactions between dynamics exist between the three levels.

Theoretically, the systemic dynamics that result in transitional change follow a typical pattern. Landscape events (like the 2011 Fukushima nuclear disaster) create pressure on socio-technical regimes which result in problems that regimes cannot solve from within. Solving these problems via incremental regime optimization will not suffice to solve these problems and creates opportunities for alternative radical innovations with the potential to overthrow the current regime. This provides opportunities for new innovations that develop in niches and are supported by social networks (often including 'regime outsiders') that are active in niche formation activities. After iterative sets of niche experiments (e.g. by organising demonstration projects), innovations mature and have the potential to gain a foothold in the existing socio-technical regime. If successful, it can eventually replace the existing regime and hence the socio-technical system as such [36]. When replacement of an existing regime by a novel regime takes place, one can speak of *system innovation*. When this concerns a radical innovation, one can speak of *transitional change*. Breakthroughs of radical innovations consequently depend on interactive systemic processes within and between the macro, meso and micro levels (i.e. between landscape events, the socio-technical regime and niches). In general, transitions are context dependent [12].

Although often used in academic studies of transitions, MLP and the related analytical frameworks that suggest 'management of transitions' (e.g. SNM, TM) have been criticised for failing to address context-related (regional and local) factors that influence the way in which innovations develop and transitions manifest (see e.g. [26,37,38]), notably addressing agency [39-41], politics and power [42-44]; the role of user groups in consumptive action [37,45]; the role of civil society in transitions [46,47]; policy [48,49]; and geographic-bound factors such as comparative institutional, socio-economic or infrastructural advantages that differ substantially from one geographic administrative jurisdiction to another [50]. When analysing energy transitions at a regional scale, identifying these factors is important because one needs to first understand the context, culture and historical pathways that explain the establishments of key institutions, interstakeholder power relations and comparative institutional (dis-)advantages. Based on this perception, the next step is then to understand the functioning of the socio technical systems, before one attempts to plan and 'manage' transitional change. Based on this criticism, Geels recently [51] furthered the conceptualization of energy transitions by incorporating agency, power and politics. In relation to attaining low-carbon societies, he [52] states that there should be more academic attention to resistance by incumbent regime actors to fundamental change than continuing to overemphasise the potential of a great many 'green' innovations. Regarding future research agendas, this should be understood as shifting away from studying niche development to studying regime power and politics as defensive mechanisms to prevent radical change from happening. Hence, the unit of analysis should be the regime, not the niche. In this paper, we adopt Geels' view, although our initial focus is on a particular niche (decentralised CHP). However, the main part of our analysis addresses regime forces (see the 'Methods' section).[e]

Furthermore, the paper contributes to the academic literature on the energy transition (strategies towards attaining low-carbon goals) in cities (e.g. [53 55]; and the thematic issue of which this paper is part [56]).

Methods

This paper presents the case study of the Ruhr Valley, which will be analysed in two ways, in which the results of the first analysis will be used as the input for the second analysis. First, a stakeholders' analysis will be conducted to identify key stakeholders, their positions in the regional energy system and barriers they experience in relation to the uptake and integration of decentralised CHP units. Second, a systemic analysis will be conducted using the MLP framework. Results from the stakeholders' analysis (notably identification of regime barriers) will be used as input for the analysis with MLP.

Case selection

The Ruhr Valley was selected for its urban and industrial regional character. Theoretically speaking, these characteristics would be expected to benefit the take up of energy efficiency technological pathways, in particular decentralised CHP. This makes our case different to other areas in Germany which have adopted renewable energy technologies on a large scale (in particular wind and solar energy) because they are more rural and hence spatially more suited to spur production of green electricity. Thus, case selection follows the criterion of the regional socio-economic and spatial character. Selecting the Ruhr Valley as a regional case study should be seen as selecting an extreme case[f] [57] for Germany as no other region compares readily to the Ruhr area. So, hypothetically, the highly industrial and urban character of the region, with potential for use of residual heat and high local demand for heat and power, should favour uptake and integration of decentralised CHP. As this study presents a (single) case study, external validity (and hence representativeness) is low. Nonetheless, potentially regions that also feature the previously mentioned characteristics can readily be compared with the Ruhr area. However, this is not the objective of this paper which only studies the one region.

Stakeholders' analysis

A stakeholder analysis is conducted to generate knowledge about the stakeholders involved with decentralised CHP 'so as to understand their behaviour, intentions, interrelations and interests; and for assessing the influence and resources they bring to bear on decision-making or the implementation process' [58], p. 338. The stakeholder analysis consisted of the identification of stakeholder characteristics, views and experienced barriers *vis-à-vis* the uptake and integration of decentralised CHP. Data collection concerned 11 qualitative - that is face to face, in-depth - interviews[g] with representatives from different stakeholder groups relevant to the uptake of decentralised CHP. A semi-structured questionnaire was used during the interviews. Questions raised addressed the status quo of decentralised CHP uptake in the Ruhr Valley, the future role of this technology in the regional energy system, interaction with district heating, experienced barriers, niche development, policy support strategies and (other) stakeholders' influence on development of this technology in the Ruhr Valley region. Addressing these issues, in particular, the barriers that hinder the uptake and integration of decentralised CHP provide key information necessary for understanding the forms of resistance exercised by the incumbent socio-technical energy regime [1].

Systemic analysis of the status quo using the MLP framework

In this paper, decentralised CHP is treated as a 'niche' (or technology with the potential to become a radical innovation) using the MLP conceptualization, although we argue that this technology has the potential to become a radical innovation and stimulate the (sustainable) transition in the energy system. In order to analyse niche development landscape events, regime factors and niche development will be identified. In common with MLP, the focus of our analysis will be *identification* of regime forces that tend to block decentralised CHP niche development.

Data used for conducting the analysis included the results from the stakeholder analysis (to a large extent this had identified the barriers) and from an analysis of documents. Next, the data was categorised into three clusters representing the three MLP levels of landscape (macro level), regime (meso level) and niche (micro level). After which, the interlinkages between the three levels were identified.

MLP is used a lot to analyse historical cases [26]. However, in this paper, we use MLP as an analytical tool to identify factors (and their interlinkages) that hinder or support development of a contemporary niche. Hence, MLP is used to assess the status quo of regime forces that influence the niche development of decentralised CHP.

Results

The results of the Ruhr Valley case study are presented in two parts. First, the results of the stakeholder analysis are presented, in terms of stakeholders' views, positions, interrelations and experienced barriers *vis-à-vis* the uptake of decentralised CHP. After which, the results of the systemic analysis using MLP are presented.

Results of stakeholders' analysis

Table 2 presents an overview of the stakeholder characteristics and views regarding the uptake of decentralised CHP. In the table, stakeholders, their positions, interrelations with other stakeholders, opinions towards the uptake of this technology and experienced barriers are presented. Table 2 reveals that consumers - be they private households, the commercial sector or housing associations - are sceptical towards implementation. In the next section, barriers experienced by stakeholders will be discussed in more detail.

Perceived drivers and barriers by stakeholders

Several drivers and barriers regarding uptake and integration of decentralised CHP can be identified. They can be categorised as: investments, regional characteristics, information deficit and lack of awareness, policy-related barriers, district heating and lack of market services. They will now be discussed in more detail.

Investments

Decentralised CHP units are still considered relatively *expensive* (especially upfront investments). Payback periods are considered quite long. Linked to this, another barrier mentioned is the lack of knowledge about business models in relation to contracting or leasing. Return on investment is considered rather low. This is related to legal framework schemes changing frequently and the feed-in tariff for electricity generation, which is becoming unpredictable. Interviewees often draw comparison to the uptake of solar PV in the country. They claim that 'Feed-in tariffs for PV-electricity decreased dramatically in recent years and continue to do so on a monthly basis'. This creates investment uncertainties. Nonetheless, most interviewees believe that current policy and funding conditions are beneficial in respect of the uptake and integration of decentralised CHP.

Among the driving factors mentioned by interviewees increasing energy prices were frequently mentioned, which were linked to the aspiration of establishing 'energy autarkies' and not being dependent anymore on the large-scale energy producers. This idea is currently receiving more attention from different end-user groups. Another driving factor (but linked to the former) that could spur the development of decentralised CHP is the increasing popularity among citizens and local business companies for making local investments. The interviewees observe a trend that 'goes local'. People wish to invest in projects or organisations they can trust and relate to; often in close proximity to their homes. Decentralised CHP plants would offer such investment opportunities for local communities. Citizens' local cooperatives or local business in collaboration with private investors jointly provide the necessary financial means to collectively design, build and operate (renewable) energy production plants, for example wind farms. This example makes clear that our interviewees believe that financial barriers could be potentially be resolved by collective local investments, especially if they are based on popular support by local communities, notably local *Bürgerenergiegenossenschaften* (authors' translation: citizens' energy cooperatives).

Regional characteristics

The Ruhr Valley has several specific characteristics that contribute to this area being rather unique. It has a large stock of old buildings, many having a status as a historical monument. Our interviewees view this as a positive reason for installing decentralised CHP units. They state that it is often more economic than retrofitting old buildings (by applying thermal insulation). In the case of conservation of monuments (especially in old fashioned

Table 2 Stakeholders characteristics and views regarding uptake of decentralised CHP

Stakeholder	Position and function	Interrelation with other stakeholders	Opinion towards uptake of decentralised CHP	Experienced barriers
Energy supply companies (e.g. RWE/E.ON)	Produce and supply energy and operate some of the local heating grids via long-term concessions.	Supplies heat and power to end consumer via DSO. Heavily regulated by government.	Negative	Combining sufficient regional and local heat demand, failing centralised heating plants (momentarily absence of heat supply).
Local governments/cities	Run local action plans to support uptake of decentralised CHP. Own public utility companies that in turn operate local heating grids.	Serves the common interest. Owns public utility companies. Relations to other stakeholders via local decentralised CHP action plan.	Positive/neutral	Financial deficits, lack of information, lack of central government (climate) policy to support decentralised CHP, uncertainty related to elections, and lack of policy focus.
Providers of decentralised CHP	Produce and sell decentralised CHP units and systems.	Provide CHP unit to end users, housing corporation or private contractor to exploit the CHP unit.	Positive	Existing district heating grid, lack of information, lack of central government (climate) policy goals to support decentralised CHP, strong coal and gas lobby in NRW policymaking venues, elections and lack of policy focus.
CHP unit system installers	Provide installation and maintenance services that are used by end consumers.	Installs and maintains CHP units as commissioned by end users, housing corporation or private contractor to exploit the CHP unit.	Neutral, albeit hardly aware of benefits and potential	Status of a historical monument of buildings and its legal implications, lack of information, elections and lack of policy focus, lack of market (support) services, due to small size little innovation receptive capacity.
Private contracting parties	Operate commercial decentralised CHP units and make contracts with end users.	Contract with end users or housing corporation.	Positive	Lack of information, many barriers from other stakeholders indirectly harm contracting parties.
Consumers (e.g. households)	Use of energy produced by decentralised CHP units (heat and/or electricity).	Contractually related with CHP unit installer, contracting party, (in some cases) housing corporations, consultants, and the DSO. Might be connected to local government via decentralised CHP action plan (e.g. recipient of subsidy).	Sceptical, albeit hardly aware	High upfront investments, lack of financial governments support, monumental status of buildings, existing district heating grid, lack of information, habits of using energy equipment, transaction costs that go with CHP unit registration and using subsidies, elections and lack of policy focus.
Social housing corporations	Adopt, own or rent and operate decentralise d CHP units from which the produced energy is to be used by tenants of the housing corporation.	Contractually related with CHP unit provider, CHP unit installer, contracting party, DSO and local government, energy company.	Positive	Lack of information, transaction costs that go with CHP unit registration and using subsidies, social housing corporations legal entitlements and limitations.
Consultancy/advisory agencies	Provide advice clients seeking information on adopting, using and finance of decentralised CHP.	With commissioner, typically local governments, housing corporations, DSOs, some private households and companies.	Positive	Lack of information, many barriers from other stakeholders indirectly harm contracting parties.
Distribution system operator			Depending on ownership	
Owned by municipality (public utility company; PUC)	Public utility companies operate part of local heating grids.	Local government, end users, housing corporations, centralised energy company. Local government owns PUC.	Positive	Lack of information, concessions preventing ownership of all local heating grids.
Owned by RWE/E.ON	Large energy companies possess long-term concessions to operate a part of the heating grids	Idem. However, energy company operates grid via concessions, not the PUC.	Negative	Potential ending of concession contracts. Loss of local heating grid monopolies.

mining settlement sites), basic thermal insulation cannot simply be applied since it would damage the fabric of the building and spoil the appearance. Moreover, our interviewees state that there is a good potential to use decentralised CHP units in urban districts that are not connected to district heating grids.

A second condition that could favour the uptake of decentralised CHP technology is the regional high-population density and having several big cities sited in close proximity. These conditions allow for a high-heat production potential in the region at close proximity to sites of high heat (and electricity) demand by end users. The important issue here is combining sufficient regional and local heat demand, since CHP plants can only operate economically and efficiently if a significant demand for heat can be ensured and transportation infrastructure of heat is not too long when taking into account potential heat loss.

Third, from an infrastructural point of view, the Ruhr Valley is characterised by an extensive existing district heating grid through which heat is already distributed to many city districts. This grid has however also been listed by the interviewees among the barriers to further uptake and development of decentralised CHP since the current district heating grid supplies sufficient amounts of heat already (from centralised industrial heat production plants). However, the existing grid offers important preconditions for low-carbon heat supply. The interviewees advocate exploring ways to create interconnections between the existing district heating grids and decentralised CHP units. In relation to this claim, one of our interviewees stated: 'It is not the grid that hinders the energy system transition in Germany. It is the energy production utilities that are centralized, large-sized and inflexible plants. Moreover, there is more potential for decentralised units to be integrated (better) into the current district heating system'.

The fourth regional-specific factor, and barrier, is the difficult financial situation in which the majority of the municipalities in the Ruhr Valley now find themselves. They have limited financial means and therefore are not in the position to allocate sufficient budget to promote the uptake and integration of decentralised CHP. Consequently, it is not possible to implement substantial earmarked policy support measures to convince local building owners in making appropriate upfront investments. Next to awareness-raising campaigns and other 'soft policies' local - non-governmental - building owners are left to finance investments in CHP themselves, unless they are able to attract other financial sources for making investments. Another negative consequence from having limited budgetary means is that municipalities have only a limited opportunity to become local 'launching customers'.

Lack of awareness and information deficit

All interviewees mention the important influence of information provision and awareness raising on the uptake and integration of decentralised CHP in the Ruhr Valley by local stakeholders. They agree, however, that decentralised CHP technology is not familiar among its potential target groups (i.e. private home-owners, housing corporations and companies looking for office space). CHP units are considered as 'not visible' by end users, as they are located in places where people do not notice them, for example in cellars. Interestingly, technological equipment suppliers and service companies (such as those responsible for installation) are often not familiar with the technology. Interviewees claim that it is, to a large extent, this information deficit that is responsible for the lack of decentralised CHP usage in the Ruhr Valley. Several interviewees state that promoting this technology is not easy as the concept is perceived as 'rather difficult to understand'. Moreover, the technology is considered as rather 'complex', notably by those in society who do not engage with energy technology regularly.

Another barrier mentioned by respondents concerns the habits and behaviour of end users. In particular, it is difficult to change habits. However, once end users are familiar with using an energy technology, it is difficult to convince them to switch to using a new (energy) technology (e.g. decentralised CHP) since it requires moving away from a proven technology the user is familiar with. The interviewees suggest that it is possibly easier to convince target groups, when promoting the technology, by mentioning the particular benefits of CHP that would appeal to them. However, this strategy would require data about the target group's heat profiles and that decentralised CHP energy production can be tailored to their specific needs. Unfortunately, there appears to be little information available on individual end-user heat profiles.

Interviewees mentioned pilot projects as a positive driving factor to encourage the uptake and development of decentralised CHP. Pilot projects receive a lot of public attention. It creates awareness and helps familiarise potential adopters of the technology. If the results of a pilot project are positive, then a basis for trust and further experimentation is likely to have been created. Decentralised CHP then becomes known to the potential end users as a more or less 'proven technology'. According to the interviewees, trust in the technology and its installers is of particular importance. It is not unreasonable to assume that one would only invest and adopt, or participate in collectively buying, a decentralised CHP unit if one is convinced that this technology is proven and has long-term added value both financially and environmentally.

Policy barriers

Current Federal and State policy frameworks that focus on supporting decentralised CHP technology are viewed by the interviewees as having been well formulated. Essentially, they are considered to create sufficient framework conditions to foster uptake. Several funding instruments are deemed useful and considered as positive driving factors. However, some interviewees believe that the amount of funding offered (for example by feed-in tariffs and subsidies through grants) is not sufficient to promote the uptake. This response seems to be related to the situation in many municipalities in the Ruhr Valley which are suffering from stretched financial budgets (related to the economic crisis). For this reason, municipalities find it difficult to co-fund investments with potential local adopters.

A second barrier mentioned by the interviewees relates to the way in which climate policies have been formulated. Interviewees considered these policies overly ambitious with too long a time horizon as an often cited criticism. As one interviews states: 'Thinking in "big steps" is not appropriate if one strives to really attain climate mitigation and energy transition goals set'. 'Setting policy goals that have to be attained in fifteen, twenty or even forty years fail to create responsibilities as politicians and public officials who are currently in power will – most likely - not be in power anymore once these policy objectives have been or should have been met.' Another interviewee comments underline this point: 'Policymakers hide behind long term goals so that they cannot be held responsible once short term policy goals have not been met'. Hence, in general, climate mitigation policy goals are seen as unrealistic, lacking attention to situational settings, and henceforth fail to address realistic, feasible opportunities to promote the use of decentralised CHP.

Third, complex policy framework conditions are seen as a hindrance factor *vis-à-vis* the uptake and integration of decentralised CHP. For instance, one interviewee states: 'Once one applies for a financial support scheme one needs to fill in and register a decentralised CHP unit in order to be entitled to receive payments.' This is perceived as a difficult and time-consuming task. Thus, it is considered challenging to comply with all the necessary formalities. This situation applies to both private homeowners and housing corporations. It is with the latter where the greatest potential for adopting decentralised CHP units lies. These corporations are in a position where they can sell both heat and electricity directly to their tenants. However, the contemporary legal framework hinders the housing corporations in becoming energy producers. Opinion is divided between stakeholder groups about whether or not this regulation is the main barrier that prevents housing corporations from installing decentralised CHP units.

A fourth policy-related barrier is the strong coal and gas lobby in the Ruhr area. Policymakers and lobbyists are considered to have very close working relations. The influence that is exerted by the coal and gas lobby on policy making is seen as a strong hindrance to policy-making *vis-à-vis* supporting the uptake and integration of decentralised power production in general and renewable energy-based technologies in particular.

A fifth policy barrier mentioned is the 2013 national elections.[h] Prior to the 2013 elections, interviewees acknowledge it is difficult to foresee which parties will be involved in the new government coalition, and whether the current policies and programs will be maintained. Interviewees claim that it is necessary to wait and see whether developments after the elections will affect current policy framework conditions, especially the long-term goals and the decentralised CHP support policies. As a consequence, investment decisions are postponed until after the elections have taken place and a new government coalition is in power and has published its policy strategies.

District heating

As mentioned above, the existing district heating grid is seen as a barrier to the uptake and integration of decentralised CHP. Expectations are that several (notably gas-fired) power plants will have to shut down in the near future. This already resulted in widespread uncertainties among stakeholders regarding the continuation of heat supply to the district heating grid. One interviewee underlines this by stating that 'A difficult situation needed to be overcome when it turned out that the coal power plant "Datteln 4" with a capacity of 1,050 MW would not start operating in time (if at all). Due to planning mistakes the power plant could not start feeding heat into the district heating grid. As a consequence an alternative solution had to be found'. This event led to strong public protests. Such situations - creating tensions to the current hegemonic energy regime in the region - can consequently spur the development of decentralised energy production (including by small- and medium-sized CHP units) to ensure the continued availability of local heat production. The emphasis on 'decentralised and local' is considered important since it creates independence from the large-scale heat producers from whom security of supply cannot be expected anymore which represents an undermining of trust considered as a factor in the stability of the dominant energy regime.

The role of municipal public utility companies (PUCs) can be seen as a potential influential factor in this respect. In cities that own PUCs, the interviewees observe a large potential to install decentralised CHP units: 'Municipalities could potentially start projects in which public grids connect to decentralised CHP units and their end users.'

However, ownership of district heating grids by PUCs is not found in all cities in the Ruhr Valley. In some cities, energy companies like RWE and E.ON have district heating grid concessions. This means that municipalities which want to support the integration of decentralised CHP in district heating systems will be dependent on the willingness of energy companies to see this scale of technology as part of their technology mix. This dependency of the municipalities on a stakeholder which has not shown enthusiasm about decentralised systems is viewed as a substantial barrier by the interviewees.

Lack of market services

A lack of companies providing services regarding installation and maintenance of decentralised CHP units is perceived as an important barrier by the interviewees. Two reasons were advanced to explain this view. First, installation workers lack proper instructions on how to install and maintain decentralised CHP units. Without these instructions, they simply cannot offer these services to customers, particularly in a manner that is likely to create trust. However, producers of CHP units and installers do offer courses and training sessions about installation and maintenance of the decentralised units. The difficulty is that such courses need to fit into the day-to-day working of these (typically small-sized) companies. Usually, they have little time or spare staff available (if at all) to participate in these courses, particularly if they are offered during working hours. A related problem in this regard is that this type of small-sized service companies lacks the capacity to specialise in developing services regarding installation and maintenance of decentralised CHP units. This results in a situation where very few (specialised) people need to cover an increasing range of specialised services related to particular technological products.

Results of systemic analysis using MLP

The stakeholders' analysis presented an overview of barriers that can be seen as regime resistance forces. Major identified barriers relate to (i) lack of financial means for making investments, (ii) lack of awareness regarding the benefits of decentralised CHP to potential users, (iii) the presence of centralised district heating systems, (iv) lack of sufficient policies supporting diffusion of decentralised CHP units, (v) legal stipulations from social housing policies that prevent housing cooperatives from becoming energy producers, (vi) district heating systems owned by public and private owners (via concessions contracts), (vii) lack of skilled service-providing companies and (viii) vested interests by the coal and gas industry. Many of the identified barriers seem interrelated, especially those related to policies promoting decentralised CHP units and the required investments and training of skilled personnel. In MLP, the identified barriers fall within the following regime forces: energy markets, user preferences, (sectoral) policies and industry. We will now present the results of the MLP analysis, using the MLP categorization of the macro, meso and micro levels.

Macro level: landscape factors

Although there are many regime barriers (i.e. the problems presented in the previous section), there are also many ('landscape') developments that could potentially offer windows of opportunity to encourage the uptake of decentralised CHP technology. The security of energy supply is no longer taken for granted. Recent problems with heat supply and expectations of rising energy prices can be considered to have played a role in local communities being stimulated to organise their own energy supply with the emphasis on local provision, particularly at the district level. This can be seen as part of a 'decentralisation' trend in Germany [59] in which citizens want to move away from a large centralised organisation of utilities. In that sense, decentralised CHP technology offers a promising alternative, as both heat and power can be produced efficiently at the district level. Moreover, it also allows for the use of renewable energy sources (which appeals to citizens who prefer 'green' energy sources to fossil ones). Connecting to the district heating systems could offer interesting opportunities for upscaling decentralised CHP use (e.g. from the building level to the district level). However, this requires that concession holders of the district heating grid allow for such use and do not only use it but also sell excess heat to industry and the large fossil-fuelled power plants. The ending of district heating systems' exploitation concessions in 2015 offers a window of opportunity for municipalities, local communities and perhaps citizen's energy cooperatives to consider decentralised CHP. Policy frameworks, in particular the *Energiewende* program, can influence the policy agendas of NRW and the Ruhr Valley municipalities including ownership of the heating grids. Nonetheless, another landscape event, the economic crisis, has had a direct negative financial impact on the capability of governments to allocate budgets for investments in the niche development of decentralised CHP units.

Meso level: regime factors

Currently, potential users (particularly social housing corporations and private homeowners living in multistorey buildings) are hardly aware of the technology's benefits. When users are aware of the benefits, they appear not to be willing to make the investment required to install decentralised CHP units. This latter situation might be linked to the decision-making practices of building owners. First, private homeowners need collective action in order to decide on investments and plans towards installing CHP units. This is time consuming with high-transaction costs. Second, regulations do not allow housing corporations to become energy producers. In addition, decentralised CHP does not seem to match with current user preferences and habits concerning (common) energy consumptive practices. Without social housing policy changing, an important stakeholder is left without institutional allowance to even start using (medium-sized) decentralised CHP units. This situation creates the impression that there is, as yet, little market demand. The installation sector being not willing to invest in specialising in decentralised CHP technology and therefore lacking service provision to accommodate potential users of the technology is also problematic.

The situation with these two stakeholder groups' lack of enthusiasm for decentralised CHP can be linked to the absence of effective support policies, in particular policy instruments to support potential users in making upfront investments. As a consequence, potential users adopt a 'wait and see' strategy and delay making investments until sufficient, trustworthy government programs are in place. As a consequence, current conditions for creating a decentralised CHP (niche) market in the Ruhr Valley are limited. It seems that decentralised CHP still represents a niche in an infant development stage: take off and diffusion into larger market segments is currently not taking place.

Micro level: niche development

Within the Ruhr Valley, a protected space in which room for experimentation with decentralised CHP is created is the 'Innovation City Ruhr' in the city of Bottrop. The latter was selected by the State of NRW to become a 'climate friendly model city' in the Ruhr Area. The aim is to transform a city district into an energy efficient area, which includes the construction of a decentralised CHP unit based on renewable energy sources, and in so doing provides a best practice example for the larger Ruhr Valley region.

In addition, the policies designed by the nine Ruhr Valley municipalities to support decentralised CHP uptake show a wide scope of actions of how this 'niche' can be further developed. Actions planned by Ruhr area municipalities differ in various ways. Whereas some municipalities only strive to install decentralised CHP in public buildings, others target local stakeholders, like households and commercial firms. Moreover, municipalities deploy different policy actions, such as constructing apartment block-heating installations, integration of decentralised CHP units in urban development planning and establishing contracts with private home owners for the provision of heat. Decentralised CHP is also seen as a means to feed in decentralised-produced heat into the existing district heating grids. Studies regarding the technical and economic potential or feasibility of decentralised CHP have been prepared in municipalities, and plans for constructing decentralised CHP units are prepared. The ways the city governments address the uptake and integration of decentralised CHP show little coherence: there is little alignment in the formulation of goals and coordination policy actions between Ruhr Valley cities. From a systemic - strategic niche management - perspective, the proposed actions look somewhat promising on the one hand, but rather uncoordinated and little aligned on the other hand. Moreover, there seems not to be sufficient shielding and nurturing of the 'decentralised CHP niche'. The same can be said in respect of the proposed actions to destabilise the incumbent regime structure. Nonetheless, as mentioned before, several landscape events are creating tensions in the regime which could create momentum for further niche development. In the end, however, municipalities cannot bring about systemic change alone. Although cities in the Ruhr Area would like to contribute to climate protection goals with their own utility companies and the possession of the electricity grid, they are depending on other stakeholders, like energy companies and distributed system operators. Many of the district heating concession contracts will end in 2015 which opens up opportunities for regime change. Energy companies are said (according to our interviewees) to use gaps in the energy anti-monopoly regulations to hamper this development. Therefore, a change of regulatory schemes is deemed a necessary requirement to create a level playing field [60].

Discussion

The *Energiewende* does not necessarily only need to rely on supporting green electricity production. Actors in urban environments like the Ruhr Valley could also embrace harvesting the potential of energy efficiency goals, *in casu* adopting decentralised CHP technology. The results, however, show that many barriers have been identified that block the uptake and integration of this technology in practice.

The stakeholder analysis proved useful in giving us more insight into the 'agency' factor of the Ruhr Valley energy system. In line with [38-40], we feel that providing more insight into agency (in our case, by conducting stakeholder analysis prior to using/applying MLP) provides/adds complementary functions in a methodological sense. It contributes to identifying regime barriers and interlevel system dynamics. We leave it to scholars in the field of sustainable transitions to judge whether this can be concerned a methodological 'novel insight'.

Furthermore, we feel that paying attention to specific regional characteristics of the Ruhr Valley is worthwhile as well. Without understanding the industrial character, the availability of district heating infrastructures (and its ownership contracts via concessions), the influence of the gas and coal lobby, the large-scale of dwellings owned by social housing corporations and the decentralised governments who can afford little budget to formulate and implement significant niche supporting programs, it is difficult to understand why the Ruhr Valley's energy system is not responding positively to the uptake and integration of decentralised CHP. Hence, in line with [50], we argue that paying special attention to geographically bound comparative (dis-) advantageous circumstances matters in understanding energy system dynamics. Furthermore, we agree with [48,49] that policies also matter and therefore deserve sufficient analytical attention. If policies that intend to support niche developments are non-aligned and are not coordinated with sectoral policies (e.g. social housing policies addressing housing corporations' entitlements in not becoming energy producers), niche development (e.g. uptake of decentralised CHP technology) is not expected to take off [52]. In addition, in the absence of using 'systemic instruments' that exceed local authorities' action plans and jurisdictions, this is not expected to happen either [48]. In conclusion, we agree with Geels [52] that the understanding of power relations at the regime level is critical to the further understanding of greening energy systems. In our opinion, this especially applies to urban, industrial and energy-intensive regions.

Next to addressing the use of MLP and reflecting on the added value of the theoretical contributions to elaborate on MLP, TM and SNM, as mentioned in the academic literature (see the paragraph in the 'Theoretical framework: using the multilevel perspective to analyse regional energy transitions' section), we also want to address the commonalities that our analysis show with results from empirical studies addressing barriers that impede the adoption of (green) energy innovations in city districts. They concern difficulties experienced when trying to have such innovations adopted in existing urban areas, such as the lack of finance, lack of trust and little effective (local) government policies to persuade public (social housing associations) [61,62] and private homeowners [63] to adopt measures to increase the energy efficiency of dwellings. As shown in the Ruhr Valley case study, the lack of market demand and the lack of incentives from the regulatory framework fail to stimulate change in the energy regime. The situation resembles the status quo in (existing) urban contexts in other Western Europe with a lack of diffusion of 'green' energy innovations (see e.g. [64]). Learning from these experiences and taking these lessons into account can be useful for those interested in studying and formulating policies to stimulate adoption of these kinds of innovations (notably addressing energy efficiency), which is now seen as a key challenge in light of the German *Energiewende* [22].

Generalisation of the results from this study is difficult since we focus on only one case (the Ruhr Valley region). The study, however, can be replicated in other regions. These regions should, however (at least) match some of the main characteristics of the Ruhr Valley region: highly industrialised, energy intensive and urbanised. Studying the results and system innovation indicators of theoretical interest in a set of multiple (comparable) cases would allow for more systematic research and would contribute to the academic literature.

Conclusions

Energy-intensive regions and energy efficiency in buildings are considered to have taken a backseat in Germany's *Energiewende* program. Recently, the importance of these issues has received scholarly attention [23]. In this paper, we contribute to this agenda. The main question in this paper is: *Which factors block the uptake and integration of decentralised CHP in the German Ruhr area's energy system?*

Following a stakeholder analysis, and a systemic analysis using MLP, the results show that the incumbent regime provides substantial barriers that prevent large-scale uptake and integration of decentralised CHP. These barriers relate to lack of financial means for making investments, unawareness and information deficit regarding the benefits of decentralised CHP to potential users, the presence of centralised district heating systems, lack of sufficient policies supporting diffusion of decentralised CHP units, legal stipulations from social housing policies that prevent housing cooperatives from becoming energy producers, district heating systems owned by public and private owners (via concessions contracts), lack of skilled service-providing companies and vested interests by the coal and gas industry. Moreover, many of the mentioned factors seem interrelated, especially those linked to policy and finance for upfront investments.

Although the study presented in this paper was conducted in a region with rather unique characters, it would be worthwhile to further the research agenda on the uptake and integration - and more in general: niche development - of decentralised CHP or other decentralised energy production technologies - in regions. Future academic work in this field would particularly be useful if comparative regional studies are conducted. Our paper has also demonstrated the added value of conducting stakeholder analysis as a complementary method to the use of MLP, notably as a way that agency can be explored. Moreover, we argue that conducting a stakeholder analysis prior to studying sustainable transitions in case study research designs using the more commonly used theoretical frameworks and heuristic tools in this disciplinary field (e.g. MLP or SNM) contributes in a methodological sense to the way transition studies address agency. For this reason, we encourage future case studies on sustainable transitions to adopt stakeholder analysis methodology.

Endnotes

[a][19] EU Tech Energie & Management (2008) NRW-Klima2020 - Beitrag Nordrhein-Westfalens zur Erreichung des nationalen Klimaschutzziels. Aachen

[b]There is one exception: the city of Duisburg. The reason is that the district heating grid already supplies heat to the inhabitants of the municipality. The city already goes beyond the national objective to raise the share of electricity produced in CHP plants to 25%.

[c]For example, in 2012, the municipality of Oberhausen was the most indebted municipality in Germany [15].

[d]We would like to stress, however, that in the context of the Ruhr area there is no common conceptualization of 'decentralised CHP'. From our interviews with local practitioners, we conclude that there seem to be multiple understandings of 'decentralised CHP' varying among (types of) stakeholders and contexts.

[e]For this reason, we decided to include conducting a stakeholders' analysis in our research design. Stakeholder analysis reveals much of existing regime barriers, agency relations, power and politics.

[f]Here we assume a relationship between the spatial type of region and the uptake of CHP. For classifying a region, we assume an ordinal scale which on the extreme can be classified as 'rural/non-industrial' and on the other side as 'urban/high industrial'. Hence, the Ruhr Valley region can be discerned as an 'extreme case' [58] since it is characterised by its very urban and highly industrial nature.

[g]The interviewees were directors of environmental and climate protection units in city governments, energy consultants at the regional and local level (including a representative from the National CHP Association), managers of energy suppliers, a representative of a public utility company, representatives of an energy agency and a research institute, and a CHP project manager from city government.

[h]Data were collected in the weeks prior to the September 2013 elections.

Competing interests

The authors declare that they have no competing interests.

Authors' contributions

All authors were involved in drafting the manuscript. All authors read and approved the final manuscript.

Acknowledgements

The authors would like to thank the Wuppertal Institute for Climate, Environment and Energy and particularly Dr. Johannes Venjakob for his contribution. The institute hosted the main researcher and supported the identification of the stakeholder network and the organisation of interviews. We also like to thank three independent reviewers for assessing the manuscript and providing us with suggestions for improvement.

Author details

[1]Department: Nature and Environment, Bonner Str. 100, 42697 Solingen, Germany. [2]Department of Governance and Technology for Sustainability (CSTM), Institute for Governance and Innovation Studies (IGS), University of Twente, Drienerlolaan 5, 7522, NB Enschede, The Netherlands.

References

1. Jacobssen S, Johnson A (2000) The diffusion of renewable energy technology: an analytical framework and key issues for research. Energy Policy 28:625–640
2. Loorbach D, Verbong G (2012) Governing the Energy Transition, Reality, Illusion or Necessity. Routledge, New York, London
3. Westner G, Madlener R (2011) Development of cogeneration in Germany. A mean-variance portfolio analysis of individual technology's prospects in view of the new regulatory framework. Energy 36:5301–5313
4. Agentur für Erneuerbare Energien (2012) Deutschlands Informationsportal für Erneuerbare Energien. Berlin.
5. Ministerium für Klimaschutz, Umwelt, Landwirtschaft, Natur- und Verbraucherschutz des Landes NRW (2013) Kraft-Wärme-Kopplung. Impulse für Energiewende "Made in NRW". Düsseldorf https://broschueren.nordrheinwestfalendirekt.de/broschuerenservice/energieagentur/kraft-waerme-kopplung/1550. Accessed on 16 July 2013
6. DSGV&VKU (2012) Gemeinsam für die kommunale Energiewende. Auf dem Weg in eine klimafreundliche Zukunft, Berlin
7. Presse- und Informationsamt der Bundesregierung (2011) Deutschlands Energiewende. Ein Gemeinschaftswerk für die Zukunft, Berlin
8. Hoppe T, Coenen F (2011) Creating an analytical framework for local sustainability performance: a Dutch case study. Local Environ 16(3):229–250
9. Truffer B, Coenen L (2012) Environmental innovation and sustainability transitions in regional studies. Reg Stud 46(1):1–21
10. EU Tech Energie & Management (2008) NRW-Klima2020 – Beitrag Nordrhein-Westfalens zur Erreichung des nationalen Klimaschutzziels. Aachen
11. Rip A, Kemp R (1998) Technological change. In: Rayner S, Malone EL (eds) Human Choice and Climate Change. Battelle, Columbus, pp 327–399
12. Geels F (2002) Technological transitions as evolutionary reconfiguration processes: a multi-level perspective and a case study. Res Policy 31:1257–1274
13. Regionalverband Ruhr (2013) Regionalverband Ruhr – Aufgaben und Verbandsgebiet., Essen http://www.metropoleruhr.de/regionalverband-ruhr/ueber-uns/gebiet-aufgaben.html. Accessed on 28 May 2013
14. Ruhrgebiet Regionalkunde (2013) Die administrative Gliederung des Ruhrgebietes., Essen http://www.ruhrgebiet-regionalkunde.de/grundlagen_und_anfaenge/kohle/kohle.php?p=2 = 0 Accesed on 28 May 2013
15. Geographie Infothek (2012) Infoblatt Strukturwandel im Ruhrgebiet. Stuttgart http://www2.klett.de/sixcms/list.php?page=geo_infothek&miniinfothek=&node=Ruhrgebiet&article=Infoblatt+Strukturwandel+im+Ruhrgebiet. Accessed on 22 May 2012
16. Beggs C (2011) Energy: Management, Supply and Conservation. Spon Press, London/New York
17. Dena – German Energy Agency (2013) Combined heat and power generation. Berlin http://www.dena.de/en/projects/energy-systems/combined-heat-and-power-generation.html Accessed on 26 June 2013
18. Dena – German Energy Agency (2013): Kraft-Wärme-Kopplung. Berlin http://www.effiziente-energiesysteme.de/themen/kraft-waerme-kopplung/heraus-forderungen.html Accessed on 26 June 2013
19. BMU - Bundesministerium für Umwelt, Naturschutz und Reaktorsicherheit (2009) Langfristszenarien und Strategien für den Ausbau erneuerbarer Energien in Deutschland. Leitszenario, Berlin
20. Eikmeier B, Klobasa M, Toro F, Menzler G (2011) Potenzialerhebung von Kraft-Wärme-Kopplung in Nordrhein-Westfalen, Bremen
21. BET – Büro für Energiewirtschaft und technische Planung GmbH (2013) Perspektiven der Fernwärme im Ruhrgebiet bis 2050. Aachen
22. Gawel E, Lehmann P, Korte K, Strunz S, Bovet J, Köck W, Massier P, Löschel A, Schober D, Ohlhorst D, Tews K, Schreurs M, Reeg M, Wassermann S (2014) The future of the energy transition in Germany. Energy Sustainability Soc 2014(4):15

23. Kemp R (1994) Technology and the transition to environmental sustainability. Futures 26:1023–1046
24. Geels FW, Schot J (2010) The dynamics of sociotechnical transitions – a sociotechnical perspective. In: Grin J, Rotmans J, Schot J (eds) Transitions to Sustainable Development. Routledge, New York, pp 9–101
25. Geels F (2004) From sectoral systems of innovation to socio-technical systems. Insights about dynamics and change from sociology and institutional theory. Res Policy 33:897–920
26. Markard J, Raven R, Truffer B (2012) Sustainability transitions: an emerging field of research and its prospects. Res Policy 41(6):955–967
27. Smith A, Stirling A, Berkhout F (2005) The governance of sustainable sociotechnical transitions. Res Policy 34:1491–1510
28. Unruh GC (2000) Understanding carbon lock-in. Energy Policy 28:817–830
29. Markard J (2011) Transformation of infrastructures: sector characteristics and implications for fundamental change. J Infrastructure Syst (ASCE) 17:107–117
30. Verbong G, Geels F (2007) The ongoing energy transition: lessons from a socio-technical, multi-level analysis of the Dutch electricity system (1960–2004). Energy Policy 35(2):1025–1037
31. Smith A, Raven R (2012) What is protective space? Reconsidering niches in transitions to sustainability. Res Policy 41:1025–1036
32. Kemp R, Schot J, Hoogma R (1998) Regime shifts to sustainability through processes of niche formation: the approach of strategic niche management. Tech Anal Strat Manag 10(2):175–198
33. Raven RPJM (2005) Strategic niche management for biomass: a comparative study on the experimental introduction of bioenergy technologies in the Netherlands and Denmark. PhD thesis, TU/E Eindhoven
34. Rotmans J, Kemp R, van Asselt M, Geels F, Verbong G, Molendijk K (2000) Transities & transitiemanagement. De casus van een emissiearme energievoorziening, Rotterdam
35. Loorbach DA (2007) Transition management: new mode of governance for sustainable development. PhD thesis. Erasmus University, Rottredam
36. Geels FW, Kemp R (2005) Transitions, transformations and reproduction: dynamics in socio-technical systems. Paper presented at the DRUID Tenth Anniversary Summer Conference, Copenhagen Denmark
37. Shove E, Walker G (2007) CAUTION! Transitions ahead: politics, practice and sustainable transition management. Environ Plann A 39:763–770
38. Van den Bergh JC, Truffer B, Kallis G (2011) Environmental innovation and societal transitions: introduction and overview. Environ Innov Soc Trans 1(1):1–23
39. Raven RPJM, Verbong GPJ, Schilpzand WF, Witkamp MJ (2011) Translation mechanisms in socio-technical niches: a case study of Dutch river management. Tech Anal Strat Manag 23:1063–1078
40. Farla J, Markard J, Raven R, Coenen L (2012) Sustainability transitions in the making: a closer look at actors, strategies and resources. Technol Forecast Soc Chang 79(6):991–998
41. Markard J, Truffer B (2008) Actor-oriented analysis of innovation systems: exploring micro-meso level linkages in the case of stationary fuel cells. Tech Anal Strat Manag 20:443–464
42. Grin J (2010) Understanding transitions from a governance perspective. In: Grin J, Rotmans J, Schot J (eds) Transition to Sustainable Development. Routledge, New York/London, pp 221–337
43. Avelino F, Rotmans J (2009) Power in transition: an interdisciplinary framework to study power in relation to structural change. Eur J Soc Theory 12(4):543–569
44. Avelino F (2011) Power in transition: empowering discourses on sustainability transitions. PhD thesis, Erasmus University, Rotterdam
45. Shove E, Walker G (2010) Governing transitions in the sustainability of everyday life. Res Policy 39:471–476
46. Seyfang G, Smith A (2007) Grassroots innovations for sustainable development: towards a new research and policy agenda. Environ Polit 16(4):584–603
47. Arentsen M, Bellekom S (2014) Power to the people: local energy initiatives as seedbeds of innovation? Energy Sustainability Soc 4:2
48. Smits R, Kuhlmann S (2004) The rise of systemic instruments in innovation policy. Int J Foresight Innov Policy 1(1):4–32
49. Weber M, Rohracher H (2012) Legitimizing research, technology and innovation policies for transformative change. Res Policy 41:1037–1047
50. Coenen L, Benneworth P, Truffer B (2012) Towards a spatial perspective on sustainability transitions. Res Policy 41:968–979
51. Geels FW (2011) The multi-level perspective on sustainability transitions: responses to seven criticisms. Environ Innov Soc Trans 1(1):24–40
52. Geels FW (2014) Regime resistance against low-carbon transitions: introducing politics and power into the multi-level perspective. Theory, Culture & Society doi:10.1177/0263276414531627
53. Bulkeley H, Betsill M (2003) Cities and Climate Change: Urban Sustainability and Global Environmental Governance. Routledge, New York London
54. Bulkeley H (2013) Cities and Climate Change. Routledge, New York London
55. Hoppe T, van den Berg MM, Coenen FHJM (2014) Reflections on the uptake of climate change policies by local governments: facing the challenges of mitigation and adaptation. Energy Sustainability Soc 4:8
56. Hoppe T, van Bueren E (2014) Governing the climate challenge and energy transition in cities. Energy, Sustainability and Society, Thematic Series., http://www.energsustainsoc.com/series/GCCE
57. Gerring J (2007) Case Study Research. Principles and Practices, Cambridge
58. Varvasovszky Z, Brugha R (2000) How to do (or not to do) … A stakeholder analysis. Health Policy Plan 15:338–345
59. Oteman M, Wiering M, Helderman JK (2014) The institutional space of community initiatives for renewable energy: a comparative case study of the Netherlands, Germany and Denmark. Energy Sustainability Soc 4(1):11
60. Berlo K, Wagner O (2013) Harter Gegenwind bei der Rekommunalisierung. AKP 4:22–23
61. Hoppe T, Bressers H, Lulofs K (2010) Energy Conservation in Dutch Housing Renovation Projects. In: Martens P, Chiang CT (eds) The Social and Behavioural Aspects of Climate Change. Greenleaf, Sheffield, pp 68–95
62. Hoppe T (2012) Adoption of innovative energy systems in social housing: lessons from eight large-scale renovation projects in The Netherlands. Energy Policy 51:791–801
63. Murphy L, Meijer F, Visscher H (2012) A qualitative evaluation of policy instruments used to improve energy performance of existing private dwellings in the Netherlands. Energy Policy 45:459–468
64. Faber A, Hoppe T (2013) Co-constructing a sustainable built environment in the Netherlands—Dynamics and opportunities in an environmental sectoral innovation system. Energy Policy 52:628–638

10

Digestion of bio-waste - GHG emissions and mitigation potential

Jaqueline Daniel-Gromke*, Jan Liebetrau, Velina Denysenko and Christian Krebs

Abstract

Background: For a precise description of the emission situation of the anaerobic digestion (AD) of the separately collected organic fraction of household waste (bio-waste), only a few data are available. The paper presents the greenhouse gas (GHG) emissions measured at 12 representative AD plants treating bio-waste. The results of the emission measurements were used to assess the ecological impact of bio-waste digestion and to describe possible mitigation measures to reduce the occurring GHG emissions. With respect to the climate protection, a quantitative assessment of the emissions of energy generation from biomass and biological waste treatment is important. Biogas plants need to be operated in a way that negative environmental effects are avoided and human health is not compromised.

Methods: GHG balances were calculated based on the measured emissions of the gases methane, nitrous oxide, and ammonia of bio-waste AD plants. The emission analysis supports the reduction of GHGs in biogas production and contributes to a climate-efficient technology.

Results: The results show that GHG emissions can be minimized, if the technology and operation of the plant are adjusted accordingly. The open storage of active material (e.g., insufficient fermented residues from batch fermentation systems), open digestate storage tanks, missing acidic scrubbers in front of bio-filters, or insufficient air supply during the post-composting of digestate can cause relevant GHG emissions.

Conclusions: Consequently avoiding open storage of insufficient fermented residues and using aerated post-composting with short turnover periods, smaller heaps, and an optimized amount of structure (woody) material can reduce GHG emissions.

Keywords: Anaerobic digestion; Emission measurement; Greenhouse gas balance; GHG mitigation

Background

Gaseous emissions are of great importance referring to the operation of biogas plants because they can affect the safety, the greenhouse gas (GHG) balance, and the economy of plants significantly. Depending on the used technology and the kind of operation, GHG emissions like methane, nitrous oxide, and ammonia are occurring. Methane emissions dominate GHG emissions of biogas plants.

Due to the global warming potential (GWP) of 25 relative to carbon dioxide [1], methane emissions have a strong effect on the climate change. Leakages, process disturbances, and unavoidable emissions during operation can influence the total GHG performance of the biogas plant negatively. Regarding measured emissions of biogas plants in operation, only a small number of detailed studies are available.

In former studies, the overall emissions of biogas plants usually have been estimated by assumptions, e.g., '1 % of diffuse methane emissions from the components of anaerobic digestion (AD) plants like digester, pipes,' etc. (e.g. [2,3]). However, in the recent years, several studies estimated methane emissions from biogas plants (e.g. [4-9]). Most of the published studies analyzed agricultural AD plants; if waste treating plants were investigated, only a few AD components were monitored as summarized by Dumont et al. [10]. Due to the fact that there are only few data describing the emission situation of AD plants based on bio-waste, in the study described here, 12 representative bio-waste treatment plants with AD process as part of

* Correspondence: jaqueline.daniel-gromke@dbfz.de
Department Biochemical Conversion, DBFZ Deutsches Biomasseforschungszentrum gGmbH, Torgauer Straße 116, 04347 Leipzig, Germany

the overall operation were analyzed. The overall objective of the study was a detailed analysis of GHG emissions generated from biogas production from bio-waste. This paper presents the results of a comprehensive measurement for GHG emissions at bio-waste digestion plants in operation during a long-term period of 3 years. Representative bio-waste digestion plants have been selected, and all relevant components of the process chain were investigated during two periods of a week per year on each of the selected plants to identify the main emission sources and the quantity of the emissions. The results of the emission measurements were implemented in an ecological assessment focused on GHG balances. The results of the examined biogas plants allow a description of possible mitigation measures to reduce GHG emissions. The results bring new aspects into the actual data base to support the assessment of environmental impacts of bio-waste digestion. Thus, the tests on practice biogas plants with respect to the whole process chain allow an optimization of the process in terms of reducing any identified emissions.

In Germany, approximately 9 million tons of bio-waste and green waste per year were collected separately in 2011 [11]. Most of this collected bio-waste and green cuts are used in composting processes. About 1.15 million tons of bio-waste per year and 0.05 tons of green cuts per year are used for digestion in biogas facilities [12]. By the end of 2013, there have been about 130 plants generating biogas from organic waste in operation. Compared to agricultural biogas plants, there is a higher share of dry fermentation processes in AD plants based on bio-waste. About one half of the bio-waste digestion plants are operated as dry fermentation plants in Germany, whereas half of the dry fermentation plants are operated discontinuously (batch system). Currently, there are 25 batch systems based on bio-waste in operation [13]. Due to the robustness of the process and the possibility to treat substrates which are hardly pumpable and contain disturbing materials (e.g., stones, metals, glass), the use of batch systems in case of dry fermentation processes of bio-waste is increasing. In the future, it will be more important to exploit additional potentials in the field of organic waste and residues from industry and municipalities. In the field of municipal bio-waste, the exploitation of additional potentials is in progress. The amount of municipal bio-waste that is available for digestion in biogas plants will increase considerably within the next years. Currently, a considerable trend to digestion of bio-waste and green waste, often integrated as so-called upstream systems into existing composting plants, can be assessed.

Methods

Twelve biogas plants were selected for the detection of plant-based emissions of methane (CH_4), nitrous oxide (N_2O), and ammonia (NH_3). Based on the measured emission rates, GHG balances in compliance with the analysis of GHG credits (e.g., for biogas production, fertilizer, and humus effect of fermentation products and composts) were prepared. Thus, the electricity production and heat utilization of biogas as well as the credits of the various fermentation residues were analyzed to estimate the specific GHG performance of the investigated facilities. Finally, the measurements with respect to mitigation of GHG emissions were analyzed and described.

Investigated biogas plants

The emission analysis includes four continuously operated wet fermentation plants (continuous stirred-tank reactor, CSTR), five continuous dry fermentation plants (plug-flow fermenter), and three batch fermentation processes (discontinuous operation, 'garage style' digesters). Table 1 shows the investigated 12 AD plants based on bio-waste with their specific characteristics. Table 2 presents the amount and kind of substrate treated at the bio-waste facility. The treated bio-waste is used completely for digestion in AD plant nos. 2, 4, and 5. Most AD plants operate with partial stream digestion of bio-waste. In these plants, just the bio-waste from separate collection is used for fermentation, whereas the green cut and structure (woody) material is added after digestion within the composting process.

AD plant nos. 1, 2, and 12 were operated with open, unaerated post-composting processes. AD plant no. 3 had a covered but no enclosed composting steps. In AD plant no. 4, larger quantities of sludge from wastewater treatment were treated. Thus, primarily liquid digestate was generated. The small amounts of solid digestate were stored on site and were used for external composting. The solid digestate of AD plant no. 5 were stored open after separation. Post-composting processes with active ventilation (pressure ventilation) and enclosed composting systems were used at AD plant nos. 7, 9, and 10. A defined step of aeration in which the air is integrated into the exhaust gas treatment (bio-filter) was considered at plant no. 10.

All investigated biogas facilities operated with bio-filters as gas treatment. However, most of plant operators did not use acidic scrubbers at biogas facilities. Only four of 12 plants operated with acidic scrubbers, and the proper operation was not always ensured. Five plants used the bio-filter combined with humidifier. The exhaust gas should be treated with acid scrubbers to deposit NH_3 and minimize N_2O formation in the bio-filter (e.g., plant nos. 5 and 9). It should be recognized that there were also diffuse emission sources which were not collected by bio-filters (e.g., open doors of delivery hall at AD plant nos. 6 and 7; post-composting at AD plant nos. 8, 9, 11).

Table 1 Characteristics of investigated AD plants based on bio-waste

Plant no.	Installed electrical capacity kW_{el}	Kind of fermentation[a]	Temperature[b]	Range of temp. in °C[c]	Mode of operation	HRT in days[d]	Residues storage tank	Post-composting[e]	Type of aeration (post-composting)	External heat utilization[f]
1	630	Wet	M		Multi-stage	8	Covered	x	Open, unaerated	
2	536	Wet	T		Multi-stage	20	Covered	x	Open, unaerated	x
3	986	Wet	M		Single-stage	17	Open			
4	1200	Wet	M	37-40	Multi-stage	25	Open, covered			
5	1790	Dry	T		Single-stage	25	Gas-proof covered			
6	1413	Dry	T	55	Single-stage	21	Covered	x	Open, unaerated	
7	816	Dry	T		Single-stage	28		x	Enclosed, aerated (pressure ventilation)	
8	625	Dry	T	55	Single-stage	14	Gas-proof covered	x	Enclosed, aerated and unaerated (pressure ventilation)	x
9	640	Dry	T		Single-stage	21	Covered	x	Enclosed, aerated (pressure ventilation)	
10	625	Batch	M	37-39	Single-stage	28		x	Enclosed, aerated (pressure ventilation)	x
11	680	Batch	M	40-42	Single-stage	21		x	Open, aerated, enclosed	x
12	370	Batch	M	40-42	Single-stage	21		x	Open, unaerated	x

[a]Wet = wet fermentation, dry = dry fermentation, batch = batch system (discontinuous). [b]M = mesophilic, T = thermophilic. [c]According to the information of plant operators (if available). [d]Hydraulic retention time. [e]x = post-composting process. [f]x = external heat utilization.

Table 2 Amount and kind of treated substrate of investigated bio-waste facilities

Plant no.	Total amount of substrate input treated at facility t/a (fresh matter)	Amount of input for AD t/a (fresh matter)	Percent share of AD (mass) related to total amount treated at facility	Kind of substrate[a]
1	83.840	32.000	38	Bio-waste, green cut
2	10.062	9.865	98	Bio-waste, catering waste, green cut
3	34.976	25.702	73	Bio-waste
4	35.388	35.388	100	Sludge from wastewater treatment, catering waste, food waste, manure
5	33.130	33.130	100	Bio-waste, catering waste
6	29.900	26.910	90	Bio-waste, green cut
7	35.450	17.725	50	Bio-waste
8	20.000	12.000	60	Bio-waste, green cut
9	23.000	17.250	75	Bio-waste, green cut
10	36.000	18.000	50	Bio-waste, green cut
11	73.333	22.000	30	Bio-waste, green cut
12	13.333	12.000	90	Bio-waste, green cut, food-waste

[a]Bio-waste = bio-waste from separate collection.

Often, digestate - whether separated or not separated - is stored open temporarily or for longer periods. Four of the seven examined plants which stored liquid digestate or process waters used covered storage tank (AD plant nos. 4, 5, 8, and 9). Two plants (nos. 5 and 8) with gas-proof covered storage tank are able to use the exhaust gas by involving into the CHP.

Emission measurements

There are in general two methods to determine the emissions of a large industrial facility or areas with diffuse emission sources. One way is to attempt to capture the overall emissions of the facilities by means of concentration measurements in the surroundings and the application of inverse dispersion models [7] or radial plume mapping [14]. These methods allow the determination of the overall emissions of a large area with uncertain sources of emission. They do not allow the localization of single sources and allocation of a certain quantity to them. However, for further efficient measures to reduce emissions, it is very important to identify and quantify the emission sources on site. For this reason, the methods used focus on the identification and quantification of single sources [5].

The emission analysis included two measurement periods in each plant (each 1 week in 2010 and 2011), in which all plant components from substrate delivery to storage of digestate and composting were investigated. The measured emissions of both periods were averaged. Several sampling points at AD plant and compost heaps were examined. Following the inspection of the biogas facilities on site, potential significant emission sources within the process chain were identified. The following emission sources were investigated: delivery and conditioning of substrate (material handling), storage of fermentation residues (digestate), fermenter, before and after exhaust gas treatment (acid scrubber and bio-filter), and exhaust of CHP unit (combined heat and power plant) as well as post-composting process of digestate. The emission measurements focused on the emission detection at the AD plant and post-composting processes - not the utilization of biogas in CHP units. Therefore, not all CHP were measured. With respect to the total GHG balance, the production as well as the utilization of biogas in CHP is important. Thus, an average of CHP emissions was considered (see 'Emissions from CHP'). For the emission measurements of the composting process, four or five sections of the windrow were selected for each measurement period, which differed in time of composting resp. age of rotting material.

According to the characteristics of the gases, the applied measurement techniques were adjusted. Leakage detection techniques were used to find the critical spots within the process; open and closed domes were used to determine the main emission sources. Regarding the methods of emission measurements, there are differences between captured and diffuse emission sources. Accordingly, different measurements for emissions from encapsulated areas (e.g., delivery hall with collection of exhaust) and diffuse emission sources during several measured periods were used. Waste treatment facilities often have gas collection systems that collect air from the captured process steps and deliver the gas after a cleaning stage into the atmosphere. In most cases, the cleaning step is a bio-filter. Because of that, in all investigated AD plants, the exhaust streams before and after treatment by bio-filters were examined. Depending on the plant system, further sampling points were analyzed. In case of encapsulated emission sources, the exhaust air flow was examined directly. Thereby, the volume flow and mass concentration within the investigated pipelines were determined. The volume flows were measured with vane anemometers. The quantity of the emission source was calculated from the concentration difference and the flow rate of the blower by using the following equation [5].

$$F = Q * \rho * (c_{out} - c_{in}) \quad (1)$$

F, emission flow rate (mg/h); *Q*, air flow rate (m^3/h); ρ, density of the target gas (kg/m^3); c_{out} exhaust gas concentration (mg/kg); c_{in}, background gas concentration (mg/kg).

Emissions of post-composting with active aeration (e.g., actively ventilated tunnel or container systems) were measured by using encapsulated areas with air extraction. In case of open windrows composting without active aeration, a wind tunnel as emission measurement was used. An air flow was generated by using a ventilator. The measurement methods, techniques, and technical guidelines used for the determination of emission concentrations are shown in Table 3. CH_4 was detected by gas chromatography with a flame ionization detector (FID), N_2O by gas chromatography, and NH_3 by absorption in an acid solution. The sampling for the determination of CH_4 and N_2O was carried out by a measuring gas line which is connected to a gas analysis with online data collection. The sampling for the determination of NH_3 occurs directly at the tunnel exit. The sample gas is guided without gas cooling through two wash bottles filled with sulfuric acid. Further information according to the methods of emission measurement at biogas plants are published in [4].

Residual gas potential

The residual gas potential of digestate from anaerobic treatment of bio-waste was considered. The gas potential can be analyzed by different temperature levels as described by [16]. The temperature of the stored digestate has a great influence on the emissions. Laboratory tests within the studies of [17] and [18] showed that depending on the temperature of the digestate during storage, the emission potential can be significantly reduced. In [5], it is shown that the average CH_4 potentials obtained at 20°C represent 39% of the CH_4 potential obtained at 39°C. According to [17], the CH_4 production at a temperature of 25°C is reduced to 40–50% of the value obtained at 37°C and at 10°C, the CH_4 production goes down to even 1% [5].

In this study, the residual gas potential of digestate was determined at a temperature of 38°C. The digestate samples were taken directly after the fermentation step and - in case of separation of digestate - after separation (see AD plant nos. 1, 2, and 7). With these samples, batch experiments were carried out according to the German technical guideline VDI 4630 [19]. Finally, relative residual gas potentials with respect to the used fresh matter were determined using the following assumptions: average CH_4 yield of 74 m^3 CH_4 (STP) per metric ton fresh matter bio-waste, 10% degradation of fresh matter by the fermentation stage, and a separation ratio of 20% solid digestate to 80% liquid digestate.

Assumptions - GHG balances

Based on a survey of plant operator, additional emission-related data (e.g., energy demand, amount, and kind of heat utilization) were collected to prepare the GHG balance of each plant. For the total GHG balances, the emissions as well as credits for the kind of products (combined heat and electricity from biogas; fertilizer and humus supply from fermentation residues) were considered. The overall GHG performance of each AD plant included in particular the following: GHG emissions according to the measured components of AD plant, calculated emissions of the electricity demand

Table 3 Measurement methods, techniques and technical guidelines for the determination of emissions at the investigated AD plants [15]

Compound	Kind of determination	Measurement methods	Measurement techniques	German technical guideline used for the calculation of emissions
Total carbon	Continuously, online data	Flame ionization detection (FID)	Bernath Atomic 3006	VDI 3481 - 3, VDI 3481 - 4, DIN EN 12619, DIN EN 13526
Methane	Continuously, online data	Infra-red (IR) method	ABB Advance Optima URAS 14	
Nitrous oxide	Continuously, online data	Infra-red (IR) spectroscopy	ABB Advance Optima URAS 14	DIN EN ISO 21258
Methane	Discontinuously, laboratory analysis	Gas chromatography (GC) with autosampler	Sampling with evacuated vials	DIN EN ISO 25139
Nitrous oxide	Discontinuously, laboratory analysis	Gas chromatography (GC) with autosampler	Sampling with evacuated vials	VDI 2469 - 1
Ammonia	Discontinuously, laboratory analysis	Wet chemical methods with sulfuric acid	Sampling with Desaga-pump and 2 wash bottles	VDI 3496 - 1

(AD plant and CHP), calculated emissions during the application of the fermentation residues, credits for the electricity production from biogas (substitution of fossil electricity supply), credits for the utilization of exhaust heat (substitution of fossil heat), and credits for the use of fermentation products (substitution of fossil fertilizer and peat, humus effects).

The considered GHG emissions for all processes of bio-waste digestion were converted into CO_2 equivalents (CO_2-eq) by using characterization factors.

The following factors according to the GWP for a 100-year time period were stated: $CO_2 = 1$, $CH_4 = 25$, $N_2O = 298$ [1]. With respect to the NH_3 emissions, it is assumed that 1% of the NH_3 is converted to N_2O emissions [1].

As a functional unit of GHG balances, 'ton input bio-waste treated at facility (fresh matter)' was used. This unit included the total amount of waste treated at the facility (bio-waste and green waste - if any) - not only the amount of bio-waste in the fermentation process. In few biogas plants, municipal bio-waste from separate collection and green waste from gardens and parks were treated, but only the bio-waste is used in the step of digestion. After the fermentation process, the digestate is often combined with the green cuts within the post-composting process. Thus, the measured emissions of post-composting processes based on the treated waste at the facility in total.

In addition to the measured GHG emissions of the AD plants, further assumptions to calculate the GHG performance were considered.

Emissions from CHP

Due to the fact that not all CHP units were measured, an average emission value for the CHP is assumed. According to measurements of gewitra (personal communications), the median of CH_4 and N_2O emissions of 161 measured CHP units in the range from 300 to 1,000 kW_{el} were determined with 1,760 g CH_4 per ton of bio-waste and 2.1 g of N_2O per ton of bio-waste treated at the facility. Considering the emission factors [1] for N_2O (298) and CH_4 (25), a GWP of 44.6 kg CO_2-eq per ton of bio-waste was estimated for all CHP units.

The energy demand of the investigated biogas plants was determined according to the data of plant operators. It was estimated to cover the electricity demand by using external electricity from the grid. The electricity production in Germany in 2011 produced in average 559 g CO_2-eq per kWh_{el} [20].

Electricity production

The electricity production from biogas replaces fossil fuels and can be considered as credit [21]. The amount of credit for the electricity production depends on the amount of produced electricity referring to the data of plant operators. The electricity mix of Germany in 2011 with 559 g CO_2-eq per kWh_{el} [20] was assumed to calculate the credit of electricity production.

Heat utilization

The exhaust heat of electricity generation in CHP units can - if used - substitute heat production based on fossil fuels [21]. The avoided GHG emissions of fossil heat supply by providing heat for external utilization (e.g., district heating, drying process) was stated as heat credits. The amount of heat credit may vary depending on the amount of heat and type of fossil heat, which is replaced in the specific case. With regard to the substitution of fossil heat, an average of the specified external heat mix of 291 g CO_2-eq per kWh_{th} [21] was used to calculate the heat credits.

Digestate - fertilizer and humus effects

Depending on the kind of digestate, respectively, the kind of treatment of the fermentation residues (e.g., with/without separation, with/without post-composting after fermentation process), different utilization pathways of digestate have been considered. According to the kind of digestate (finished compost, fresh compost, liquid fermentation residues, solid digestate), different GHG emissions can be saved and considered to the GHG balances as credits (Table 1). Referring to the kind of digestate, the following credits were determined: substitution of mineral fertilizer (nitrogen, phosphorus, potassium), substitution of peat (only in case of finished compost), humus accumulation (carbon-sink), and humus reproduction (i.e., for maintaining soil fertility).

According to the nutrient content (i.e., nitrogen, phosphorus, potassium amounts) of investigated digestates, the production of mineral fertilizer can be substituted and is stated in GHG balances as credit. The following emission factors for the production of mineral fertilizer were assumed according to [22]: 6.41 kg CO_2-eq per kg nitrogen (N), 1.18 kg CO^2-eq per kg phosphorus (P_2O_5), and 0.663 kg CO_2-eq per kg potassium (K_2O).

Humus effects of digestate at investigated AD plants were considered if applied on agricultural land. To evaluate the humus effects of fermentation residues, estimations according to [23] were used. That means, for the amount of finished compost, 20% substitution of peat and 80% agricultural use, thereof 20% of humus accumulation and 80% of humus reproduction was assumed. For the scenario of humus, reproduction was stated - in contrast to [23] - that the substitution of straw is considered and the credits for the fermentation of straw with recirculation of the digestate can be estimated. The humus reproduction (i.e., for maintaining soil fertility) of digestates depends on the content of dry matter and organic dry matter as well as the degrading stability of organic dry matter. Data regarding the humus reproduction of digestate from AD based on bio-waste are not available. The humus reproduction of digestate of investigated AD plants was calculated. The characteristics (e.g., dry matter, organic dry matter, amount of nutrients especially nitrogen) of each digestate were determined based on the 1-year certificate of digestate referring to the quality assurance of the Federal Compost Association.

According to the kind of digestate, the substitution effect compared to straw was analyzed. Therefore, the amount of straw was calculated which might be used for biogas production if the application of digestate on agricultural land is assumed. Differed to the kind of digestate, the amount of straw per ton of digestate (fresh matter) was calculated as follows: 2.11 (finished compost), 1.82 (fresh compost), 0.91 (digestate with post-composting), and 0.15 (liquid digestate). The electricity production of the assumed biogas production due to the fermentation of straw was considered as credit for humus reproduction of digestate.

The substitution of peat was estimated only in case of finished compost. According to the assumptions in [24], 1 kg dry peat (respectively, 2 kg fossil carbon dioxide) is replaced by 1 kg compost (organic dry matter). Referring to the humus accumulation (carbon sink) of composted digestate, the amount of organic carbon (Corg) as published in [23] was assumed as follows: 21.6 kg Corg per ton of digestate for fresh compost and 64.5 kg Corg per ton of digestate for finished compost. In consideration of the stoichiometric ratio of Corg relative to CO_2, 1 kg Corg can fix 3.7 kg CO_2.

Application of digestate

The application of digestate on agricultural land can cause N_2O emissions as well as NH_3 emissions [25]. With respect to the NH_3 emissions, it was assumed that 1% of the NH_3 is converted to N_2O emissions [1].

Results and discussion

GHG emissions

Various fermentation processes such as wet fermentation, dry fermentation, and batch fermentation were analyzed according to the emission situation. The results show that the emissions are dominated not by the kind of the fermentation process or the technology but by the manner of plant operation.

Figure 1 shows the measured emissions of CH_4, N_2O, and NH_3 (converted to carbon dioxide equivalents) of the investigated AD plants. The range of determined plant emissions varied between 40 and 320 kg CO_2-eq per ton of bio-waste. The detailed presentation on the type of GHGs shows that the CH_4 emissions - except for plant no. 6 - dominate the indicated GHG equivalents of biogas facilities.

Important sources of GHG emissions were identified. The component-specific GHG emissions of the bio-waste digestion plants are presented in Figure 2.

Especially, the inadequate aeration directly after fermentation (in order to interrupt the methanogenic activity) processes as well as unaerated or less aerated post-composting processes caused extremely high GHG emissions (see plant no. 1, no. 2, or no. 12). In case of some of the investigated biogas plants, the emissions of post-composting are summarized in the amount of 'emissions after bio-filter' (e.g., AD plant no. 10). The overall emissions of AD plant no. 10 was quite low because all parts of the fermentation and post-composting process were totally encapsulated.

Furthermore, AD plant no. 6 showed higher NH_3 emissions due to the drying of digestate at higher temperature and higher pH value. In this case, the existing downstream acidic scrubber was out of operation during the measurements. The operation of the bio-filters can also be problematic; extremely wet bio-filters for example can cause additional CH_4 production as observed at AD plant no. 8.

Finally, on almost all AD plants, emission sources were identified whose intensity can be reduced if the state-of-the-art treatment technology was used (e.g., acid scrubber before bio-filter, aeration of post-composting). The results show that the open storage of fermentation residues (with or without separation step) should be avoided. In addition to unaerated post-composting processes and open storage of active material (e.g., solid digestate), the CHP was one of the most important sources of CH_4.

According to the measured residual gas potential of digestate, a wide range from 4 to 23% was determined. Ten of 12 samples of digestate of investigated AD plants showed a relative residual gas potential higher than 10%. A high relative residual gas potential means insufficient fermentation of the substrate. The residual gas potential of bio-waste digestion achieved the same range as agricultural AD plants which were operated as single stage processes, whereas in comparison to agricultural biogas plants with multi-stage process, the determined CH_4 potential of fermentation residues from bio-waste digestion provides basically higher values. Table 4 shows the gas potential of the investigated bio-waste plants compared to the gas potential of agricultural biogas plants as published in [17]. According to [17] where agricultural AD plants were investigated, discontinuous systems (batch) and single-stage systems have shown the highest residual gas potential. Moreover, multi-stage systems of agricultural AD plants achieved less than half of the residual gas potential of single-stage plants [17]. The results of [17] stated that single-stage processes achieve higher residual gas potential due to their generally shorter retention time. With respect to the investigated bio-waste AD plants

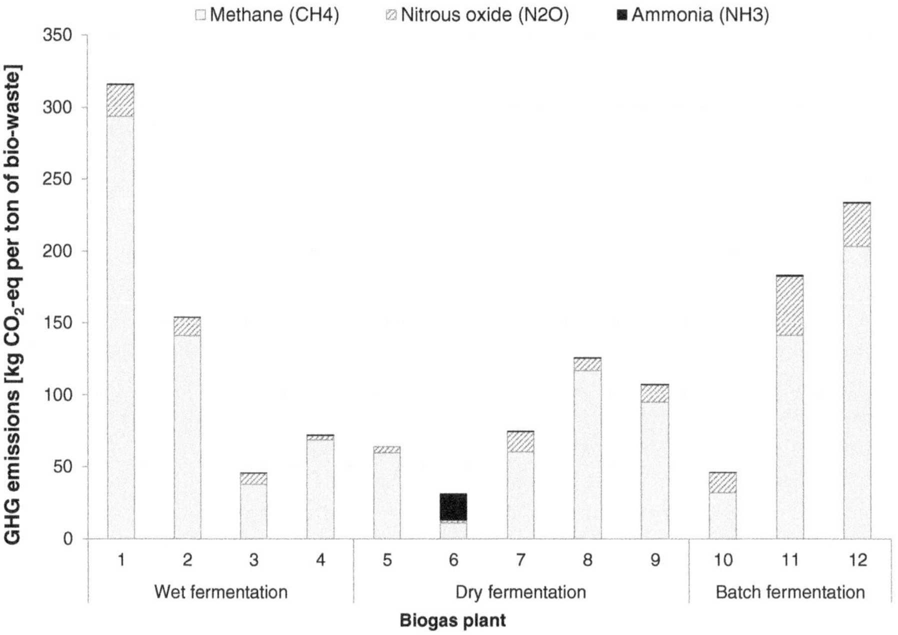

Figure 1 GHG emissions of the investigated biogas facilities (bio-waste) differed to the kind of GHG emission. The measured emissions of investigated AD plants are presented in kg CO_2-eq per ton of bio-waste differed to the kind of GHG emission and kind of fermentation process. Methane and nitrous oxide emissions = direct GHG emissions, ammonia = indirect GHG emissions.

hydraulic retention times (HRTs) ranged from 1 to 4 weeks. However, due to a great variability of other process parameters, the results do not give a clear answer regarding the estimation that lower HRT corresponds to lower gas potential (see Tables 5 and 6).

GHG balances

The overall GHG balance of the investigated AD plants depends on the measured GHG emissions on the one hand (see 'GHG emissions') and on the credits for the generated products (e.g., combined heat and electricity

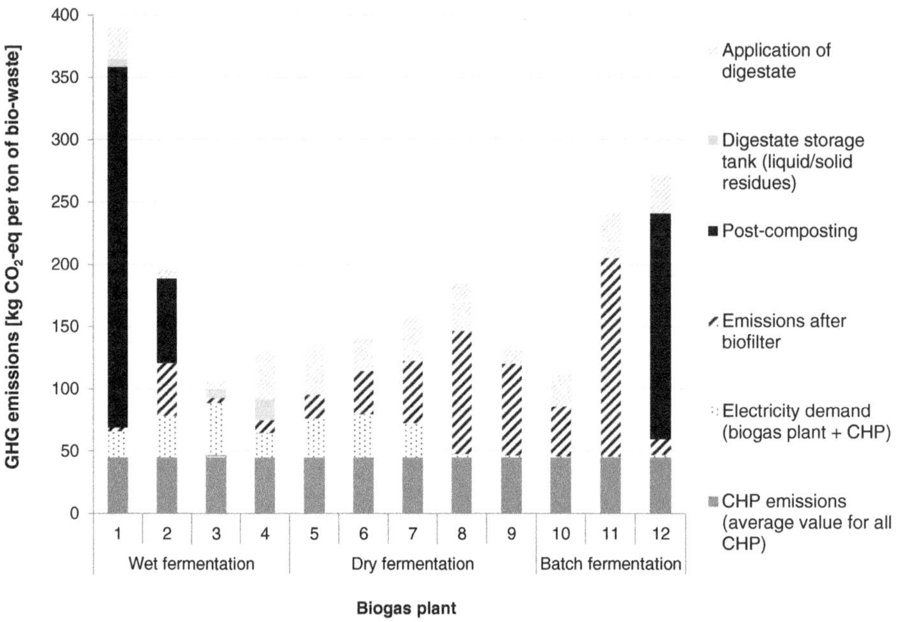

Figure 2 GHG emissions of bio-waste digestion plants differed to kind of plant components. The GHG emissions are presented in kg CO_2-eq per ton of bio-waste differed to the kind of fermentation process (wet, dry, or batch fermentation) and the main emission sources within the process chain.

Table 4 Investigated AD plants differed to kind of digestate and considered GHG credits (marked with 'x')

	Kind of digestate	Solid digestate				Liquid digestate	
		Substitution of mineral fertilizer	Substitution of peat	Humus accumulation	Humus reproduction	Substitution of mineral fertilizer	Humus reproduction
1	Fresh compost	x	-	x	x	-	-
2	Finished compost	x	x	x	x	-	-
3	Solid digestate (separated)	x	-	-	x	-	-
4	Liquid digestate	x	-	-	x	x	x
5	Separated digestate without post-composting	x	-	-	x	x	x
6[a]	Solid digestate	x		-	x	-	-
7	Fresh and finished compost	x	x	x	x	-	-
8	Fresh and finished compost	x	x	x	x	x	x
9[b]	Finished compost	x	x	x	x	x	x
10	Finished compost	x	x	x	x	-	-
11	Fresh and finished compost	x	x	x	x	-	-
12	Finished compost	x	x	x	x	-	-

[a]Assumption according to plant no. 5 (no data available). [b]Assumption according to plant no. 8 (no data available).

from biogas; fertilizer and humus supply from fermentation residues) on the other hand. The calculated GHG credits according to the AD plant concept are presented in Figure 3.

Finally, the highest amount of GHG credits of humus reproduction can be expected from composted digestate. In general, the following order of humus reproduction can be assumed: post-composted digestate (finished and fresh compost) > solid digestate > liquid digestate. In case of finished compost, additional GHG credits for the substitution of peat (by application in soil producing facilities, e.g.) can be considered.

If external heat (generated by the electricity production of CHP unit) is utilized, credits for avoided fossil heat production optimize the GHG balances as well (see plant no. 12). Nevertheless, in most cases (besides plants nos. 1, 7, 10, and 11), the credit for the electricity production based on biogas which was given for the substitution of fossil fuels dominates the GHG credits.

The total range of GHG balances (including credits) varied between −49 and 323 kg CO_2-eq per ton of bio-waste due to different plant concepts and measured emissions (see Figure 4).

Moreover, the emissions of each component have been set in relation to the amount of produced electricity in order to get an emission value according to the energy output (g CH_4/kWh_{el}). Compared to an assumed electricity mix in Germany (559 g CO_2-eq per kWh_{el} according to [20]), 8 of 12 AD plants show even lower values.

Table 5 Residual gas potential in percent related to the methane production

	Bio-waste AD plants 38°C	Agricultural AD plants 37°C	
		Single-stage	Multi-stage
Average	14.1	9.7	5.1
Min	3.6	3.2	1.2
Max	23.4	21.8	15

Investigated bio-waste AD plants in comparison to agricultural AD plants according to Weiland et al. [17] and temperature level for determination of gas potential.

Overall discussion of results gained in this study

The problem of increased emissions is not the anaerobic process itself, but a non-optimal after-treatment of the digestate. In general, the emission situation is not uniform; the plants show very different emission rates. The total emissions from AD plants no. 3, no. 6, and no. 10 were quite lower than the remaining. However, even those plants showed considerable potential for optimization. The best overall result of the analyzed AD plants belonged to a biogas facility with no external heat utilization and below average credits for digestate. It can be stated that all investigated biogas facilities showed potential for optimization. Often, there are no incentives for a sufficient utilization of waste with respect to high CH_4 yields or reduction of emissions, due to the fact that the running costs of waste facilities has to be financed by the producers of the waste paying for the waste disposal. Moreover, there are no strict regulations to avoid uncontrolled emissions as for agricultural biogas plants for energy crops and for co-digestion of waste. Therefore, waste treatment plants show relevant potentials for optimization.

AD plant no. 12 showed that very high emissions can be covered by a very good energy concept combined with a good utilization of fermentation residues. The bad overall

Table 6 Hydraulic retention time and residual gas potential of investigated AD plants

Plant no.	HRT in days[a]	Relative residual gas potential in percent[b]		Kind of fermentation[c]	Mode of operation	Separation of digestate
		Minimum	Maximum			
1	8	11	15	Wet	Multi-stage	Yes
2	20	19	19	Wet	Multi-stage	Yes
3	17	-	-	Wet	Single-stage	No
4	25	-	-	Wet	Multi-stage	No
5	25	-	-	Dry	Single-stage	Yes
6	21	15	21	Dry	Single-stage	No
7	28	4	11	Dry	Single-stage	No
8	14	23	23	Dry	Single-stage	No
9	21	17	17	Dry	Single-stage	No
10	28	12	17	Batch	Single-stage	No
11	21	-	-	Batch	Single-stage	No
12	21	6	6	Batch	Single-stage	No

[a]Hydraulic retention time in days. [b]Residual gas potential of digestate according to the input material (fresh matter based). [c]Wet = wet fermentation, dry = dry fermentation, batch = batch system (discontinuous).

GHG balance of AD plant no. 1 evidences how certain factors may interact the GHG performance negatively. In this case, extremely high emissions occurring from the post-composting process and very low electricity generation caused high GHG emissions in total. Inadequate digestion of the substrate caused not only low gas production, respectively, electricity generation but also high emissions during the post-composting process of digestate.

Regarding the GHG credits, the highest importance of an efficient fermentation had the production of energy. A high share of electricity generation led to high GHG credits. As far as the utilization of exhaust heat of electricity production was possible, it had also a positive influence on the GHG performance of the AD plant. Moreover, the use of digestate showed positive effects on the GHG balances. In addition to the nutrient effect through the utilization of the fermentation residues as a fertilizer (substitution of mineral fertilizer), GHG emissions can be saved due to the humus effect of digestate. Especially, composted digestate like fresh and finished compost contributed to the humus accumulation

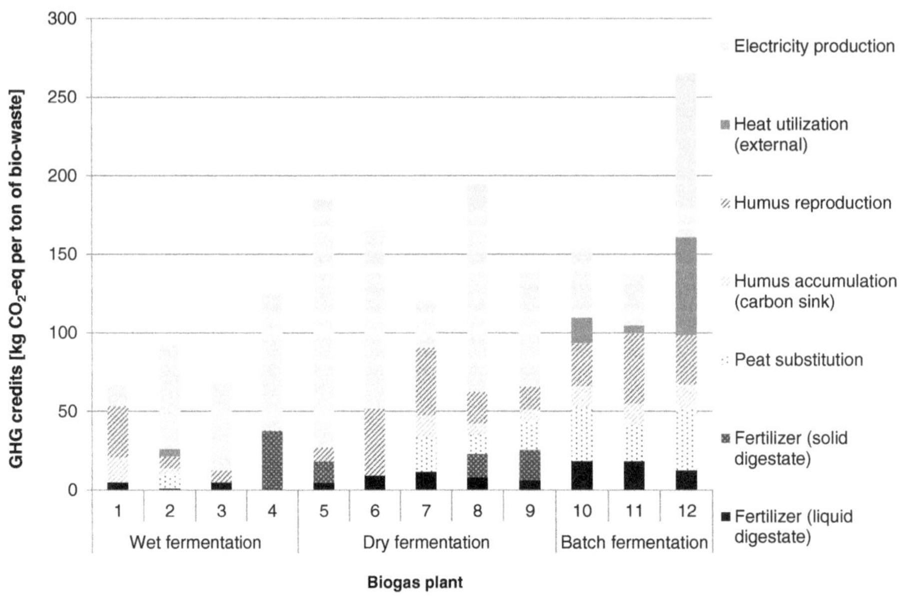

Figure 3 GHG credits of investigated bio-waste digestion plants. GHG credits depend on the amount of energy production or heat utilization (substitution of fossil electricity/fossil heat production) as well as the kind and amount of digestate (substitution of fertilizer, substitution of peat, humus effects).

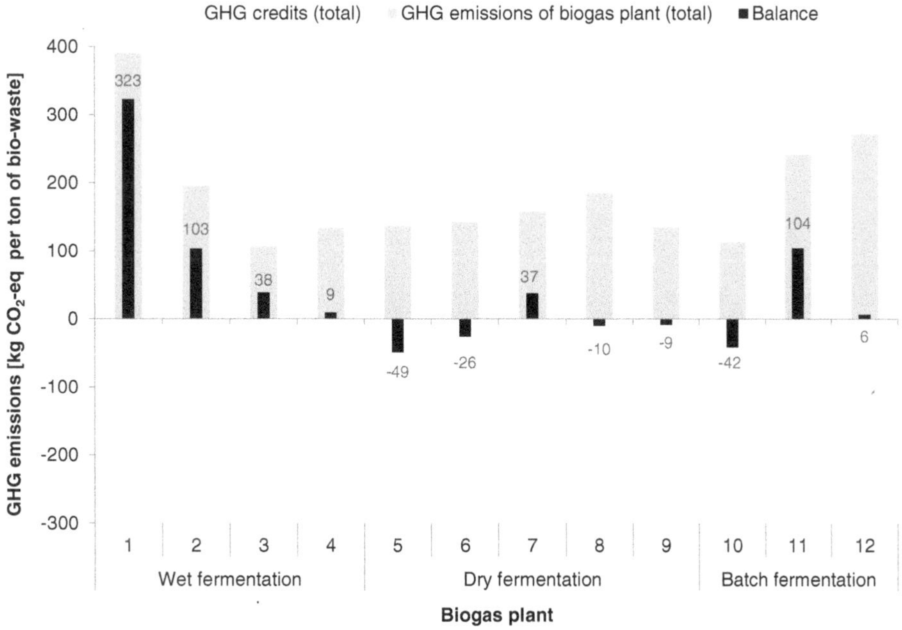

Figure 4 Total GHG balance of bio-waste digestion plants with GHG emissions of AD plant and GHG credits. The balance as a result of total GHG emissions of AD plant and total GHG credits (*black column*).

(carbon sink) and the humus reproduction of digestate. Compared to the production of fresh or finished compost digestate without post-composting process, which is used within the agriculture directly, less GHG credits were given. However, the risk of high emissions during the post-treatment of the fermentation residues was avoided.

The following measures are able to reduce GHG emission of bio-waste digestion: intensive aeration of the (solid) digestate after fermentation; gas-tight storage tank for fermentation residue and integration into biogas utilization; avoidance of any open storage of digestate and fermentation residues; and small, aerated compost windrows combined with sufficient structural materials and frequent turnover as well as the use of acidic scrubbers in front of the bio-filter.

With respect to the development of methodology of emission measurements and the standardization of procedure for the determination of emissions on biogas plants, further investigations are necessary. Further scientific data about the current emission situation and the ongoing development as well as reliable measurement methods are required to determine the CH_4 emissions from the plants in operation today. In this regard, the reliable measurement of stationary and diffuse emission sources is of high importance. Uncertain are the emissions sources that are not coupled to the gas system of the plant, but still cause GHG emissions as stated in [10]. As one example, no assessment of emissions from pressure relief valves could be carried out as part of this study. Concerning the emissions, the treatment and evaluation of temporary occurring emissions caused by certain operational conditions are still unclear. Moreover, the further development of ecological assessment of biogas pathways with respect to the humus effects of digestate in comparison to other pathways is of great importance.

Conclusions

Based on the emission measurements, significant sources of emissions were identified. The results show that GHG emissions can be minimized, if the technology and operation of the plant are adjusted accordingly. Basically, the kind of operation of the plant and the handling of digestate determine the amount of GHG emissions. The overall GHG balances of the investigated AD plants depend on the measured emissions as well as the amount of credits for the generated products (e.g., combined heat and electricity from biogas; fertilizer and humus effects from fermentation residues). The consideration of GHG credits can optimize the overall GHG performance of the biogas facilities.

Abbreviations

AD: anaerobic digestion; C: carbon; CHP: combined heat and power unit; CH_4: methane; CO_2: carbon dioxide; CO_2-eq: carbon dioxide equivalent; Corg: organic carbon; GHG: greenhouse gas; GWP: global warming potential; K_2O: potassium oxide; kW_{el}: kilowatt (electrical); kWh_{el}: kilowatt hours (electrical); kWh_{th}: kilowatt hours (thermical); N: nitrogen; NH_3: ammonia; No.: number; N_2O: nitrous oxide; STP: standard temperature pressure; t: metric ton.

Competing interests

The authors declare that they have no competing interests.

Authors' contributions

JDG carried out the total GHG balances including GHG credits and conceived of the papers structure. JL and CK conducted the emission analysis regarding the biogas plant and the components. JL, CK, and JDG described the mitigation potential. VD was responsible for the survey of plant operators and the assessment of these data. All authors drafted, read, and approved the final manuscript.

Acknowledgements

The DBFZ investigated the emission situation of biogas plants based on bio-waste in Germany within a research project in cooperation with gewitra mbH and Dr. Reinhold & Kollegen during 2009 to 2012. Gewitra mbH led the measurements on AD plants and post-composting processes. Dr. Reinhold & Kollegen Potsdam was involved into evaluation of humus effects of digestate at investigated AD plants in strong cooperation with our colleague Dr. Walter Stinner. The authors would like to thank Dr. Carsten Cuhls, Birte Mähl, and Dr. Jürgen Reinhold for their cooperation as project partners as well as the German Federal Ministry for the Environment, Nature Conservation, and Nuclear Safety for the financial support of this project.

Dedication

This publication is dedicated to Prof. Andreas Zehnsdorf on the occasion of his 50th birthday.

References

1. IPCC (2007) Climate change 2007, Fourth Assessment Report (AR4) of WG 1 Intergovernmental Panel on Climate Change (IPCC). Cambridge University Press, Cambridge and New York
2. Vogt R (2008) Basis data for GHG balances of process chains of biogas and compilation of new balances. German edition: Basisdaten zu THG Bilanzen für Biogas-Prozessketten und Erstellung neuer THG-Bilanzen, Report on behalf of E.ON Ruhrgas AG. Institut für Energie- und Umweltforschung Heidelberg GmbH, Heidelberg
3. Müller-Langer F, Rönsch S, Weithäuser M, Oehmichen K, Scholwin F, Höra S, Scheftelowitz M, Seiffert M (2009) Economical and ecological assessment of substitutes for natural gas based on energy crops. German edition: Ökonomische und ökologische Bewertung von Erdgassubstituten aus nachwachsenden Rohstoffen. Deutsches Biomasseforschungszentrum gGmbH, Leipzig
4. Liebetrau J, Clemens J, Cuhls C, Hafermann C, Friehe J, Weiland P, Daniel-Gromke J (2010) Methane emissions from biogas producing facilities within the agricultural sector. Eng Life Sci 10:1–5
5. Liebetrau J, Reuschel C, Clement J, Friehe J, Weiland P (2013) Analysis of greenhouse gas emissions from 10 biogas plants within the agricultural sector. Water Sci Technol 67(6):1370–1379, doi:10.2166/wst.2013.005
6. Petersson A (2012) The Swedish Voluntary system for control of methane emissions. http://www.iea-biogas.net/case-studies.html?file=files/daten-redaktion/download/case-studies/Case%20Study%20Sweden.pdf. Accessed 23 Jul 2013
7. Flesch T, Desjardins R, Worth D (2011) Fugitive methane emissions from an agricultural biodigester. Biomass Bioenergy 35(9):3927–3935
8. Woess-Gallasch S, Enzinger P, Jungmeier G, Padzinger R (2007) Greenhouse gas emissions from biogas plants, Germanth edn. Report for Landesenergieverein Steiermark, Graz
9. Amon B, Kryvoruchko V, Amon T, Zechenmeister-Boltenstern S (2006) Methane, nitrous oxide and ammonia emissions during storage and after application of dairy cattle slurry and influence of slurry treatment. Agric Ecosyst Environ 112:153–162
10. Dumont M, Luning L, Yildiz I, Koop K (2013) Methane emissions in biogas production. In: Wellinger A, Murphy J, Baxter D (eds) The biogas handbook - science, production and application, vol 52, Woodhead Publishing Limited. Series in Energy, Cambridge, pp 248–266
11. DESTATIS (2013) Abfall-Daten. Statistisches Bundesamt. https://www.destatis.de/DE/ZahlenFakten/GesamtwirtschaftUmwelt/Umwelt/UmweltstatistischeErhebungen/Abfallwirtschaft/Tabellen/Abfallbilanz2011.pdf?__blob=publicationFile. Accessed 20 Mrch 2013
12. Fricke K, Heußner C, Hüttner A, Turk T, Bauer W, Bidlingmaier W (2013) Vergärung von Bio- und Grünabfällen, Teil 1: Ausbaupotenzial bei der Vergärung von Bio- und Grünabfällen. Müll Und Abfall 12:628–635
13. Daniel-Gromke J, Rensberg N, Schüch A (2014) Ein Überblick - Potenziale, Techniken und Perspektiven von zukünftigen Substraten im Biogassektor mit Fokus auf die Bioabfallvergärung. In: Nelles M (ed) DBFZ-Jahrestagung 2014 Bioenergie. Vielseitig, sicher, wirtschaftlich, sauber?! 1.-2. Oktober 2014. Deutsches Biomasseforschungszentrum gGmbH, Leipzig, ISSN 2199-9384
14. Hashmonay RA, Varma RM, Modrak MT, Kagann RH, Segall RR, Sullivan PD (2008) Radial plumemapping: a US EPA test method for area and fugitive source emission monitoring using optical remote sensing. In: Kim YJ, Platt U (eds) Advanced environmental monitoring. Springer, Cham, Switzerland, pp 21–36
15. Liebetrau J, Daniel-Gromke J, Krebs C (2012) Emissionen aus Vergärungsanlagen. Erfahrungen aus dem landwirtschaftlichen Bereich und aus der Abfallbehandlung. In: Immissionsschutz. Aktuelle Entwicklungen im anlagenbezogenen Planungsprozess und Immissionsschutz. Band 3, Neuruppin, S. 337-351 – ISBN 978-3- 935317-90-0
16. Oechsner H (2013) Determination of residual gas- /residual methane potential. In: Liebetrau J, Pfeiffer D, Thrän D (eds) Method handbook - biogas as part of the funding programme "Biomass for Energy", vol 7, Report commissioned by the Federal Ministry of Environment, Nature Conservation and Nuclear Safety. Deutsches Biomasseforschungszentrum gGmbH, Leipzig
17. Weiland P, Gemmeke B, Rieger C, Plogsties V, Vogtherr J, Kissel R, Bachmaier H, Schumacher B (2009) Biogas-Messprogramm II – 61 Biogasanlagen im Vergleich. Fachagentur Nachwachsende Rohstoffe e. V. (ed), Gülzow
18. Reinhold G, Gödeke K (2011) Restgaspotential in landwirtschaftlichen Biogasanlagen. Biogas – Fachtagung Thüringen. http://www.tll.de/ainfo/pdf/bior0111.pdf. Accessed 15 Jan 2015
19. VDI (2006) VDI 4630, Vergärung organischer Stoffe, Substratcharakterisierung, Probenahme, Stoffdatenerhebung. Gärversuche, Norm, Düsseldorf
20. UBA (2012) Entwicklung der spezifischen Kohlendioxid-Emissionen des deutschen Strommix 1990-2010 und erste Schätzungen 2011. Federal Environment Agency (Umweltbundesamt), Dessau-Roßlau
21. Thrän D, Pfeiffer D (2011) Method handbook - Material flow-oriented balancing of greenhouse gas effects - methods for the determining of technology indicators, production costs, and greenhouse gas effects of projects as part of the funding programme "Biomass for Energy", vol 1, Report commissioned by the Federal Ministry of Environment. Nature Conservation and Nuclear Safety. Deutsches Biomasseforschungszentrum gGmbH, Leipzig
22. BLE (2010) Leitfaden Nachhaltige Biomasseherstellung. Bundesanstalt für Landwirtschaft und Ernährung (BLE), Bonn
23. Knappe F, Vogt R, Lazar S, Höke S (2012) Optimierung der Verwertung organischer Abfälle, Report commissioned by the Federal Environment Agency (Umweltbundesamt). Institut für Energie- und Umweltforschung Heidelberg GmbH, Heidelberg
24. Vogt R, Knappe F, Giegrich J, Detzel A (2002) Ökobilanz Bioabfallverwertung. IFEU Institut für Energie- und Umweltforschung Heidelberg GmbH, Report commissioned by the German Federal Environmental Foundation. Deutsche Bundesstiftung Umwelt DBU, Berlin
25. Wulf S, Maeting M, Clemens J (2002) Application technique and slurry co-fermentation - effects on ammonia, nitrous oxide, and methane emissions after spreading. J Environ Qual 31:1795–1801, doi:10.2134/jeq2002.1795

11

Linear programming control of a group of heat pumps

Jiří Fink[2*], Richard P. van Leeuwen[2,1], Johann L. Hurink[2] and Gerard J.M. Smit[2]

Abstract

Background: For a new district in the Dutch city Meppel, a hybrid energy concept is developed based on bio-gas co-generation. The generated electricity is used to power domestic heat pumps which supply thermal energy for domestic hot water and space heating demand of households. In this paper, we investigate direct control of the heat pumps by the utility and how the large-scale optimization problem that is created can be reduced significantly.

Methods: Two different linear programming control methods (global MILP and time scale MILP) are presented. The latter solves large-scale optimization problems in considerably less computational time. For simulation purposes, data of household thermal demand is obtained from prediction models developed for this research. The control methods are compared with a reference control method resembling PI on/off control of each heat pump.

Results: The reference control results in a dynamic electricity consumption with many peak loads on the network, which indicates a high level of simultaneous running heat pumps at those times. Both methods of mix integer linear programming (MILP) control of the heat pumps lead to a much improved, almost flat electricity consumption profile.

Conclusions: Both optimization control methods are equally able to minimize the maximum peak consumption of electric power by the heat pumps, but the time scale MILP method requires much less computational effort. Future work is dedicated on further development of optimized control of the heat pumps and the central CHP.

Keywords: Smart grid, Linear programming control, Heat pumps, Thermal storage, Load balancing, Demand side management

Background

Introduction

To reduce fossil fuel usage and increase the share of renewable energy, the Netherlands plans for 14 % renewable energy of the total energy consumption by the year 2020 and 100 % in the year 2050 [1]. However, integration of renewable energy in existing distribution grids may cause problems with power stability in some regions at times when there is a large amount of energy offered by solar PV installations or wind turbines and low demand [2]. Smart grids are seen as a possible solution and therefore the Dutch government started the Topteam Energy which defined the innovation contract smart grids [3] with the following goals:

- reduce costs for integration of renewable energy into distribution grids
- increase consumer awareness and energy savings
- contribute to a competitive energy market and consumer choice resulting in reductions in energy prices

It is widely recognized that smart grids are required to balance energy production with loads and energy storage. This is made possible by sophisticated control strategies and communication networks.

To increase the share of renewable energy in today's energy system, it is not always necessary to integrate the renewable energy into existing grids. With so-called smart microgrids, it is possible to connect or disconnect a region to or from the main distribution grid. In Meppel, a small city in the Netherlands, there is a plan to build a district (Nieuwveenselanden) with an energy system called *MeppelEnergie* [4–6] that can be defined as a smart microgrid and which has the goal to become almost 100 %

*Correspondence: fink@kam.mff.cuni.cz
[2]University of Twente, Department of Computer Science, Mathematics and Electrical Engineering, P.O. Box 217, 7500 AE, Enschede, The Netherlands
Full list of author information is available at the end of the article

renewable. The heart of the system is a combined heat and power (CHP) unit which is supplied with biogas from a nearby wastewater treatment facility. Part of the houses in the district are connected to the district heating system, while other houses have a heat pump. Both the heat for the district heating system and power for the heat pumps are supplied by the CHP. This is a typical case where a smart grid is required for load balancing purposes, otherwise the district would need frequent power exchanges with the existing power grid. Without a smart grid, this would probably involve large peak loads and strengthening of the existing grid would be required, involving high costs for new cabling and bi-directional transformers. For different cases in Germany involving regional solar PV generation, these effects are explained by Nykamp [2]. Therefore, the heat pumps should be scheduled in such a way that they only consume, if possible, the electricity produced by the CHP. If this is not possible, the remaining energy has to be bought on the electricity market at minimal cost.

The planning of a group of heating systems may have many objectives in practice. In the Meppel project, biogas is transported through a dedicated pipeline and electric energy is distributed through a private cable and converted by heating supply systems. In general, generators and transport equipment have to be dimensioned for the maximal peak consumption. Thus, the main objective is minimizing the maximal consumption which may decrease investments in the system and will also lead to maximum self consumption of the generated electricity by the biogas CHP.

This paper develops a smart grid control method with the goal to minimize peak consumption. We further demonstrate the importance of this method for the Meppel case. After a review of related work in section "Related work", a more detailed problem formulation which leads to an algorithm called global mix integer linear programming (MILP) control is given in section "Problem statement and global MILP control". As this algorithm requires a lot of computational power, we develop an algorithm called time scale MILP control in section "Time scale MILP control". The Meppel case is explained in section "Case application". The simulation results are presented in section "Results and discussion". Finally, we draw conclusions in section "Conclusions".

Related work

The problem of minimizing peak consumption of a group of heat pumps is a typical smart grid control problem. As demonstrated by Nykamp [2], integration of renewable energy (e.g., by wind turbines, solar PV, and biomass conversion) requires implementation of control strategies for load balancing and load shifting in electrical grids. This is either due to the stochastic nature of the energy production or due to supply limitations of the biomass. In the Meppel case, the supply of biogas to the CHP which generates the electricity for the heat pumps is limited. Besides that, the biogas supply is approximately constant in time and it is not possible to divert the biogas supply to other consumers. These are totally different circumstances than existing natural gas CHP systems. The natural gas supply system is widespread within the Netherlands, and for CHP's on natural gas, there are usually no limitations to size or consumption patterns.

For load balancing purposes, Stadler et al. [7] introduce three possible device control strategies for refrigerators which are also applicable for heat pumps.

1. Autonomous control is performed by the device itself, e.g., by sensing the grid frequency, interpreting the difference with a nominal frequency, and making decisions to start or stop the heat pump. In our case, we expect poor results from this type of control because there are also constraints on thermal comfort which are likely to create conflicts in time with reaching grid stability.
2. Price-based control is the dominating control strategy for scheduling devices within a smart grid. The heat pump determines an operational schedule based on dynamic price information which is sent to the heat pumps by the utility. However, to make a sensible decision about future schedules, it is required that a heat pump controller is able to predict future heat demand of the building. For this purpose, model predictive control is implemented within the heat pump controller. For climate control of a single building with a heat pump and time-varying electricity prices, model predictive control is developed and demonstrated by Halvgaard et al. [8]. For groups of devices, a smart grid control strategy based on the principles of price-based control is developed within the TRIANA smart grid control methodology by Molderink [9], Bakker [10], and Bosman [11]. Bosman demonstrates the use of this type of control for a group of microCHP's in [11], while Bakker also demonstrates results for a group of heat pumps in [10]. These examples demonstrate effectiveness of price-based control. The only minor disadvantage of price-based control is the need for an intelligent heat pump controller for each house.
3. Direct control involves making control decisions for each heat pump on the utility level. This has the advantage that the heat pump controllers can be simple. On the other hand, the utility has to solve large-scale optimization problems (in our case, minimizing peak electricity consumption) if there are a large number of heat pumps to control. Besides that, the optimization has to include heat demand predictions for each house at every time step of the

calculation. In the Meppel case, the number of houses with a heat pump connected to a single biogas CHP is limited to 135. For mathematical optimization, we estimate that this is a feasible challenge and that is why we chose this method for the Meppel case.

The mathematical background of the minimizing peak problem is presented in [12] which proves that the problem of minimizing peak is NP-complete. A somewhat similar problem was considered by Bosman et al. [13, 14] who studied a microCHP planning problem and also proved that minimizing peak is NP-complete in their model [15]. Bosman et al. [11] also present a dynamic programming algorithm for the microCHP planning problem whose time complexity is $O(T^{3C+1})$ where T is the number of time intervals and C is the number of microCHPs.

The main contribution of this paper is that we demonstrate by development of the time scale control method and the application within the Meppel case that the number of variables and equations which are part of the large-scale optimization problem on the utility level can be reduced considerably while the control objective of minimizing the peak consumption is still reached.

Methods

Problem statement and global MILP control

In this section, we present a mathematical description of the studied model and possible applications of this model. The model is set up to be generally applicable for any type of heating appliance. Typical appliances for heating water are electrical and gas heating systems, heat pumps, and combined heat and power units (microCHP). The heated water is stored in buffers to be prepared for the demand of inhabitants. In our model, a house consists of two local heating systems, one for space heating and the other for tap. A schematic overview of the model is presented in Fig. 1. It consists of:

- a supply which represents some source of energy (e.g., electricity, gas),
- a converter which converts the energy into heat (hot water),
- a buffer which stores heat for later usage, and
- a demand which represents the consumption profile of heat.

In principle, the presented model can consider arbitrary types of energy but in this paper, we use *electricity* and *heat* to distinguish consumed and produced energy. This simple model of a local heating system cannot only be applied for heating water but has many other applications in smart grids (e.g., control of fridges and freezers) and inventory management. We come back on this at the end of this section.

We consider a discrete time model for the considered problem, meaning that we split the planning period into T time intervals of the same length. We consider a set $\mathcal{C} = \{1, \ldots, C\}$ of C heating systems and a set $\mathcal{T} = \{1, \ldots, T\}$ of T time intervals. Note that the heating of a house is split into two independent heating systems (see Fig. 1). In this paper, the letter c is always an index of a heating system (either space heating or tap) and t is an index of a time interval. For mathematical purposes, we separate a heating system into a converter, a buffer, and demand; see Fig. 1. We say "a converter c," "a buffer c," or "a demand c" to refer to the devices of the heating system $c \in \mathcal{C}$.

We consider a simple converter which has only two states: In every time interval, the converter is either turned on or turned off. The amount of consumed electricity is E_c and the amount of produced heat (or any other form of energy) is H_c during one time interval in which the converter $c \in \mathcal{C}$ is turned on. If the converter is turned off, then it consumes and produces no energy. Let $x_{c,t} \in \{0, 1\}$ be the variable indicating whether the converter $c \in \mathcal{C}$ is running in time interval $t \in \mathcal{T}$.

The state of charge of a buffer $c \in \mathcal{C}$ at the beginning of time interval $t \in \mathcal{T}$ is denoted by $s_{c,t}$ which represents the amount of heat in the buffer. Note that $s_{c,T+1}$ is the state of charge at the end of planning period. The state of charge $s_{c,t}$ is limited by an upper bound U_c.

The amount of consumed heat by the inhabitants of the house from heating system $c \in \mathcal{C}$ during time interval $t \in \mathcal{T}$ is denoted by $D_{c,t}$. This amount is assumed to be given and is called the demand of heating system c. In this paper, we study offline problems, so we assume that demands $D_{c,t}$ are given for the whole planning period.

The operational variables of the converters $x_{c,t}$ and the states of charge of buffers $s_{c,t}$ are restricted by the following constraints.

$$s_{c,t+1} = s_{c,t} + H_c x_{c,t} - D_{c,t} \text{ for } c \in \mathcal{C},\ t \in \mathcal{T} \quad (1)$$

$$0 \leq s_{c,t} \leq U_c \text{ for } c \in \mathcal{C},\ t \in \{1, \ldots, T+1\} \quad (2)$$

$$x_{c,t} \in \{0, 1\} \text{ for } c \in \mathcal{C},\ t \in \mathcal{T} \quad (3)$$

Equation 1 is the charging equation of the buffer. During time interval $t \in \mathcal{T}$, the state of charge $s_{c,t}$ of a buffer $c \in \mathcal{C}$ is increased by the production of the converter which is $H_c x_{c,t}$ and it is decreased by the demand $D_{c,t}$. Equations (2) and (3) ensure that the domains of variables $s_{c,t}$ and $x_{c,t}$, respectively, are taken into account. Note that the initial state of charge $s_{c,1}$ can be fixed (e.g., by setting $s_{c,1} = \frac{U_c}{2}$).

In this paper, we consider the objective function of minimizing the peak:

$$\begin{aligned} &\text{minimize } m \\ &\quad \text{where } m \geq \textstyle\sum_{c \in C} E_c x_{c,t} \text{ for } t \in \mathcal{T} \end{aligned} \quad (4)$$

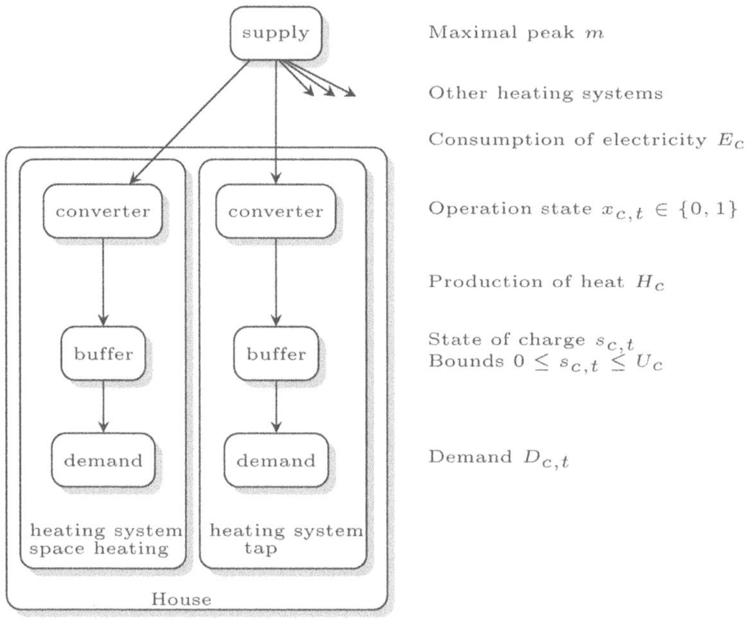

Fig. 1 Schematic picture of a house with two separated heating systems for space heating and tap

Since $E_c x_{c,t}$ is the amount of consumed electricity by a converter c in time interval t, the sum $\sum_{c \in C} E_c x_{c,t}$ is the amount of electricity consumed by all converters in time interval t. Furthermore, the inequality and the objective function (4) guarantees that the value of the variable m is the maximal consumption of electricity during one time period within the whole planning period.

In the following, we give some other possible applications of this model.

Fridges and freezers: A fridge essentially works in the opposite way than heating, so it may be modelled similarly. However, we have to be careful with the correct interpretation of all parameters. The state of charge of the buffer again represents the temperature inside the fridge, but a higher state of charge means a lower temperature. The converter does not produce heat to the fridge but it decreases the temperature inside the fridge, so the converter increases the state of charge of the buffer (fridge). The demand decreases the state of charge of the fridge due to thermal loss and usage of the fridge by humans.

Inventory: The considered heating problem is also related to inventory control problems [16]. A buffer may represent an inventory and a converter may represent orders. However, this leads to a situation, where only a limited capacity of inventory is given and it is only possible to order a fix amount of goods which is not a typical situation in inventory management.

Note that the objective function and all constrains are linear and operational state variables are binary, so constraints and the objective (1)–(4) form an instance of MILP. This instance can be solved by any MILP solver (see, e.g., [17]) and we call this approach *global MILP control.* However, as the number of binary variables may get too large for planning many houses over a long planning horizon, this method may get computationally expensive. Therefore, in the following sections, an algorithm which significantly reduces the number of variables is given.

Time scale MILP control

The method presented in the previous section creates one large instance of MILP and solves it by a MILP solver. This method gives us an optimal solution for the whole planning period but it may not be suitable for practical purposes. First, finding an optimal solution requires a lot of computational power. Next, the prediction of demand for the distant future may be very inaccurate.

Therefore, we consider an online control in which the decision which converters will be running is made only for the coming time interval. On the other hand, we cannot ignore the future completely. Indeed, we should take more care about the near future time intervals than the distant ones because the current decision has stronger impact on the near future and the prediction is in general more accurate for the near future.

As the general formal notation of time scale control may be hard to understand, we use an example to present the approach. Assume that the upcoming time interval has index 1. The decision which converters $c \in C$ will be running during the coming time interval 1 needs to be made, meaning that the values of variables $x_{c,1}$ need to be decided. Since also the influence of the very near future needs to be detailed, we also consider binary variables, e.g., for the next two time intervals, i.e., variables $x_{c,2}$ and $x_{c,3}$. For the further future, the plan does not need to be so precise, so ,e.g., another two time intervals, we relax the integral constraints, meaning that we require $0 \leq x_{c,4}, x_{c,5} \leq 1$. The reason for relaxing these variables is to decrease the number of integral variables which has the dominating effect on the computational time required to solve an MILP instance. From the practical point of view, these relaxed variables (e.g., $x_{c,4}$) can signify the probability that a converter c will run in time interval 4, and so $\sum_{c\in\mathcal{C}} H_c x_{c,4}$ is the expected demand of electricity.

Following this, for the even more distant future, we only need a rough planning. In order to explain the idea of rough planning, let us consider the state of charge equation, e.g., for time intervals $t = 8, 9$ and 10.

$$\begin{aligned} s_{c,9} &= s_{c,8} + H_c x_{c,8} - D_{c,8} \\ s_{c,10} &= s_{c,9} + H_c x_{c,9} - D_{c,9} \\ s_{c,11} &= s_{c,10} + H_c x_{c,10} - D_{c,10} \end{aligned} \tag{5}$$

We sum these equations and after simplification, we obtain

$$s_{c,11} = s_{c,8} + H_c(x_{c,8} + x_{c,9} + x_{c,10}) - (D_{c,8} + D_{c,9} + D_{c,10}) \tag{6}$$

The rough plan for converter c for time intervals $t, t+1, \ldots, t'$ is now defined by $x_{c,t..t'} = \sum_{i=t}^{t'} x_{c,i}$, that is we replace time intervals $t, t+1, \ldots, t'$ by one block of time intervals $t..t'$. During this block, converter c consumes $E_c x_{c,t..t'}$ electricity and produces $H_c x_{c,t..t'}$ heat. Using this notation, the state of charge equation for a block 8..10 of time intervals $t = 8, 9$ and 10 is

$$s_{c,11} = s_{c,8} + H_c x_{c,8..10} - D_{c,8..10} \tag{7}$$

where $D_{c,8..10} = D_{c,8} + D_{c,9} + D_{c,10}$ is the cumulative demand for time intervals 8,9 and 10. In this example, we replace the three variables $x_{c,8}$, $x_{c,9}$, and $x_{c,10}$ by one aggregated variable $x_{c,8..10}$ which is constrained by bounds $0 \leq x_{c,8..10} \leq 3$. In this way, we can cover a longer planning horizon without requiring too many variables in an MILP instance. Furthermore, note that only the sum of demands $D_{c,8..10}$ for distant future time intervals is important. The practical consequence is that the time where a significant demand occurs does not have to be predicted precisely, e.g., in the morning, it is sufficient to predict the amount of hot water demand for evening showers but the exact time when inhabitants will take a shower can be approximated.

In our example, we consider rough planning variables $x_{c,6..7}$, $x_{c,8..10}$, and $x_{11..15}$. In summary, all state of charge equations are

$$\begin{aligned} s_{c,t+1} &= s_{c,t} + H_c x_{c,t} - D_{c,t} \text{ for } t = 1, \ldots, 5 \\ s_{c,8} &= s_{c,6} + H_c x_{c,6..7} - D_{c,6..7} \\ s_{c,11} &= s_{c,8} + H_c x_{c,8..10} - D_{c,8..10} \\ s_{c,16} &= s_{c,11} + H_c x_{c,11..15} - D_{c,11..15} \end{aligned} \tag{8}$$

for every $c \in \mathcal{C}$.

The capacity constraints of buffers remain the same, so

$$L_{c,t} \leq s_{c,t} \leq U_{c,t} \text{ for } t \in \{1,2,3,4,5,6,8,11,16\} \tag{9}$$

The operational constraints of converters now are

$$\begin{aligned} & x_{c,t} \in \{0,1\} && \text{for } t \in \{1,2,3\} \\ 0 \leq\ & x_{c,t} \leq 1 && \text{for } t \in \{4,5\} \\ 0 \leq\ & x_{c,6..7} \leq 2 \\ 0 \leq\ & x_{c,8..10} \leq 3 \\ 0 \leq\ & x_{c,11..15} \leq 5 \end{aligned}$$

and the objective is

$$\begin{aligned} &\text{minimize } m \\ &\text{where} \quad m \geq \textstyle\sum_{c\in\mathcal{C}} E_c x_{c,t} \quad \text{for } t \in \{1,2,3,4,5\} \\ &\qquad\quad 2m \geq \textstyle\sum_{c\in\mathcal{C}} E_c x_{c,6..7} \\ &\qquad\quad 3m \geq \textstyle\sum_{c\in\mathcal{C}} E_c x_{c,8..10} \\ &\qquad\quad 5m \geq \textstyle\sum_{c\in\mathcal{C}} E_c x_{c,11..15} \end{aligned}$$

This instance of MILP problem can be solved by any MILP solver. The values of variables $x_{c,1}$ of an optimal solution are used to determine which converters $c \in \mathcal{C}$ should run in the coming time interval 1. For the next time interval 2, a similar instance of an MILP problem is created by shifting indices of time intervals by one and refining the predicted demands $D_{c,t}$.

There is no general rule how time intervals should be split into blocks since it is strongly influenced by the particular case studies. In this study, we use one specific choice to study the potential of this approach.

Case application

Both methods of direct control (global and time scale MILP control) are applied to a specific case involving 135 houses of the Meppel project. Each house is equipped with two heat pumps, one for domestic hot water and one for space heating. The applied energy system in Meppel is explained in detail in [18]. For the present paper, we investigate the quality of load balancing of heat pump electricity demand. We compare both types of direct control with results of reference PI control that is determined by each heat pump by itself, based on signals from house and storage thermostats. To generate heat demand profiles, the following approach is followed:

- develop a thermal model to determine house space heating demand
- apply measured weather data (we investigate 1 week with a high space heating demand)

- define various typical house and household profiles to generate a variety of space heating and domestic hot water demand profiles
- simulate space heating demand of the various households
- determine reference PI control results
- input domestic hot water and space heating demands into the global and time scale MILP control algorithms and determine results

In the next sections, the individual steps of the approach are explained in more details.

Space heating thermal model

Suitable methods to determine space heating demand are listed in [19] and include modelling of thermal network, radiant time series, and transfer function methods. In the following, we develop a so-called grey-box model, i.e., a simplified model of a house based on the thermal network approach. An advantage of this approach is that dynamic heat demand of a house is described by only a few differential equations which are easy to integrate into smart grid control algorithms. However, model parameters may not correlate very well with true physical characteristics of the concerned building. For the purpose of the present paper, absolute accuracy of the space heating demand model is of less importance, hence we determined model parameters by physical estimation and not, e.g., by applying identification techniques using simulated or measured data. Validation of accuracy of the applied space heating demand model is part of our future work. The thermal network representation of the simplified model is shown in Fig. 2. For the indoor part, the model contains lumped thermal masses for the interior zone and floor structure. We assume an air-based heating system which gives heating input directly into the zone. Towards the exterior, there are two lumped thermal masses, one for the interior walls and one for the exterior walls of the house envelope. The main heat loss contributions to the ambient are through the envelope walls, windows, and roof.

The model equations are formulated as follows:

$$\begin{aligned}
C_{\text{wo}} \cdot \frac{dT_{\text{wo}}}{\text{dt}} &= \frac{(T_a - T_{\text{wo}}) \cdot A_{\text{wall}}}{R_{wo-a}} + \frac{(T_{\text{wi}} - T_{\text{wo}}) \cdot A_{\text{wall}}}{R_{wi-wo}} \\
C_{\text{wi}} \cdot \frac{dT_{\text{wi}}}{\text{dt}} &= \frac{(T_{\text{wo}} - T_{\text{wi}}) \cdot A_{\text{wall}}}{R_{\text{wi-wo}}} + \frac{(T_z - T_{\text{wi}}) \cdot A_{\text{wall}}}{R_{z-\text{wi}}} \\
C_f \cdot \frac{dT_f}{\text{dt}} &= \frac{(T_z - T_f) \cdot A_f}{R_{z-f}} + \frac{(T_{cs} - T_f) \cdot A_f}{R_{f-\text{cs}}} \\
C_z \cdot \frac{dT_z}{\text{dt}} &= D_{c,t} + q_{\text{vent}} + q_{\text{inf}} + q_{\text{gain}} + \frac{(T_f - T_z) \cdot A_f}{R_{z-f}} \\
&\quad + \frac{(T_{\text{wi}} - T_z) \cdot A_{\text{wall}}}{R_{\text{wall}}} + \frac{(T_a - T_z) \cdot A_{\text{windows}}}{R_{\text{windows}}} \\
&\quad + \frac{(T_a - T_z) \cdot A_{\text{roof}}}{R_{\text{roof}}}
\end{aligned} \tag{10}$$

The terms in these equations are explained in Table 1. Besides heat loss to the ambient, infiltration- and ventilation-associated heat losses (q_{inf} and q_{vent}) are determined by the following equations:

$$\begin{aligned}
q_{\text{inf}} &= \phi_{\text{air,inf}} \cdot \rho_{\text{air}} \cdot c_{p,\text{air}} \cdot (T_a - T_z) \\
q_{\text{vent}} &= \phi_{\text{air,vent}} \cdot \rho_{\text{air}} \cdot c_{p,\text{air}} \cdot (T_a - T_z) \cdot (1 - \gamma_{\text{hr}})
\end{aligned} \tag{11}$$

The used terms are also explained in Table 1. Ventilation air flow values $\phi_{\text{air,vent}}$ in m^3/h are defined by schedules given in section "House and household case information". Infiltration air flows $\phi_{\text{air,inf}}$ are related to leakages of the building and are assumed as constant values. We defined some variations including heat recovery ventilation with a certain heat recovery efficiency γ_{hr}.

Besides heat losses, there are scheduled heat gains (q_{gain}) due to resident and electric appliance dissipations. The applied schedule is given in section "House and household case information". For the present paper, solar energy absorbed by the interior is neglected because we study a relatively cold and cloudy week.

The heat demand $D_{c,t}$ is calculated by solving the model equations including a simple zone temperature control. For this, setpoint schedules for the zone temperature are defined in section "House and household case information". The demand $D_{c,t}$ for every house c and time interval t is determined by the following rules:

- If the zone temperature T_z equals the setpoint, then the demand $D_{c,t}$ is the amount of energy which the heating system has to generate to keep the zone temperature constant.
- If the setpoint is increased, then the demand $D_{c,t}$ is increased to raise the zone temperature in a given warmup speed $\frac{dT_z}{\text{dt}}$.
- If the zone temperature T_z is above the setpoint (e.g., due to decreasing the setpoint or natural heating by internal gains), the demand $D_{c,t}$ is the minimal non-negative amount of energy which keeps the zone temperature above the setpoint.

Application of weather data

As we explained in the previous section, we limited our investigation to heat loss due to temperature differences between the zone temperature and ambient temperatures. Effects of solar gains and wind speeds on the heat demand are excluded. Weather data containing hourly average ambient temperatures are obtained from the website of the Dutch national weather institute ([20]). We choose data of weather station Hoogeveen which is close to Meppel. We investigate the coldest week of 2012 as this week results in a relatively high heat demand for space heating.

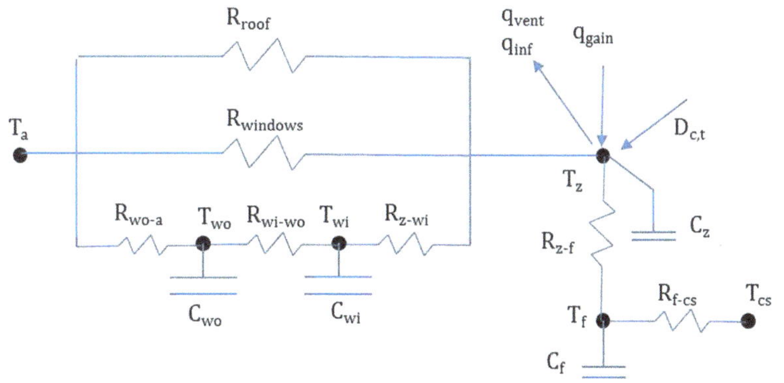

Fig. 2 Applied thermal network model

House and household case information

We consider a total of 135 households and we define three types of households (see Table 2) living in semi-detached and detached houses. Houses will be built in three construction phases and later phases will have a better thermal insulation due to tightening regulations. In this study, we assume that in each phase, 30 semi-detached and 15 detached houses will be built. Rc values of semi-detached houses in the respective construction phases are 3.5, 5.0, and 7.5 m^2K/W and the Rc values of detached houses are 5.0, 7.5, and 10.0 m^2K/W.

Table 1 Nomenclature energy system characterization

Term	Signification
T_a	Ambient temperature
T_z, C_z	Zone (room) temperature and thermal capacity
T_{wo}, C_{wo}	Outside wall temperature and thermal capacity
T_{wi}, C_{wi}	Inside wall temperature and thermal capacity
T_f, C_f	Zone floor temperature and thermal capacity
T_{cs}	Cellar or creeping space temperature
R_{roof}	Thermal resistance of roof
R_{window}	Thermal resistance of windows
R_{wo-a}	Thermal resistance between outside wall and ambient
R_{wi-wo}	Thermal resistance between outside and inside wall
R_{z-wi}	Thermal resistance between zone and inside wall
R_{z-f}	Thermal resistance between zone and floor
R_{f-cs}	Thermal resistance between floor and concrete structure
q_{vent}	Ventilation heat flow
q_{inf}	Infiltration heat flow
q_{gain}	Internal gain heat flow
$D_{c,t}$	Heating load flow (heating Demand)
$\phi_{air,inf}$	Infiltration air flow $[m^3/h]$
$\phi_{air,vent}$	Ventilation air flow $[m^3/h]$
ρ_{air}	Air density
$c_{p,air}$	Specific heat capacity of air

We define a lower setpoint for the zone temperature (18 °C) during working hours and night and a higher setpoint temperature (20 °C) otherwise. We also define domestic hot water demand for morning, afternoon, and evening peaks. See Table 2 for schedules of the temperature setpoints and energy demands for hot water.

Simulation results of heating demand

The model equations given in section "Space heating thermal model" are solved using a 15-min time step which is required as a minimum time step for controlling the thermal storage and to calculate heat pump operation times. To obtain 15-min weather data, we applied linear interpolation of the 1-h data values.

We assume a heat pump coefficient of performance (COP) of 4.5 for space heating and 2.5 for domestic hot water generation during the simulation. With that, we obtain as average electricity input 116.4 kW for space heating and 19.3 kW for domestic hot water, leading to 135.7 kW total average electricity demand. If the MILP control performs well on minimizing peaks, we expect control schedules for the heat pumps to give results close to this average electricity demand.

Results and discussion

In Fig. 3, we show the total electric energy consumption of the three methods of control (reference control, global MILP control, and time scale MILP control). The reference control, which resembles PI control, results in a dynamic electricity consumption with many peak loads on the network, which indicates a high level of simultaneous running heat pumps at those times.

As seen on Fig. 3, both methods of MILP control of the heat pumps lead to a much improved, almost flat electricity consumption profile. It is also demonstrated that the result of time scale MILP control is very close to global MILP control. The average for both is 135.6 kW. The peak reduction compared to reference control is 98 % for global MILP control and 96 % for time scale MILP control.

Table 2 Types of households, the number of household types in both types of houses, schedules of higher temperature setting, and hot water demands

Type of household		Young couple	Young family	Elderly people
Number of persons in a household		2	4	2
Number of houses	semi-detached	27	54	9
	detached	12	27	6
Higher setpoint	weekdays	17–22	8–22	10–23
	weekend	9–23	9–23	10–23
Hot water on weekdays	morning	15 MJ	8 MJ	4 MJ
	afternoon	0 MJ	4 MJ	4 MJ
	evening	20 MJ	24 MJ	20 MJ
Hot water on weekend	morning	8 MJ	4 MJ	4 MJ
	afternoon	4 MJ	4 MJ	4 MJ
	evening	24 MJ	32 MJ	24 MJ

If we consider this almost equal performance and take into account the much reduced computational effort of time scale MILP control, we prefer this method for future algorithm development.

The energy system described in section "Introduction" contains a central CHP. The obtained flat electricity demand profile of the heat pumps may be generated entirely by the central CHP but this depends on the possibility to simultaneously store or use the produced heat by the CHP in a connected district heating system. Part of future work is to study possible CHP control schemes more closely, focusing on stable CHP operation, optimal sustainability, and optimal profitability in relation to varying external grid prices.

Another important quality aspect of the heat pump operation schedules obtained from the optimized control is the experienced thermal comfort within the houses. This is influenced by the upper bound U_c of the state of charge $s_{c,t}$ (Eq. (3)). In Fig. 4, we show the total electric energy consumption of the heat pumps sorted from highest to lowest consumption. The MILP control results in a more horizontal distribution than the reference control. In fact, depending on the value of the upper bound U_c, a whole continuum of slopes between

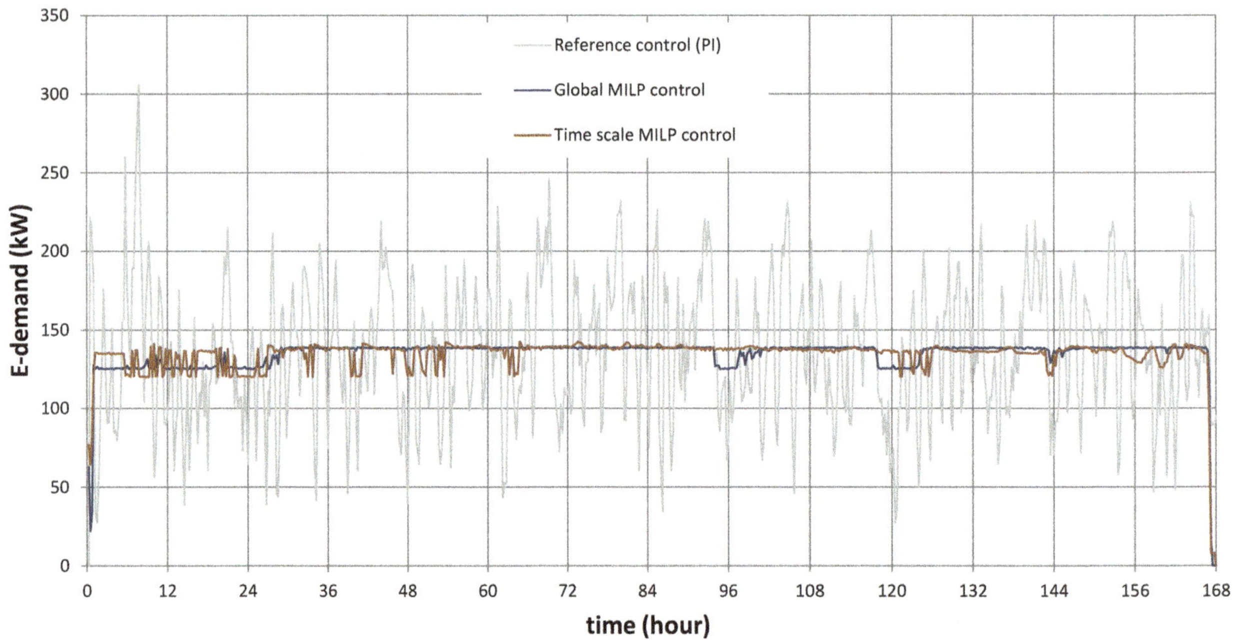

Fig. 3 Case results

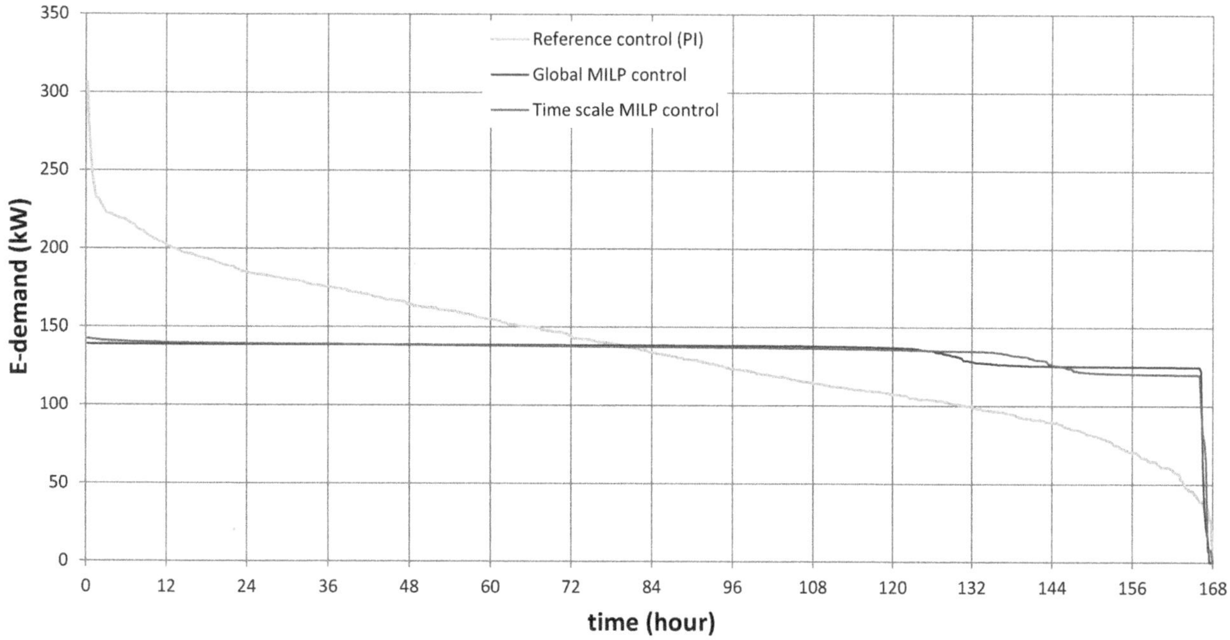

Fig. 4 Sorted case results

the slope of the reference control and MILP control is possible. For the reference control, the value of U_c is approximately zero, i.e., the interior temperature equals the defined setpoints in time, except for the warming up and cooling down times when the setpoint is increased or decreased from a previous state. During those times, the interior temperature needs some time to change due to thermal mass effects within the house. Part of our future research is to investigate the quality of the interior temperature in relation to the value of the upper bound U_c.

Conclusions

In this paper, we investigate how well a group of heat pumps for 135 different households can be controlled to minimize peaks within the electricity distribution network by two methods of direct control. The significance of this research is that this type of control enables integration of renewable energy within a microelectricity grid. Without this control, power consumption peaks would occur frequently and this results into higher investments needed for network strengthening.

From simulations with a simplified thermal network model of the houses, we obtained heat demand data for space heating. For the simulations, we defined occupancy schedules and different insulation properties of the houses. We also defined domestic hot water demand profiles for three types of households. We compare the total electricity consumption of all the heat pumps for the following control methods: reference (PI) control where each heat pump determines itself when to switch on or off, and two forms of direct control: global MILP control and time scale MILP control. The latter has the advantage of a much reduced scale for solving the optimization problem.

In the defined case of 135 houses, MILP control decreases electricity peaks by 96 % compared to reference control. The difference between global and time scale MILP control is small. As time scale MILP control is computationally much more efficient, we propose to use time scale MILP control. The influence of the chosen time scaling on computational time and the quality of the achieved result is part of future work. Although the results look promising, due to the high level of peak reduction, future work will be dedicated to investigate the resulting thermal comfort as a result of the obtained heat pump planning. Also part of future work is to investigate methods of reaching social fairness in heat pump planning and integrating this work into the TRIANA smart grid control method [21] to investigate possible CHP control schemes with a focus on stability, sustainability, and profitability in relation to varying electricity grid prices.

Competing interests

The authors declare that they have no competing interests.

Authors' contributions

JF developed the mathematical optimization modeling part, and RVL developed the thermal modeling part of this research. RVL defined the case and case-related parameters. The simulation environment was developed by JF. JF and RVL together analyzed the simulated data and drafted the manuscript. JH and GS critically reviewed the paper and gave final approval for the paper to be published. All authors read and approved the final manuscript.

Authors' information
Jiří Fink was affiliated to the STW project i-Care 11854.

Acknowledgements
The authors would like to thank the Dutch national program TKI-Switch2SmartGrids for supporting the project Meppelenergy and the STW organization for supporting the project i-Care 11854.

Author details
[1]Saxion University of Applied Sciences, Sustainable Energy Group, P.O. Box 70.000, 7500 KB, Enschede, The Netherlands. [2]University of Twente, Department of Computer Science, Mathematics and Electrical Engineering, P.O. Box 217, 7500 AE, Enschede, The Netherlands.

References

1. Meer Duurzame Energie in de Toekomst. visited October 2014. http://www.rijksoverheid.nl/onderwerpen/duurzame-energie/meer-duurzame-energie-in-de-toekomst. Accessed 26 Oct 2015
2. Nykamp S (2012) Integrating renewables in distribution grids—storage, regulation and the interaction of different stakeholders in future grids. PhD thesis, University of Twente (2013). doi:10.3990/1.9789036500579
3. Innovation Contract Smart Grids. visited October 2014. http://topsectorenergie.nl/wp-content/uploads/2013/10/InnovatieContract-Smart-Grids-2012.pdf. Accessed 26 Oct 2015
4. Nieuwveense Landen. visited March 2014. http://www.meppelwoont.nl/nieuwveense-landen/. Accessed 26 Oct 2015
5. RVO Meppel heats new housing development with biogas. visited March 2014 (2012) http://www.rvo.nl/sites/default/files/Meppel%20heats%20new%20housing%20development%20with%20biogas.pdf. Accessed 26 Oct 2015
6. Meppel Energie. visited March 2014. http://www.meppelenergie.nl/nieuwveense-landen. Accessed 26 Oct 2015
7. Stadler M, Krause W, Sonnenschein M, Vogel U (2009) Modelling and evaluation of control schemes for enhancing load shift of electricity demand for cooling devices. Environ Modelling & Softw 24(2):285–295
8. Halvgaard R, Poulsen NK, Madsen H, Jorgensen JB Economic model predictive control for building climate control in a smart grid. Proceedings of 2012 IEEE PES innovative Smart Grid Technologies (ISGT)
9. Molderink A (2011) On the three step methodology for smart grids. PhD thesis, University of Twente. doi:10.3990/1.9789036531702
10. Bakker V (2011) Triana: A control strategy for smart grids: forecasting, planning and real-time control. PhD thesis, University of Twente. doi:10.3990/1.9789036533140
11. Bosman M (2012) Planning in smart grids. PhD thesis, University of Twente. doi:10.3990/1.9789036533867
12. Fink J, Hurink JL (2015) Minimizing costs is easier than minimizing peaks when supplying the heat demand of a group of houses. Eur J Oper Res 242:644–650
13. Bosman MGC, Bakker V, Molderink A, Hurink JL, Smit GJM (2009) The microCHP scheduling problem. In: Proceedings of the Second Global Conference on Power Control and Optimization, PCO 2009, Bali, Indonesia. AIP Conference Proceedings Vol. 1159. pp 268–275
14. Bosman MGC, Bakker V, Molderink A, Hurink JL, Smit GJM (2011) Controlling a group of microCHPs: planning and realization. In: First International Conference on Smart Grids, Green Communications and IT Energy-aware Technologies, ENERGY 2011, Venice, Italy. pp 179–184
15. Bosman MGC, Bakker V, Molderink A, Hurink JL, Smit GJM (2010) On the microCHP scheduling problem. In: Proceedings of the 3rd Global Conference on Power Control and Optimization PCO, 2010, Gold Coast, Australia. AIP Conference Proceedings Vol. 1239. pp 367–374
16. Axsäter S (2006) Inventory control. International series in operations research and management Science, Vol. 90. Springer, New York
17. Cook WJ, Cunningham WH, Pulleyblank WR, Schrijver A (2011) Combinatorial optimization, Vol. 33. Wiley. com, Rhode Island
18. van Leeuwen RP, Fink J, de Wit JB, Smit GJ (2015) Upscaling a district heating system based on biogas cogeneration and heat pumps. Energy, Sustainability Soc 5(1):16. doi:10.1186/s13705-015-0044-x
19. Mitchel JW, Braun JE (2013) Principles of heating, ventilation, and air conditioning in buildings. Wiley, New York
20. KNMI weather records station Hoogeveen. http://www.knmi.nl. Accessed 26 Oct 2015
21. Toersche HA, Bakker V, Molderink A, Nykamp S, Hurink JL, Smit GJM (2012) Controlling the heating mode of heat pumps with the TRIANA three step methodology. In: Innovative Smart Grid Technologies (ISGT), 2012 IEEE PES. pp 1–7. http://ieeexplore.ieee.org/iel5/6170475/6175527/06175662.pdf

12

"Trust" and "confidence" as socio-technical problems in the transformation of energy systems

Christian Büscher* and Patrick Sumpf

Abstract

Background: For the development of a sustainable energy system, the public's contribution to the sustainment of the system's current operations will become vital. The public's role is expected to change from that of passive service abiders to active service providers, as anticipated by visions of prosumers, smart grids, demand side management, virtual power plants, and electric vehicle management. Smart technology and new business models will increase system complexity and the experience of uncertainty and risk for all parties. *Actionability* of the public's participation in the opportunities arising from the transformation of energy systems will become a problem.

Methods: With a qualitative functionalist method, we analyze the prerequisites of the public's participation: the capacity of social mechanisms, such as trust, for the solution of social problems. Functional problems affect the continuation of communication, i.e., the possibility of the sustainment of social reality. We isolate three case studies that illustrate concrete effects of trust, distrust, and confidence on three parties, respectively: consumers, investors, and those affected by grid extensions.

Results and conclusions: We argue that the social mechanisms of trust and confidence are more vital for consumers, investors, and those affected than previous research has noticed. These mechanisms determine the achievement of sustainability in energy systems by sustaining actionability as a favorable expectation (trust, confidence), or an unfavorable expectation (lack of trust, distrust). Even lack of trust and distrust can be valuable contributors on the road to sustainability, since they uncover flaws and setbacks of a fledgling smart grid.

Keywords: Smart grid, Complexity, Acceptance, Uncertainty, Risk, Prosumer, Consumers, Investors, Those affected, Decision-making

Background

The transformation of energy systems (ES) into smart grids—particularly in the domain of electricity—is not only a matter of technological proficiency. In fact, the far-reaching objectives and visions of systemic transformation require major qualitative changes in the relationship between the public and the ES [1–3]. "Smart" technologies and behavior, like smart grids, smart markets, or demand side management (DSM), are supposed to support the implementation of multiple alternative energy sources, in order to increase efficiency and sustainability of the provision, distribution, and consumption of energy.[1] From the systems perspective, the potential of technological developments cannot be exploited without the public's active participation: "Consumers represent a cornerstone in the fulfillment of goals such as energy efficiency and savings, by their active participation through their actual utilization of the system" [4].

Proponents of smart grids assume that the public will adapt to the new technological reality once the opportunities become clear. Scientists, politicians, and economists call on the ability of potential users to rationally calculate risks and chances.[2] To expose, analyze, and assess the consequences of these underlying premises is the motivation of our investigation. They rest on nothing less than the broad-scale modification of public attitudes toward the following:

* Correspondence: buescher@kit.edu
Karlsruhe Institute of Technology (KIT), Institute for Technology Assessment and Systems Analysis (ITAS), Karlstr. 11, Karlsruhe 76133, Germany

- New technical devices which are "intelligent, self-healing, autonomous machines" [5–7], into whose operation the majority of the public has no insight [8]
- Opaque markets in which rules of fairness cannot be enforced for certain, as it is unclear "who reaps the benefits, who bears the risks and burdens" [9]
- The complex of public administration and governmental agencies, which supervise the development of smart infrastructures, and on which the public has to rely [10–12]

The implementation of these changes will not be straightforward. In general, social scientists assume that the major obstacle for the implementation of a new technology is *social acceptance*. In the last decade, we have seen extended research on the matter of acceptance and energy system transformation: general assessments of regional situations like the USA [9]; the differentiation of dimensions of acceptance (socio-political, community, or market [13] dimensions); acceptance of institutional frameworks like socio-economic conditions or the consequences of innovation [14]; investigations of criteria determining acceptance (including criteria like community ownership [15], participatory exercises [16], perceived justice in planning procedures [17], values associated with large technical systems [18]); and many case studies on wind energy in general and offshore wind, smart grids, transmission lines, etc. However, this branch of research—just like most research on socio-technical systems—investigates causes for the achievement or non-achievement of positively valued developments and outputs (like public support for new technologies in particular and the ES transformation in general [19]); it investigates goal attainment following specific quotas of renewable energy or improvements to the ES concerning greater customer orientation or more sustainable, ethical, and effective developments [20]. In this sense, lack of acceptance is conceptualized mainly as a deficit that has to be overcome through educating the public, and it is assumed that once education is achieved, the related problems will probably vanish.

Contrary to the belief in promoting the public's understanding of science, everyday social life needs a "reduction of complexity" [21] to get things done; it needs "voluntary blindness." A "large measure of trust is necessary if we are to cooperate, manage the uncertainties and confront the risks of working towards a sustainable energy future" [22]. In truth, acceptance does not necessarily lead to the active involvement of persons; it may just result in passive tolerance.

As an alternative to deficit models (lack of acceptance or trust) and factor theories (causes for acceptance or trust), we introduce the application of functionalist methods [23] to the specific case of smart grid technologies for future energy provision. We are searching for the social conditions which *render unlikely commitments likely*. This search does not imply any evaluation of the outcome of such commitments. Rather, the application of this method should provide a deeper understanding of the underlying problem—free of preconception—before discussing policy options.

The alternative concept to acceptance is that of *the relation of trust and confidence*. *Trust* is a mechanism of complexity reduction, because it enables action in spite of uncertainty about the future [21, 24–26]. As a consequence, trust helps execute actions that would otherwise not have been effective [21], and it builds a foundation for the implementation of *action potential* [27]. *Confidence* is a behavior which is complementary to trust, and it is directed at abstract entities like infrastructures, politics in general, or markets. Given the need to reduce complexity and to absorb uncertainty, we assume that trust in systems and their respective decision programs, as well as confidence in the overall transformation of the systems, are necessary *preconditions for actionability*.[3]

Moreover, we argue that trust and confidence are interdependent. They systematically reinforce one another, and only if a certain threshold of trust and confidence is reached will people commit themselves to actively participate in the transformation process. Lack of trust may enforce a passive behavior; distrust may enforce autonomy, or the search for alternatives. These patterns are visible in energy-related roles such as consumers, investors, and those affected by grid extension.

Trust and confidence address specific functional problems which arise in socio-technical constellations like the ES. The task of this paper is to investigate what socio-technical problems can be identified ("Socio-technical problems" section) and how social mechanisms like trust and confidence contribute to their solutions—without claiming that those solutions are genuinely positive for the overall development of future energy provision or society as a whole. We have to bear in mind that *overconfidence* and *blind trust* may be as damaging as the lack of trust.

Methods

Accordingly, this paper will present a sociological discussion on social mechanisms to cope with the increasing degree of complexity resulting from the ES transformation. For this purpose, theoretical arguments from sociological approaches, addressing functional problems of control, change, and actionability, will be combined with empirical insights. The empirical data has been extracted from 90 to 120 min, semi-standardized interviews conducted with eight leading experts from various branches of the German energy sector. They include major technology companies, transmission service operators, federal and local business associations, and consumer and environmental associations.[4] The narratives of these experts, combined with extensive document reviews, provide the main source for our descriptions of the German ES visions. Moreover, they guided our research toward the inclusion of complexity and the resulting challenges in the three case studies. All in all—referring to the functionalist method—we follow one of the leading methodical guidelines of technology assessment: to *expose the premises* of prospective technology programs, to analyze their secondary problems, and to bring all societal consequences to full display.

For a more thorough understanding of the societal dimension of ES transformation, we will expose structural, institutional, and operational problems ("Socio-technical problems" section). We will discuss how, on the operational level, problems of acting and decision-making under uncertainty have their roots in trust, lack of trust, or distrust. To illustrate the plausibility of our theoretical arguments and to illustrate concrete challenges, we explore trust issues in the German energy sector ("Trust and distrust as drivers of stability and change in ES transformation" section). Three case studies will be laid out, with consideration of the different situations which the public finds itself in, already, and more so in the future: consumers, investors, and those affected ("Results: three case studies about smart grids—consumers, investors, and those affected" section). Finally, we will discuss briefly how the existing "architecture of trust" may be modified and what consequences this can have on the development toward sustainability ("Discussion: an "architecture of trust" for ES" section).

Socio-technical problems

The ES interrelates technical and social processes to solve specific problems, in order to provide expected outputs. Prevailing research on socio-technical systems emphasizes either *structural* aspects or *institutional* aspects. While these aspects are crucial, we will also draw attention to *operational* aspects, thereby dealing with problems of action and decision-making. Since every socio-technical system poses a "linked series of socio-technical problems" [28], we propose that *socio-technical systems are expectation complexes of services (institutional aspect), to which technical and social systems contribute (structural aspect), to solve from moment-to-moment socio-technical problems (operational aspect).* We will discuss structural and institutional changes which lead to complexity. Complexity results in the anticipation of contingency, which itself results in decision-uncertainty and risk. Problems of actionability concern—mainly but not exclusively—the operational dimension. In the following, we discuss socio-technical problems and why trust and confidence are requisite social mechanisms for the development of future smart energy infrastructures.

Structure and control

Previous research on the emergence of socio-technical systems [29, 30], on the governance of large technical systems [31, 32], or on socio-technical infrastructure systems [28, 33, 34] analyzes the emergent qualities of entities wherein social and technical elements interrelate. It conceives of socio-technical systems as open systems or networks of heterogeneous elements, held together by a purpose: that of providing energy, transportation, water, or worldwide data exchange. It assumes that technical operations and social actions are functionally complementary [35, 36], and it focuses on antagonistic developments, stress, or breaches. Usually, one finds exogenous (environmental) or endogenous (systematic) factors triggering changes of the system's characteristics, which then influence the quality of the infrastructure service [37]. This makes future states or behavior of systems harder to predict (indeterminateness).

Concepts of socio-technical systems highlight the multiple, varying relations between heterogeneous elements, indicating a high level of "organized complexity" [38]. For the German ES Transformation, complexity is all-embracing: Different types of power plants (for conventional and renewable energy sources) are connected to the network with transmission lines, distribution grids, and smart devices which contribute to the service of the ES. Moreover, different actors of the ES (administrative bodies, companies, communities, groups, and private persons) are interrelated through rules, contracts and markets, and regulations [33]. Control of such interwoven networks becomes vital.

Experts already face multiple challenges due to the changing structure of infrastructure systems; they are particularly challenged by the "complexity of these systems and the related emergent behaviors which may arise in collective ways, difficult to predict from the superposition of the behavior of the individual elements of the system" [39]. Given that the visionaries of smart grids propose even more integration of Information and Communications Technology (ICT), the problem of non-knowledge about the behavior of smart grids may progressively become *the* most significant issue for all involved, for operators, supervisors, investors, "prosumers", and citizens [40]

Institutions and change

Several concepts concerned with socio-technical systems accord an important role to institutions [41] and to processes of institutionalization [36, 42]. Recently, *transition research* has invoked the structural and institutional features of socio-technical systems with crucial infrastructure (such as energy, water, railroads, and telecommunications). A very prominent feature of transition research is the analysis of the relations of *stability* (configuration, structure, and institution) and of *change* (co-evolution, structuration, and institutionalization). Within the framework of transition research, the multi-level perspective addresses socio-technical transition as a function of stability and change caused on three analytically distinct levels: *regime, niche,* and *landscape*. The regime is the dominant structure within a socio-technical system. In a regime, a multitude of actors and organizations is tightly interwoven into a network of mutual dependencies held together through formal and informal relations, e.g., through contracts and trust (respectively). The regime determines social relations by virtue of institutionalized expectations, such as cognitive rules of scientific observation, agreed upon knowledge, established technical paradigms, and belief systems [35, 36]. In abstract terms, institutions entail that not every event, action, or decision is similarly possible, legitimate, or probable. *Structures and institutions are necessary features of social life to provide orientation and to enable action.*

The energy infrastructure in Germany is a highly regulated complex, with strongly institutionalized networks of incumbent actors, but it is now in flux because of the German ES Transformation. This transformation is executed as a real-time experiment of a socio-technical nature, comprising experiments with technical and social aspects (e.g., regulations or consumer behavior). "Research on energy has increasingly turned society into a laboratory – one in which the energy user and non-scientist can potentially play an active part in the experiment" [40]. To control such a transformation, a balance between tight experiment-reality couplings (which enables innovative, realistic, close-to-the-market benefits) and loose couplings (which disturb the system's operation as little as possible) is required. To foster social change, one needs to sustain action orientation with stable generalized expectations, while increasing learning capacities with less "resilient" institutions [43]. Nonetheless, due to the inevitable non-knowledge about the success of the German ES Transformation, the public may be opposed to a learning experience, and such opposition will, presumably, prevent their active involvement [40].

Operation and intransparency

The problems of control and change affect the operational dimension of the ES. Social operations include acting and decision-making. In socio-technical systems, operations are conditioned by technical determinism (1) and social dynamics (2). These are very distinct modi operandi.

1. In this paper, we refer to a *technical* operation on a physical, chemical, or biological basis, as a "functioning simplification in the medium of causality" [44]. In the sense of causal determination of technical operations, technology exonerates from contingency—if it works. One does not need to know how technical devices operate, but with some experience, one can expect a certain output on a regular basis. This idea can be illustrated with the expression "trivial machine," coined by Heinz von Foerster [45]. Trivial machines are those machines which use *a set of rules* in order to transform certain facts into other facts (for example, heat from combustion of coal boils water in a boiler to produce steam, which expands across a number of blades in a turbine, leading to rotation and, subsequently, to the generation of electricity). "Non-trivial machines," on the contrary, change the rules on how certain facts are transformed into other facts *of their own accord*.[5] Therefore, we understand technology in connection with regularity of behavior; provided a machine functions, it will operate in accordance with the actors' stable expectations of what will happen when the machines operates.[6]
2. We refer to a *social* operation as communication (the basic operational element of social systems) based on meaning. Unlike technical operations, social operations entail the possibility of *surprise*. Successful communication is, first and foremost, *unlikely*. The synthesis of a threefold selection of information, utterance, and understanding constitutes communication. Selectivity implies the potential to surprise in every situation anew: People are always free to say "yes" or "no" to any kind of communication offer. Paradigmatically, this problem has been addressed by concepts emphasizing that social interaction is a precarious affair because of the freedom other actors possess. The social condition for collective actions, i.e., cooperative behavior (as pointed out by rational choice theory), is one important concept here [46]; the "problem of double contingency" is another [47].

From a functional perspective, sociological theory seeks to explain why all the occurrences we take for granted, because we observe them on a regular basis, are happening despite their indeterminateness (contingency, uncertainty, risk). Unlike technical operations and routinized, quasi-automatic action, social realities emerge when actors relate to each other via communication and when the threefold selection of information, utterance, and understanding creates meaningful distinctions which link communication to an identifiable unit (a social system) distinct from all other events (in the system's environment). "Such a model allows for a complex interplay of what information is being transferred, how it is transferred (whether intended or not), and then how it is received (which itself reinstitutes the communicative process by re-conceptualizing information)" [48]. Social realities have to solve problems of contingency, e.g., the surprise of unexpected events, so that structures and institutions can emerge and sustain themselves. Norms, rules, routines, values, etc., are contingent in their origins, but once they are established, they offer stable expectations. However, communication, as the basic element of social systems, always inherits the moment of surprise. Every social situation is comprised of events which bring about a new combination of determinacy *and* indeterminacy [47]—which is (or should) not be the case for technology—where only determinacy is in order.

Structural coupling of technology and communication

Since there is a sharp distinction between technical operations and social operations, the interrelations of the socio-technical should not be approached in terms of functional equivalency (like in the metaphor of the "seamless web") but in terms of *structural coupling*. A structural coupling implies that while technology is a topic of (or stimulates) communication, it does not determine, overlay, or substitute communication, because the type of operation is distinct: "The technical network of energy flow is completely neutral to communication; in other words, information is produced outside the network [...]. Causal relations between technological physics and communicated information are freed of overlap and take the form of structural coupling" [49].

The case of ICT illustrates this structural coupling clearly, for in spite of tremendous developments (speed, volume, and accessibility) in electronic data processing, social actors remain dependent on interpretation and choice in order to exploit the technological capacities. The information value of electronic data processing is determined by the processing of meaning by psychic or social systems. Actors are confronted with an enormous asymmetry between the visible, symbolic tokens on the interface of a system and the machines lying hidden behind the interface. Behind it are these programmed and trivial, but very complicated, machines which execute algorithms that greatly exceed the data processing capacities of human cognition. "Although they [computers] are manufactured and programmed machines, such computers function intransparently for consciousness and communication through structural couplings. Strictly speaking they are invisible machines" [49]. The structural coupling of technical and social realities produces both relief and new forms of stress. Unless uncertainty is absorbed by social mechanisms like trust and confidence, actionability cannot be sustained (Table 1).

Trust and distrust as drivers of stability and change in ES transformation

In the operational dimension of socio-technical systems, trust is a prototypical problem concerned largely with time. Similarly, in Simmel's "philosophy of money," trust implies a "quasi-religious faith" which is required to cope with non-knowledge of the future [50, 51]. Trust empowers persons to "act as if the future was certain" [21]. Recent research on "process views of trusting" [52] highlights the temporal dimension of trust. Trust and time are long-standing relatives. When building stable expectations of future developments, people have to commit themselves to a "leap of faith" [51], and this leap has become a focal point for research on trust.

According to Lewis and Weigert ([53], p. 25), the most "forceful" theoretical advances on conceptualizing trust in the past decades have been conducted by Möllering [26, 51, 54]. Möllering refers to Simmel's leap as the act of "suspending irreducible social vulnerability and uncertainty as if they were favourably resolved" ([54], p. 356). This theory on the act of "suspension," which is largely shared in the international research community, tries to account for the irrationality of trust that upholds favorable expectations even in the face of possible disappointment—*the perception of risk* [55]. In particular, this interdisciplinary approach attempts to conceptualize trust in a more holistic fashion than single disciplines—such as economics. It identifies the core of the trust problem in the *irreducibility of uncertainty* (as quoted above). This corresponds to the functionalist view of upholding actionability through trusting.

Comparing his own approach with game theory models of trust and comparable rational choice concepts, Möllering ([54], p. 356) explains that "The need to find alternative explanations is evident, because rationalistic explanations regularly face the paradox that they are either explaining trust away or explaining everything but trust." In this way, rational choice models of trust usually refer to measurability, or certainty, or cost versus benefit analyses. Yet these aspects are only one part of what trust is concerned with; they are merely concerned with *reason* ([54], p. 357). As a result of this limited view, game theorists who are

Table 1 Three dimensions of socio-technical problems

Dimension	Socio-technical problems		Solutions
	Stability	Change	
Structure	Control of system performance, security	Complexity	Governance
Institution	General orientation, reliable outputs	Surprise	Experiments, learning
Operation	Actionability, participation	Intransparency	Confidence, trust

concerned with trust suffer from the assumption that it is knowledge or complete information that arouses trust or distrust. Möllering, on the contrary, alludes to the role of non-knowledge—absorbing uncertainty and risk—as a central variable in the trust process. Even in promising research on "the role of trust in the transition to sustainable energy" [22], the capacity of rational choice and other fledgling trust concepts is overstretched in that "all are based on a common foundation: trust is a feeling or belief that someone (or some institution) will act in your best interest" ([22], p. 2615). Not only is the crucial element of *suspension*—the core of studying trust—missing in this definition; moreover, this definition limits the possible references of trust to micro-scale units such as "someone (or some institution)." In contrast, the approach presented in this paper relies on trust and confidence in abstract systems in a very specific manner [55, 56] that we will further unfold in the following sections.

While trust is already significant in many areas of the energy sector (e.g., investment decisions, market transactions, power plant and network operations), it will probably become even more important in future constellations, if the public's inclusion is required. The transformation of an ES cannot happen thus unless all actors modify their attitudes (or adopt new ones) with respect to new technologies, regulatory frameworks, and market opportunities. Such a transformation of attitude cannot happen through personal relationships; instead, trust in abstract systems is required [56]. Everyone affected by the transformation of the ES is forced to evaluate if the changing system will perform as expected. The function of trust is to enable engaging in new forms of interaction, which the system will offer as arising opportunities. However, evaluation must happen prior to monitoring whether the desired outcomes materialize. Therefore, the problem of intransparency is not only characterized by the trust addressees' degrees of freedom or the possibility of technical failure. Foremost, we should consider the necessity to momentarily close the temporal gap to an open future by a "leap of faith" [26, 50]. Trust is the grease which softens the frictions of constant rational calculation on what can "go wrong."

Trust is intrinsically interwoven in the *sustainment* of present operations:

> Trust can only be secured and maintained in the present. Neither the uncertain future nor even the past can arouse trust since that which has been does not eliminate the possibility of the future discovery of alternative antecedents. [...] The basis of all trust is the present as an unbroken continuum of changing events, as the totality of states in respect to which events can occur [21]

Analyses of system transformation hinge on the temporal horizon chosen; e.g., German scenarios of possible energy generation and consumption paths and respective addressees of trust vary according to whether they concern the next year, the next 5 years, or the next 30 years. Visions of potential smart grid components concern its final state, which is temporally located far away, at approximately 2050, according to the federal government [1]. Hence energy practitioners describe some envisioned components, such as virtual power plants, as "fantasies."[7] The following section discusses visions of future ES, particularly the German ES Transformation ("Smart grids" section). Then, the concept of trust will be clarified further ("Familiarity, confidence, and trust in complex systems" section) and the reader will discover three pivotal trust challenges for smart grids ("Results: three case studies about smart grids—consumers, investors, and those affected" section).

Smart grids

The German ES Transformation is largely a development toward a smart grid. Nearly every significant political party refers to smart grid development, while the government funds large-scale testing with smart grid components, such as "E-Energy" [2], and promotes the development of an "Internet of Energy" [57]. Although a smart grid (understood as a mere effectuation of energy consumption) is conceivable without a large integration of renewable energy sources (RES), the German ES Transformation relies on innovations in this area to mitigate ensuing volatilities.

The smart grid cannot function successfully without "appropriate" smart meter *usage* and economic *activity*, as determined in prevalent visions. Hence, the success of the smart grid is very dependent on the behavioral conformity of prosumers, because actions (trust), the absence of actions (lack of trust), and counteractions (distrust) of individual smart meter users and small-scale electricity vendors in emerging energy markets may have significant cumulative effects on the overall rationality—and therefore success—of the ES. We assess the relevance of trust as a social mechanism in terms of its *impact* as an *action-enabling* mechanism for electricity consumers confronted with unfamiliar complexity and facing unprecedented decision-making and reflexivity in smart grids [58]. We emphasize the *consequences* of trust in the sense that a given action (smart meter usage, economic activity) actually will occur, or remain absent, because of a lack of trust.

These expectations imply a *qualitative* change in the underlying prerogatives of the energy sector. An active trust in technology (smart meters), markets (reputation management, sanction systems), and organizations (supervision, trust intermediaries) will become indispensable for the prospective ES to work, given the critical role of decentralized consumption and feed-in units for system performance. Mere acceptance of infrastructure plays a smaller role.

Hence, the German ES is moving toward a smart grid model based upon large-scale consumer involvement[8] (c.f. how four out of five smart grid components affect the average electricity customer (Table 2)). Since this transformation will probably be most significant for the average, private consumers, this paper focuses primarily on their broad household usage.

Familiarity, confidence, and trust in complex systems

Even though past ES were technically complicated [39], electricity *consumption* was quite simple and rarely led, in the OECD world, to users' disappointment in the reliability of an ES. In Germany, which possesses a particularly reliable power grid, an uninterrupted electricity supply is expected, and consumers have built a high degree of *familiarity* in interacting with electricity services. Familiarity indicates a low level of risk reflection on technology interaction [9] and a high level of *confidence*, thus resulting in a very routine mode of interaction. When consumers' positive experiences with technological services are constantly reaffirmed, familiarity and stable expectations are built, which can eventually obscure huge technological infrastructure systems in an unnoticeable latency [28]. Although confidence is related to trust, it is not bound to decision-making, because disappointments in confidence are attributed to external factors rather than to one's own decisions [55].

Like trust, distrust is an action-enabling strategy. Distrust is not the opposite of trust, but its functional equivalent. Distrust reduces complexity into a narrow action corridor by making a few actions probable and certain others highly unlikely [21]. Distrust provokes counterstrategies by consumers who wish to circumvent the distrusted situation or object, thus aspiring autonomy, searching for alternatives and perhaps even boycotting. Unlike a lack of trust, which can hinder action, distrust results in a *mobilization of action potential*.

In the past, consumers could afford to distrust both the regime operating the ES and its technology (e.g., nuclear plants) without causing an overarching, systemic consequence.[10] Since smart grids depend on the behavior of single users, they are more vulnerable to cumulated effects of massive simultaneous action—and such action can fall outside the expected behavioral patterns. As large technical systems become interactive technical systems with direct consumer involvement through ICT and market participation, the systems' invisibility will dissolve. Smart grids will contain far more "access points," i.e., intersections (smart meters, market organizations, electric vehicles) where consumers can do "facework" [56] with the impersonal ES. This increases the system's complexity and the potential for uncertainty and incalculable consumer action, reinforcing the system's foundation on trust instead of control.

In the German ES Transformation, complexity is an issue in its own right, depicted by experts as the central challenge which they face. Our interviewees explicitly stated that the complexity has a direct bearing on their work. They claimed that a high proportion of non-knowledge exists and pointed out the fragility of interdisciplinary knowledge between engineering (energy technology), computer science (grid communication), and social science (consumer behavior). They portrayed the German ES Transformation as a large real-time experiment rather than a planned project—an appraisal which substantiates the impact of complexity.

Table 2 Five components of smart grids [7]

Domains of smart energy technology				
DSM	Virtual power plants	Electric vehicle management	Energy prosumers	Self-healing networks

From left to right: ICT, increasing connectivity/communication, (distributive) grid extension, decentralization, implementation of "artificial intelligence"

Results: three case studies about smart grids—consumers, investors, and those affected

As trust reduces complexity and thus enables action under future uncertainty, trust unfolds its vital effect in the face of non-knowledge and becomes manifest in the risk-taking act [59]. We shall illustrate, with case studies about smart grid actors, the growing need for distrust, and trust as a "certainty equivalent" [21], to manage rising degrees of contingency, uncertainty, and risk, particularly from a consumer perspective.

Consumers

Electricity has been available to consumers as a quasi-invisible force [58] and, for decades, as an unquestioned resource [60]. The currently passive consumption will increasingly become a matter of liability and daily attention, as the organization of generation, transport, and distribution of electricity is transformed. For the integration of volatile RES, "smart technology" is envisioned to help align consummation during phases of low or high RES output. Proponents of "smart meters" claim that customers can be encouraged to save energy (because smart meters offer transparency about quantity and costs), while operators can monitor and control appliances of customers (because smart meters provide data, identity, and location of home appliances), and control their maximum load demand during peak load (because supply can be disconnected via smart meter if customers exceed their quotas) [61]. According to the technical premises of dominating visions, "smart technology" not only results in more efficient use of energy, it will also provide "grid-supportive" measures and thereby help maintain grid stability and security of supply [2, 62].

In sharp contrast to the traditional supply-driven consumption, smart grids require consumers who actively adapt their consumption behavior through smart meters allowing for DSM, new tariff systems, mobile applications, etc. Consumers may decide to use electrical appliances (e.g., heat pumps, air-conditioning, and washing machines) and discharge vehicle batteries to the grid only at determined times, responding to signals securing the demand-supply equilibrium [2, 7, 63].

As choice always involves uncertainty, consumers' decisions expose the system to trust and distrust vulnerabilities. A basic choice consists in whether to comply with the expectations of engineers, operators, and/or supervisors. Consumers may reject the effort of rational calculation: They may not gain familiarity and fail to gather information by comparing possibilities and prices; they may reject the delegation of control to machines, operators, and supervisors. Yet delegating one's power to make decisions, even partially (e.g., to software agents in smart meter appliances), is itself a decision.

With a view to encouraging consumers' compliance, psychologists research the regulation of incentives (e.g., online platforms, price signaling, and smart meter setup). In many European countries [3], including Germany (E-Energy projects [2]), large-scale regulatory experiments are carried out. The fact that incentives are heavily researched is a strong indicator for the social instability caused by the freedom of choice which consumers are likely to have. Therefore, incentive mechanisms also create risk. Incentives can fail and lead collective consumer behavior into unanticipated directions. If consumer behavior becomes a critical factor for system operation, then the risk of cumulated effects (like collective distrust against the new grid or parts of it) will rise. During the initial phases of transition in the Netherlands and California (USA), distrust was manifested by protest against the installation of smart meters [4, 12].[11]

In 2011, the US Government Accountability Office [64] concluded in its report on Electricity Grid Modernization that "consumers are not adequately informed about the benefits, costs, and risks associated with smart grid systems." As smart grids enable the transmission of large amounts of data about household activity, fear rises about privacy and customer security with regard to the unauthorized use or modification of sensitive "personal data" (that is, recorded data attributed to a person, e.g., name, contact, biographical data like individual preferences, transaction history, or recorded activities), or of information derived from such data (including the compilation of individual profiles or scores) [10]. Some experts anticipate "nothing short of a paradigm shift in industry from the current hardware centric focus on system adequacy and reliability, towards the inclusion [of] a more directly consumer-oriented view of security" [11]. However, even if governments attempt to enforce the implementation of security standards and the industry promises to incorporate these standards, the problem of "authenticity" arises as soon as trust is lacking and the attribution of motives is prevalent. The EU commission has mandated the European Standardization Organization (ESO) to develop security standards for smart meters. Commercial interests dominate, since the ESO relies on the expertise of commercial parties. Hoenkamp and Huitema claim that "These [standards] will be useless if the public rejects them" ([12], p. 5).

The public's fears could sabotage the envisaged overall rationality of the system. Comparable "emergent effects" of cumulative, collective behavior have been observed in other sectors, such as the rejection of E10 biofuel in Germany,[12] sudden investment withdrawals in the financial sector [65], and periodic, or even constant, distrust in the food sector, as with *Enterohaemorrhagic Escherichia Coli* bacteria (EHEC) in Europe [66] or milk powder in China.[13]

Investors

As mentioned above, smart grid visions encompass numerous actors engaged in the energy trade. In some countries and regions, this process is well under way; particularly, small and decentralized RES units with self-generated electricity are increasing in Germany, Denmark, or California (USA). It is expected that electricity loads will serve self-supply and assist with grid stability and, moreover, cause new markets to flourish, so that every actor equipped with technology for photovoltaics, wind, combined heat and power, or geothermal applications, etc., can act as an electricity vendor for "virtual power plants" and "smart markets" ([2, 7], e.g., [63, 67]).

The desired "prosumers" are expected to consume and produce electricity simultaneously, making them both consumers and entrepreneurs. Even at present, such developments and prospects are significant, as can be illustrated by Siemens research: The company created a web portal, with concomitant mobile applications, to explore the potential of virtual power plants in Germany and in European cross-border conglomerates [68]. Its main objective was to form stronger power plants that every participant could benefit from. Therefore, Siemens set up cloud computing networks between local providers and various electricity generators who were willing to combine their electricity loads with those of the other actors. The means to achieve this were "virtual," the actors being "plugged" together through exposition of available load information (kWh, gigawatt) in the cloud. In this way, virtual power plants emerge through ICT as a fresh new field of energy commerce.

Like the global financial system [65], the ES may see the establishment of rating agencies as "guardians of trust" which become crucial to the formation of market networks. According to smart grid visions, institutions which provide recommendations can increase the probability of market interaction, by generating a reputation or signaling certain qualities [2, 7, 69]. They can be certification agencies, market intermediaries, or online platforms.

Since the smart grids' implementation depends, at least partly, on consumers' attitudes, we need to determine the qualities of these attitudes more precisely. We can determine these attitudes by using the analytic distinction between trust and confidence: An actor must, for example, explicitly *trust* in the price stability of her *chosen* supplier's electricity, whereas she must implicitly feel *confident* about the stability of *general* market operation, in how responsibilities are allocated, how security of supply is maintained, and how legal security is guaranteed in the emerging business world.

For active prosumers to be endowed with actionability in this diffuse new system, trustful decision-making can only be embedded in a climate of general confidence in (the parts of) the overall system. This *interdependence between confidence and trust* can be pictured as a relation of mutual spirals, where, progressively, one either reinforces the other (virtuous circle), or one weakens the other (vicious circle). Together, confidence and trust determine the future state of the system [55]. In a virtuous circle, confidence and trust mutually stabilize each other, creating an atmosphere of business optimism and leading to a massive exploitation of opportunities in technology and markets.

The scenario of a vicious circle is particularly relevant during the mid-term phase of the German ES Transformation. Investment in RES (which may be provided by small private investors like average consumers, but also by local providers and larger industries) is crucial for smart grids to evolve. Business representatives and scientific experts insist that confidence in a stable legal framework and financial returns for the investments in RES in the coming years are necessary. However, some actors doubt whether these investments are being made, because uncertainty in the business seems quite acute (according to an interviewed expert from a business association). Certain actors' distrust extends to skepticism of the system's ability to maintain security of supply; some electricity-intensive companies have even begun building their own power plants (according to an interviewed expert from academia).

A lack of trust among consumers and investors is not necessarily a negative development, nor does our discussion intend to promote some sort of "trust creation" for smart grid implementation. It may well be that involved actors are right to distrust—time will show. We merely wish to point out that the expectations of a majority of smart grid practitioners, promoters, and stakeholders do not take into sufficient account the functions and consequences fulfilled by trust, lack of trust, and distrust (e.g., [1, 3]). Indeed, *distrust* can serve an important function of learning and being conscious and critical of relevant changes in technology or policies. Notably, distrust can counter an excessive accrual of trust, after the equilibrium of trust and distrust has tipped toward the trust side, adding up to "blind trust" (which in everyday life we may call "carelessness").

Those affected

In the foreground of the debate on transforming ES into smart grids and thereby integrating RES are appeals to consider the "public acceptance of emerging energy technologies" [9]. These technologies include smart grid facilities, grid extensions, and economic or legal measures that can accompany these technological changes. Most debates about public acceptance are centered on such infrastructure issues. Whenever a grid extension is planned, certain actors must decide on its localization. These siting choices are often accompanied by protest and/or resistance movements ("not in my backyard" problems).[14] In Germany, such NIMBY problems are often dealt with politically, by application of legal (participatory or planning) procedures. When someone is exposed to phenomena such as grid extension or power plant sites near their residential area, they are *affected* by decisions taken elsewhere with little or no own decision-making involved, and thus, they slip into the role of a "personally affected party" in a technology conflict [70].

In Germany, the public debate about siting decisions, NIMBY problems, and, more generally, contentious technologies has a long-standing history which has encouraged actors and observers (politicians, scientists, business persons, civil associations, and non-profit organizations) to declare "acceptance" the social cornerstone of the German ES Transformation.[15]

Indeed, acceptance of infrastructures is crucial, as two projects in Germany exemplify: First, the *SuedLink* project is supposed to transport electricity from offshore wind parks on the northern German coast toward southern industrial recipients. Planning considers technical and economic criteria, as well as objections from those affected by the project; several participatory exercises have been implemented. Although initial propositions of the possible route tried to exclude cities, towns, hospitals, industrial sites, nature reserves etc., more than 3000 objections have so far been articulated and publicly discussed.[16] Also, planning is influenced by several elections in the affected states or municipalities. Political parties openly demand postponement of the project because it interferes with local electoral campaigns and engenders strong opposition.[17] Second is the *South-East Direct Current Passage* project: it concerns transmission lines from eastern to southern Germany. The government claims that energy generated from RES will be transported; however, at least a significant part of the public does not believe these claims, accusing the government and the responsible business companies of intending to build a route for the transportation of coal electricity, since the route is planned to begin in *Saxony-Anhalt*—a brown coal mining area. As protesters remain outraged about the proposed route to the southern region of Bavaria, the project had to be put on hold.[18]

These siting decisions are *static* and *singular*: "Static" indicates an asymmetry in the relationship between the affected actors and the technology or procedure, for, unlike the actors, neither the technology nor the procedure reacts to communication. "Singular" emphasizes the one-sided character of the decisions: Either one accepts, or one refuses, the infrastructure—there is no middle ground. Ordinary spheres of social interaction in the smart grid, though, require more active participation than "merely" being affected as outlined above; they require constant *decision-making* and decisions are facilitated by using trust as a social mechanism to reduce complexity. Even in the extraordinary situation where interaction concerns infrastructure, and where acceptance (understood as civil tolerance of infrastructure by absence of protest) is the preponderant form of social coordination, trust still plays an important role.

The relation between the concepts of trust and acceptance has not been clarified sufficiently by the social sciences. Yet with reference to Kohring [71], one could identify the very foundation of technology acceptance in strategies of attributing trust. With regard to the case of the German ES Transformation, one can draw two conclusions on the interplay of trust and acceptance: Firstly, as acceptance allows actors to primarily accept shared rules on (legal) proceedings in the place of shared outcomes [72], trust is the mechanism which compensates for this lack of factual knowledge. Since attending actual decision-making processes is reserved to elected politicians, trusting that a decision was taken based on solid reasoning and a competent assessment of the future is simplified by accepting that the decision to accomplish the German ES Transformation has been reached through a politically legitimate procedure. The fact of not-knowing the exact manner in which the decision has been reached is bridged by trust, provided one "accepts" the set of rules according to which the decision has been negotiated. In this way, siting decisions and NIMBY problems, as primary examples of acceptance conflicts, are always accompanied by trust in legal (decision-making, participatory, and planning) procedures.

Moreover, a functional shift of acceptance in the German ES Transformation case can be expected: In earlier periods of ES development, acceptance issues limited to particular region proceedings (e.g., in a certain federal state) were a cause in their own right. Today, the German ES Transformation is observed by the entire nation, and the public's observation of regional acceptance can serve as an indicator of *system trust building*. To be more precise, the degree of *confidence* can be associated with a public observation of (successful or failing) acceptance creation in regions where grid extensions actually take place and must be dealt with by personally affected fellow citizens.

From a temporal perspective, these developments indicate that (i) processes of being affected and investing trust in procedures mainly occur during current grid extension projects; (ii) from a mid-term perspective, these processes occur along the transformation period; and (iii) in the smart grid of the far future, the significance of genuine acceptance phenomena concerning physical infrastructures is likely to vanish.

Discussion: an "architecture of trust" for ES

Over time, consumers, investors, and other stakeholders' attitudes toward the ES have coalesced into expectations which are—and will continue to be—challenged. These expectations are meanwhile inherent to the structures and institutions of the ES; it shapes their manifestation, impact, and degrees of stability and change. The extent and manner in which elements of a system can relate to each other are built into structures, and, therefore, structures determine (in technical systems), or condition (in social systems), the *processes* of each system. When we encounter structures, we can develop expectations of what will happen in the future. Moreover, after a longer period, we increasingly experience *familiarity* upon acting with, or within, structures.

Based on this sense of familiarity, structures support an architecture of trust [73] within which actors can identify persons or objects to whom or which they can address their expectations with respect to the ES (*addressees*). The public can identify certain institutions as representatives of "the energy system," e.g., energy providers (who maintain price stability), public agencies for technical supervision (who ensure security of supply), and sanction agencies (who enforce regulations) [56]. As addressees, these institutions represent mutually shared expectations to which services and outputs are attributed [82], either when reassurances about fulfillment of expectations are necessary (e.g., "electricity supply is safe") or when disappointments occur, be they singular or continuous (e.g., "the engineers will manage/not manage the transition"). Notwithstanding actors' familiarity with structures and institutions of the ES, their expectations are challenged at every moment in which they deal with the system and its perceived representatives. Their expectations are continuously confirmed, negated, or modified in concrete situations during the operation of the ES. It is this type of *sustainment*—driven by trust, distrust, and confidence—that will crucially determine the potentials of sustainability in the future system.

For consumers, this means a redirection of familiarity toward either actively trusting and running the risk of being overconfident or creating collective distrust. Either investors trust and therefore invest in fledgling smart grid technology, thus contributing to an investment climate, or investors lack the confidence to support the ES, given the instability of legal frameworks. Finally, those personally affected by spatial planning issues have to accept local decision-making while, more importantly, investing trust in planning procedures. For the success of the ES on a national level, the local settlement of NIMBY conflicts is an essential means of building system trust. Meanwhile, as our case studies demonstrate, distrust by citizens in the form of protest, investment holdbacks, or technology refusal may even provide learning opportunities. Since every act of trust holds a risk [21, 73] and since the German ES suffers many flaws, setbacks, and risks during the transformation, choosing to *balance trust and distrust* may be the better option on the road toward sustainability.

From a functional perspective, it is inevitable that we attribute trust, in order to reduce complexity, to make selections, and to build up actionability, for no one can distrust everything simultaneously. The framework of a trust architecture highlights prominent trustees in the social environment, so that selections are limited to certain potential addressees. This leads to various reconfigurations of trust and distrust, action and counteraction, affirmation, negation, apathy, etc. According to Luhmann [47], one can direct expectations at *persons* (e.g., opinion leaders), *roles* (e.g., engineers), *programs* (e.g., energy transformation policy), and *values* (e.g., data privacy, sustainability). Within the framework for the ES evolution, these addressees of trust and distrust will determine the success and failure of smart grids. At this point, we cannot present methods for managing the systemic consequences of trust, distrust, and confidence; we have to settle for demonstrating their status as solutions to socio-technical problems of ES transformation.

Conclusion

Concepts of trust, distrust, and confidence offer valuable insights into the relations of various actors of the socio-technical ES. With these insights, we can challenge some assumptions of technological visions. Technical innovation and institutional change cannot, alone, guarantee active participation by the public. Given that familiar structures and institutions of the ES are changing, we need to ascertain a future architecture of trust which offers all participants certain addressees for their attribution of trust. The authors aim at conducting further empirical research to investigate *processes* of trust and distrust attribution.

Endnotes

[1]The debate concerns not only the provision of electricity but also broader infrastructures, including technologies for heating and cooling and technologies using various energy sources such as gas. However, in the following, we concentrate on the transformation of the electricity domain, because not only do we find the most drastic changes in this area but also many innovations of exploiting various energy carriers align to the electricity system, like geothermal electricity.

[2]Research on the expectations about new smart technologies is sometimes misleading [74]; similarly, rationally calculating cost and benefits is also prone to lead to mistakes.

[3]Discussions about "empowerment" also point in the direction of the capacity to act [75, 76].

[4]In Germany, such "associations" are often comparable to American non-profit organizations.

[5]Likewise, Ashby [77] characterizes machines as "law-abiding," because no machine can be called "self-organizing," as it cannot automatically change its way of organization.

[6]Normal Accident Theory shows how non-linear operations in tightly coupled structures inevitably lead to failures and accidents and therefore to negative surprises [78].

[7]Data from our interview with a representative of a German energy business association.

[8]This assumption was strongly corroborated at a smart grid congress in Karlsruhe, Germany, in 2014, where a central message was "the client as [a] central future asset" [79].

[9]Examples of low-risk reflection include interaction with cell phones, public transport, and cars.

[10]The consequences of a "crisis of trust" can be more severe for the actionability of specific organizations than for consumers: "For some [organizations], the crisis of trust is especially acute, coloring and constraining virtually all their actions and choices" [80].

[11]See, for example, Felicity Barringer: "New Electricity Meters Stir Fears." *The New York Times*, 30.01.2011, or initiatives like www.stopsmartmeters.org.

[12]See Hughes D' Arcy: "Biofuel scepticism prompts German summit." *The Guardian*, Accessed 07.03.2011.

[13]See Qiang Zhang: "Distrust of Chinese milk persists despite Fonterra case." *BBC News China*, 13.08.2015 (http://www.bbc.co.uk/news/world-asia-china-23582282; accessed 16.02.2015).

[14]*Siting* relates to processes surrounding decision-making about the regional and local spaces and corridors to be used for infrastructure installations. NIMBY opposition can be a reaction to it.

[15]See [1, 2, 57, 81].

[16]See "Dokumentation des Dialogprozesses vor der Antragsstellung." TenneT TSO GmbH (http://suedlink.tennet.eu/bundesfachplanung/antrag-6/dialogprozess.html; accessed 05.02.2015).

[17]See Nicolai Kwasniewski: "Gleichstromtrasse Suedlink: Planung und Verlauf der Stromautobahn." *Spiegel Online*, 05.02.2014.

[18]See Markus Szymanski, Mike Balser, and Marlene Weiß: "Der Stress mit den Stromtrassen"; *sueddeutsche.de*, 30.07.2014.

Competing interests
This work is the result of research conducted within the Helmholtz Initiative ENERGY-TRANS (www.energy-trans.de/english/index.php). The authors declare that they have no competing interests.

Authors' contributions
The article was jointly prepared by both authors. Both authors read and approved the final manuscript.

Authors' information
Christian Büscher is a senior researcher at the Institute of Technology Assessment and Systems Analysis (ITAS) at Karlsruhe Institute of Technology (KIT). He holds a Doctoral Degree in Sociology (Dr. Phil.) from Darmstadt Technical University and a diploma in Sociology (Diplom-Soziologe) from the University of Bielefeld. His research interests are in the area of technology assessment and risk and environmental sociology. His list of publications includes papers in peer-reviewed journals and in edited books.
Patrick Sumpf is a research associate at the Institute of Technology Assessment and Systems Analysis (ITAS) at Karlsruhe Institute of Technology (KIT). He holds a B.A. in Political Science and an M.A. in Political Communication, both from Bielefeld University, Germany. He currently is a doctoral candidate at Mannheim University, Germany. His work focuses on trust research.

Acknowledgements
The authors wish to thank the editors of this journal for their useful suggestions. Discussion with colleagues from the ENERGY-TRANS project helped us clarify arguments. We are especially grateful for the revisions by Mira Klemm.

References
1. BMWi/BMU (2011) The Federal Government's energy concept of 2010 and the transformation of the energy system of 2011. Berlin
2. B.A.U.M. Consult (2012) Smart Energy made in Germany. Interim results of the E-Energy pilot projects towards the Internet of Energy. Munich/Berlin
3. Giordano V, Gangale F, Fulli G, Jiménez MS (2011) Smart Grid projects in Europe. JRC Reference Reports by the Joint Research Centre of the European Commission
4. AlAbdulkarim L, Lukszo Z, Fens T (2012) Acceptance of privacy-sensitive technologies: Smart Metering Case in the Netherlands. Third International Engineering Systems Symposium (CESUN), Delft University of Technology, Delft
5. Amin M (2001) Toward self-healing energy infrastructure systems. IEEE Comput Appl Power 14:20–28. doi:10.1109/67.893351
6. Ma Y, Zhou L, Tse N, Osman A, Lai LL (2009) An initial study on computational intelligence for smart grid. In: Proceedings of the 2009 International Conference on Machine Learning and Cybernetics., pp 3425–3429
7. Ramchurn S, Vytelingum P, Rogers A, Jennings N (2012) Putting the "smarts" into the smart grid: a grand challenge for artificial intelligence. Commun ACM 55:86–97
8. Wissner M (2011) The smart grid—a saucerful of secrets? Appl Energy 88: 2509–2518
9. Kasperson RE, Ram BJ (2013) The public acceptance of new energy technologies. Daedalus 142:90–96
10. Cavoukian A, Polonetsky J, Wolf C (2010) SmartPrivacy for the smart grid: embedding privacy into the design of electricity conservation. Identity Inf Soc 3:275–294. doi:10.1007/s12394-010-0046-y
11. Pearson ILG (2011) Smart grid cyber security for Europe. Energy Policy 39: 5211–5218. doi:10.1016/j.enpol.2011.05.043
12. Hoenkamp RA, Huitema GB (2012) Good standards for smart meters. In: European Energy Market (EEM), 2012 9th International Conference on the European Energy Market.10.1109/EEM.2012.6254820
13. Wüstenhagen R, Wolsink M, Bürer MJ (2007) Social acceptance of renewable energy innovation: an introduction to the concept. Energy Policy 35:2683–2691. doi:10.1016/j.enpol.2006.12.001
14. Wolsink M (2012) The research agenda on social acceptance of distributed generation in smart grids: renewable as common pool resources. Renewable Sustainable Energy Rev 16:822–835. doi:10.1016/j.rser.2011.09.006
15. Musall FD, Kuik O (2011) Local acceptance of renewable energy—a case study from southeast Germany. Energy Policy 39:3252–3260. doi:10.1016/j.enpol.2011.03.017
16. Ciupuliga AR, Cuppen E (2013) The role of dialogue in fostering acceptance of transmission lines: the case of a France–Spain interconnection project. Energy Policy 60:224–233. doi:10.1016/j.enpol.2013.05.028
17. Zoellner J, Schweizer-Ries P, Wemheuer C (2008) Public acceptance of renewable energies: results from case studies in Germany. Energy Policy 36: 4136–4141. doi:10.1016/j.enpol.2008.06.026
18. Künneke R, Mehos DC, Hillerbrand R, Hemmes K (2015) Understanding values embedded in offshore wind energy systems: toward a purposeful institutional and technological design. Environ Sci Policy 53:118–129. doi:10.1016/j.envsci.2015.06.013
19. acatech (2011) Akzeptanz von Technik und Infrastrukturen. Anmerkungen zu einem aktuellen gesellschaftlichem Problem. Deutsche Akademie der Technikwissenschaften acatech, München
20. Ribeiro PF, Polinder H, Verkerk MJ (2012) Planning and designing smart grids: philosophical considerations. IEEE Technol Soc Mag 31:34–43. doi:10.1109/MTS.2012.2211771
21. Luhmann N (1979) Trust and power. Wiley, Chichester
22. Bellaby P, Eames M, Flynn R (2010) The role of "trust" in the transition to sustainable energy. Energy Policy 38:2613–2614. doi:10.1016/j.enpol.2009.03.066
23. Knudsen M (2010) Surprised by method—functional method and systems theory. Forum Qualitative Sozialforschung/ Forum: Qualitative Social Research 11
24. Kohring M (2004) Vertrauen in Journalismus: Theorie und Empirie. UVK-Verl.-Ges, Konstanz
25. Strulik T (2004) Nichtwissen und Vertrauen in der Wissensökonomie. Campus-Verl, Frankfurt [u.a.]
26. Möllering G (2006) Trust: reason, routine, reflexivity, 1st edn. Elsevier, Amsterdam [u.a.]
27. Strulik T (2006) Rating agencies, ignorance and the knowledge-based production of system trust. In: Strulik T, Willke H (eds) Towards a Cognitive Mode in Global Finance. The Governance of a Knowledge-based Financial System. Campus, Frankfurt am Main, New York, pp 239–255
28. Edwards PN (2004) Infrastructure and modernity: force, time, and social organization in the history of socio-technical systems. In: Misa T, Brey P, Feenberg A (eds) Modernity and Technology. MIT Press, Cambridge, London, pp 185–225
29. Hughes TP (1983) Networks of Power: Electrification in Western Society: Baltimore 1880–1930. The Johns Hopkins University Press, Baltimore
30. Hughes TP (1987) The evolution of large technological systems. In: Bijker WE, Hughes TP, Pinch TJ (eds) The Social construction of technological systems: new directions in the sociology and history of technology. MIT Press, London. pp 51–82
31. Mayntz R (1993) Grosse technische Systeme und ihre gesellschaftstheoretische Bedeutung. Kölner Zeitschrift für Soziologie und Sozialpsychologie 45:97–108
32. Mayntz R (2009) The changing governance of large technical infrastructure systems. In: Über Governance. Institutionen und Prozesse politischer Regelung. Campus, Frankfurt am Main, New York, pp 121–150
33. Edwards PN, Jackson SJ, Bowker GC, Knobel CP (2007) Understanding infrastructure: dynamics, tensions, and design. DeepBlue, Ann Arbor
34. Jackson SJ, Edwards PN, Bowker GC, Knobel CP (2007) Understanding infrastructure: history, heuristics and cyberinfrastructure policy. First Monday 12
35. Geels FW (2004) From sectoral systems of innovation to socio-technical systems: insights about dynamics and change from sociology and institutional theory. Res Policy 33:897–920

36. Smith A, Stirling A, Berkhout F (2005) The governance of sustainable socio-technical transitions. Res Policy 34:1491–1510. doi:10.1016/j.respol.2005.07.005
37. Künneke RW (2008) Institutional reform and technological practice: the case of electricity. Ind Corporate Change 17:233–265
38. La Porte TR (1975) Organized social complexity: explication of a concept. In: Organized social complexity: challenge to politics and policy. Princeton Univ. Press, Princeton, pp 3–39
39. Kröger W, Zio E (2011) Vulnerable systems. Springer, London, Dordrecht, Heidelberg, New York
40. Gross M, Mautz R (2015) Renewable energies. Routledge, London, New York
41. Carlsson B, Stankiewicz R (1991) On the nature, function and composition of technological systems. J Evol Econ 1:93–118. doi:10.1007/BF01224915
42. Fuenfschilling L, Truffer B (2014) The structuration of socio-technical regimes—conceptual foundations from institutional theory. Res Policy. doi:10.1016/j.respol.2013.10.010
43. Strunz S (2014) The German energy transition as a regime shift. Ecol Econ 100:150–158. doi:10.1016/j.ecolecon.2014.01.019
44. Luhmann N (2005) Risk—a sociological theory. Aldine Transactions, New Brunswick (USA), London (UK)
45. von Foerster H (1993) KybernEthik. Merve, Berlin
46. Elster J (1979) Ulysses and the Sirens. Studies in rationality and irrationality. Cambridge University Press, Cambridge
47. Luhmann N (1995) Social systems. Stanford University Press, Stanford
48. Khan SR (2014) "Humans cannot communicate"—unravelling the mental knots of Niklas Luhmann. Contemp Sociology 43:49–55. doi:10.1177/0094306113514538a
49. Luhmann N (2012) Theory of society—volume 1. Stanford University Press, Stanford
50. Simmel G (1978) The philosophy of money. Routledge & Kegan Paul, Boston
51. Möllering G (2001) The nature of trust: from Georg Simmel to a theory of expectation, interpretation and suspension. Sociology 35:403–420. doi:10.1177/S0038038501000190
52. Möllering G (2013) Process views of trusting and crises. In: Bachmann R, Zaheer A (eds) Handbook of Advances in Trust Research. Edward Elgar, Cheltenham, pp 285–305
53. Lewis JD, Weigert AJ (2012) The social dynamics of trust: theoretical and empirical research, 1985–2012. Social Forces 91:25–31. doi:10.1093/sf/sos116
54. Möllering G (2006) Trust, institutions, agency: towards a neoinstitutional theory of trust. In: Bachmann R, Zaheer A (eds) Handbook of Trust Research. Edward Elgar, Cheltenham, pp 355–376
55. Luhmann N (1988) Familiarity, confidence, trust: problems and alternatives. In: Gambetta D (ed) Trust. Making and Breaking Cooperative Relations. Basil Blackwell, Oxford [u.a.], pp 94–107
56. Giddens A (1990) The consequences of modernity. Polity Press, Cambridge
57. Appelrath HJ, Kagermann H, Mayer C (2012) Future energy grid: Migrationspfade ins Internet der Energie. acatech, Cambridge
58. Shove E, Warde A (1998) Inconspicuous consumption: the sociology of consumption and the environment. www.lancaster.ac.uk/fass/resources/sociology-online-papers/papers/shove-warde-inconspicuous-consumption.pdf, retrieved 16.09.2015
59. Skinner D, Dietz G, Weibel A (2013) The dark side of trust: when trust becomes a "Poisoned Chalice". Organization. doi:10.1177/1350508412473866
60. Fekete A (2011) Common criteria for the assessment of critical infrastructures. Int J Disaster Risk Sci 2:15–24. doi:10.1007/s13753-011-0002-y
61. Depuru SSSR, Wang L, Devabhaktuni V (2011) Smart meters for power grid: challenges, issues, advantages and status. Renewable Sustainable Energy Rev 15:2736–2742
62. Amin SM, Wollenberg BF (2005) Toward a smart grid. IEEE Power Energy Magazine 3:34–41.
63. Amin SM, Giacomoni AM (2011) Smart grid-safe, secure, self-healing. IEEE Power Energy Magazine 10:33–40. doi:10.1109/MPE.2011.943112
64. GAO (2011) Electricity grid modernization: progress being made on cybersecurity guidelines, but key challenges remain to be addressed. U.S. Government Accountability Office, Washington, D.C
65. Shapiro SP (1987) The social control of impersonal trust. Am J Sociol 93:623–658
66. Sumpf P (2013) Mechanismen der Vertrauens(de)konstitution in öffentlichen Krisen - Das Beispiel EHEC O104:H4. In: Medien müssen draußen bleiben! Wo liegen die Grenzen politischer Transparenz? Frank & Timme, Berlin, pp 217–235
67. BNetzA (2011) "Smart grid" und "smart market". Eckpunktepapier der Bundesnetzagentur zu den Aspekten des sich ändernden Energieversorgungssystems. Bundesnetzagentur, Bonn
68. Jopp K (2014) Virtuelle Kraftwerke in der Cloud. Energy 20:23–25
69. Orwat C (2011) Systemic risks in the electric power infrastructure? TATuP 20:47–55
70. Luhmann N (1990) Technology, environment and social risk: a systems perspective. Organ Environ 4:223–231. doi:10.1177/108602669000400305
71. Kohring M (2001) Vertrauen in Medien - Vertrauen in Technologie. Akademie für Technikfolgenabschätzung, Baden-Württemberg
72. Vollmer H (1996) Akzeptanzbeschaffung: Verfahren und Verhandlungen. Zeitschrift für Soziologie 25:147–164
73. Strulik T (2011) Vertrauen. Ein Ferment gesellschaftlicher Risikoproduktion. Erwägen, Wissen, Ethik 22:239–251
74. Krishnamurti T, Schwartz D, Davis A, Fischhoff B, de Bruin W, Lave L, Wang J (2012) Preparing for smart grid technologies: a behavioral decision research approach to understanding consumer expectations about smart meters. Energy Policy 41:790–797. doi:10.1016/j.enpol.2011.11.047
75. Verbong GPJ, Beemsterboer S, Sengers F (2013) Smart grids or smart users? Involving users in developing a low carbon electricity economy. Energy Policy 52:117–125. doi:10.1016/j.enpol.2012.05.003
76. Geelen D, Reinders A, Keyson D (2013) Empowering the end-user in smart grids: recommendations for the design of products and services. Energy Policy 61:151–161. doi:10.1016/j.enpol.2013.05.107
77. Ashby WR (2004) Principles of the self-organizing system. E:CO 6:102–126
78. Perrow C (1984) Normal accidents. Living with high-risk technologies. Basic Books, New York
79. Khala N, Sumpf P, Büscher C (2014) "Wir sind das Smart Grid" – Zu einer aktuellen Debatte unter InformatikerInnen. Bericht vom 2. Kongress Energie + Informatik – Dezentrale Energie smart verknüpft. TATuP 23:118–121
80. LaPorte TR, Metlay DS (1996) Hazards and institutional trustworthiness. Public Administration Rev 56:341–347
81. VDE (2008) Smart Distribution 2020. Virtuelle Kraftwerke in Verteilungsnetzen. Technische, regulatorische und kommerzielle Rahmenbedingungen. Verband der Elektrotechnik Elektronik Informationstechnik e.V. (VDE), Frankfurt am Main, Cambridge
82. Malle BF (1999) How people explain behavior: a new theoretical framework. Pers Soc Psychol Rev 3:23–48. doi:10.1207/s15327957pspr0301_2

Danger explodes, space implodes: the evolution of the environmental discourse on nuclear waste, 1945–1969

Judi Pajo

Abstract

Background: The ways in which the public understands nuclear waste affect nuclear waste policy and the actual disposal of nuclear waste. This paper traces the origins and the evolution of the public understanding of nuclear waste through an analysis of a sample of the historical record of the public discourse on the topic.

Methods: This paper employs sociocultural anthropology methods innovatively: rather than emphasizing the data collection aspect of participant observation, it emphasizes interpretive discourse analysis, sampling the archives of the *New York Times* as a de facto ethnographic repository of the U.S. discourse on nuclear waste spanning the years 1945–1969.

Results: Whereas no evidence of a public concept for nuclear waste is found in the chronologically first item of this study's dataset, the chronologically last item implies a public conceptualization of nuclear waste that appears not too different from the one encountered today.

Conclusions: The paper illuminates especially the evolution of two key threads of environmental consciousness in the U.S. public discourse: as waste grew ever more harmful in the public understanding, the space available for disposing of it shrank and eventually disappeared.

Keywords: Nuclear waste disposal, Public discourse, Radioactive contamination, Environmental consciousness, *New York Times*

"[S]ome kind of national burying place will be needed.... It has been proposed ... to cast the atomic waste coffins in heavy concrete and lower them into one of the great deeps of the ocean or shoot them off the earth in a rocket that would become a satellite in outer space" [1].

Background

The problem of nuclear waste disposal

"America's nuclear waste management program is at an impasse" states the 2012 report submitted to the U.S. Secretary of Energy by the Blue Ribbon Commission on America's Nuclear Future[1]. Indeed, over 165 million pounds of high-level radioactive waste wait for permanent disposal in the U.S.; most is spent nuclear fuel from the energy industry[2], and the rest originates from legacy national defense activities and from the navy[3]. While decay requires tens of thousands of years[4], the temporary arrangements under which the bulk of this inventory is currently stored, at hundreds of locations dispersed throughout the U.S.[5], were not designed for permanent storage and are not deemed safe for the permanent storage of high-level nuclear waste[6]. Yet radioactive waste has been accumulating there ever since the nuclear energy industry took off in the 1950s and then grew in the 1960s. Some of the high-level nuclear waste currently under temporary storage dates as far back as the Second World War.

The last event of major significance in a decadelong and conflict-ridden journey towards a permanent disposal solution to the nation's nuclear waste problem was the halting of work on the targeted site at Yucca Mountain in Nevada in 2008. The Blue Ribbon Commission (BRC) characterized this as "the latest indicator of a [nuclear

Correspondence: jpajo@pace.edu
Sociology/Anthropology Department, Dyson College, Pace University, 41 Park Row, Room 725, New York, NY 10038, USA

waste management] policy that has been troubled for decades and has now all but completely broken down" [2]. There is little dispute that the key reason underlying nuclear waste's unfolding complex story of "controversy, litigation, and protracted delay" lies with grave public concerns over "adequate protection of public health and safety and the environment" [2]. The BRC assesses that "finding sites where all affected [...] are willing to support or at least accept a [permanent nuclear waste storage] facility has proved exceptionally difficult" [2]. Put in other words, the U.S. public, in virtual unanimity, views nuclear waste as hazardous to human health and to the natural environment and does not deem safe any of the known methods for disposing of it.

Yet inability to reach agreement about a site of permanent disposal does not make nuclear waste disappear. In fact, the U.S. inventory of nuclear waste continues to grow: some four to five million pounds of high-level radioactive waste are generated by the nuclear energy industry annually[7]. Piling up nuclear waste in temporary storage facilities poses as great, if not greater, threats to the natural environment, to public health, and to national security as does disposing of such waste under permanent arrangements. In addition, the current impasse threatens the viability of a significant source of energy, making visible one aspect of the complex body of obstacles that the development of non-carbon sources of energy needs to overcome: at present, nuclear power supplies around one fifth of the electricity in the U.S.; with the caveats of solving accident vulnerability and waste disposal issues, nuclear energy may represent an alternative to the continuing global rise in carbon emissions.

The public conceptualization of nuclear waste

At present, the opposition to hosting a nuclear disposal site usually appears to be considered a given. Yet a range of explanations coexist as to the origins of the public conceptualization of nuclear waste[8].

One common and seemingly commonsensical suggestion is that today's public concerns over the safety of nuclear waste disposal started with the popular political movements of the 1960s and 1970s. These movements have been historically well documented [3] and continue to draw significant scholarly interest into the present [4, 5]. At the same time, they are also generally viewed as springing out of certain remarkable deeds of some remarkable individuals. The biologist and social critic Rachel Carson, for example, is often credited with having "laid the foundation for the modern environmental movement" [6]. "Senator Gaylord Nelson created Earth Day as a way to force this issue onto the national agenda," reads a page from the website of the Environmental Protection Agency; "Twenty million Americans demonstrated in different U.S. cities, and it worked!" [7].

Another line of argument specifically relates the American public's negative perception of all things nuclear to the history of nuclear weapons [8], including weapons testing[9]. This view focuses on the devastating aftermath of nuclear explosions as well as, in addition, on a series of visible individual accidents such as the most recent meltdown at Fukushima, Japan in the year 2011 [9].

In yet another portrayal, the cultural origins of "nuclear fear" may be rooted more profoundly in our shared myths if not even human consciousness. It is some kind of deep mythical fear, in other words, that is then stoked by both the imagery of nuclear warfare and the visibility of nuclear energy disasters such as the recent Fukushima meltdown or Chernobyl and Three Mile Island before it [10, 11].

Given that the public conceptualization of nuclear waste underlies the political impasse that in effect perpetuates the storage of radioactive waste under temporary arrangements and threatens the viability as well as future of a significant energy source, it may be instructive to inquire into it more closely. Where does the public conceptualization of nuclear waste really come from?

Nuclear waste in public discourse

To understand the public conceptualization of nuclear waste, this paper takes the route of empirical investigation, while recognizing that a number of routes are possible and promising[10]. It adapts key methodological tools from sociocultural anthropology to produce a sample of the records of the U.S. public discourse on nuclear waste and then subjects that sample of records to interpretive discourse analysis.

The records in the dataset produced for the study span the years 1945 1969, the timeframe corresponding to what this researcher earlier identified as the first wave of public discourse on nuclear waste[11]. The core reason for limiting the investigation to this timeframe has to do with the emergence and the shaping of the notion of nuclear waste in public discourse. Regarding starting at 1945, as discussed in the researcher's earlier study, there is no evidence of a notion of nuclear or atomic or radioactive waste in the public discourse prior to that year. Regarding the 1969 cutoff, the researcher's earlier study also indicated that, by the time President Nixon took office in early 1969, the concept of nuclear waste in the public discourse had already assumed the familiar shape in which it is encountered today: the second wave of public discourse characterized nuclear waste primarily in terms of its threats, dissociated from the benefits of nuclear exploitation, and as carrying risks that cannot be eliminated.

Methods

Designing an anthropological lens to look at public discourse on nuclear waste

To trace the evolution of the environmental consciousness on nuclear waste in the U.S., this paper employs the research and analytical methods of sociocultural anthropology. The methodology is fitting to the task for anthropologists have honed their research to the very aims of mapping out and understanding what at various points in the discipline's history has been seen as cultural structure, grammar of culture, cultural meaning, or simply culture.

Yet it is not possible to conventionally employ the ethnographic toolkit for meaningfully studying the environmental consciousness on nuclear waste in the U.S. The key reason lies with the method of participant observation, which within the discipline of sociocultural anthropology is widely considered to be the vital core of ethnographic research. In its conventional application, ethnographic participant observation requires the researcher personally to take part, indeed to immerse her own self, in most facets of human activity in the setting that she investigates. But a researcher cannot travel back in time to the U.S. of the years 1945–1969 when the environmental consciousness regarding nuclear waste took shape. Neither can a researcher personally participate in the range of human activities that take place in a setting the size of the U.S., at least not in the same way that she can participate in the lives of the people on a small island.

The methodology of ethnographic participant observation was originally developed for making sense of cultures that appeared rather entirely alien when encountered by the West, cultures of peoples not known to "us," cultures "without history," or, in better words, cultures that were encountered in the absence of a written record of documentation recognizable in the West [12]. Indeed, many an ethnographic project traditionally started with the researcher's learning of a native tongue. In contrast, this study's object of focus is far from alien and has been extensively documented. An abundance of textual and audio-visual records of what was said about nuclear waste disposal in the U.S. between 1945 and 1969 can be found across numerous physical libraries, archives, and electronic databases: news reporting and commentaries across a range of media outlets, as well as innumerable governmental reports and scholarly studies, not to mention popular literature across different nonfictional and fictional genres as well as works of visual art and moving pictures.

In order to design an anthropological lens that could be useful for looking at the evolution of environmental consciousness on nuclear waste in the U.S., this researcher interpreted the core of ethnographic participant observation to lie not with personal immersion in the life of a community bound to a physical setting but rather with the method's blending of rigorous empirical observation with a shifting of perspective from the observer to the subject of observation. Instead of emphasizing the convention of the data collection aspect of the participant observation method, the anthropological lens re-designed in this way focuses on identifying and sampling data that represent an equivalent to traditional ethnographic fieldnotes.

Such re-design yields the additional benefit of improving the scope of data under investigation. The emphasis that conventional anthropological research places on a researcher's personal immersion in fieldwork and on a holistic approach to observation clearly aims at broadening the body of empirical observations. In furtherance of the same direction, the painstaking documentation of the researcher's holistic observations in fieldnotes aims to produce as extensive a record of data for subsequent analysis and interpretation as possible. Nevertheless, the researcher's personal participation in the collection of data necessarily limits the scope of such data strictly to what one individual can produce within exacting frames of time. Even in the best of implementations of participant observation, the voluminous fieldnotes and records produced through ethnographic research represent narrowly limited data samples.

In contrast, this study's empirical focus, the body of the environmental discourse on nuclear waste in the U.S., challenges the researcher not with the production of ever more data. Rather, the challenge is how to sample the multiple and rich sources of records in a way that the dataset produced, while representative, and while holistic in scope so that potentially significant findings are not hindered by exclusion of potentially significant variables from the start, would also meaningfully lend itself to qualitative analysis and in-depth interpretation in the best of the anthropological tradition.

Sampling environmental discourse in the archive of the *New York Times*

The challenge was resolved with identification of the archive of the *New York Times* as the source for sampling the environmental discourse on nuclear waste between the years 1945 and 1969.

Given this timeframe, a newspaper archive is an especially appropriate source from which to draw a dataset that would address a question such as where the concept of nuclear waste disposal comes from. Arguably, the extent to which newspapers reflect the public discourse and influence the public discourse is commensurate with at least the extent to which their physical printed copies reach into the hands of readers[12]. This study's timeframe appears to have also been the heyday of newspapers in

America: their circulation grew steadily after the Second World War and it was not until the 1970s that newspaper circulation fell below the number of U.S. households [13, 14].

Among the large body of U.S. media outlets, the *New York Times* stands out for its sizable national circulation as well as for its editorial ambitions of covering "All the News That's Fit to Print." During the timeframe that is the focus of this study, the pages of the *New York Times* recorded the voices of a great number of individuals: not only reporters and editors but also readers whose published addresses and feedback to the newspaper regarding what they read in its pages and what they perceived to happen in the world around them, though editorially controlled, represent a sampling of voices from a substantial pool. So the *New York Times* is best positioned also for serving as a proxy for the knowledgeable and articulate individual that conventional ethnographic research designates as "key informant". Throughout the years under investigation, the *New York Times* pursued the role of the nation's "newspaper of record," and, arguably, it accordingly influenced numerous other media organizations throughout the nation. This special standing is highlighted by a fact of symbolic pertinence to this study: it was a reporter of the *New York Times*, William L. Laurence, who the U.S. government drafted in 1945 to write the official history of the atomic bomb project.

Regarding the technical aspects of sample selection, the records that constitute the dataset used for the analyses and interpretations outlined in the following pages were drawn from the electronic archives of the *New York Times* through keyword searches: they were selected for containing in their headings and/or subheadings one of the keywords "atomic," "radioactive," or "nuclear" in combination with the keyword "waste," and they were all published between the years 1945 and 1969[13]. The electronic archives characterize these items as "articles," "front page articles," "editorials," and "letters to the editor."

The dataset contains 73 records in total[14]. Placed in a historical chronology, 17 of these items correspond to the presidency of Harry Truman, 1945–1953; 44 to the presidency of Dwight Eisenhower, 1953–1961; 6 to the presidency of John F. Kennedy, 1961–1963; and 6 to the presidency of Lyndon B. Johnson, 1963–1969. There is some significance to the fact that the vast majority of these records correspond to the Eisenhower presidency: while President Truman and President Eisenhower each served the same amount of time in office, the *New York Times* published well over twice as many items fitting this study's sampling criteria during the Eisenhower presidency. The rate of publication for such items then fell under President Kennedy and all but disappeared under President Johnson.

Interpretive discourse analysis

The findings outlined in the following pages were reached through subjecting this dataset to what may be called interpretive discourse analysis. Given the researcher's prior work with these discourse records as part of a larger dataset[11], the present analysis represents a re-visitation of the data that focused specifically on addressing the question of the conceptualization of nuclear waste. Specifically, the researcher freshly read each one of the items in the dataset, with which she was already familiar from previous readings and analyses, now literally to answer the question: How is nuclear waste conceptualized in this item? The answer to that question could be a direct quotation from the item, a paraphrase, or a verbalization of what appeared to be implied in specific discourse segments in the item. The last, an interpretation, counts for how the researcher most typically answered the research question at the conclusion of such re-reading of a dataset item. Oftentimes, the answer reached after visiting a record, or a tentative articulation of a finding, prompted yet another visitation of a previously visited record or even pool of records. The goal of such repeated interpretation was to unearth the meanings that underlay the literal level of discourse in a given record and that also permeated the larger sample of records. Focusing on the underlying shared meanings would have also helped counter for biases that would arguably be present in records that were editorially pre-selected. This analysis was less interested in the literal level of discursive statements possibly influenced by "for" or "against" editorial and/or political positions, and rather more interested in teasing out the shared underlying understandings that, regardless of the political position an individual had assumed or would eventually assume, defined the very topic of nuclear waste in the public discourse of the time under investigation. The quotations and illustrations in the following section are not exhaustive; other segments could be cited, and the entire dataset is available from the author upon request.

Results

A significant finding of this research concerns the evolution of the public conceptualization of nuclear waste during 1945–1969. Whereas no evidence of a public concept for nuclear waste is found in the chronologically first item of this study's dataset, the chronologically last item implies a public conceptualization of nuclear waste that appears not too different from the one encountered today. The interpretive analysis of the items in the study's sample identified four key conceptual stages in this evolution: in the first stage, nuclear waste was introduced to public discourse in a tentative definition; in the second, public discourse agnostically assessed nuclear waste; during the third stage, a view of nuclear waste crystallized; by the fourth, nuclear waste appeared to be resolutely defined.

An additional significant finding is that this process was shaped largely by the mutual evolution of two conceptual threads. The first is the hazard perspective: over the course of the years under investigation, nuclear waste was framed in the public discourse in terms of danger—increasingly and with increasing certainty. The second thread is the vanishing of physical space for safe disposal: as nuclear waste was ever more markedly understood in terms of threat to human health and to the natural environment, the space available for disposing of it came to be conceptualized as ever more limited and remote—until it eventually disappeared.

The following paragraphs periodize the evolution that brought the public conceptualization of nuclear waste into the shape in which it is encountered at present and also highlight some of the terms in which the hazard definition of nuclear waste and the shrinking of disposal space for it were articulated, in tandem, as they evolved.

Discussion

Periodizing the evolution of the public conceptualization of nuclear waste

Plotting the distribution of dataset items by year of publication reveals an approximate correspondence between the stages of conceptual evolution named above and the annual rates at which the sampled records of the public discourse on nuclear waste made the pages of the *New York Times*. Interestingly, at least three of the four stages through which the public conceptualization of nuclear waste evolved appear to correspond to nearly normal distributions with clear peaks. The first stage, which started in the year 1948 and continued through the year 1951, saw the peak in the rate of publication of items that fit the sampling criteria of this study sometime between the years 1949 and 1950. The records interpreted here as indicative of the second stage of evolution of public discourse on nuclear waste were published in the *New York Times* between the years 1953 and 1956; the rate of their publication peaked in the year 1955. The third stage, which can be placed within the years 1958 and 1963, reached the peak of publications in 1959. It may be slightly a stretch to view the years 1966 and 1967 as the flat peak of the period 1964–1969; more appropriately, the fourth stage of evolution may be viewed as corresponding in the distribution chart not to a curve but rather to a flat without a peak: indeed, the fourth stage represents the end of the evolution of the public concept of nuclear waste.

The entire sample's pinnacle of publication rate, reached in the year 1957, lends itself to two complementary views. In the first, the year 1957 marks the transition from the second to the third stage of conceptual evolution: the interpretive analysis reveals that it was essentially in this year that the public discourse re-oriented its focus from the earnest assessment of the potential dangers of nuclear waste to crystallizing the notion of nuclear waste as irredeemably dangerous. Complementing this watershed view, 1957 may also be seen as the peak of the 11-year timeframe 1952–1963, a longer stage of conceptual evolution, in this complementary view, that started with an agnostic assessment of nuclear waste and concluded with a crystallization of the public view of nuclear waste as irredeemably a threat.

1945, before nuclear waste

The first time that the notion of waste appears in proximity to the notion of atomic reaction in this dataset is in the remarkable firsthand account of the bombing of Nagasaki in 1945. William L. Laurence, the only reporter to be part of the military mission on August 9, subtitled his September 9 piece "A City Laid Waste by World's Most Destructive Force." The waste of nuclear, in other words, was the devastation that followed the explosion of the bomb. Deploying the "World's Most Destructive Force" had to do with putting an end to war. So the bomb was to be admired: "It is a thing of beauty to behold, this 'gadget.' It its design went millions of man-hours of what is without doubt the most concentrated intellectual effort in history." Clearly, as of 1945, public discourse must not yet have indexed the connotations of nuclear that dominate public discourse today.

Stage one, tentative definitions

To the extent that the historical archive of the *New York Times* may serve as a proxy for a chronological record, "atomic waste" in the sense of "waste materials from atomic energy plants" was formally introduced to the U.S. public discourse on April 15, 1948. As discussed in this researcher's earlier study, this was the era of the science and technology paradigm of public discourse on nuclear waste[11]. Covering an expert conference session on "the peace-time problems of the disposal of radioactive waste" qualified as "the first on that subject ever incorporated in a general safety meeting," the report quotes government technocrats arguing for this "serious hazard of the future" to "be put on the same agenda with problems of hotel and restaurant safety, traffic, fire prevention and home accidents." Some effort, in other words, appears to have been deemed necessary to draw proper attention to the topic of atomic waste. But the war was over and peacetime was entitled to attend to its own challenges. The headline of the quoted record speaks of "safety menace" and further on the record relates that even though "sufficient dilution of some radioisotopes in water renders them presently harmless, it is questionable whether radioactive wastes can be made safe for disposal in a drainage system in the ocean merely by aqueous dilution."

Remarkably, the records from the years 1948 through 1951 indicate that the most significant terms in which nuclear waste would eventually come to be defined and is presently understood were introduced into the public discourse already in this first stage: the key notion of a potential hazard that is difficult for humans to sense yet may constitute a safety concern; the recognition that gradations in the level of radioactivity would correspond to gradations in the threat that radioactivity poses to human health; the notion that certain methods of disposal and certain places of disposal may amplify or mitigate that threat as well as carry different costs. Even the idea of one single national disposal site was already introduced in this stage. Titled "Atomic 'Cemetery' Needed for Waste," a report from July 29, 1950 argues that "some kind of national burying place will be needed for the lethal substances;" this would be superior to disposing of nuclear waste by dumping it in the ocean, conditional on a plausible scenario: "[i]f fish ate the material, scientists fear it might find its way into food used by humans." Clearly, a route for radioactive materials to enter the human food supply was already thought about.

Yet nuclear waste as well as the threats associated with it appear to have felt distant and abstract upon their initial introduction into public discourse—especially so when compared to how they would come to be articulated in the later stages through which the concept of nuclear waste evolved into the current public definition. At this first stage, the threat of nuclear waste was located persistently in a loosely specified future. For example, a report from February 13, 1949 that refers to information issued by the Atomic Energy Commission regarding land burial in containers comments that, though such disposal "may be good enough for the present," "the time will surely come when atomic power plants will be generating electricity and when it will be necessary to get rid of their radioactive wastes in a less cumbersome way."

"Cumbersome" is also a telling word: often in this first stage, the threat of nuclear waste seemed to be viewed as minor and dismissible; there was no dispute to the assumption that proper scientific and technological effort would make it disappear. Titled "Atomic Waste Dumped in Sea," a brief report of May 22, 1949 quotes the British Ministry of Supply at face value in explaining that nuclear waste dumped into sea was "only slightly radioactive and the amount too small 'to have any harmful effect on fish or on human life.'" On September 9, 1950, reporting on the success of laboratory experimenting that "may be a solution to the problem of getting rid of radioactive wastes from atomic research laboratories," a report optimistically titled "New Method Handles Radioactive Waste" concludes that "it is believed that [the experiments] would be applicable to larger plants, which might handle thousands of gallons of wastes daily." Better yet, a report from February 11, 1951 declares, in its title already, that "Dangerous Radioactive Wastes, Which Now Are Buried, May Have Many Industrial Uses." Of those "many" envisioned uses, the record specifies "the sterilization of food and pharmaceuticals."

Importantly, during this first stage of the evolution of the concept of nuclear waste in the public discourse, the space available for disposing of nuclear waste was envisioned as virtually unlimited. This is how a brief report of August 7, 1950 describes disposal at the Brookhaven National Laboratory in Long Island: "radioactive material [...] is collected in small tanks after it goes into the drains. If it 'cools off,' it goes on into sewers. Any accidental seepage [...] can be detected quickly [...] Gas used to cool off the atomic pile is highly diluted before being released into the air." The water, the ground, and the air, indeed all of the planet must have been assumed to cache disposal possibilities.

All in all, as of 1951 nuclear waste was, in the words of two reports published that month, "a nuisance." And any of the multiple themes through which that nuisance was introduced into the public discourse seemed just as likely as any of the others eventually to dominate how public discourse would define nuclear waste: nuclear waste could turn out to become a matter primarily of threat and of danger just as much as it could turn out to become a matter of industrial cost efficiency just as much as it could turn out to become a matter of triumph of the scientific and technological effort over the wonders of the (manipulated) natural world.

Stage two, focal points emerge

Theme wise, the records from the year 1953 through the year 1956 have a mode: 6 out of the 14 touch on the theme of putting nuclear waste to one or another beneficial use. A triumphant headline of January 27, 1954, for example, announced that "Electricity Is Made From Atomic Waste." Penned by none other than the famed William L. Laurence, the report detailed at some length the technical backend of "[a]n atomic battery that converts atomic energy directly and simply into small but usable quantities of electrical energy." It was a genuine technological achievement, "[t]his direct conversion for the first time in history of nuclear energy into electricity." To relate its significance, Laurence quotes David Sarnoff, head of the Radio Corporation of America and one of the most prominent industry figures of the time, to say that the technology underlying the prototype battery "may prove to be as significant as Edison's conversion of electricity into light." That the generation of electricity might have been among the most prominent of the potential uses of nuclear waste that were being pursued is supported by a brief record from May 18, 1955. Reporting on "a young Swiss student," just "18 years old," who had "designed a tiny self-powered electronic tube" that "would use built-in radioactive material from the atomic trash pile," the record noted that the achievement had won the student various prestigious accolades and concluded that it would eventually "assur[e] him a top college scholarship."

But electricity was not the only potential use that was envisioned for nuclear waste. Turning on its head, the potential military threat that the nuclear waste was sometimes seen to constitute, a record from May 20, 1954 entertained the idea that "the radioactive by-products of atomic fission [...] have a potential military utility": they could be used to create "a lethal radioactive 'line' along a frontier, behind a river, across a peninsula, that would deny an area to the enemy." So it was with regret for the report to note that radioactive wastes were "too small in quantity today to irradiate anything but a small area." And it was with hope to conclude that "a large industrial complex dependent upon commercial atomic power would add daily significant amounts of radioactive waste products to the Nation's chemical armory."

The entertainment of scenarios such as a radioactive Maginot line, as well as the numerous technological pursuits of potential uses for nuclear waste, seems to have been manifestations of an overall effort to deny nuclear waste an existence as waste. That theme was articulated pithily in a record from May 8, 1956. Titled "Atomic Waste No Problem Yet, Strauss Says at Generator Site," the report cited the head of the U.S. Atomic Energy Commission while he, together with his British homologue, toured "this country's first full-scale atomic power project": Lewis L. "Strauss said the term 'atomic waste' is a misnomer".

Importantly, denying nuclear waste an existence as waste was not denial of the harms and hazards of radioactivity[15]. On the contrary, the sample indicates that the danger presented by nuclear waste was submitted to public comprehension. For example, readers learned that, while the amount of nuclear waste in existence was relatively small, the anticipated expansion of the nuclear energy industry would expand it in the future.

Significantly, scientific delimitations of danger were given prominent attention. In a brief science note of June 26, 1955, an oceanographer opposed the dumping of nuclear waste in the ocean; the argument was that, over time, such practice "would have an unpredictable influence on the whole biological cycle of the sea." Some months later, on December 17, a substantive record spoke of scientific consensus on the point that "dispersal to the ocean cannot be the ultimate answer to radioactive waste disposal," on the grounds that, assuming dumping over five decades, "a highly radioactive form of strontium, known as Strontium 90, would be enough to contaminate 5 percent of the entire ocean volume." Readers were also to learn that "[i]n the body, strontium behaves much like calcium, with a specific affinity for bone, and is quite effective in producing bone cancer."

The key reference of the scientific delimitation of danger would indeed become human health. The second stage's most stern warning on "the biological effects of radiation" came on June 17, 1956. It was launched with the epic remark that "mankind must make up its mind whether a world movement for the control of atomic energy is to be organized or whether the human race must be slowly extinguished centuries hence." Singling out the "genetic dangers" posed by the radioactive byproducts of the energy industry, the record visualized "the perils that attend the peaceful uses of atomic energy" with "a four-leafed clover" and "a two-headed calf" and commented that a "legless animal has no chance of survival in the wilderness."

Such strong visual language notwithstanding, it would be incorrect to conclude that the second stage of evolution of the notion of nuclear waste in the public discourse concluded with a firm definition of nuclear waste as an irredeemable hazard. Indeed, a mere fortnight after the publication of the report quoted above, on July 1, 1956, one of the science notes carrying the initials of the very author of that article now reported on an achievement "in disposing safely of radioactive waste". But it is significant that, in this second stage, public discourse came to delimit the danger of radioactive waste in terms of science and with reference to human health.

It is also significant that the scientific consensus that had developed against ocean dumping put a crucial limitation to the space that, up until that point, was assumed available for the disposal of nuclear waste. No longer was there a public understanding of the planet as rife with possibilities for the disposal; the ocean was no longer to be used. Within less than 3 months from the first reporting of the oceanographer's argument against dumping came a revisitation of one earlier dumping report. Whereas on January 19, the *New York Times* had commented regarding the British dumping of 1500 tons of radioactive waste in the Atlantic Ocean that "[e]laborate precautions were taken to assure that the waste containers do not disintegrate until their contents have become harmless," on September 18 readers were reminded of "the consignment of 1,500 tons that was dropped into the sea last January" and were now to learn that "during the operation one container fell from a derrick with the result that the outer concrete casing broke."

1957, shifting from seeking solutions to seeking protection

This study's sample indicates that 1957 must have been a crucial year in the evolution of the notion of nuclear waste in the public discourse. That nuclear waste must have drawn intense interest is suggested even by the frequency of the coverage of the topic in the *New York Times*, the highest of the years 1945 through 1969. Indeed, revisiting the nuclear waste topic was common in 1957. For example, the inauguration of a research laboratory at a commercial petroleum company in Illinois that would "put to work" radioactive waste originated some 1500 miles away "from an atomic reactor in Idaho" was the theme of two consecutive records, on April 10 and April 14, 1957. Or, the conclusions of a report of the National Academy of Sciences regarding the safety of disposal procedures received coverage on May 1 and then again on May 5, 1957. Then in July, one case of dumping in the vicinities of New York City prompted three separate sampled records.

The three records of July 1957 also throw light on the shifting of focus that marks 1957 as the year in which public discourse moved from articulating nuclear waste in terms of multiple searches for utilization solutions to articulating it in terms of a pragmatic pursuit of protection from danger. On July 13, a brief record informed the readers that "a tanker carrying a load of radioactive sodium waste down the Hudson River en route to a dumping point 150 miles southeast of New York" was "expected to pass under the George Washington Bridge [on that same day of the reporting] and reach the Battery early at about noon." Nothing had yet happened, in other words, yet the fact that nuclear waste would be passing through common geographic points of reference clearly commanded attention. Then, on July 15, readers learned that an accident had befallen the shipment: "one of about twenty-five yellow, airtight receptacles dumped into the ocean" had "failed to sink." It was actual news this time, and the lengthier report announced it with alarm: "[a] steel container filled with radioactive sodium waste is floating somewhere in the Atlantic Ocean about 185 miles southeast of New York and is a 'menace to navigation.'" By the following day, July 16, the incident had also prompted "An Analysis of the Radioactive Waste Problem and Moves by U.S. to Meet It." This record designated the incident to be an illustration for "one of the increasing problems of the nuclear age" and argued for "most detailed precautions [...] to be taken about the disposal of radioactive waste materials; otherwise careless disposal procedures can be as damaging to a community as fall-out from a nuclear weapon." Despite its even greater length, the record spent one single sentence to relate that the lost container "was [...] later found."

The attention was justified neither by the shipment nor by the incident itself. It was remarkable for a shipment of just 25 nuclear waste containers of unspecified weight to be the topic of a report in the *New York Times* while uneventfully underway, especially in light of the fact that, as it would surface in a September 5 record that scrutinized the operations of a small company dedicated to the dumping of nuclear waste "somewhere at sea about twenty-seven miles off the Massachusetts coast," that one company had carried out more than 500 dumping trips by July 1957 and its capacities allowed for "more than 200 containers [to] be shipped to sea at a time." Moreover, while the incident of the one container failing to sink was announced with alarm, the lengthiest of the records related to the topic scattered only three words to relate that the lost container "was [...] later found."

The prompt for the extraordinary attention had to be something else. The sample indicates that it was a feeling of threat. That by 1957, nuclear waste must have been consensually perceived as threatening was already indicated in that year's earliest sampled record. The sole 1957 record to report on the success of yet another effort to channel nuclear waste into something of use, this time "a tiny atomic battery" that would power small devices "for almost limitless periods of time," the January 30 record addressed the safety issue already in its subtitle: "Device Using Waste of Reactor Is Called Safe as House Key."

But while the claim of the successful utilization appears to have been simply ignored, the safety issue continued its move towards the very center of public attention. On March 12, 1957, reporting on "the 1957 nuclear congress, supported by twenty-four scientific and engineering societies," the record zoomed in for "the dangers to the public from the spread of the atomic energy industry." Under the subtitle "Engineer Group Urges Plans by States to Control Wastes in Air and Streams From Nuclear Industry Plants" this record set the need for protection from the "long radioactive life of some of the wastes" into the "hundreds of years." This was a considerable augmentation of danger given that the highest previously cited number of radioactivity had been 33: a July 1, 1956 science note had explained that "Cessium has a half-life of thirty-three years, meaning that in thirty-three years it looses half of its radioactivity." By June 5, attention was focused on the "long range plans" of the U.S. Food and Drug Administration "to safeguard the country's future food supply against radioactivity." Headlined "U.S. Combatting Food Radiation," this record expressed the certainty of the "threat of radioactivity to food" in the words of a government official: "'there is little doubt that a material amount of radioactivity will spill over into the farmland and water supplies of this nation.'"

So the government was already taking action; the record reported that "large-scale analyses to determine safe radiation tolerances were already under way."

Whereas the first record of the year 1957 reported on the search for a beneficial utilization of nuclear waste, the year's last record was about a different kind of technological pursuit: "the feasibility of injecting radioactive wastes into the ground without contaminating drinking water or creating other health hazards." So 1957 can also be seen as a watershed year: the search for utilization solutions was now the past; the future would be about protection from danger.

Stage three, definition crystallized

The records from the years 1958 through 1963 imply a crystallizing if not crystallized public definition of nuclear waste. Much of what took place with regard to nuclear waste between 1958 and 1963, as well as the ways in which that action was put into words, appear to spring from an understanding of nuclear waste as a grave danger.

The coverage of an occurrence in France in October 1960 is illustrative of how widely shared this understanding was. On October 11, the first of three consecutive records on the matter introduced the "intention" of the French Atomic Energy Commission "to sink 6,500 barrels of atomic waste in the sea between Antibes and Corsica Oct. 20" in terms of an international disagreement: "Prince Rainer III of Monaco wrote to President de Gaulle today to question the wisdom of a plan to dump radioactive waste in the Mediterranean." Then, the record went on to discuss the dynamics of "several protests from citizens on the French Riviera" and quoted one local official as saying "people here are legitimately alarmed" and as anticipating that "tourists would be afraid to swim in the Mediterranean." Minute details of the grassroots opposition to the dumping within the city of Nice, as well as an exchange between the city's mayor and the head of the French Atomic Energy Commission, were laid out in a record published on the next day, October 12, 1960. This second record highlighted "a rare manifestation of political unity" against the dumping: "Gaullist and Conservative City Councillors joined Communists in threatening to strike if the Government did not reconsider." The topic of a third item of coverage became the fact that the French government did reconsider. The record of October 13, 1960 speaks of "victory" that was "temporary" yet at the same time also "initial." This third record used the dismissive tone of the French atomic governing body to highlight the public understanding of nuclear waste as a threat: the agency's statement was quoted in referring to the anti-dumping campaigners as "persons who, lacking sufficient information, are currently expressing their fears." The record concluded by undermining that dismissiveness through lining the sovereign of Monaco up on the side of the anti-dumping campaign: "it was learned on good authority that Prince Rainer, now in Paris, has secretly visited President de Gaulle to express opposition to the dumping project."

That these public understandings of nuclear waste would have been shared this side of the Atlantic, is suggested by a record that was published on April 22, 1962. In reporting the decision of the Atomic Energy Commission to deny "a Houston company's application to dispose of low-level radioactive wastes in the Gulf of Mexico," the Associated Press cited an internal "finding" of the commission as ground for the action: "the granting of a license to dispose of packaged low-level radioactive waste in the Gulf of Mexico would have serious harmful effects on our friendly relations with Mexico."

The data indicate that this stage of the American public's understanding of nuclear waste was well recognized by elected representatives. Three records from these years specifically mention congressional or federal inquiries of one form of another: on March 18, 1958, upcoming "hearings on the disposal of industrial radioactive wastes" were reported in the *New York Times*; on June 28, 1959, readers learned that four federal agencies had requested a scientific investigation of "the problem of radioactive waste disposal into Atlantic and Gulf of Mexico coastal waters;" by September 3, 1959, "[a] joint Congressional subcommittee warned [that existing protection from radioactive waste] might run out as the peaceful use of the atom was expanded." To spell out the distinctions between high-level and low-level nuclear waste, this last record paraphrases the report presented to the U.S. Congress in explaining that "[h]igh-level waste could be more than a million times as dangerous to life and resources as low-level waste" and that "[h]igh-level waste is so dangerous that experts have not yet decided whether it would be safe to cast it 150 miles off shore into 1,000 fathoms of water (6,000 feet) and leave it there."

Nuclear waste would have been suspected as a cause of harm to the degree that incidents involving seemingly anything nuclear drew attention. An uneventful accidental leaking of "[a] small amount of radioactive material [...] from an exhaust stack at Oak Ridge National Laboratory" became news by the next day; the November 13, 1959 record informed that, despite the absence of "hazard to the public," employee cars "were washed, if necessary, to prevent spread of the material [...] in minute quantities." "Blasts in Boston Bring Atom Scare" reads the headline of a record from August 10, 1960 that reports on "three explosions [...] on an open pier about 200 feet from radioactive atomic waste; the fire authority declared "no fall-out danger" however, and the readers were additionally assured that "[t]he radioactive material on the pier was in thick concrete casks." A record of November 16, 1960 that reported the accidental death of a nuclear scientist at Shirley, Long Island "when a huge truck loaded with used atomic reactor fuel skidded on wet pavement and struck his car" must have felt the need to address the understanding of nuclear waste as danger already in its subheading: "Car Is Struck by Truck Full of Radioactive Waste–None Leaks or Spills."

By this stage of its evolution, public discourse on nuclear waste centered on getting rid of it. Whereas only 1 of the 25 records of this timeframe focused on the beneficial utilization of nuclear waste, the disposal theme was present in each one of the records. Indeed, even that one December 10, 1961 report on certain buoy "tests on the use of nuclear waste products as a source of electric power for lighted aids to [sea] navigation" touched on disposal in commenting that the tests were part of the broader search effort "by the A.E.C. for the use of waste atomic materials [that] are now buried as a rule for disposal."

The records overwhelmingly indicate that, in the public understanding, the space available for the disposal of radioactive waste had shrunk significantly. A brief Reuters news of April 9, 1958 reported on governmental efforts at the United Nations for "regulations to prevent pollution of the high seas or the air above" involved in radioactive waste disposal. Covering the study of a "panel of oceanographers" that was "asked to suggest disposal sites," a record from June 28, 1959 reported that "[a]fter intensive study of local oceanographic conditions, the panel selected twenty-eight possible disposal sites [that included] every major seaport area from Boston, Mass., to Corpus Christi, Tex." But on July 14, 1959, a record reporting on the finding by the National Science Foundation that "a fleet of 300 nuclear-powered ships could safely dispose low-level radioactive wastes in the oceans 'without undue hazard to human health'" concluded with a remark about the suggestion "that there were about twenty sites in the coastal waters of the Atlantic and Gulf of Mexico that could be used safely for the disposal of low-level radioactive wastes:" that suggestion had "drawn strenuous objections from some of the localities near the proposed dumping grounds."

The public sentiment of this stage might have been best captured in a sentence from a research report quoted in the *New York Times* on January 24, 1958: "[s]afe disposal means that the waste shall not come in contact with any living thing."

Stage four, immune to denial

The fourth and last stage of the evolution of the concept of nuclear waste in the U.S. public discourse is recorded in the dataset in a relative paucity of items: for all of the 6 years 1964–1969, there were only three records from the year 1966 and another three from the year 1967 that fit the study's sampling criteria; the other 4 years have no records in the dataset.

Though all six of the sampled records support the interpretation that the public concept of nuclear waste, which had already crystallized before 1964, was now growing immune to denials of danger, only one of them provides such support explicitly. For all the denials of the dangers of nuclear waste and for all the shunning of the topic of hazard in the five other records from this stage, the record of November 2, 1967 documents how the Maltese representative to the United Nations, one Dr. Arvid Pardo, gravely warned "that continued indiscriminate dumping of radioactive waste into the ocean might prevent man from utilizing the incalculable food and mineral resources of the sea bottom."

The other five records support the interpretation indirectly. They do so first of all by documenting the strategies of an underlying effort to quell what must have been perceived as a prevailing public understanding of nuclear waste as dangerous. One such strategy was avoidance of the topic of waste hazards. For example, two records concerning waste reprocessing and ocean dumping overseas, one from July 8, 1966 involving 13 European countries and the other from July 30, 1967 involving Japan, steered completely clear of the topic of danger. And on December 28, 1967, an item of Reuters news marked several months of delay in informing the readers of the previous summer's dumping in "the Eastern Atlantic" of "[e]leven thousand tons of solid radioactive waste." This one record also documented the denial strategy in action: it quoted the European Nuclear Energy Agency in giving assurance that the dumping would "represent no risk either to man or marine organisms."

Yet the most telling support for the interpretation that the public conceptualization of nuclear waste was now growing immune to denials of danger came from the manner in which such denials were sometimes articulated. A May 23, 1966 record, boldly titled "No Hazard Found in Atomic Wastes," opens with the case of one doctor who "can hardly abide the sight of whitefish, no matter how nicely prepared" because "for a year he has eaten a half pound of it every Friday, freshly caught downstream from the great atomic energy works at Hanford, Wash." The record hypothesizes, "[h]e probably carries the heaviest body burden of radioactive zinc of any man in history," yet "[h]e is not, however, in much danger. His burden of 'hot zinc' is still less than one-tenth that considered permissible for the general public." But if nuclear waste was generally perceived as safe, why would anyone have needed to go through such an unusual symbolic experiment?

The prompt for this report was an international symposium of the International Atomic Energy Agency in Europe which was then again covered on May 29, 1966. Not only does the second report recapitulate the case of the radioactively contaminated yet harmless fish, but it generalizes that the scientific evidence discussed at the symposium translated in "almost uniformly good" news: "[t]here was no evidence that individuals in any part of the world had ingested enough of the debris to suffer harm, even by the most conservative health standards." And though admitting that "radioactivity seems sinister to the public" and that "[t]he idea of radioactive waste in food is alarming," this record denied the dangers of radioactivity with some striking sentences: "[e]very day [... w]ater drawn from the Columbia River to cool the great reactors at Hanford, Wash., is returned to the river with more than 60 different kinds of radioactive element. Those that survive the trip down the river are carried far out to sea, deposited on the ocean floor, or are picked up by oysters, clams, mussels and other creatures that delight the palates of those living in that region. Whenever residents of Chattanooga, Tenn., take a drink of water from the municipal supply, they imbibe a little bit of radioactive waste from Oak Ridge National Laboratory, far up the Tennessee and Clinch Rivers."

If not sarcastic, this bucolic depiction of the delights of radioactively contaminated victuals indicates at least that the understanding of nuclear waste as hazard was pervasive to the degree of prompting one to conceptualize its entering the human body. That even an explicit denial of danger cannot but reveal how profoundly the idea of danger[16] was engaged suggests that the evolution of the concept of nuclear in the public discourse had been concluded.

Conclusions

It was after 1945 that public discourse in the U.S. wedded the notion of atomic with the notion of litter to denote the byproducts of the exploitation of nuclear reactions for energy production. And it was before 1969 that this new notion of nuclear waste was associated with devastating consequences to human health and to the natural environment to a point that, at present, puts a consensus for disposing of nuclear waste out of practical reach. When sampled and analyzed through an adaptation of anthropological methods, this critical period's record of the public discourse on nuclear waste suggests that the view of nuclear waste evolved over a number of conceptual stages—from an initial introduction as a byproduct that was suspected of potential distant threats, to an assessment of the nature of that threat, to a subsequent crystallization of the view of nuclear waste as danger, to the present view of a grave and implacable hazard to human health and to the natural environment that cannot be entitled to a physical presence.

In other words, the current public understanding of nuclear waste neither just appeared out of the blue at a specific point in time, as it is sometimes assumed, nor had at its start the same shape in which it is encountered today. In addition, though the conceptualization of residues from the exploitation of nuclear reactions took shape at the same time that the American public grew aware of the devastation that followed the atomic bombing in Japan, the data investigated here show that nuclear waste was initially viewed as a peacetime concern and in contrast to military concerns: tellingly, the very introduction of nuclear waste into public discourse appealed for public attention in the context of the war having ended. Finally, though this research cannot directly assess the proposition that nuclear fear is rooted deeply in the human psyche, the data investigated here show that nuclear waste was not always feared in the same seemingly implacable way in which it appears to be feared at present. Instead, to draw a metaphor from the context of nuclear reactions, it was between the years 1945 and 1969 that the nucleus of danger in the public view of nuclear waste exploded, and then, as the awareness of the hazard that nuclear waste presents to human health and to the environment mushroomed, the physical space for disposing of nuclear waste imploded.

Endnotes

[1]Early in the year 2010, President Obama requested the formation of a commission "to conduct a comprehensive review of policies for managing the back end of the nuclear fuel cycle," known as the Blue Ribbon Commission on America's Nuclear Future, or BRC. The quotation is the very heading of the executive summary of the BRC's final report [2].

[2]In its 2012 report, the Blue Ribbon Commission cited 75,000 metric tons as the nation's then-total inventory waiting for permanent disposal [2]. According to the Nuclear Energy Institute, or NEI, the current inventory of high-level waste and spent nuclear fuel that are the byproduct of the nuclear energy industry is 71,780 metric tons [15]. One metric ton is equal to approximately 2200 lb.

[3]These figures do not include low-level radioactive waste such as from healthcare and other sources.

[4]For example, plutonium, a form of transuranic waste, has a half-life of 24,000 years, meaning that half of the original amount remains radioactive after 24,000 years [16].

[5]It is stored at well over 100 commercial nuclear reactor sites where the waste was generated, including sites of reactors no longer in operation, as well as at over 200 sites managed by the Department of Energy, or DOE.

[6]"[M]any [commercial] nuclear plants have run out of pool storage capacity," according to the NEI. "At these plants, used fuel is stored above ground" [17]. The storage arrangements for what is known as "defense waste" are described by an observer as "aging tanks larger than most state capitol domes" some of which "threaten groundwater and waterways" [18].

[7]The NEI [19] and the BRC [2] cite the range in metric tons: 2000 to 2300 metric tons per year.

[8]There is a need, even before attempting a taxonomy of these explanations for the origins of the public conceptualization of nuclear waste, to note that the key concepts driving this research (concepts such as "public understanding" and even "the public" itself, or concepts such as "discourse" and "evolution") are not universally self-explanatory but demand, as one peer reviewer put it, conceptual and theoretical clarifications. Acknowledging that concepts of this order have long been at the center of debates of a philosophical nature in sociology, political science, cultural studies, etc. [20, 21], this present paper does not side with or dispute any specific theoretical position, nor does it aim to contribute to such debates. Instead, it uses "public," "discourse," and "public discourse" heuristically, for the purposes of the research effort outlined. The author acknowledges that familiarity with the theoretical contexts in which such concepts are usually debated may aid a reader of the paper in better contextualizing the empirical and interpretive findings and conclusions presented here.

[9]A body of writing exists in anthropology on the complex legacies of nuclear weapons testing [22].

[10]Responding to peer review commentary, the author recognizes that this route is but one of the possible routes for undertaking this investigation. Indeed, other methodological perspectives exist, rooted in a range of scholarly disciplines, that also promise to help illuminate the important institutional and political forces, as well as the significant historical events that helped shape the public conceptualizations of nuclear waste we encounter in the U.S. at present [23, 24]. Moreover, as this project's approach may emphasize "the American perspective," perhaps by the very nature of its topical focus, the author acknowledges that the evolution of the public conceptualization of nuclear waste, including that within the U.S. that is the focus of this study, would owe also to other perspectives that originated and developed outside of the U.S. In addition, nuclear waste policy and its history, both within the U.S. and internationally, are subjects of substantial bodies of scholarly writing that conceptualize nuclear waste from a range of angles, including particular sites [25], the siting process as such [26], comparative international perspectives [27], and more.

[11]Pajo J (n.d.) The Two Waves of Public Discourse on Nuclear Waste in the U.S. since 1945.

[12]The ways in which a broad range of forces outside of the media are related to the rise of public conceptualizations may themselves pose questions that may be simply too numerous and complex to be sufficiently addressed within the frame of a research paper such as this one. Responding to peer review commentary, the author acknowledges the importance to this paper's topic of such complex relationships between "institutional forces," "elite groups," "military-industrial," "international actors," and others.

[13]Following the recommendation of one of the journal's peer reviewers, the author experimented with broadening this pool of sampling keywords as follows: one additional search was run in which the keywords "radiation," "fission," and "fusion" were searched in combination with the keyword "waste;" another additional search was run in which the keywords "pollution," "emissions," and "runoff" were searched instead of the keyword "waste." The rationale for this effort was to see whether the initial sample would be thus augmented substantially. The sample could indeed be augmented, but by a total of only 12 additional items. These were triggered as follows: the keyword "pollution" produced seven additional items, the keyword "radiation" produced four additional items, and the keyword "emissions" produced one additional item. The keywords "fission," "fusion," and "runoff" did not pull any additional items from the database. Of the 12 items, two were simply reports of overseas sentiments regarding testing. Reading all 12 items triggered through this subsequent sampling effort appeared to support the interpretations presented in this paper. However, given that these 12 items were not part of the original sample, they are not discussed here at length.

[14]Four of the 77 records that were selected by the electronic database based on these criteria were later excluded from the sample due to the fact that, despite containing valid combinations of the selective keywords, they were topically not about nuclear waste: the keyword "waste" was in those cases used in its "misspending" connotation.

[15]The only exception to the absence of denial of the harms of nuclear waste during these years lies with one single record, which was retained in the sample despite not being directly about nuclear waste. On July 31, 1955, on the occasion of the tenth anniversary of the bombing of Hiroshima, Robert Trumbull, head of the *New York Times* bureau in Tokyo, wrote: "the stories of how the A-bomb killed the soil, so that plants will not grow in it, are false. Trees are green wherever they stand, some in 'ground zero' of the big blast. If one makes the trip through a bewildering maze of alleys to where the

medieval Hiroshima Castle stood in architectural beauty until the big explosion, the foundations that are left will be found covered with greenery." Further on, the article targeted the human health topic directly: "according to scientific analyses, the emission of deadly gamma rays and neutrons in the atomic blast over [Hiroshima caused much less human damage than the force and heat released. As for the 'after-effects' of radiation, it is calculated on the basis of careful observation that those who survive the initial impact of such an explosion have a pretty good chance of living a normal life thereafter."

[16]This nicely illustrates the validity of Mary Douglas's classical paradigm of pollution and danger: danger is very great indeed when such matter-out-of-place as nuclear waste enters some place where it especially does not belong and which especially matters: the human body [28].

Competing interests
The author declares that she has no competing interests.

Authors' contributions
The author prepared the manuscript. The author read and approved the final manuscript.

Authors' information
The author is Assistant Professor of Anthropology at Pace University.

Acknowledgements
The author would like to thank Dagmar Fiedler, Editorial Manager, and the two anonymous reviewers at *Environment, Sustainability and Society* for helpful and insightful commentaries. The author also acknowledges a Summer 2014 Faculty Research Grant from Dyson College, Pace University that helped write this paper and thanks Maria Iacullo-Bird for commentary on the research. A Visiting Scholarship at the New York University Center for European and Mediterranean Studies and the resources of the Bobst Library were also of help. At NYU, Larry Wolff fostered the intellectual environment that inspired the author to explore historically. Finally, the author thanks Erind Pajo for engaging commentary throughout the process.

References

1. NYT (1950) Atomic 'Cemetery' needed for waste. New York Times. Washington, DC. 29 Jul 1950: p. 15
2. BRC (2012) Report to the secretary of energy. Blue Ribbon Commission on America's Nuclear Future., http://energy.gov/sites/prod/files/2013/04/f0/brc_finalreport_jan2012.pdf 5 Oct 2015
3. Rudig W (1990) Anti-nuclear movements: a world survey of opposition to nuclear energy. Harlow, Longman
4. Karapin R (2009) On his book Protest Politics in Germany: Movements on the Left and Right since the 1960s. Rorotoko., 15 Jul 2009 http://rorotoko.com/interview/20090715_karapin_roger_protest_politics_germany_movements_left_right/ 5 Oct 2015
5. Hughes M (2012) Reason, emotion, pressure, violence: modes of demonstration as conceptions of political citizenship in 1960s West Germany. Ger Hist 30(2):222–246
6. UNESCO (2012) The legacy of Rachel Carson's Silent Spring. United Nations Educational, Scientific and Cultural Organization., http://www.unesco.org/new/en/natural-sciences/ioc-oceans/about-us/special-events/the-legacy-of-rachel-carsons-silent-spring/ Accessed 22 Aug 2014
7. EPA (2014) EPA history: Earth day. Environmental Protection Agency., http://www2.epa.gov/aboutepa/epa-history-earth-day Accessed 2 Mar 2015
8. Mahaffey J (2010) Atomic awakening: a new look at the history and future of nuclear power. Pegasus, New York
9. Mahaffey J (2014) Atomic accident: a history of nuclear meltdowns and disasters. Pegasus, New York
10. Weart SR (1988) Nuclear fear: a history of images. Harvard University Press, Cambridge
11. Weart SR (2012) The rise of nuclear fear. Harvard University Press, Cambridge
12. Wolf ER (1982) Europe and the people without history. University of California Press, Berkeley
13. Davenport JS (1949) Newspaper circulation: backbone of the industry. W MC Brown Company, Dubuque
14. Bogart L (1989) Press and public: who reads what, when, where and why in American newspapers, 2nd edn. Lawrence Erlbaum Associates, Hillsdale
15. NEI (2014) On-site storage of nuclear waste. Nuclear Energy Institute., http://www.nei.org/Knowledge-Center/Nuclear-Statistics/On-Site-Storage-of-Nuclear-Waste Accessed 15 Aug 2014
16. EPA (2014) Transuranic radioactive waste. Environmental Protection Agency., http://www.epa.gov/radiation/docs/radwaste/402-k-94-001-tru.html#source Accessed 20 Aug 2014
17. NEI (2014) Used nuclear fuel storage. Nuclear Energy Institute., http://www.nei.org/Issues-Policy/Nuclear-Waste-Management/Used-Nuclear-Fuel-Storage Accessed 18 Aug 2014
18. Alvarez R (2014) A primer: military nuclear wastes in the United States. Bulletin of the Atomic Scientists., 24 Feb 2014 http://thebulletin.org/primer-military-nuclear-wastes-united-states 5 Oct 2015
19. NEI (2014) Fact sheets. Nuclear Energy Institute., http://www.nei.org/Master-Document-Folder/Backgrounders/Fact-Sheets/Safely-Managing-Used-Nuclear-Fuel Accessed 22 Aug 2014
20. Habermas J (1991) The structural transformation of the public sphere: an inquiry into a category of bourgeois society (trans: Burger T). The MIT Press, Cambridge
21. Habermas J (1985) The theory of communicative action (Volume 1): reason and the rationalization of society (trans: McCarthy T). Beacon, Boston
22. Johnston BR (ed) (2007) Half-lives and half-truths: confronting the radioactive legacies of the cold war. School for Advanced Research Press, Santa Fe
23. Stewart RB (2011) Fuel cycle to nowhere: U.S. law and policy on nuclear waste. Vanderbilt University Press, Nashville
24. de la Bruheze AAA (1992) Political construction of technology: nuclear waste disposal in the United States, 1945–1972, Dissertation, Universiteit Twente
25. Ringius L (2001) Radioactive waste disposal at sea: public ideas, transnational policy entrepreneurs, and environmental regimes. The MIT Press, Cambridge
26. Jacob G (1990) Site unseen: the politics of siting a nuclear waste repository. University of Pittsburgh Press, Pittsburgh
27. Berkhout F (1991) Radioactive waste: politics and technology. Routledge, London
28. Douglas M (1966) Purity and danger: an analysis of the concepts of pollution and taboo. Routledge & Kegan Paul, London

14

Root biomass and soil carbon response to growing perennial grasses for bioenergy

Leonard C. Kibet[1*], Humberto Blanco-Canqui[1], Robert B. Mitchell[2] and Walter H. Schacht[1]

Abstract

Background: Dedicated bioenergy crops such as switchgrass (*Panicum virgatum* L.), miscanthus [*Miscanthus x giganteus* (Mxg)], indiangrass [*Sorghastrum nutans* (L.) *Nash*], and big bluestem (*Andropogon gerardii Vitman*) can provide cellulosic feedstock for biofuel production while maintaining or improving soil and environmental quality. To better understand bioenergy crop effects on soils, we studied changes in soil properties of a Tomek silt loam under inorganic fertilization of switchgrass after 4 years and warm-season grass monocultures and mixtures after 6 years in eastern Nebraska.

Methods: The first experiment had two study factors: two switchgrass harvest dates (August and November) and nitrogen (N), phosphorus (P), and potassium (K) fertilization rates. Nitrogen fertilizer levels (0, 60, and 120 kg N ha^{-1}) were the main plots, while P levels (0, 22, and 44 kg P ha^{-1}) were the split plots and K levels (0, 11, and 22 kg K ha^{-1}) were the split-split plots. The second experiment included six bioenergy feedstocks comprised of four monocultures [switchgrass (cv. Shawnee and an experimental strain tracked as Kanlow N1), indiangrass (Chief), and miscanthus (Mxg)] and two mixtures [big bluestem (Goldmine) + indiangrass (Warrior) + switchgrass (Shawnee) and big bluestem (Bonanza) + indiangrass (Scout) + switchgrass (Shawnee)]. Soil samples were analyzed for root biomass, soil organic C (SOC), total N, bulk density, aggregate stability, and pH.

Results: In the first experiment, inorganic fertilization and harvest dates had no effect on switchgrass root biomass, SOC pools, soil aggregate stability, and other properties. In the second experiment, cumulative root biomass under Chief indiangrass monoculture was lower than that under other grass monocultures and mixtures except miscanthus. These results suggest that inorganic fertilization and harvest dates do not affect soil properties in the short term, but Chief indiangrass monoculture may have lower root biomass than other grasses.

Conclusions: Overall, fertilization management did not induce changes in root biomass and soil properties, but Chief indiangrass monoculture had lower cumulative root biomass compared with mixtures and switchgrass monocultures, suggesting that cultivar selection will affect root biomass accumulation. Further monitoring is needed to determine long-term changes in root biomass and soil properties under these bioenergy crop systems.

Keywords: Bioenergy, Perennial warm season grasses, Monocultures, Mixtures, Root biomass, Soil organic carbon, Total nitrogen

* Correspondence: lkibet2@unl.edu
[1]Department of Agronomy and Horticulture, University of Nebraska-Lincoln, Lincoln, NE 68583, USA
Full list of author information is available at the end of the article

Background

Dedicated bioenergy crops such as native perennial warm-season grass monocultures and mixtures may provide abundant biomass for biofuel production. However, their effects on soil ecosystem services have not been widely documented under different soil types, management scenarios, and climatic conditions. Although perennial bioenergy crops generally improve soil properties such as soil organic C (SOC), and soil structural properties compared with row crops [3, 9], little information is available on the differences in soil properties between perennial grass monocultures and mixtures and among inorganic fertilization levels. This knowledge is crucial for the development of sustainable bioenergy cropping systems as soil properties are an integral component of soil productivity.

Perennial warm-season grasses may improve soil properties by virtue of their fibrous and extensive root network. Warm-season grass species and varieties may, however, differ in their root biomass and distribution. Some studies have suggested that species in diverse communities might alter the depth distribution of roots in response to the density and uniqueness of neighboring roots [9, 28, 32]. In pot experiments, von Felten and Schmid [48] found that mixtures of grass species increased root biomass in deeper soil depths than monocultures. Also, Cong et al. [8] reported that a mixture of eight grasses that consisted of four grass species (*Agrostis capillaris* L., *Anthoxanthum odoratum* L., *Festuca rubra* L., and *Holcus lanatus* L.) and four forbs (*Centaurea jacea* L., *Leucanthemum vulgare* Lam., *Plantago lanceolata* L., and *Rumex acetosa* L.) increased root biomass more than their monocultures (6.85 versus 4.33 Mg ha^{-1}).

One potential impact of bioenergy production systems is SOC accumulation. This is important because SOC can enhance soil ecosystem services and increase soil productivity [2]. Perennial grasses may have different impacts on above- and belowground biomass production and SOC accumulation, depending on grass species, varieties, and mixtures, but such impacts have not been widely studied. Accumulation of C and N in the soil can be complex and depends on many variables such as soil type, management practices, residue quality, precipitation input, and harvest treatments [1, 10]. For example, soils with low initial SOC concentration may have greater potential to accumulate SOC under perennial energy crops.

Perennial grass monocultures and mixtures may affect SOC and other soil properties differently. Blanco-Canqui [4] reported that mixtures of perennial warm-season grasses have the potential to store more SOC than monocultures. In a 12-year study in Minnesota, Fornara and Tilman [12] reported that high-diversity mixtures of perennial grass species can store, on average, 500 % more SOC and 600 % more N than monocultures of the same species. In an 11-year study, perennial warm-season grass mixtures increased SOC stocks by 18 % and N stocks by 16 % compared with monocultures of the same species [8]. The large increase of SOC under perennial mixtures may be due to the greater grass diversity, more abundant root biomass, and deeper distribution of C-enriched roots relative to monocultures. In a 3-year study, Bonin et al. [6] showed, however, no effect of multiple species mixtures of perennial warm-season grasses on SOC, suggesting that perennial mixtures may not increase SOC stocks in the short term relative to monocultures.

Additionally, since aboveground biomass is harvested for bioenergy, growing perennial grasses may deplete soil nutrients. Nutrient removal from the soil may vary because of differences in biomass production, cultivars, weather, soil type, and management practices [42]. For example, Newton et al. [36] showed that grasses have greater ability to absorb soil nutrients than broadleaf forages, implying that grasses differ in their potential to remove nutrients from the soil. Also, in a study comparing nutrient removal rates by perennial grasses, giant reed (*Arundo donax* L.) had the highest aboveground biomass yield and the highest nutrient (N, P, and K) concentrations and removal rates compared to switchgrass, giant miscanthus, weeping lovegrass [*Eragrostis curvula* (Shrad.) Nees], kleingrass (*Panicum coloratum* L.), and Johnsongrass [*Sorghum halepense* (L.) *Pers.*] [23], suggesting that aboveground biomass is correlated with soil nutrient removal.

Production of perennial warm-season grasses for bioenergy consumes soil nutrients and needs constant supplementation of nutrients through inorganic fertilization [23]. Inorganic fertilization increases aboveground biomass production of perennial warm-season grasses [19, 25, 47]; however, its effect on root biomass, SOC stocks, and other soil properties is unclear [19]. Root biomass response to N application is variable. Ma et al. [28] found that fertilization (224 kg N ha^{-1}) of 4-year-old switchgrass stands in Alabama had no effect on root biomass relative to control (0 kg N ha^{-1}). Garten et al. [15] also reported no effect of N fertilization on switchgrass root biomass in a 5-year study. In contrast, Heggenstaller et al. [19] found that high rates of fertilization (220 kg N ha^{-1}) reduced root biomass in 3- or 4-year-old stands of switchgrass in Iowa. Similarly, fertilization-induced changes in SOC storage are inconsistent. Previous studies indicated that SOC increased [20, 39], decreased [24, 30], or were unaffected [29, 33] following fertilization of perennial grasses.

The objectives of this study were to assess the following: (1) the impact of different rates and mixtures of N, P, and K fertilization and harvest date on switchgrass root biomass, SOC, total N, and other soil properties, and (2) soil-profile distribution of root biomass, SOC, bulk density, and total N under perennial warm-season grass monocultures and mixtures. Our first hypothesis was that fertilization of switchgrass increases root biomass, SOC, and total N pools relative to non-fertilized plots. Our second hypothesis was that perennial warm-season grass mixtures increase root biomass, SOC, and total N pools compared with warm-season grass monocultures. Perennial warm-season grasses need to be in production for at least 5 years to be economically feasible [37]. Therefore, our study was designed to assess soil response to growing perennial grasses during the above timeframe. Soil response to perennial grasses in the short term (<5 years) may not be measurable until the grasses reach their full potential at about 5 years [37].

Methods

Study site

The study was conducted on two sites seeded to perennial warm-season grasses at the University of Nebraska-Lincoln's Agricultural Research and Development (ARDC) Center near Mead, NE (41.2261° N, 96.4883° W). The soil for both sites is classified as Tomek silt loam (Fine, smectitic, mesic Pachic Argiudolls). The Tomek series consists of very deep well-drained soils formed in loess with clay content between 24 and 35 %, while the sand content is <10 % in the A and AB horizons. Experiment I was established in 2006 and was seeded to switchgrass (cv. Shawnee) monocultures and had two study factors: two biomass harvest dates (August and November) and N, P, and K fertilization rates. This study was seeded at a rate of 320 pure live seed (PLS) per meter square. The main plot was 3.3 by 3 m, while the split-split plots were 1.1 by 3 m. Treatments were applied to plots in a randomized complete block, split-split-plot design with three replications. Fertilizer treatments were applied beginning in 2010, 4 years after field establishment. The main plot treatment was a level of N fertilization (0, 60, and 120 kg N ha^{-1}). The split plot treatment was a level of P fertilization (0, 22, and 44 kg P ha^{-1}), and the split-split plot treatment was a level of K fertilization (0, 11, and 22 kg K ha^{-1}). Nitrogen was applied as urea (46-0-0), P was applied as triple superphosphate (0-16-0), and K was applied as potassium sulfate (0-0-50). All fertilizer treatments were broadcast with a drop spreader. Plots were harvested at a 10 cm cutting height with a small-plot flail harvester (Carter Manufacturing, Brookston, IN) either at anthesis (near August 1) or after a killing frost each year by harvesting in the direction of the split-split plot treatment.

Experiment II was established in 2009 and was seeded with 29 perennial grass monocultures and mixtures and replicated four times in a randomized complete block design. Our study evaluated soil properties for only six of the 29 high-yielding grasses (four monocultures and two mixtures). The monocultures evaluated were switchgrass (cv. Shawnee), switchgrass (Kanlow N1), indiangrass (Chief), and miscanthus (Mxg). The mixtures included grass mixtures of [big bluestem (Goldmine) + indiangrass (Warrior) + switchgrass (Shawnee)] (Mixture 1) and [big bluestem (Bonanza) + indiangrass (Scout) + switchgrass (Shawnee)] (Mixture 2) in four field replicates. The mixtures will hereafter be referred to as mixture 1 and mixture 2 for discussion purposes. Experimental units were 1.1 × 3-m plots seeded to the monocultures or the mixtures at a rate of 320 PLS per square meter. The miscanthus used was the Illinois *Miscanthus x giganteus* clone started from rhizomes, grown in a greenhouse, and transplanted into the plots from pots at 12 plants per plot. Immediately after planting or transplanting, all plots were sprayed with 0.24 L ha^{-1} of Paramount (quinclorac: 3, 7-dichloro-8-quinolinecarboxylic acid) plus 0.95 L ha^{-1} atrazine (2-chloro-4 ethylamino-6-isopropylamino-s-triazine) to control grassy weeds. Broadleaf weeds were sprayed as needed, typically with 2, 4-D (2, 4-dichlorophenoxyacetic acid) at 0.95 L ha^{-1}. No fertilizer was applied in the seeding year, but all plots were fertilized with 112 kg N ha^{-1} as urea (46-0-0) after growth initiation in April or May of each year. Harvesting was done as described for experiment I (N, P, and K study) above. Data on aboveground biomass yield for both experiments (I and II) will be reported in a companion paper. The focus of this paper was on root biomass and soil properties.

Soil sampling

Under experiment I, we measured root biomass, SOC, total N, bulk density, soil aggregate stability, and pH. Under experiment II, we measured root biomass, SOC, total N, and bulk density. We included soil aggregate stability and pH measurements in experiment I to study possible effects of inorganic fertilization on these two properties. Previous studies on inorganic fertilization of row crops reported decreased pH and mixed effects on aggregate stability with fertilization [5, 26].

Soil samples were collected in June 2014 from experiment I and June 2015 from experiment II. In experiment I (N, P, and K study), four soil cores were collected from each plot using a 3-cm diameter hand probe for the 0–10- and 10–20-cm depths. Two core samples were composited for soil analyses, and the other two cores were composited for root biomass analysis. In experiment II (monocultures and mixtures), six soil cores (4.57 cm diameter) per plot were collected using a tractor-mounted Giddings probe (Giddings Machine Company, Inc. Windsor, CO). We collected three soil cores from the center of the plants and three between plants to a depth of 100 cm per plot and sliced at 0–10, 10–20, 20–40, 40–60, and 60–80 and 80–100 cm depths. The six cores were composited, placed in paper bags, air-dried for 72 h, and analyzed. One hundred grams of soil without roots was carefully extracted from the composite sample for the analyses of SOC and total N concentration. The remaining soil sample was used for root biomass analysis.

Measurement of soil properties

Root biomass was analyzed on air-dried soil samples for both experiments. The samples were weighed, soaked, and roots rinsed in tap water over a 0.5-mm sieve. Sand and debris were hand-sorted from root material while rinsing in tap water. Roots including crown, rhizomes, and fine roots recovered were placed in an aluminum pan and oven-dried at 60 °C for root biomass determination [32].

Soil samples collected for experiment I were analyzed for wet aggregate stability using the wet sieving method described by Nimmo and Perkins [35]. Soil samples were passed through 4.75- and 8-mm sieves to obtain soil aggregates with diameters between 4.75 and 8 mm. Fifty grams of aggregates were weighed, placed on top of a nest of sieves with 4.75- and 0.25-mm openings, immersed in water, and saturated by capillarity for 10 min before wet sieving for another 10 min. A mechanical sieving device was used to oscillate the column of sieves with a stroke of 3 cm. Aggregates remaining in each sieve were transferred to pre-weighed beakers and dried at 105 °C. Amount of water-stable aggregates was expressed as the percentage of soil remaining on the sieve after 10 min relative to the initial mass of soil used. The amount of water-stable aggregates was corrected for sand content and reported as macro- and micro-aggregates. Aggregates with diameters >0.25 mm were classified as macro-aggregates, while aggregates with diameters <0.25 mm were classified as micro-aggregates [45].

All other soil samples were air-dried and sieved (2 mm) before analysis. Soil organic C and total N concentrations were determined on soil ground on a roller mill before analysis by dry combustion using a LECO CN 2000 analyzer (Leco Corp., St Joseph, MI; [34]). Baseline data on SOC concentration were not available for either experiment. Soil particle-size distribution was determined using the hydrometer method [17]. Soil particle-size distribution did not significantly differ among treatments and between the two experiments. Averaged across both experiments and treatments, soil had 15 % sand, 36 % clay, and 49 % silt for the 0- to 10-cm depth and was classified as silty clay loam. The bulk density was determined using the core method for each soil depth [38]. Soil pH was measured on soil and water in a 1:2 ratio using Thermo Orion 525A+ pH/mV meter (Thermo-electron Corporation, MN). The mixture was allowed to stand for at least 1 h, stirred every 10 to 15 min to allow the pH of the soil slurry to stabilize before measuring pH [44].

Data were analyzed by SAS PROC UNIVARIATE to test their normality and equal variance distribution. Root biomass data required logarithmic transformation to achieve normality. Data were reported as geometric means for discussion purposes. Analysis of variance in PROC MIXED in SAS [43] and least significant difference test were used to assess treatment differences. Fertilization effects were analyzed by harvest date and soil depth. Nitrogen, P, and K fertilization levels were the fixed variables, while replications were the random variables in the analysis. Differences among treatments were studied at the 0.10 probability level.

Results and discussion

Inorganic fertilization (experiment I)

Root biomass

Inorganic fertilization of switchgrass had no effect on root biomass. Mean root biomass across N, P, and K fertilization rates was 3.87 ± 1.59 Mg ha^{-1} (mean ± SD) for the 0- to 20-cm depth. These results did not support our hypothesis stating that inorganic fertilization will increase root biomass of switchgrass. The lack of differences in root biomass may be due to the following: (1) the relatively short period of fertilization (4 years) and (2) the use of a highly productive soil in this experiment, which possibly reduced the effects of fertilization on root biomass. We hypothesize that inorganic fertilization would have significant effects on root biomass if switchgrass were managed in marginally productive lands with low organic matter content.

Previous studies of fertilization on root biomass have reported similar findings to our study. For example, after 5 years, N fertilization of switchgrass did not affect root biomass in a study in western Tennessee [15]. Similarly, after 4 years, N fertilization did not affect root biomass of perennial warm-season grasses across three sites in Ohio [21]. Also, in another study in western Tennessee, N fertilization of switchgrass reduced root biomass in the spring but had no effect in the fall after 4 years [22]. Some studies have indicated that addition of fertilizers to highly productive soils can enhance biomass production without changing the root systems [46]. Previous studies and the present study suggest that inorganic fertilization of switchgrass may not generally increase root biomass relative to control without fertilization in the short term.

Soil organic carbon and other soil properties

Inorganic fertilization of switchgrass and harvest dates (August and November) had no effect on SOC pools and other soil properties. Because trends in mean SOC concentrations and pools were similar, only SOC pools are presented. Mean SOC across fertilization levels was 24.3 ± 3.7 Mg C ha^{-1} at the 0- to 10-cm depth and $27.9.3 \pm 3.4$ Mg C ha^{-1} at the 10- to 20-cm depth. While the lack of fertilization effect may not be surprising, we expected that high fertilization rates would increase SOC pools by increasing root biomass production, but that was not the case. No differences in SOC among N, P, and K fertilization rates are likely due to no differences in root biomass amount.

Similar to our study after 9 years of another switchgrass experiment at ARDC, fertilization and harvest date had variable effects on SOC [13]. Previous studies have found small or no increase in SOC pools with fertilization [21, 27]. Corresponding with SOC, inorganic fertilization did not affect soil C:N ratio. Average C:N ratio across N, P, and K fertilization levels was 10.69 ± 2.01 ppm at 0- to 10-cm depth and 11.74 ± 1.55 ppm at 10- to 20-cm depth. Some studies have indicated that N fertilization of switchgrass decreases root C:N ratio [15, 19] by increasing organic matter decomposition, thereby reducing SOC concentration.

Also, inorganic fertilization did not affect soil bulk density and soil pH. Bulk density averaged across fertilization levels ranged from 1.32 to 1.45 g cm^{-3} at the 0- to 10-cm depth and 1.54 to 1.64 g cm^{-3} at the 10- to 20-cm depth and the mean pH was 5.92 ± 0.21 at the 0- to 10-cm depth and 6.11 ± 0.44 at the 10- to 20-cm depth. Under row crops, inorganic fertilization often reduces soil pH through the release of H^+ [26], but in our study, under perennial grasses, soil pH was unaffected by fertilization.

Wet aggregate stability

Similarly, inorganic fertilization of switchgrass did not affect soil aggregate stability. The amount of macro-aggregates among fertilizer levels ranged from 53 to 84 % at the 0- to 10-cm depth. In row crops, N fertilization negatively affects wet aggregate stability in some soils. For example, Blanco-Canqui et al. [5] reported that in the 7.5- to 45-cm soil depth, soil aggregate stability decreased with an increase in N application. The decreased aggregate stability with fertilization suggests that the use of N fertilizers, particularly those that release NH_4^+, can reduce soil aggregation by dispersing colloids and secondary particles [18] and reducing the amount of microbial biomass [26].

Published studies on the effects of switchgrass fertilization on soil aggregation are few. In Ohio, 4 years of N fertilization of switchgrass at 202 kg N ha^{-1} reduced the amount of macro-aggregates at the 0- to 5-cm depth compared with non-fertilized plots [22]. The highest N application rate in the latter study was 202 kg N ha^{-1}, whereas our maximum rate was 120 kg N ha^{-1}, suggesting that higher rates of N application than that used in our study may reduce soil aggregate stability.

Root biomass of monocultures and mixtures (experiment II)

Root biomass

Perennial grass monocultures and mixtures did not affect the amount of root biomass when data were analyzed by depth interval (Table 1). However, cumulative root biomass under Chief indiangrass monoculture for the 0- to 100-cm depth was significantly lower than under other grass monocultures and mixtures except miscanthus (Table 1). The indiangrass monoculture had 1.8 times lower (10.63 Mg ha^{-1}) root biomass than the average across all other treatments except miscanthus (19.31 Mg ha^{-1}; Table 1). The order of cumulative root biomass was the following: mixture 1 = mixture 2 = Shawnee switchgrass = Kanlow N1 switchgrass ≥ miscanthus ≥ Chief indiangrass.

Results showed that mixtures and monocultures that included switchgrass varieties have greater root biomass than the Chief indiangrass monoculture. Switchgrass as a species is considered to be the most adaptable to diverse regions with variable soil and climatic conditions

Table 1 Impact of growing perennial warm-season grasses in monocultures and mixtures on root biomass on a Tomek silt loam in eastern Nebraska after 6 years of management

Treatments	Root biomass (Mg ha^{-1})						
	Depths (cm)						
	0–10	10–20	20–40	40–60	60–80	80–100	0–100
Grass mixture 1	11.98	1.54	2.34	1.15	1.32	0.58	18.54a*
Grass mixture 2	11.64	1.34	2.38	1.73	0.87	1.06	18.17a
Shawnee	14.41	1.84	1.61	1.13	0.65	0.81	20.27a
Kanlow N1	12.47	2.53	2.78	1.59	1.05	0.89	20.27a
Miscanthus (Mxg)	9.21	2.22	3.78	0.70	0.66	0.74	16.79ab
Chief	6.82	1.29	1.38	1.12	0.71	0.50	10.63b
LSD 0.10	6.70	1.32	2.44	0.84	0.49	0.86	1.59

The means are back-transformed log values. Perennial warm-season grass monocultures [switchgrass (cv. Shawnee, Kanlow N1), indiangrass (Chief) and Mxg] and grass mixture 1 [big bluestem (Goldmine) + indiangrass (Warrior) + switchgrass (Shawnee)] and grass mixture 2 [big bluestem (Bonanza) + indiangrass (Scout) + switchgrass (Shawnee)]

*Numbers followed by different letters within a column show significant differences among treatments at $P \leq 0.10$

in the USA including the Midwest [31]. Based on the results, some varieties of indiangrass such as Warrior and Scout may be grown in mixtures of perennial grasses for enhanced root biomass relative to Chief indiangrass alone. While our experiment did not have Warrior and Scout indiangrass varieties grown alone to compare with the Chief indiangrass monoculture, our results appear to suggest that indiangrass grown in monocultures may likely fill less of the soil profile with roots than indiangrass grown in mixtures. Results also suggest that root characteristics may differ among indiangrass varieties.

Previous studies under different types of grasses, mostly cool-season grasses, have found that root biomass may not always differ among grass species or between monocultures and mixtures. For example, in two studies after 3 and 4 years, differences in root biomass among grass species mixtures and monocultures were not significant in a study in Germany [16, 40]. However, mixtures increased root biomass relative to monocultures after 11 years, suggesting that differences may develop in the longer term [40]. The total soil-profile root biomass in switchgrass monocultures, in this study, was comparable with that reported in a 3-year study of switchgrass for the 0- to 90-cm depth in southern Washington [7] and a 3-year study of switchgrass for the 0- to 110-cm depth in central North Dakota [14]. In our study, the lack of significant differences between mixtures and monocultures may be due to similar rooting patterns across the species studied.

Soil organic carbon and other soil properties

Soil organic C (Fig. 1) and total N (Table 2) pools did not differ at any soil depth among grass monocultures and mixtures after 6 years of management. Cumulative SOC pool for the 100-cm depth was 98.09 ± 8.82 Mg ha^{-1} and cumulative total N pool was 10.74 ± 0.74 Mg ha^{-1}. Results did not support our hypothesis that mixtures will improve SOC and other soil properties compared with monocultures after 6 years. As in experiment I, the soil under experiment II was highly productive, which may have reduced treatment effects. Our results are consistent with two previous studies of warm-season grasses. After 3 years, SOC pools among a switchgrass monoculture, a 4-grass species mixture, and a 10-species mixture of native grasses were similar in Virginia [6]. Also, after 11 years, big bluestem, indiangrass, and little bluestem [*Schizachyrium scoparium* (*Michx*)] did not increase SOC relative to C3-grass species in Michigan [41], suggesting that even after a long period of time, SOC pools among perennial grasses may not differ.

The soil C:N ratio among grass monocultures and mixtures did not differ, which suggests that soil organic matter decomposition rates among the grass treatments studied were similar. Averaged across grass treatments, C:N ratio was 8.97 ± 3.41 for the 0- to 100-cm depth. Nitrogen fertilization often reduces soil C:N ratio in croplands, but studies on the effect of warm-season grass monocultures and mixtures on C:N ratio are not available to compare with our results.

The SOC pool (98.09 ± 8.82 Mg C ha^{-1}) at 0 to 100 cm in this study after 6 years was similar to that (116.6 ± 59.4 Mg C ha^{-1}) found in a nearby switchgrass experiment after 9 years [13]. These results appear to indicate that, after 6 years, grass monocultures and mixtures stored similar amounts of SOC to that under switchgrass monocultures after 9 years. Also, after 5 years, SOC concentration among switchgrass, miscanthus, and

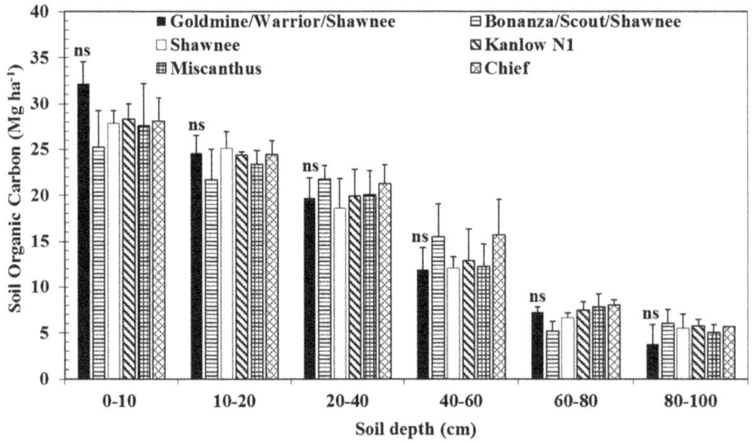

Fig. 1 Soil-profile organic C distribution under perennial warm-season grass monocultures [switchgrass (cv. Shawnee, Kanlow N1), indiangrass (Chief) and Mxg] and mixtures [(big bluestem (Goldmine) + indiangrass (Warrior) + switchgrass (Shawnee) and big bluestem (Bonanza) + indiangrass (Scout) + switchgrass (Shawnee)] on a Tomek silt loam in eastern Nebraska after 6 years of management. No significant differences (ns) were detected on SOC among grass monocultures and mixtures. *Error bars* are standard deviation values

Table 2 Total nitrogen response to perennial warm-season grass monocultures and mixtures on a Tomek silt loam in eastern Nebraska after 6 years of management

Treatments	Total nitrogen (Mg ha^{-1})					
	Depths (cm)					
	0–10	10–20	20–40	40–60	60–80	80–100
Grass mixture 1	3.03 ns	2.67	2.01	1.56	1.02	1.6
Grass mixture 2	2.41	2.11	1.85	1.41	1.05	0.84
Shawnee	2.64	2.46	1.93	1.21	1.15	0.75
Kanlow N1	2.63	2.48	2.25	1.45	1.16	1.27
Miscanthus (Mxg)	2.63	1.95	1.99	1.41	1.03	0.81
Chief	2.49	2.6	2.09	1.77	1.23	1.55
LSD 0.05	0.58	0.69	0.52	0.47	0.49	0.89

Perennial warm-season grass monocultures [switchgrass (cv. Shawnee, Kanlow N1), indiangrass (Chief) and Mxg] and grass mixture 1[big bluestem (Goldmine) + indiangrass (Warrior) + switchgrass (Shawnee)] and grass mixture 2 [big bluestem (Bonanza) + indiangrass (Scout) + switchgrass (Shawnee)]. Means among grass treatments were not significantly different (ns)

big bluestem monocultures did not differ in a study in Kansas [11], suggesting that perennial warm-season grasses regardless of species may have similar potential in accumulating SOC in this region. Also, grass monocultures and mixtures did not affect soil bulk density. Bulk density averaged across grass monocultures and mixtures was 1.40 ± 0.09 g cm^{-3} at 0 to 10 cm, 1.61 ± 0.13 g cm^{-3} at 10 to 20 cm, 1.60 ± 0.07 g cm^{-3} at 20 to 40 cm, 1.62 ± 0.12 g cm^{-3} at 40 to 60 cm, 1.61 ± 1.10 g cm^{-3} at 60 to 80 cm and 1.70 ± 0.07 g cm^{-3} at 80 to 100 cm.

Conclusions

Four years of inorganic (NPK) fertilization of switchgrass in a silty loam in eastern Nebraska did not affect root biomass and soil properties. Similarly, 6-year-old stands of simple mixtures of big bluestem, indiangrass, and switchgrass had no significant impact on the SOC pool, total N, and C:N ratio compared with monocultures. However, Chief indiangrass had lower root biomass compared to all other grasses except miscanthus. As mentioned earlier, aboveground biomass production of perennial warm-season grasses often reaches full potential in about 5 years [37]. Thus, our results suggest that there could be limited or no impacts of inorganic fertilization and mixtures on root production and soil properties within the above timeframe and soil type. We hypothesize that the lack of root biomass and soil properties' response to fertilization may be attributed to the use of a highly productive soil. The lack of differences in root biomass among the four warm-season grasses (big bluestem, indiangrass, switchgrass, and miscanthus) suggests that these grasses have similar rooting characteristics except Chief indiangrass. Further study is warranted to test the longer term effects of fertilization and mixtures on root production and soil properties.

Competing interests
The authors declare that they have no competing interests.

Authors' contributions
LCK and HB-C collected, prepared, analyzed the soil samples and prepared the manuscript. RM and WS designed and managed the field experiments and contributed to the critical review and inputs into the manuscript. All authors read and approved the final manuscript.

Acknowledgements
This research was supported by funding from the North Central Regional Sun Grant Center at South Dakota State University through a grant provided by the US Department of Agriculture under award number 2010-38502-21861.

Author details
[1]Department of Agronomy and Horticulture, University of Nebraska-Lincoln, Lincoln, NE 68583, USA. [2]Grain, Forage, and Bioenergy Research Unit USDA-ARS, University of Nebraska-Lincoln, Lincoln, NE 68583, USA.

References

1. Adler PR, Del Grosso SJ, Parton WJ (2007) Life-cycle assessment of net greenhouse-gas flux for bioenergy cropping systems. Ecol Appl 17:675–691
2. Blanco-Canqui H, Shapiro CA, Wortmann CS, Drijber RA, Mamo M, Shaver TM, Ferguson RB (2013) Soil organic carbon: the value to soil properties. J Soil Water Conserv 68:129A–134A
3. Blanco-Canqui H, Gilley JE, Eisenhauer DE, Jasa PJ, Boldt A (2014) Soil carbon accumulation under switchgrass barriers. Agron J 106:2185–2192
4. Blanco-Canqui H (2010) Energy crops and their implications on soil and environment. Agron J 102:403–419
5. Blanco-Canqui H, Ferguson RB, Shapiro CA, Drijber RA, Walters DT (2014) Does inorganic nitrogen fertilization improve soil aggregation? Insights from long-term tillage experiments. J Environ Qual 43:995–1003
6. Bonin CL, Lal R, Tracy BF (2014) Evaluation of perennial warm-season grass mixtures managed for grazing or biomass production. Crop Sci 54:2373–2385
7. Collins HP, Smith JL, Fransen S, Alva AK, Kruger CE, Granatstein DM (2010) Carbon sequestration under irrigated switchgrass (Panicum virgatum L.) production. Soil Sci Am J 74:2049–2058
8. Cong W, Ruijven J, Mommer L, De Deyn BG, Berendse F, Hoffland E (2014) Plant species richness promotes soil carbon and nitrogen stocks in grasslands without legumes. J Ecol 102:1163–1170
9. de Kroon H, Hendriks M, van Ruijven J, Ravenek FM, Padilla F, Jongejans E, Visser W, Mommer L (2012) Root responses to nutrients and soil biota: drivers of species coexistence and ecosystem productivity. J Ecol 100:6–15
10. Del Grosso SJ, Mosier AR, Parton WJ, Ojima DS (2005) DAYCENT model analysis of past and contemporary soil N_2O and net greenhouse gas flux for major crops in the USA. Soil Tillage Res 83:9–24
11. Evers BJ, Blanco-Canqui H, Staggenborg SA, Tatarko J (2013) Dedicated bioenergy crop impacts on soil wind erodibility and organic carbon in Kansas. Agron J 105:1271–1276
12. Fornara DA, Tilman D (2008) Plant functional composition influences rates of soil carbon and nitrogen accumulation. J Ecol 96:314–322
13. Follett RF, Vogel KP, Varvel GE, Mitchell RB, Kimble J (2012) Soil carbon sequestration by switchgrass and no-till maize grown for bioenergy. Bioenergy Res 5:866–875
14. Frank AB, Berdahl JD, Hanson JD, Liebig MA, Johnson HA (2004) Biomass and carbon partitioning in switchgrass. Crop Sci 44:1391–1396
15. Garten CT Jr, Brice DJ, Castro HF, Graham RL, Mayes MA, Phillips JR, Post WM III, Schadt CW, Wullschleger SD, Tyler DD, Jardine PM, Jastrow JD, Matamala R, Miller RM, Moran KK, Vugteveen TW, Izaurralde RC, Thomson AM, West TO, Amonette JE, Bailey VL, Metting FB, Smith JL (2011) Response of "Alamo" switchgrass tissue chemistry and biomass to nitrogen fertilization in West Tennessee, USA. Agric Ecosyst Environ 140:289–297

16. Gastine A, Scherer-Lorenzen M, Leadley PW (2003) No consistent effects of plant diversity on root biomass, soil biota and soil abiotic conditions in temperate grassland communities. Applied Soil Ecol 24:101–111
17. Gee GW, Or D (2002) Particle-size analysis. p. 255–293. In: Dane JH, Topp GC (eds) Methods of soil analysis. Part 4. SSSA Book Ser. 5. SSSA, Madison
18. Haynes RJ, Naidu R (1998) Influence of lime, fertilizer, and manure applications on soil organic matter content and soil physical conditions: a review. Nutr Cycling Agroecosyst 51:123–137
19. Heggenstaller AH, Moore KJ, Liebman M, Anex RP (2009) Nitrogen influences biomass and nutrient partitioning by perennial, warm-season grasses. Agron J 101:1363–1371
20. Hyvonen R, Persson T, Andersson S, Olsson B, Agren GI, Linder S (2008) Impact of long-term nitrogen addition on carbon stocks in trees and soils in northern Europe. Biogeochemistry 89:121–137
21. Jung JY, Lal R (2011) Impacts of nitrogen fertilization on biomass production of switchgrass (*Panicum virgatum L.*) and changes in soil organic carbon in Ohio. Geoderma 166:145–152
22. Jung JY, Lal R, Jastrow JD, Tyler DD (2011) Nitrogenous fertilizer effects on soil structural properties under switchgrass. Agric Ecosyst Environ 141:215–220
23. Kering MK, Butler TJ, Biermacher JT, Guretzky JA (2012) Biomass yield and nutrient removal rates of perennial grasses under nitrogen fertilization. Bioenerg Res 5:61–70
24. Khan SA, Mulvaney RL, Ellsworth TR, Boast CW (2007) The myth of nitrogen fertilization for soil carbon sequestration. J Environ Qual 36:1821–1832
25. Lemus R, Parrish DJ, Abaye O (2008) Nitrogen-use dynamics in switchgrass grown for biomass. Bioenergy Res 1:153–162
26. Liebig MA, Varvel GE, Doran JW, Wienhold BJ (2002) Crop sequence and nitrogen fertilization effects on soil properties in the western Corn Belt. Soil Sci Soc Am J 66:596–601
27. Lu M, Zhou XH, Luo YQ, Yang YH, Fang CM, Chen JK, Li B (2011) Minor stimulation of soil carbon storage by nitrogen addition: a meta-analysis. Agr Ecosyst Environ 140:234–244
28. Ma Z, Wood CW, Bransby DI (2000) Impacts of soil management on root characteristics of switchgrass. Biomass Bioenergy 18:105–112
29. Ma Z, Wood CA, Bransby DI (2000) Soil management impacts on soil carbon sequestration by switchgrass. Biomass Bioenergy 18:469–477
30. Mack MC, Schuur EAG, Bret-Harte MS, Shaver GR, Chapin FS (2004) Ecosystem carbon storage in arctic tundra reduced by long-term nutrient fertilization. Nature 431:440–443
31. McLaughlin SB (1992) New Switchgrass Biofuels Research Program for the Southeast, pp. 111-115. In: Proceedings of the Annual Automotive Technology Development Contractors' Coordination Meeting. SAE International, Warrendale
32. Mommer L, van Ruijven J, de Caluwe H, SmitTiekstra AE, Wagemaker CAM, Ouborg NJ, Bogemann GM, van der Weerden GM, Berendse F, de Kroon H (2010) Unveiling below-ground species abundance in a biodiversity experiment: a test of vertical niche differentiation among grassland species. J Eco 98:1117–1127
33. Neff JC, Townsend AR, Gleixner G, Lehman SJ, Turnbull J, Bowman WD (2002) Variable effects of nitrogen additions on the stability and turnover of soil carbon. Nature 419:915–917
34. Nelson DW, Sommers LE (1996) Total carbon, organic carbon and organic matter: laboratory methods. In: Sparks DL et al (eds) Methods of soil analysis. Part3. SSSA Book Ser 5. SSSA, Madison, pp 961–1010
35. Nimmo JR, Perkins KS (2002) Aggregate stability and size distribution. In: Dane JH, Topp GC (eds) Methods of soil analysis. Part 4. SSSA Book Ser. 5. SSSA, Madison, pp 317–327
36. Newton GL, Bernard JK, Hubbard RK, Allison JR, Lowrance RR, Gascho GJ, Gates RN, Vellidis G (2003) Managing manure nutrients through multi-crop forage production. J Dairy Sci 86:2243–2252
37. Perrin R, Vogel K, Schmer M, Mitchell R (2008) Farm-scale production cost of switchgrass for biomass. Bioenerg Res 1:91–97
38. Peterson RG, Colvin LD (2002) Sampling. In: Dane JH, Topp GC (eds) Methods of soil analysis. Part 4. Agron. Monogr. 5. SSSA, Madison
39. Pregitzer KS, Burton AJ, Zak DR, Talhelm AF (2008) Simulated chronic nitrogen deposition increases carbon storage in northern temperate forests. Glob Chang Biol 14:142–153
40. Ravenek JM, Bessler H, Engels C, Scherer-Lorenzen M, Gessler A, Gockele A, De Luca E, Temperton VM, Ebeling A, Roscher C (2014) Long-term study of root biomass in a biodiversity experiment reveals shifts in diversity effects over time. Oikos 123:1528–1536
41. Reich PB, Knops J, Tilman D, Craine J, Ellsworth D, Tjoelker M, Lee T, Wedin D, Naeem S, Bahauddin D, Hendrey G, Jose S, Wrage K, Goth J, Bengston W (2001) Plant diversity enhances ecosystem responses to elevated CO_2 and nitrogen deposition. Nature 410:809–812
42. Robinson DL (1996) Fertilization and nutrient utilization in harvested forage systems—southern forage crops. In: Joost RE, Roberts CA (eds) Nutrient cycling in forage systems., pp 65–92
43. SAS Institute (2012) The analyst applications. SAS Inst, Cary
44. Thomas GW (1996) Soil pH and soil acidity. In: Sparks DL (ed) Methods of soil analysis. Part 3. SSSA Book Ser. 5. SSSA, Madison, pp 475–490
45. Tisdall JM, Oades JM (1982) Organic matter and water stable aggregates in soils. J Soil Sci 33:141–163
46. Van Noordwijk M, Brouwer G (1997) Roots as sinks and sources of nutrients and carbon in agricultural systems. In: Brussaard L, Ferrera-Cerrato R (eds) Soil ecology in sustainable agricultural systems. Adv Agroeco CRC-Press, Boca Raton, pp 71–89
47. Vogel KP, Brejda JJ, Walters DT, Buxton DR (2002) Switchgrass biomass production in the Midwest USA: harvest and nitrogen management. Agron J 94:413–420
48. Von Felten S, Schmid B (2008) Complementarity among species in horizontal versus vertical rooting space. J Plant Ecol 1:33–41

Legume-based mixed intercropping systems may lower agricultural born N_2O emissions

Mehmet Senbayram[1,2*], Christian Wenthe[1], Annika Lingner[1], Johannes Isselstein[3], Horst Steinmann[4], Cengiz Kaya[2] and Sarah Köbke[5]

Abstract

Background: The area used for bioenergy crops (annual row crops (e.g., wheat, maize), herbaceous perennial grasses, and short-rotation woody crops (e.g., poplar)) is increasing because the substitution of fossil fuels by bioenergy is promoted as an option to reduce greenhouse gas (GHG) emissions. However, biomass used for bioenergy production is not per se environmentally benign, since bioenergy crop production is associated with negative side effects such as GHG emissions from soil (dominated by N_2O). N_2O emissions vary greatly in space and time; thus, direct comparison of soil N_2O fluxes from various agro-ecosystems is certainly crucial for the assessment of the GHG reduction potential from energy crops.

Methods: Therefore, our study aimed to evaluate the two different agro-ecosystems (cropland and agro-forestry) cultivated in central Germany for their environmental impact. In a 1-year field experiment, we compared N_2O fluxes from cropland (non-fertilized wheat, N-fertilized wheat, non-fertilized faba bean, and wheat mixed intercropping with faba bean) and agro-forestry (non-fertilized poplar, N-fertilized poplar, non-fertilized Robinia, and poplar mixed intercropping with Robinia) as a randomized split-block design.

Results: Rainfall at the field site was slightly over average during the period from 1 April to 1 July in 2014 (201 mm rain) and considerably below average during the same period in 2015 (100 mm rain). Cumulative mean N_2O fluxes were up to five fold higher in agro-forestry than in arable crop treatments during 2014 growing period. We hypothesized that the difference in N_2O emissions when comparing arable land and agro-forestry was mainly due to the limited water and nutrient uptake of plantations during the first year. Among the arable crops (wheat, N-fertilized wheat, wheat mixed intercropped with bean, and bean), seasonal and annual N_2O emissions were highest in soils when faba bean was grown as a mono-crop. On the other hand, cumulative mean N_2O fluxes were 31 % lower ($p < 0.05$) when faba bean mixed with wheat than in soils planted with N-fertilized wheat.

Conclusions: The latter clearly suggests that using legume crops as intercrop or mixed crop in wheat may significantly mitigate fertilizer-derived N_2O fluxes and may be an effective proxy for increasing GHG emission savings for energy crops.

* Correspondence: mehmetsenbayram6@yahoo.co.uk
[1]Institute of Applied Plant Nutrition, University of Gottingen, Carl-Sprengel-Weg 1, Gottingen 37075, Germany
[2]Present Address: Department of Plant Nutrition and Soil Science, University of Harran, SanliUrfa TR-63000, Turkey
Full list of author information is available at the end of the article

Background

The observed increase in global average temperatures over the last decades is very likely due to the observed increase in anthropogenic greenhouse gas concentrations in the atmosphere. Nitrous oxide (N_2O) is a potent greenhouse gas as it absorbs long-wave radiation and contributes to the reduction of the ozone layer in the stratosphere [1]. Data from the ice-core analysis show that for thousands of years, mean atmospheric N_2O concentrations were close to 270 ppbv; however, the latter increased about 20 % in recent years [2, 3]. The Intergovernmental Panel on Climate Change (IPCC) Third Assessment Report identified microbial production of N_2O in expanding and fertilized agricultural lands as a primary driver of this increase [4]. However, large uncertainties remain on the estimates of N_2O fluxes from the biosphere [2, 5] due to complex interactions between the related processes and controls on production, consumption, and transport through the soil and on the release into the atmosphere [6, 7].

High-yielding agricultural systems have specifically high nitrogen (N) demand which cannot be supplied by soil N reserves. Therefore, additional N input is needed; however, in most agricultural land-use systems, the application of organic and inorganic nitrogen fertilizers triggers the emissions of anthropogenic N_2O emissions. The present IPCC default factor for direct N_2O emissions arising from the nitrogenous fertilizer application to managed soils is 1 %. However, soil N_2O emissions vary significantly depending on soil type, plant species, climate, crop rotation, tillage method, and fertilizer application rates [4]. At high N_2O-emitting sites, most N_2O release is characterized by short peak emissions (up to 90 % of the annual emissions) connected mainly to (i) precipitation events and change in soil moisture [8–10], (ii) N fertilization [11–13], (ii) freeze-thaw cycle [14], and (iv) soil tillage [15]. Such peak N_2O emissions from soils are highly variable in space and time; thus, measuring and quantifying variance in N_2O emissions are rather difficult, and there are only a few field experiments available allowing long-term comparison of various crops and other factors [16]. It is absolutely crucial to simulate such N_2O peak events and to identify the driving factors in order to be able to develop mitigation options.

The main microbial reactions involved in the production of N_2O are nitrification (oxidation of NH_4^+ to NO_3^-) and denitrification (reduction of NO_3^-, via NO, N_2O, to N_2). There is now growing evidence that denitrification (bacterial or fungal) is the dominating process responsible for N_2O losses from agricultural soils [8, 10, 17]. Therefore, in many agricultural field studies, N_2O emission events are found in periods when high mineral nitrogen concentrations in soils coincide with high soil moisture (increasing the rate of denitrification) [7, 18]. The relevance of soil moisture can also be seen in field studies with similar climatic, soil, and substrate conditions, where N_2O emissions nevertheless show significant inter-annual variability caused by the differences in soil water content [19]. These differences can usually be deduced from the weather records as they are related to the differences in the input of water through rainfall [18]. But the soil-water balance is also strongly influenced by the offtake of water through evapotranspiration. Crop plants play an important role, as they may transpire 500 to 600 mm of water per growth cycle which is 30 to 90 % of the precipitation input in that period. In this respect, crops differ substantially, not only in the total amount of water that is transpired but also in the growth and transpiration pattern in the course of the year [19, 20].

Biofuels are often called "CO_2 neutral" in the sense that CO_2 which is emitted in the course of their combustion has previously been fixed from the atmosphere via photosynthesis during plant growth. Many industrialized countries have established ambitious policy targets and often offer financial incentives to stimulate the production or use of bioenergy. The main reasons for the promotion of biofuel production are that it is made from renewable resources (organic manures, plant materials, food waste), that it is expected to have no or even positive effect on the atmospheric greenhouse gas balance, and that it may reduce the dependency on fossil fuel [21]. However, biomass used for bioenergy generation is not per se environmentally benign, since its production is inevitably associated with negative side effects such as GHG emissions or N leaching. Soil N_2O emissions are likely to be the dominating greenhouse gas emissions associated with bioenergy crop production [22]. It has been reported that the production and use of biofuels compared with the use of conventional fossil fuels may lead to a reduction or even increase in the total greenhouse gas emissions (72 to 107 %), depending on the type of bioenergy crop used and combustion technology chosen [23]. Here, the authors showed that N_2O would typically make up 10 to 80 % of the total greenhouse gas emissions in the biofuel production chain.

Intercropping, defined as any system of multiple cropping within the same space can be used as an alternate bioenergy cropping system [24]. The intercropping of cereals with legumes is particularly common, and introducing N_2-fixing legumes into cereal-based crop rotations may reduce synthetic mineral N-fertilizer use and thought to mitigate N_2O fluxes. A reduction of N_2O in tree-based intercropping systems has been reported [25]. In contrast, in a review study, Rochette and Janzen [26] concluded that legumes can produce substantial N_2O emissions. The cultivation of N_2-fixing legume species (e.g., faba bean (*Vicia faba L.*) as arable crop or Robinia

(*Robinia pseudoacacia*) as woody plant) could stimulate N_2O emissions simply by increasing N input to soils, thus providing additional substrate for nitrification and denitrification [27]. Authors reported that faba bean could release 13 % of their fixed N as rhizodeposition [28]. Furthermore, denitrification by N-fixing bacteria can be another source of N_2O in legume domains [29].

As discussed above, N_2O emissions vary greatly in space and time; thus, the direct comparison of soil N_2O fluxes from various agro-ecosystems is certainly necessary and only very few studies are available on this subject. Therefore, our study aimed to evaluate two different agro-ecosystems (cropland and agro-forestry) cultivated in central Germany for their environmental impact. In a 1-year field experiment, we compared N_2O fluxes from cropland (non-fertilized wheat, N-fertilized wheat, non-fertilized faba bean, and wheat mixed intercropping with faba bean) and agro-forestry (non-fertilized poplar, N-fertilized poplar, non-fertilized Robinia, and poplar mixed intercropping with Robinia) as a randomized split-block design. Therefore, the objectives of the present study were (i) to obtain year-round N_2O emission data in various agro-ecosystems (cropland and agro-forestry) in central Germany and (ii) to provide more knowledge on how a legume-based crop production system may affect N_2O budget.

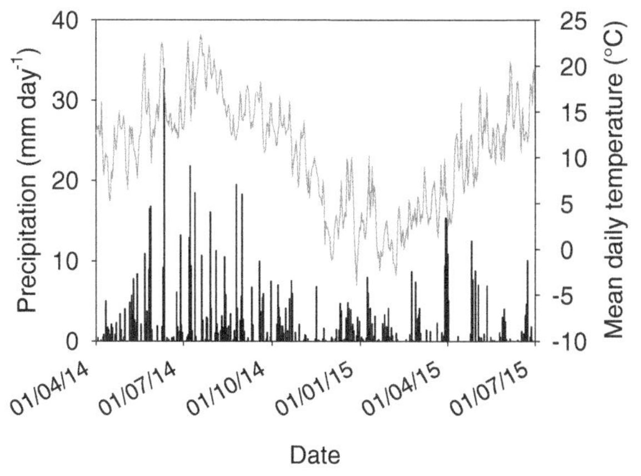

Fig. 1 Mean daily air temperature [°C] at canopy level (*gray lines*) and daily precipitation [mm day^{-1}] (*black bars*) at experimental site Reinshof, Göttingen, in central Germany during the investigation period (April 2014–July 2015)

Methods

Field experiment

A field experiment was conducted at the research farm Reinshof (51.49° N, 9.93° E, 150 m asl) of the Georg-August-University Göttingen, Germany. The continental climate leads to an average precipitation of 651 mm and a mean temperature of 9.2 °C. The rainfall and temperature at field site during the investigation period can be found in Fig. 1. Here, annual mean temperature and cumulative rainfall were 10.1 °C and 677 mm in 2014, respectively. The soil was classified as Haplic Luvisol according to the FAO classification system. At 0–30 cm soil section, the soil contained 15 % clay, 73 % silt, and 12 % sand with a pH of 6.7, and 0.1 % total N and 1.0 % total organic carbon content [30].

The 1-year field experiment (part of a large field trial) was set up in April 2014 with the aim of comparing N_2O fluxes from the cropland (non-fertilized wheat (WT), N-fertilized wheat (NWT), non-fertilized faba bean (FB), wheat mixed intercropping with faba bean (WFB)), agro-forestry (non-fertilized poplar (PL), N-fertilized poplar (NPL), non-fertilized Robinia (RB), and poplar mixed intercropping with Robinia (PRB)) as a randomized split-block design. In the first year of the study, spring wheat (*Triticum aestivum L.)* cultivar Tybald and faba bean (*V. faba L.)* cultivar Fuego were sown at the same time at the optimum planting date for spring wheat (see Table 1 for details). Agro-forestry treatments consisted of poplar (*Populus* "Hybride 275") (PL); poplar mixed with Robinia (*R. pseudoacacia* "HKG 81901") (PLR); Robinia (RB); and N-fertilized poplar (PLN). In the agro-forestry treatments, all trees were planted with 1 tree m^{-2} on 30 April 2014. Plot size was 5 × 5 m for forest and 3 × 9 m for arable treatments with three replications for each treatment. After the harvest of crop land in 2014, winter wheat (*T. aestivum* "genus"), wheat intercropped with winter bean (*V. faba* "S_004"), mono-winter bean, and N-fertilized wheat were seeded on 28 October 2014. Seeding density in the mono-cropped winter wheat plots was 320 plants m^{-2}, inter-cropped plots were alternately seeded with wheat density of 160 plants m^{-2} and a bean density of 20 plants m^{-2}, and mono-bean plots had a density of 40 plants m^{-2}. Granular N fertilizer in the form of calcium-ammonium-nitrate was applied at a rate of 80 kg N ha^{-1} as a single dressing to the soil surface on 15 May 2014 and 25 March 2015 in the respective treatments.

Soil mineral N

For the analysis of soil mineral N, soil from 0–15 cm depth was sampled extracted with a 0.0125 M $CaCl_2$ solution (1:5 *w*/*v*) and shaked for 1 h. The extracts were then filtered with Whatman 602 filter paper and stored at −20 °C until analysis. The extracts were analyzed colorimetrically for the concentrations of NO_3^- and NH_4^+ using the San++ continuous flow analyzer (Skalar Analytical B.V., Breda, The Netherlands).

Trace gas flux measurement

After the N application, gas samples were taken daily for a period of 1 week, followed by intervals of 2–3 days

Table 1 Sowing dates and fertilizer application rate of each treatment

Abbreviation	Crop	Fertilizer (kg CAN–N ha^{-1})	Stand	Date of sowing spring and winter crops
WT	Spring wheat/winter wheat	None	Mono	25 March 2014/28 October 2014
NWT	Spring wheat + N/winter wheat + N	80	Mono	25 March 2014/28 October 2014
WFB	Mixed intercropping (wheat and faba bean)[a]	None	Mixed intercrop	25 March 2014/28 October 2014
FB	Field beans[a]	None	Mono	25 March 2014/28 October 2014
PL	Poplar	None	Mono	30 April 2014
NPL	Poplar	80	Mono	30 April 2014
PRB	Mixed planting Poplar/Robinia	None	Mixed planting	30 April 2014
RB	Robinia	None	Mono	30 April 2014

[a]Information was given for both spring and winter crops

until the end of the vegetation period and once per week during the winter period using the closed chamber method described by Hutchinson and Mosier [31]. On each of 24 plots, basal rings made of polyvinyl chloride (PVC) (height 10 cm, diameter 60 cm) were pressed 5 cm into the soil. For the measurements, PVC chambers with an inner diameter of 60 cm and a height of 30 cm were put onto the basal rings, and tightened by a butyl rubber band all around the junction. The chambers were closed for 40 min within the period of 10.00–14.00 h at each sampling day. Gas samples were taken from the chamber atmosphere at 0–20–40 min after closing the chamber using pre-evacuated 12-ml glass vials (Labco, Crewe, UK). The top of the PVC chambers were covered with white polystyrene to reflect solar radiation and have temperature stability within the chamber.

Concentrations of N_2O were analyzed by a Bruker gas chromatography system (456-GC, Bruker, Billerica, USA) by deploying an electron capture detector (ECD) for N_2O. Operating conditions for the GC were as follows: injector temperature 95 °C, column temperature 85 °C, and detector temperature 320 °C. Samples were introduced using a Gilson auto-sampler (GX-281) (Gilson Inc., Middleton, WI, USA). Data processing was performed using the CompassCDS (vers. 3.0) software.

Statistics

Daily N_2O flux rates for dates between sampling dates were calculated using linear interpolation, and annual cumulative N_2O emissions were calculated as the sum of all daily flux rates for the vegetation period of each crop and the entire investigation periods during March 2014–August 2015. We used general linear model and Tukey's test for pairwise comparison of cumulative N_2O flux between the treatments within each year. Statistical analyses were done using SPSS version 13.0.

Results and discussion

Rainfall and crop production

Rainfall at the site Reinshof was slightly over average during the period from 1 April to 1 July in 2014 (201 mm) and considerably below average during the period from 1 April to 1 July in 2015 (100 mm). Therefore, the 2014 spring period was much wetter than spring 2015. Overall, 2014–2015 winter period was quite mild with no significant freeze-thaw event.

Soil mineral N

The time course of the soil mineral-N concentrations are shown in Fig. 2. In soils planted with arable crops, soil NH_4^+ concentrations during the investigation period remained rather low (below 10 kg N ha^{-1} at 30 cm layer) in non-fertilized treatments. The application of mineral-N fertilizer in NWT treatment caused a slight increase in topsoil NH_4^+ concentrations for a short time period which decreased rapidly to the background concentrations within a week. The low concentration of soil NH_4^+ even in N-fertilized treatments (in the form of calcium-ammonium-nitrate) can be attributed to a rapid nitrification as soil texture and pH (6.7) serve ideal conditions for nitrification. In all treatments, there was a significant increase in topsoil NH_4^+ concentrations in spring 2015 regardless from the N application. Early spring period in 2015 was reasonably dry compared to the same period in 2014. Therefore, the latter can be attributed to the processes related to the soil wetting after a long dry period (such as in spring 2015) which can accelerate N release from the mineralization of soil organic matter immediately after rewetting [32].

Overall treatments, soil NO_3^- concentrations in 0–15 cm soil segment varied between 10 and 85 kg NO_3^-–N ha^{-1}. Here, NO_3^- was generally the dominant soil N form and highly variable when sampled soon after additions of fertilizer N. In all soils, concentrations of NO_3^- in the 0–15-cm layer decreased over time with the largest

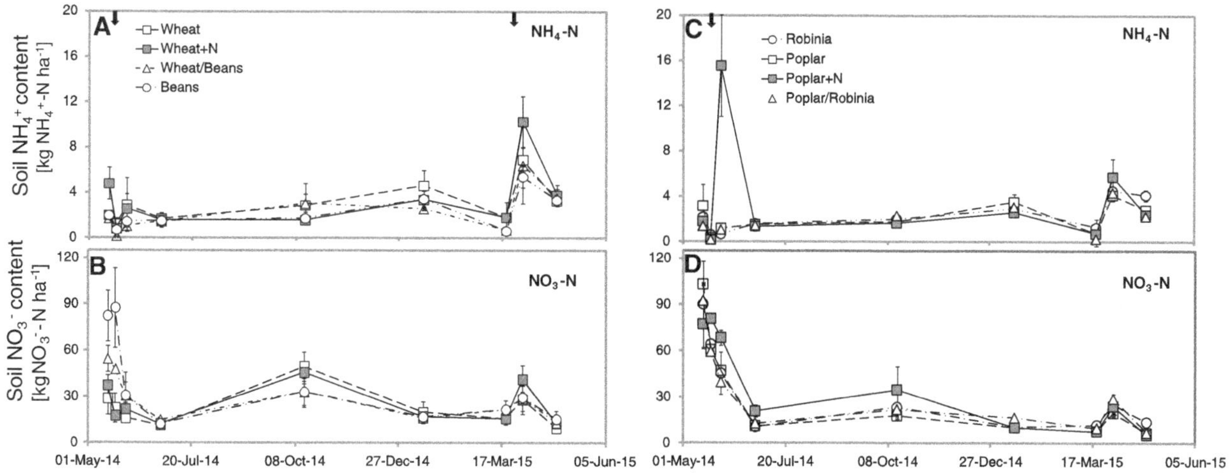

Fig. 2 Top soil (0–15 cm soil section) NH_4^+ (**a**, **c**) and NO_3^- (**b**, **d**) content of the non-fertilized mono-crops (wheat (WT), faba bean (FB), poplar (PL), and Robinia (RB)), N-fertilized mono-crops (80 kg N ha^{-1} in the form of calcium-ammonium-nitrate; wheat (NWT) and poplar (NPL)) and non-fertilized mixed crops (wheat with faba bean (WFB) and poplar with Robinia (PRB)). *Vertical arrows* above the charts indicate application dates of mineral fertilizers. *Error bars* show the standard error of the mean of each treatment ($n = 3$)

decrease found in the arable crops specifically in WT and NWT treatments. During the vegetation period in 2014, soil NO_3^- content was generally higher in agro-forestry soils than in arable soils. Plant nutrient and water uptake was expected to be higher in cropland compared to the young agro-forestry treatments in 2014 due to small size and low growth rate of young trees. Thus, more rapid depletion of soil mineral N in arable crops than agro-forestry treatments can mainly be attributed to the differences in plant N uptake.

In arable land stand, soil NO_3^- concentrations were clearly higher (significant in 2014, $p < 0.01$) in FB than in other non-fertilized treatments during the vegetation period (Fig. 2a, b). For legume crops, inputs of biologically fixed N largely supplement to the uptake of soil mineral N to meet crop N demand. Thus, the legume species also take up mineral N from soils for growth before fixing additional N. The preferential use of soil mineral N helps explain why there is also significant depletion of soil NO_3^- in FB treatment [33]. In a review study, authors reported that average 41 % (for chickpea), 65 % (for faba bean), and 66 % (for field peas) of N that were present in legumes were derived from soil N [33]. However, slightly higher soil NO_3^- concentration in FB treatments than non-legume soils suggests that there were still reasonably more NO_3^- available for potential denitrification losses during the legume-growing season.

Seasonal N_2O emissions

In both years, flux data indicate that N_2O emissions were dominated by specific event periods (Fig. 3). Overall, maximum daily emissions of N_2O in the early summer period in 2014 were 0.16 ± 0.07 and 0.04 ± 0.01 kg N_2O–N ha^{-1} day^{-1} in agro-forestry and arable land treatments, respectively. In a 2-year field study, Lebender et al. [12] observed similar flux rates over a nearby site with similar soil conditions and agricultural practices (wheat and spring barley). Maximum N_2O emissions measured in agro-forestry treatments (0.16 ± 0.07 kg N_2O–N ha^{-1} day^{-1}) have been usually observed in young agro-forest ecosystems [7, 27]. Almost all significant N_2O fluxes occurred as daily peak N_2O emissions and were measured only during the early summer period in 2014. The importance of these peak emissions in early summer period on the annual budget of N_2O emissions highlights the necessity of continuous flux monitoring to accurately determine the N loss from agro-ecosystems specifically in spring and early summer seasons [7, 20].

In 2014, N_2O emissions gradually decreased to the background levels (below 10 g N_2O–N ha^{-1} day^{-1}) from the months of May to July and remained in background levels until March 2015 (Fig. 3). Interestingly, the latter was less than 0.01 kg N_2O–N ha^{-1} day^{-1} during the early summer period in 2015 for all treatments (including N-fertilized treatments). As seen in Fig. 1, early summer period in 2014 was relatively wet (April–June, 201 mm rainfall) compared to the same period in 2015 (April–June, only 100 mm rainfall). Therefore, we may attribute higher daily N_2O fluxes in 2014 than in 2015 (in early summer period) to the differences in mineral N and moisture content of the soil. Mineral N content of all soils in June 2014 was almost similar as compared to the same period in 2015, whereas N_2O fluxes were still about 10-fold higher in June 2014 than in June 2015. In this context, we may conclude that soil moisture seems to be the major

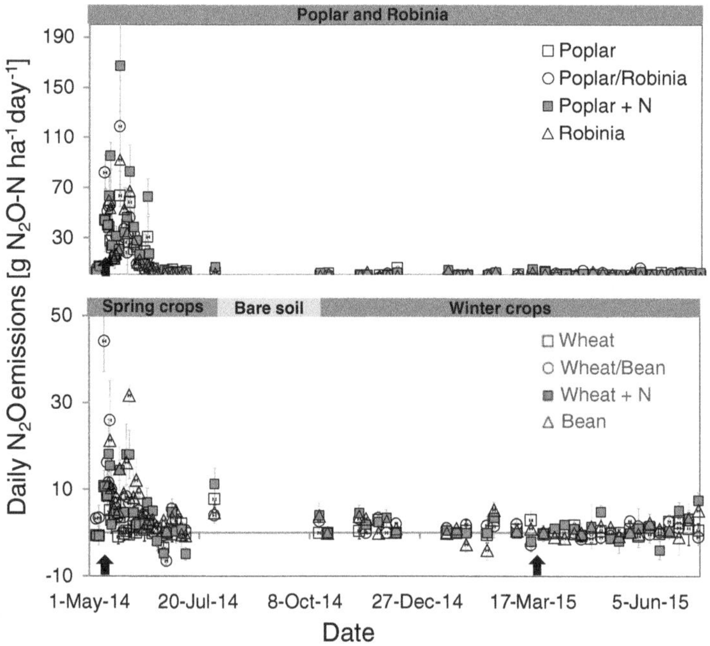

Fig. 3 Daily N_2O fluxes of the non-fertilized mono-crops (wheat (WT), faba bean (FB), poplar (PL), and Robinia (RB)), N-fertilized mono-crops (80 kg N ha^{-1} in the form of calcium-ammonium-nitrate; wheat (NWT) and poplar (NPL)), and non-fertilized mixed crops (wheat with faba bean (WFB) and poplar with Robinia (PRB)). *Vertical arrows* under the charts indicate application dates of mineral fertilizers. *Horizontal bars* above the charts indicate growing periods of crops. *Error bars* show the standard error of the mean of each treatment ($n = 3$)

driving factor of higher N_2O emissions in the 2014 summer period than in 2015.

All the abovementioned factors (e.g., high moisture, high soil temperature in early summer, and moderate or high NO_3^- content of the soil) are known to trigger specifically the denitrification rate in soils. Thus, we may speculate that denitrification (fungal or bacterial) was the potential key source of measured large N_2O fluxes in the 2014 early summer period. Our earlier report supports this hypothesis in which sandy loam soil was incubated under laboratory conditions, and similar to the field experiment, large N_2O peak events were observed immediately after rewetting of the soil. Here, a stable isotope-labeling study clearly showed that denitrification was the major source (over 90 % of emitted N_2O) of large N_2O peaks that occurred in wet seasons [8, 19]. Furthermore, there is now a growing evidence that fungal denitrification may be the key process producing N_2O in such situations rather than bacterial denitrification [17]. Surely, more research is needed to reveal (i) the dominant processes and (ii) key microbial or fungal strains producing N_2O under these specific conditions (especially during early summer period).

Effect of plant species on N_2O emissions

Mean cumulative N_2O fluxes during the vegetation period in 2014 were 152 ± 58, 217 ± 29, and 441 ± 10 g N_2O–N ha^{-1} in WT, WFB, and FB treatments, respectively. Among the non-fertilized arable crops (wheat, wheat mixed intercropped with bean, and bean), N_2O emission over the 2014 growing seasons is highest in soils when faba bean (FB) was grown (Fig. 4). Introducing N-fixing legumes into cereal-based crop rotations may reduce synthetic mineral-N fertilizer use and thought to mitigate N_2O fluxes. However, the present study clearly showed that when faba bean was grown as a mono-crop, N_2O fluxes were about threefold higher compared to WT treatment. In contrast to the present study, authors reported that growing season N_2O emissions from N_2-fixing legumes are significantly lower than from non-legumes and are often comparable to unfertilized background emissions [26, 34]. In line with the present study, Rochette and Janzen [26] (in a review study) concluded that legumes can produce substantial N_2O emissions. They speculated that the main source of N_2O emissions from soils planted with N_2-fixing legumes during the vegetation period may be attributable to the N release from root exudates and/or from the decomposition of dead root residues. An alternative process that may contribute to the latter would be the N_2O emission during the N_2-fixation process in the nodules where N_2 is fixed. Authors reported that several Rhizobium species in the free-living forms or in legume roots can denitrify NO_3^- and release N_2O from active

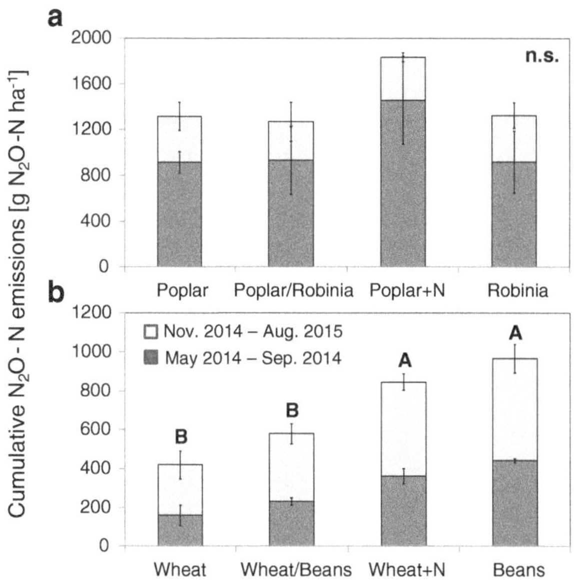

Fig. 4 Cumulative mean N_2O emission rates of the non-fertilized mono-crops (wheat (WT), faba bean (FB), poplar (PL), and Robinia (RB)), N-fertilized mono-crops (80 kg N ha^{-1} in the form of calcium-ammonium-nitrate; wheat (NWT) and poplar (NPL)) and non-fertilized mixed crops (wheat with faba bean (WFB) and poplar with Robinia (PRB)). Upper graphs (**a**) show the agro-forestry treatments, and lower graphs (**b**) show the crop domains during the period from May–September 2014 (vegetation period of spring crops) and November 2014 to August 2015. *Error bars* show the standard error of the mean of each treatment ($n = 3$). Means within each crop followed by the same letter (*A*, *B*) are not significantly different ($p > 0.05$). *n.s.* not significant

nodules most likely to prevent excess NO_3^- that inhibits the activity of N_2-fixing enzymes [35].

In agro-forestry treatments, both daily and cumulative N_2O emissions did not differ among each other in the 2014 growing period. Here, mean seasonal N_2O emissions (during the growing season of arable crops) were 1121 ± 161, 1102 ± 159, and 1052 ± 266 g N_2O–N ha^{-1} in PL, PRB, and RB treatments (no significant difference), respectively. The cumulative mean N_2O fluxes during the 2014 growth period were considerably higher in agro-forestry than in arable crop treatments. During the first year, young plantations in agro-forestry domains generally have limited N and water uptake, while wheat and faba bean as arable crops are at their most productive growth stage specifically during May and June (growth rates are almost at their maximum during this period). Here, soil conditions seem to be more favorable specifically for denitrification in agro-forestry treatments than in soils planted with arable crops that may explain large N_2O emissions [27]. The cumulative mean N_2O emissions in agro-forestry treatments were about fivefold higher than in both WT and WFB treatments, and the latter was still more than twofold higher when compared to FB plots. In line with the present data set, authors reported large N_2O fluxes after conversion of pastures lands [7] or grasslands [36]. Here, authors attributed large N_2O fluxes to the soil disturbance associated with tillage and cultivation that accelerate soil organic matter decomposition and microbial activity (nitrification and denitrification) leading to N_2O emissions. In the present experiment, soil tillage has been done almost at the same time for all treatments. Thus, we may speculate that the difference in N_2O emissions when comparing arable land to agro-forestry was mainly due to differences in water and nutrient uptake of plant species. Water and nutrient demand of young plantations during the first year are generally low which in return may cause more favorable conditions specifically for denitrification and N_2O losses from denitrification. Overall, we can therefore summarize that direct plant effect seems to be one of the key variables that regulates N_2O losses from soils. Zona et al. (2013) concluded that vegetation uptake of NO_3^- together with water ultimately may reduce the anaerobic volume of soils and may lower both denitrification rate and product stoichiometry of denitrification (lower $N_2O/N_2O + N_2$ ratio; meaning reduced N_2O and enhanced N_2 production) in agricultural soils [8, 17].

Effect of mineral N supply on N_2O emissions

Although it was not the main goal of the present experiment to study the effect of mineral-N addition on N_2O fluxes, we added fertilizer N to mono-crop wheat (80 kg N ha^{-1}; calcium-ammonium-nitrate) in parallel plots to be able to compare N_2O fluxes from soils planted with N_2-fixing plants (faba bean mono-culture or faba bean intercropped with wheat) with fertilized and non-fertilized wheat soils. N fertilization during the first year of new agro-forestry plantations is also not a common practice. However, N doses similar to the arable treatments were applied at the same date in order to be able to gain better scientific knowledge about the dominant factors regulating N_2O fluxes in agro forest ecosystems. Expectedly, in all N-fertilizer treatments, N_2O fluxes increased immediately after fertilizer application, however, only in 2014 (wet early summer) but no response observed in 2015 (dry early summer). The latter clearly suggests that environmental factors specifically soil moisture was the most dominant factor in 2014 that leads to relatively high N_2O fluxes and without sufficient moisture or rainfall, fertilizer application alone does not affect N_2O fluxes significantly, e.g., in 2015. In line with the present study, authors also reported that the application of organic or inorganic fertilizers affects N_2O fluxes only in wet seasons but not in dry years [10, 17, 19].

Overall, the application of nitrogenous fertilizer affected N_2O fluxes predominantly in cropland soils and had limited impact in agro-forestry soils. Here, cumulative N_2O fluxes (from May to September 2014) were 46 and 121 % higher in fertilized than in non-fertilized agro-forestry (non-fertilized poplar vs. fertilized poplar) or in cropland treatments (non-fertilized wheat vs. fertilized wheat), respectively. The latter clearly suggests that N_2O fluxes were more dependent on soil mineral-N content in arable crops than in agro-forestry most likely due to greater competition between plants and N_2O-producing soil microorganisms in cropland than in agro-forestry soils. Fertilizer-derived cumulative N_2O emissions (emission factor) during the period from April to December 2014 were 0.21 and 0.45 % of applied N in wheat and poplar soils, respectively. Measured emission factors were significantly lower than what IPCC predicts (1 %; [4]). However, the latter was similar to what we reported in our previous study for central and northern Germany ([12, 19]). For operational reasons in the present IPCC protocol, the N_2O emission factor was set to 1 % for all fertilizer N regardless of crop or soil type. Large variations in N_2O emissions from different agricultural systems due to differences in management, climate, and soil type are very well known. Low N_2O emission factors in the present experiment and in our previous reports suggest that in the future, different emission factors should be considered at least for different crops or regions that account for their different risks of N_2O emissions.

Cumulative mean N_2O emission during the growing season was still 31 % higher ($p < 0.05$) in FB than in NWT treatment. In a review study, Rochette and Jansen [26] summarized that legumes can increase N_2O emissions during growth compared to evenly fertilized arable crops most likely due to the N release from the root exudates and decomposition of crop residues. Our study clearly agrees with Rochette and Jansen [26] and many others (e.g., [7, 27]) that growing legumes as mono-crop can increase N_2O fluxes compared to N-fertilized arable crops. On the other hand, seasonal N_2O fluxes were 35 % lower in WFB (wheat mixed intercropped with faba bean) than in NWT (wheat as mono-crop) treatment. The latter suggests that using legume crops as intercrop or mixed crop in wheat may significantly mitigate fertilizer-derived N_2O fluxes. However, surely more research is needed to upscale current findings due to the complexity and variability of N_2O fluxes in complex agricultural systems, e.g., mixed cropping systems.

Conclusions

The N_2O emission from soils is variable in space and time, thus measuring and quantifying variance in N_2O emissions is rather difficult, and there are only few field experiments available allowing long-term comparison of various plant species (crops, legumes, and agro-forests) simultaneously with fertilizer effects. We see three take home messages:

- Currently, biogas production from energy crops is mainly based on anaerobic fermentation of mono-crops; however, high-yielding mono-crops require high N-fertilizer input that increases the risk of N_2O losses. Present study clearly showed that mixed intercropping agricultural systems (legume and non-legume plant species) may significantly lower (about 35 %) N_2O losses compared to the N-fertilized mono-crops.
- Cumulative N_2O emissions in agro-forestry soils were about 2–5-fold higher than in cropland soils. Soil conditions in agro-forestry treatments seem to be more favorable specifically for denitrification (due to limited water and nutrient uptake of young plantations during their initial growth stage) than soils planted with arable crops that may be responsible for large N_2O emissions.
- Cumulative mean N_2O emissions during the growth period of annual crops were 31 % higher ($p < 0.05$) in soils planted with faba bean than in N-fertilized wheat. We can conclude that legumes (when grown alone) can produce substantial N_2O emissions most likely due to enhanced denitrification activity in their rhizosphere due to N-rich root exudates/dead organic matter.

Competing interests

The authors declare that they have no competing interests.

Authors' contributions

MS contributed to the acquisition of funding, supervised the field work, perform the statistics, and drafted the paper (50 %). CW as a MSc student run the experiment (70 %; gas sampling, soil sampling, and analysis), collected and analyzed the data, and drafted the paper (10 %). AL as a PhD student contributed to the field experiment (crop management, soil sampling, and data analysis), and drafted the paper (5 %). JI contributed to the acquisition of funding and design of the experiment and drafted the paper (5 %). HS contributed to the acquisition of funding, participated to the design of the experiment, and drafted the paper (5 %). CK drafted the paper and contributed to the planning of the experiment. SK as a PhD student contributed to the field experiment (crop management, soil sampling, and analysis), supervised the MSc student, and drafted the paper (25 %). All authors read and approved the final manuscript.

Acknowledgements

This study was funded by the Federal Ministry of Education and Research (FKZ 031A351A). IMPAC[3] is a project of the Centre of Biodiversity and sustainable Land Use at the University of Gottingen.

Author details

[1]Institute of Applied Plant Nutrition, University of Gottingen, Carl-Sprengel-Weg 1, Gottingen 37075, Germany. [2]Present Address: Department of Plant Nutrition and Soil Science, University of Harran, SanliUrfa TR-63000, Turkey. [3]Department Crop Science, Grassland Science, University of Gottingen, Gottingen 37075, Germany. [4]Department Biodiversity and Sustainable Land Use, University of Gottingen, Gottingen 37077, Germany. [5]Department Crop Science, Plant Nutrition and Soil Science, University of Gottingen, Gottingen 37075, Germany.

References

1. Ravishankara AR, Daniel JS, Portmann RW (2009) Nitrous oxide (N2O): the dominant ozone-depleting substance emitted in the 21st Century. Science 326:123–125. doi:10.1126/science.1176985
2. Smith KA, Mosier AR, Crutzen PJ, Winiwarter W (2012) The role of N2O derived from crop-based biofuels, and from agriculture in general, in Earth's climate. Philos Trans R Soc Lond B Biol Sci 367:1169–1174. doi:10.1098/rstb.2011.0313
3. Fluckiger J, Monnin E, Stauffer B, Schwander J, Stocker TF. (2002) High-resolution Holocene N2O ice core record and its relationship with CH4 and CO2. Glob Biogeochem Cycles 16:1010. doi:10.1029/2001GB001417
4. IPCC Fourth Assessment Report: Climate Change 2007 (AR4). https://www.ipcc.ch/publications_and_data/ar4/syr/en/contents.html. Accessed 5 Aug 2015
5. Kroeze C, Mosier A, Bouwman L (1999) Closing the global N2O budget: a retrospective analysis 1500–1994. Glob Biogeochem Cycles 13:1–8. doi:10.1029/1998GB900020
6. Davidson E (1992) Sources of nitric-oxide and nitrous-oxide following wetting of dry soil. Soil Sci Soc Am J 56:95–102
7. Zona D, Janssens IA, Gioli B, Jungkunst HF, Serrano MC, Ceulemans R. (2013) N2O fluxes of a bio-energy poplar plantation during a two years rotation period. GCB Bioenergy 5:536–547. doi:10.1111/gcbb.12019
8. Senbayram M, Chen R, Mühling KH, Dittert K (2009) Contribution of nitrification and denitrification to nitrous oxide emissions from soils after application of biogas waste and other fertilizers. Rapid Commun Mass Spectrom RCM 23:2489–2498. doi:10.1002/rcm.4067
9. Bergstermann A, Cardenas L, Bol R, Gilliam L, Goulding K, Meijide A, et al. (2011) Effect of antecedent soil moisture conditions on emissions and isotopologue distribution of N2O during denitrification. Soil Biol Biochem 43:240–250. doi:10.1016/j.soilbio.2010.10.003
10. Dobbie KE, Smith KA (2001) The effects of temperature, water-filled pore space and land use on N2O emissions from an imperfectly drained gleysol. Eur J Soil Sci 52:667–673. doi:10.1046/j.1365-2389.2001.00395.x
11. Bouwman AF, Boumans LJM, Batjes NH (2002) Emissions of N2O and NO from fertilized fields: summary of available measurement data. Glob Biogeochem Cycles 16:1058. doi:10.1029/2001GB001811
12. Lebender U, Senbayram M, Lammel J, Kuhlmann H (2014) Impact of mineral N fertilizer application rates on N2O emissions from arable soils under winter wheat. Nutr Cycl Agroecosystems 100:111–120. doi:10.1007/s10705-014-9630-0
13. Weier K, Doran J, Power J, Walters D (1993) Denitrification and the dinitrogen nitrous-oxide ratio as affected by soil-water, available carbon, and nitrate. Soil Sci Soc Am J 57:66–72
14. Morkved PT, Dorsch P, Henriksen TM, Bakken LR (2006) N2O emissions and product ratios of nitrification and denitrification as affected by freezing and thawing. Soil Biol Biochem 38:3411–3420. doi:10.1016/j.soilbio.2006.05.015
15. Metay A, Chapuis-Lardy L, Findeling A, Oliver R, Alves Moreira JA, Feller C. (2011) Simulating N2O fluxes from a Brazilian cropped soil with contrasted tillage practices. Agric Ecosyst Environ 140:255–263. doi:10.1016/j.agee.2010.12.012
16. Saggar S (2010) Estimation of nitrous oxide emission from ecosystems and its mitigation technologies preface. Agric Ecosyst Environ 136:189–191. doi:10.1016/j.agee.2010.01.007
17. Koester JR, Well R, Dittert K, Giesemann A, Lewicka-Szczebak D, Mühling KH, et al. (2013) Soil denitrification potential and its influence on N2O reduction and N2O isotopomer ratios. Rapid Commun Mass Spectrom 27:2363–2373. doi:10.1002/rcm.6699
18. Jungkunst HF, Fiedler S, Stahr K (2004) N2O emissions of a mature Norway spruce (Picea abies) stand in the Black Forest (southwest Germany) as differentiated by the soil pattern. J Geophys Res Atmos 109:D07302. doi:10.1029/2003JD004344
19. Senbayram M, Chen R, Wienforth B, Herrmann A, Kage H, Mühling KH, et al. (2014) Emission of N2O from biogas crop production systems in Northern Germany. Bio Energy Res 7:1223–1236. doi:10.1007/s12155-014-9456-2
20. Drury CF, Reynolds WD, Tan CS, McLaughlin NB, Yang XM, Calder W, et al. (2014) Impacts of 49–51 years of fertilization and crop rotation on growing season nitrous oxide emissions, nitrogen uptake and corn yields. Can J Soil Sci 94:421–433. doi:10.4141/CJSS2013-101
21. Herrmann A (2013) Biogas production from maize: current state, challenges and prospects. 2. Agronomic and environmental aspects. Bio Energy Res 6:372–387. doi:10.1007/s12155-012-9227-x
22. Cherubini F, Jungmeier G (2010) LCA of a biorefinery concept producing bioethanol, bioenergy, and chemicals from switchgrass. Int J Life Cycle Assess 15:53–66. doi:10.1007/s11367-009-0124-2
23. Smeets EMW, Bouwmanw LF, Stehfest E, Van Vuuren DP, Posthuma A. (2009) Contribution of N2O to the greenhouse gas balance of first-generation biofuels. Glob Chang Biol 15:1–23. doi:10.1111/j.1365-2486.2008.01704.x
24. Whitmore AP, Schroder JJ (2007) Intercropping reduces nitrate leaching from under field crops without loss of yield: a modelling study. Eur J Agron 27:81–88. doi:10.1016/j.eja.2007.02.004
25. Beaudette C, Bradley RL, Whalen JK, McVetty PBE, Vessey K, Smith DL. (2010) Tree-based intercropping does not compromise canola (Brassica napus L.) seed oil yield and reduces soil nitrous oxide emissions. Agric Ecosyst Environ 139:33–39. doi:10.1016/j.agee.2010.06.014
26. Rochette P, Janzen HH (2005) Towards a revised coefficient for estimating N2O emissions from legumes. Nutr Cycl Agroecosystems 73:171–179. doi:10.1007/s10705-005-0357-9
27. Liu XP, Zhang WJ, Hu CS, Tang XG (2014) Soil greenhouse gas fluxes from different tree species on Taihang Mountain, North China. Biogeosciences 11:1649–1666. doi:10.5194/bg-11-1649-2014
28. Mayer J, Buegger F, Jensen ES, Schloter M, Heß J. (2003) Estimating N rhizodeposition of grain legumes using a N-15 in situ stem labelling method. Soil Biol Biochem 35:21–28. doi:10.1016/S0038-0717(02)00212-2
29. Breitenbeck G, Bremner J (1989) Ability of free-living cells of bradyrhizobium-japonicum to denitrify in soils. Biol Fertil Soils 7:219–224. doi:10.1007/BF00709652
30. Müller E, Rottmann N, Bergstermann A, Wildhagen, H, Joergensen RG. (2011) Soil CO2 evolution rates in the field—a comparison of three methods. Arch Agron Soil Sci 57:597–608. doi:10.1080/03650340.2010.485984
31. Hutchinson GL, Mosier AR (1981) Improved soil cover method for field measurement of nitrous oxide fluxes1. Soil Sci Soc Am J 45:311. doi:10.2136/sssaj1981.03615995004500020017x
32. Xiang S-R, Doyle A, Holden PA, Schimel JP (2008) Drying and rewetting effects on C and N mineralization and microbial activity in surface and subsurface California grassland soils. Soil Biol Biochem 40:2281–2289. doi:10.1016/j.soilbio.2008.05.004
33. Unkovich M (2012) Nitrogen fixation in Australian dairy systems: review and prospect. Crop Pasture Sci 63:787–804. doi:10.1071/CP12180
34. Jeuffroy MH, Baranger E, Carrouee B, de Chezelles E, Gosme M, Hénault C, et al. (2013) Nitrous oxide emissions from crop rotations including wheat, oilseed rape and dry peas. Biogeosciences 10:1787–1797. doi:10.5194/bg-10-1787-2013
35. Rosen A, Lindgren PE, Ljunggren H (1996) Denitrification by Rhizobium meliloti.1. Studies of free-living cells and nodulated plants. Swed J Agric Res 26:105–113
36. Gelfand I, Zenone T, Jasrotia P, Chen J, Hamilton SK, Robertson GP. (2011) Carbon debt of conservation reserve program (CRP) grasslands converted to bioenergy production. Proc Natl Acad Sci U S A 108:13864–13869. doi:10.1073/pnas.1017277108

16

Local power and land use: spatial implications for local energy development

Cheryl de Boer[1*], Richard Hewitt[2], Hans Bressers[1], Patricia Martinez Alonso[2], Verónica Hernández Jiménez[2], Jaime Díaz Pacheco[2] and Lara Román Bermejo[2]

Abstract

Background: The decentralised and private nature of small-scale renewable energy development does not fit traditional models of government planning and oversight. The land use impacts related to these developments are not well understood and data is lacking related to the environmental, social and economic impacts that can occur under various scenarios.

Methods: This research note provides a literature review of the scarce information available about the spatial impacts of small-scale renewable energy and outlines the current stream of research being undertaken to address this knowledge gap. The preliminary case studies in Overijssel, the Netherlands and Navarre, Spain provide the background for understanding this complex issue, and a new integrated policy and land use model is introduced in order to combine qualitative and qualitative data that is important for understanding the dynamics of this growing field.

Results: The main difficulties in moving forward in planning for the decentralised renewable energytransition are the variation of perspectives on the attractiveness and appropriateness of urban renewableenergy (RE) development, differences in implementation processes and incentives, the dynamic nature ofthe relevant technologies and the lack of up to date information on land use.

Conclusions: Multi-functional land use is a key strategy for increasing the uptake of small-scale renewable energy but little to no data is available regarding it in European land use literature and policy. This needs to be addressed in order to enable pragmatic policies that will enable effective implementation of renewable energy.

Keywords: Renewable energy; Land use impacts; Integrated modelling

Background

Decentralised and smaller-scale approaches to renewable energy may facilitate or accelerate implementation under certain conditions. Key elements supporting this hypothesis are (1) larger projects have more visible spatial impacts and projects with high spatial impacts are slow to develop in places with strict planning regulations and where land use changes are open to public debate[1] and (2) larger projects require large investment of capital which developers may be unwilling to outlay when economic or legislative changes generate instability in the sector [1].

Some articles exist with respect to the spatial implications of land-based photovoltaic energy (PV) installations [2–4] and the landscape impacts and related implementation issues of wind turbines [5–8]. However, there are few local or sub-regional scale studies on implementation or land planning conflicts specifically due to displacement or incompatibility of uses with the development of renewable energy-related landscape features (RELF). RELF are the physical structures necessary for the deployment of renewable energy (RE). This is an important area of study since there has been a relatively recent emergence of local RE developments aligned with the transition to a more decentralised energy model. Often due to the smaller impact per installation, the typical environmental impact assessment (EIA) or related planning instruments do not apply and installations can go "unnoticed" from a planning

* Correspondence: cldeboer@gmail.com
[1]CSTM/ITC, University of Twente, Drienerlolaan 5, 7522 NB Enschede, the Netherlands
Full list of author information is available at the end of the article

perspective[2]. We focus principally on wind and PV developments since they can be achieved through both large- and small-scale activities, which are important for increasing the resilience of the energy supply in a given region or locality.

This research note corresponds to research being undertaken as part of the EU FP7 funded project COMPLEX: Knowledge Based Climate Mitigation Systems for a Low Carbon Economy. The efforts involve better understanding how contextual (qualitative and quantitative) data can be used to understand land use patterns and possible future scenarios of local renewable energy in the Netherlands and Spain. We provide here the results of our literature review of the available knowledge on spatial impacts of RE and a summary of what this means for researchers attempting to understand this dynamic field. Our main research in this project is related to the relationships between land use change, renewable energy development and policy. We dedicate this research note to increasing the visibility of how current data is limiting us in understanding this dynamic and important relationship. Secondly, we introduce very briefly the model being developed to help us understand these relationships and relevant case studies that we are applying it to in order to advance future studies in understanding the spatial impacts of RE. As such, this research will support future efforts in the fields of land use change modelling and those involved in areas of public administration research-related policy implementation.

Literature review of the spatial impacts of RE

The spatial impacts of implementing various renewable energies (RE) are complex and dependent on the technology as well as the particular context in which RELF are implemented. They may be visual, physical, functional, long-term, short-term, easily reversible or not, etc. Spatial impacts are related to the topography of the landscape, the area of land physically covered by RELF, the current land use and geography, the distance from areas of natural beauty or sensitive ecosystems and biodiversity [9]. Full attribution of spatial impacts of RELF requires an understanding of the entire implementation and usage process chain; impacts may include changes to existing infrastructure, land conversion and agricultural productivity impacts, ecosystem modifications and habitat and biodiversity reduction, aesthetic changes and adjustments to recreational potential. Some spatial impacts, such as the negative impacts of wind turbines on bird populations (ecological impacts), landscape visual quality (aesthetic impacts) and disruption to food supply from transforming agricultural land to biofuels (direct land use impacts) are widely known. Other related aspects, such as land use intensity, indirect land use change and land multi-functionality have received less attention in the literature. In the following paragraphs, we provide a review of the literature relevant to the question of spatial impacts associated with solar PV and wind energy implementation and then include a more detailed discussion of land use, the key focus of this paper. Much literature relevant to the potential spatial impacts of RE development was identified; however, significant gaps do exist. The spatial impacts of solar energy installations are generally less well studied than for other RE types. The impact of solar thermic (ST) energy is likely to be minimal, since this is mostly associated with heating systems added to buildings [2], but the spatial impact on the territory of PV installations may be quite substantial. In addition, though the literature on wind energy is abundant, few studies seem to relate to the direct land use impacts of wind installations, a key concern of our research. Overall, precise assessment of expected and real impacts was generally sparse for all RE types. One exception to this is that spatial impacts due to land use transformation from food to biofuel have been deeply studied; however, the focus on RELF requires information on the location of these "crop installations". Gathering the information related to this is seen as a particular issue in and of itself as crops for biofuels are not distinguishable from regular crops from a land use perspective. As such, we have chosen not to include biomass-related RE in this research.

Direct land use impacts

Alterations to land use can be expected to occur for all new RE installations. PV installations and wind turbines that are not located on top of already existing infrastructure will reduce the ability to use the land for other activities, though to different degrees based on the previous land uses and the type of project implemented. While for example, it could be expected that much agricultural land would be too valuable to turn into PV fields[3], new standards requiring a percentage of utility supplied power to come from renewable sources is resulting in large-scale PV projects expanding into a wide range of locations and ecosystems that were once considered uneconomic. The potential land use issues and concepts related to PV and wind installations found in the literature are highlighted below.

Construction phase impacts are relevant to the installation of PV or wind installations [3]. Large-scale wind installations together with their associated access roads and electricity supply infrastructure have major direct impacts and are compatible only with some existing uses (e.g. forestry, pasture). Literature relating to impacts of wind turbines however mostly deals with issues around ecological or visual impact of the erected turbines. Research with stakeholders to estimate the social cost of wind farm developments in Saragossa (Spain) identified loss of natural areas and the impact of access road construction as key impacts [5]. In this work, stakeholders consulted in the Spanish region of Navarre noted that the impact of access construction was one of the factors that provoked public controversy when the first wind farms were developed in the 1990s (Martínez Alonso P, Hewitt R, Pacheco JD, Román L, Hernández Jiménez V, Bressers H and de Boer C: Losing the roadmap: Renewable energy paralysis in Spain and its implications for the EU low-carbon economy, Submitted). A lack of assessment procedures established at an international level for PV is also contributing to this gap [3].

The concept of land use intensity is important when addressing the land use impacts from installation and operation of large-scale PV plants [4]. Land use intensity is a metric that expresses the quantity of land transformed relative to power output or unit of electricity generated. The land occupation relative to the time the power plant is in use is also important and should take into account the time required for the land to recover following use. While land use intensity for large-scale solar installations and coal plants was comparable in the short-term over the long-term PV installations required lower land areas for equivalent energy generation capacity [4]. These authors found that recent commercial PV power plants in the USA covered an average of 25 km^2/GWp. They estimated also that in cases where the PV systems were installed in forest areas that recovery of the forest requires an average of 10 years. Additionally, they calculate that a 30-year-old PV plant occupies ~15 % less land than a coal power plant of the same age. The study of overall land coverage of PV panels that has increased due to legislative incentives in Italy quantified the future potential impact with an assumption that between 7 and 10 m^2 of surface area is necessary to generate 1 kWp [3]. A continuing increase in the installed capacity of 10 MWp per month would then result in between 80,000 and 100,000 m^2 of additional PV land cover per month.

These studies are surprising since they challenge two key popular assumptions: (1) that solar installations always occupy far less land than coal plants—in fact, their impact is initially comparable and only begins to show advantage after ~25 years—and (2) energy generation capacity of solar is per se less intense than all fossil fuels—something that can be seen to be untrue if intensity is measured over the lifetime of the plant [4]. These impacts are related to land-based PV, not those that are implemented on top of existing infrastructure.

More generally, the available literature suggests that, besides the various negative impacts, there may be a number of potential land use benefits of PV systems. PV projects can be used to reclaim degraded land and reduce requirements of transmission lines of electricity grids, be used in scenic areas and National Parks, where the avoidance of pylons and wires is a major advantage, can be integrated into the façades of buildings, used as a cladding material for commercial buildings and provide shading and heat extraction [2]. Additionally, greater multi-functionality is generally more possible than with many other energy technologies. Placing PV panels on top of existing infrastructures rather than on forested or farmland can also prevent biodiversity loss [4]. The main negative land use impacts related to PV systems are the loss of amenity; the implementation of a PV system in once cultivable land limits the soil productivity in that area [2]. The electricity output in this case is then competing with food production and as such is similar to the dilemmas faced by energy crops [3]. Some negative visual aspects have also been noted, but there is little certainty about how these impacts are experienced.

Aesthetic impacts

The literature on aesthetic or visual landscape impacts resulting from RE installations relates mostly to wind energy as widespread deployment of wind turbines has taken place over the last 2 or 3 decades. Since this literature is extensive, the following summary is intended to be illustrative rather than comprehensive. Generally, the research on the aesthetic impacts of solar is based on solar cells that are not added to previous infrastructures. This needs to be considered when determining the relevance of these impacts to the context at hand.

"One of the most difficult-to-quantify...environmental costs is visual impacts, whether from wind, solar or hydropower developments" [5]. These authors used choice modelling techniques based on stakeholder interviews to estimate the social cost of wind farm developments in Saragossa, Spain. Respondents tended to value impacts on flora and fauna more highly than impacts to the visual landscape or to the area's rare geological heritage. This study is notable in that it recognises that wind farm developments have significant social costs and that these costs can be assessed through engagement with local stakeholders. In a study of annoyance due to wind turbine noise in Sweden, it was found that interviewees in a municipality in the south of Sweden were more likely to report annoyance due to turbine noise if they felt that turbines negatively impacted the visual landscape [10]. In general, respondents accepted the necessity of wind turbines but felt that their contribution to the landscape scenery was negative. Further, a 2007 study of renewable energy planning in the Netherlands, found the visual evaluation of the impact of wind power on landscape values to be the most dominant factor in community acceptance of such schemes [11].

Although there seems to be broad general consensus about the importance of visual impact of wind energy developments, not all authors explicitly consider it [12].

Stakeholders interviewed in this study's Navarre case considered that public controversy related to installation of wind farms in the 1990s had now mostly been resolved [13] thanks to public information campaigns and growing general awareness of environmental issues. Other parts of Spain have seen political disagreement between autonomous communities (e.g. between Castille and Leon and Cantabria) over the visual impact of wind energy installations in border areas of landscape beauty [13].

For solar energy installations [14], the aesthetic impacts of PV installations can be quantified based on four criteria: visibility, colour, fractality and concurrence between fixed and mobile panels. This study determined the overall impact by expert opinion through a Delphi procedure. Different impacts were generated by different types of plants in different types of landscapes. The proximity to urban or recreational areas was however not taken into account, and so this study can better be seen as potential aesthetic impact since frequency of impact was not considered. Analysing subjective reactions requires a cognitive study of people's preferences for PV power plants. There is yet no psychological research specific to PV power plants in the literature according to [14], and this would be extremely important for highlighting successful opportunities for implementing these projects in urban and semi-urban areas. The only empirical result related to land use impacts was that differences in size were important in the overall impact of the subjects and that both objective variables and subjective feelings were taken into account when making their judgments. A scoping survey completed in the province of Overijssel found that among those ($N = 15$) that completed the survey, the landscape implications were of lower importance in the development of PV energy than other factors such as importance for the environment and economic benefits.

Recreational impacts

Implementation of RELF may also imply changes to the land as a public amenity. Metrics for the impacts of RE installations on recreational resources have not been developed but would likely be similar to the aesthetic impacts mentioned earlier [4]. However, additional factors may need to be considered such as loss of public access. While wind farms may be accessible to the public, in most cases, PV arrays are not.

Ecological impacts

Ecological impacts relate to habitat loss or damage, such as insect and bird death from intense light generated by solar installations or wind turbines, and increased use of resources such as water. The impact of wind turbines on birds has been widely studied, e.g. [6–8], and the importance of this aspect for assessing the environmental impact of windfarms is well-recognised. In the previously mentioned study, stakeholders in Saragossa valued impacts to flora and fauna more highly than other landscape impacts [5]. The authors include resource depletion in their list of environmental impacts of PV systems, since their manufacture required scarce materials (In/Te/Ga). A 2009 study of the territorial and landscape impacts of PV systems identifies impacts from landscape fragmentation, vegetation degradation, interference with flora and fauna as well as microclimatic change caused by the daytime warming of the surface of the solar array [3]. Concerns have also been noted about impacts to wildlife, which may prohibit solar development on large areas of desert land in California [4].

In terms of habitat and biodiversity impacts, research continues on measuring habitat fragmentation and risk assessment of complex ecosystem collapse. Land use in tensity is often used as a proxy for assessment of impacts on biodiversity [4]. However, biodiversity can be directly measured by species density (recorded in the Millennium Ecosystem Assessment), and PV installations can be sited accordingly in less biodiverse areas. The water needed for cleaning solar panels is approximated at 500–1000 gallons per MWp of panels per year [4]. The hydrological footprint of large solar arrays, particularly in arid or semi-arid environments, is likely to be considerable.

The term *energy sprawl* addresses the varying spatial extents of different energy production techniques [9]. It is the product of the total quantity of energy produced annually (e.g. TW h/year) and the land use intensity of production (e.g. km^2 of habitat per TW h/year). This is used to address the potential habitat effects of energy sprawl and to show that more compact energy generation does not necessarily reduce damage to biodiversity. Particularly important is that energy production techniques can have multiple effects on biodiversity, which "operate at different spatial and temporal scales. Biodiversity impacts that are likely to scale with real impact include habitat replacement and habitat fragmentation. Further, the longevity of the impacts… also… varies" ([9]; p3).

Spatial planning for RE impacts

Spatial impacts of different types of RELF can be assessed, but no simple calculation will enable the right technology to be selected to ensure maximum capacity for minimum spatial impact. Instead impacts should be determined based on the particular characteristics of the area in question. Many different criteria for siting of RE installations need to be balanced to take into account the various types of impact from each technology. For example, while the elevation of the wind turbines increases the area of aesthetic land use impact compared to PV, the actual area occupied by PV is the main intrusion into the landscape due to their low elevation and wide expanse.

Reversibility

Some impacts resulting from installation of RELF are quickly reversible through the removal of the relevant installations, such as solar panels or wind turbines. In other cases, previous or alternative uses will only return once the previous state has been regained. Clearly, installations in urban areas on rooftops are nearly completely reversible, but large-scale land clearance for wind farms or PV arrays may take decades or centuries to recover. In general, spatial planning procedures do not deal well with the issue of reversibility, often because high value land uses such as urban development or mining are irreversible on a human time scale.

Large-scale wind and solar developments in the EU require an EIA, which includes a visual impact analysis. Though there do not seem to be many published examples (EIA studies are mainly an annex to the project itself), a GIS-based visibility catchment (viewshed) analysis would be an appropriate starting point for evaluation of visual impact. Published research exists on the environmental impacts from the manufacturing and decommissioning phases of PV power equipment; however, more studies are needed about the installation and operation phase.

Methods

Local communities can experience difficulty in deciding the siting of PV installations [3]. Innovative approaches involving participatory cartography are one way to address these difficulties [13, 15, 16]. Modelling approaches directed at RE installation location, such as multi-criteria decision-making (MCDM) procedures [17] or optimal site potential models [18], are beginning to emerge and are likely to offer a way forward with respect to the spatial allocation of RELF in line with environmental criteria. This section provides an overview of the two case studies used in this research to understand how local RE development is taking place and particularly what elements of spatial issues are developing alongside it that should be considered.

The study of local renewable energy development from a spatial perspective is a relatively new field of study. Two cases were chosen with different spatial contexts in order to develop understandings of relationships that take place in both space-scarce (the Netherlands) and space-abundant (Spain) areas. In the Netherlands, locally based solar implementation is increasing exponentially. Collective purchasing, crowd-sourcing and small-scale solar farms are becoming commonplace. The highly regulative nature of Dutch spatial planning could be used to reduce the changes seen due to solar energy. Currently, this is only being done in cases where the scale is considered to be industrial and not for personal or community use. Discussions are being held at the provincial level about how to incorporate medium-size solar installations in the landscape, but no final decisions have been made. Wind energy is highly regulated and has not achieved much in comparison to other countries in Europe. In Spain, previous advancements made through large-scale wind and solar installations have come to a standstill due to an altered funding environment at the national level. Current developments are considered to be more modest or small scale and need to address more local issues due to the lack of support from higher levels.

Connecting the types of local RE projects and the resulting implementation processes and land use impacts that have taken place in these two cases is desired in order to provide us with the necessary preliminary data to simulate and model these relationships. What is missing from the available literature is how qualitative and quantitative social factors increase or decrease the likelihood of implementation of renewable energy in different cases. In this ongoing research, we are using a specially developed integrated participatory land use model APOLUS—actor and policy land use simulator.

APOLUS is articulated as two model blocks (see Fig. 1): land demand and land use allocation are connected by a feedback loop. The land demand model block determines the amount of land to be changed or combined with RE uses which is then allocated in the land use allocation model block. Land demand is initially determined for each scenario on the basis of factors exogenous to the land allocation block, but land allocation outcomes are fed back into the demand block so that the system is modified dynamically as RE implementation progresses. To produce simulations of land use and RELF at future dates (e.g. 2020, 2050), the storyline and simulation approach is employed [19]. Storylines are developed through the participatory process and the model represents them. This enables the contextual factors related to the implementation process to be introduced into the model. Determination of RE demand for each scenario is a key challenge. As noted above, the demand model block also needs to consider multi-functionality, enabling the allocation of new RELF in existing land areas. It is anticipated that more multi-functionality of land use will take place in the Netherlands area than in Spain, so some difference in the operation of the system across the two areas is anticipated.

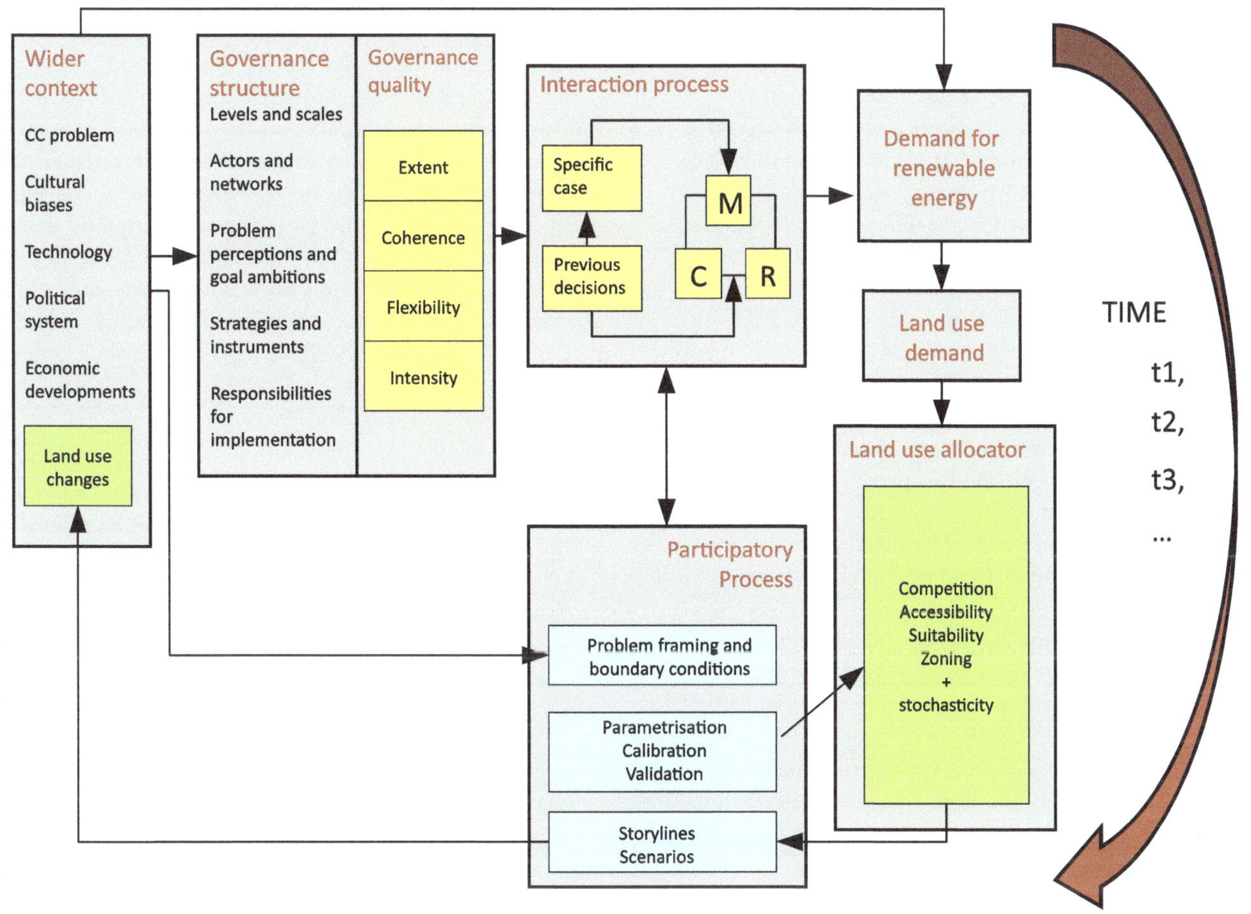

Fig. 1 Conceptual Model for APOLUS: Actor and POlicy Land Use Simulator

As such, the differences and similarities between the chosen case studies will give us a perspective on different European approaches to land use and renewable energy. The model enables us to incorporate geographical, social and economic factors into the development of future scenarios with the help of expert and local stakeholder input. Doing so at a regional/local scale in our two case studies will provide direction as to what type of spatial land use data is necessary for understanding land use changes associated with local renewable energy development.

Results and discussion

Implementation of local RE installations, as with any change to local resources or amenities, is more effectively accomplished if it is community driven and if both costs and benefits are shared across different members of society. As seen in Spain, a success story can turn into failure overnight if implementation is excessively reliant on top-down mechanisms such as legislative support and subsidy regimes. Conversely, even if a considerable amount of general support for renewable energy policy exists, at a local level, many residents feel that a renewable energy system may limit their quality of life [3]. A number of important points emerged from the preliminary case study work that need to be addressed in the spatially focused simulation model that has been developed to study these implementation processes. These are as follows:

(1) *Multi-functionality:* No clear distinction was identified between wind and PV energy with respect to multi-functionality of use. In the opinion of Spanish stakeholders, PV energy installation development implies loss of the previous land use, while wind energy was regarded as compatible with other uses. This is quite different to the current situation in the Netherlands where there is strong resistance to wind energy because it is viewed as incompatible with the landscape, yet PV on roof tops is generally accepted. Since multiple land uses are not recorded under Corine land cover data, it is not possible to verify whether there were good examples of compatibility of land use from the maps. It is very important that the simulation model take multi-

functionality into account since under some circumstances, RE must be incorporated into the existing land uses without taking them over.

(2) *Temporal resolution of innovations:* Another challenging aspect of this research is the speed at which changes are taking place and the minimal level of oversight which can occur. This is particularly visible for the diffusion of PV panels in the Netherlands. An underlying hypothesis is that new projects are influenced by the presence (or lack) of other PV panels in the area. Detecting the presence of any causal aspects to this relationship implies a need for data on a highly detailed temporal scale. However, since this level of detail is not present in the available open data, the model will be tasked with simulating this process. By generating plausible outcomes from multiple model runs, the simulation model serves as a virtual laboratory for exploring the spatial and temporal diffusion of these processes.

(3) *Path dependency*: A further challenge to the successful development of future scenarios is the underlying assumption that the past land use relationships will hold in the future (path dependency). Using past land use changes relationships as the basis for the scenarios has a number of issues. As technologies change, the underlying characteristics of RELF also change. This is true particularly with respect to the footprint and aesthetic impacts that are related to a technology. To address this, we include a certain amount of randomness into the development of scenarios and will also include the changes in potentially influential contextual factors as variables in the model. The spatial allocation component of the model and the cellular automata model of White and collaborators [20] is especially suitable for this kind of work, since small changes to model parameters can produce highly divergent outcomes as the simulation progresses, a feature known as bifurcation. Model results are thus not limited to a few scenarios (e.g. business as usual, renewables super development, continued fossil fuel dependence, etc.), but instead produce a broad range of possible options based on the various bifurcations from each scenario path. Multiple model runs can be used to assess the probability of the various outcomes. Model results can assist policymakers in assessing the likelihood of the developments having an impact in their particular area based on the timespan over which they expect their policies and plans to be relevant.

Conclusions

The need for additional and interdisciplinary studies related to the development of local renewable energy is clear. The multiple types of RELF possible will have significant and varying impacts on the speed and success of the energy transition in various places. The increasingly private and local nature of renewable energy development increases the complexity of monitoring and understanding these processes. We aim for this study to provide some insight into the gaps and lack of information available for policymakers and planners in preparing, supporting and steering the local, decentral renewable energy transition.

Endnotes

[1]The exception to this is the case where changes are required for flood safety, in which case the high level of public acceptance can slightly speed up the necessary processes.

[2]Recently, Spain approved a new law on EIA (Law 21/2013, 9th of December, of Env. Impac. Asses.) relaxing these instruments further: photovoltaic installations under 10 ha. Do not require study as well as those dedicated to self-consumption under 100 kW.

[3]See results of the land use analysis for the Navarre case study, end of Section 4.2.

Abbreviations

RE: renewable energy; RELF: renewable energy landscape features.

Competing interests

The authors declare that they have no competing interests.

Authors' contributions

CDB coordinated the development of the paper, contributed to the Dutch research and text of the paper. RH coordinated the team in Spain, contributed to the research, contributed to the text of the paper and revised the manuscript. HB contributed to the Dutch research and to the text of the paper. PMA carried out the research work in Spain, contributed to the text of the paper and revised the manuscript. VHJ, LRB and JDP contributed to the research and revised the manuscript. All authors read and approved the final manuscript.

Authors' information

This team of authors have grouped themselves together based on their collective work and interest in the effects of social factors in land use change. CDB is currently a senior researcher appointed jointly to two departments of the University of Twente—PGM (Urban and Regional Planning and Geo-Information Management) and CSTM (Department of Governance and Technology for Sustainability) at the University of Twente (UT). Her research focuses on connecting land use data and knowledge with policy relevant instruments and planning. The largest support for the land use change mapping and modelling component comes from the expertise of RH, based on his experience in participatory modelling, land use modelling and software. VHJ has a lengthy profile in community engagement and modelling which has helped to determine the stakeholder participation elements of the research. Both RH and VHJ are members of OCT (Observatory for a Culture of the Territory). The other authors (HB, VHJ, PMA, LRB, JDP) are also affiliated with these two institutions (OCT and UT) and provide expertise on policy implementation, GIS and modelling and land use change and impacts of development.

Acknowledgements
The research presented in this paper is conducted and funded from the EU FP7 project number 308601—COMPLEX. Further detailed information is available on the project website.

Author details
[1]CSTM/ITC, University of Twente, Drienerlolaan 5, 7522 NB Enschede, the Netherlands. [2]Observatory for a Culture of the Territory (OCT), C/ Duque Fernán Nuñez, 2 – 1º, 28012 Madrid, Spain.

References

1. De Boer C, Hewitt R, Bressers H, Verónica Hernández J, Martínez Alonso P, Warbroek B (2014) Stakeholder input and feedback on model development of PLUS4-CMP. Project Report for EU FP7 Programme COMPLEX Project
2. Tsoutsos T, Frantzeskaki N, Gekas V (2005) Environmental impacts from the solar energy technologies. Energy Policy 33(2005):289–296
3. Chiabrando R, Fabrizio E, Garnero G (2009) The territorial and landscape impacts of photovoltaic systems: definition of impacts and assessment of the glare risk. Renew Sust Energ Rev 13(2009):2441–2451
4. Turney D, Fthenakis V (2011) Environmental impacts from the installation and operation of large scale solar power plants. Renew Sustain Energy Rev 15(2011):3261–3270
5. Álvarez-Farizo B, Hanley N (2002) Using conjoint analysis to quantify public preferences over the environmental impacts of wind farms. An example from Spain. Energy Policy 30(2):107–116
6. Drewitt AL, Langston RH (2006) Assessing the impacts of wind farms on birds. Ibis 148(s1):29–42
7. Fox AD, Desholm M, Kahlert J, Christensen TK, Krag Petersen IB (2006) Information needs to support environmental impact assessment of the effects of European marine offshore wind farms on birds. Ibis 148(s1):129–144
8. Stewart GB, Pullin AS, Coles CF (2007) Poor evidence-base for assessment of windfarm impacts on birds. Environ Conserv 34(01):1–11
9. McDonald RI, Fargione J, Kiesecker J, Miller WM, Powell J (2009) Energy sprawl or energy efficiency: climate policy impacts on natural habitat for the United States of America. PLoS One 4(8), e6802. doi:10.1371/journal.pone.0006802
10. Pedersen E, Persson Waye K (2004) Perception and annoyance due to wind turbine noise—a dose-response relationship. J Acoust Soc Am 116(6):3460–3470
11. Wolsink M (2007) Planning of renewables schemes: deliberative and fair decision-making on landscape issues instead of reproachful accusations of non-cooperation. Energy Policy 35(2007):2692–2704
12. Aydin NY, Kentel E, Duzgun S (2010) GIS-based environmental assessment of wind energy systems for spatial planning: a case study from Western Turkey. Renew Sustain Energy Rev 14(1):364–373
13. Martínez Alonso P, Hewitt R, Vicente Guillén J, Román L, Hernández Jiménez V, Pacheco JD (2013) Climate Related Energy Developments in Spain, COMPLEX FP7 project report., Available on line at: http://www.complex.ac.uk/outreach/publications/reports/OCT-COMPLEX_Deliverable31_26122013.pdf
14. Torres-Sibille AC, Cloquell-Ballester VA, Artacho Ramirez MA. (2009) Aesthetic impact assessment of solar power plants: an objective and a subjective approach. Renew Sustain Energy Rev 13(2009):986–999
15. Lange E, Sigrid Hehl-Lange S (2005) Combining a participatory planning approach with a virtual landscape model for the siting of wind turbines. J Environ Plan Manag 48(6):833–852
16. Van Hoesen J, Letendre S (2010) Evaluating potential renewable energy resources in Poultney, Vermont: a GIS-based approach to supporting rural community energy planning. Renew Energy 35(9):2114–2122
17. Aydin NY, Kentel E, Sebnem Duzgun H (2013) GIS-based site selection methodology for hybrid renewable energy systems: a case study from western Turkey. Energy Convers Manage 70:90–106
18. Yeo I, Yee JJ (2014) A proposal for a site location planning model of environmentally friendly urban energy supply plants using an environment and energy geographical information system (E-GIS) database (DB) and an artificial neural network (ANN). Applied Energy 119:99–117
19. White R, Engelen G, Uljee I (1997) The use of constrained cellular automata for high-resolution modelling of urban land-use dynamics. Environment and planning B 24:323–344
20. Alcamo J (2008) The SAS approach: combining qualitative and quantitative knowledge in environmental scenarios. Environmental futures: The practice of environmental scenario analysis 2:123–150

17

Biogas and nutrients from blackwater, lawn cuttings and grease trap residues—experiments for Hamburg's Jenfelder Au district

S. Hertel*, P. Navarro, S. Deegener and I. Körner

Abstract

Background: The project KREIS focuses on a new combination of renewable energy provision with innovative wastewater treatment, called the "Hamburg Water Cycle®" (HWC) which will be applied in Hamburg's neighbourhood Jenfelder Au. HWC includes a separate collection of rainwater, greywater and blackwater. Vacuum toilets are used to concentrate the blackwater. Biogas will be produced from the blackwater in an anaerobic digestion process together with co-substrates. The blackwater will be transported to the anaerobic pre-treatment facility via a vacuum system. Construction of water systems started in 2013, and commercialization of houses is planned to be finished in 2018.

Methods: The article focuses on research work accompanying the demonstration project. Blackwater and the co-substrates, lawn cuttings and grease trap residues from restaurants and canteens will be considered as bioresources, not as residues. To evaluate the utilization efficiency, three investigation steps were carried out: inventory to determine substrate quantities and qualities, anaerobic digestion to determine biogas production, and evaluation of digestate utilization options.

Results: The daily amount of blackwater in Jenfelder Au is calculated to be about 12 m^3 (dry matter (DM) 0.6 %; organic dry matter (oDM) 65 % DM; nitrogen (N) 28 % DM; phosphorus (P) 2.7 %). To increase the biogas production, co-substrates will be added. Grease trap residues (averages: DM 2 %; oDM 85 % DM; N 2.5 % DM; P 0.6 % DM) and lawn cuttings (averages: DM 30 %; oDM 80 % DM; N 2.6 % DM; P 0.3 % DM) were selected. The inventory study showed a sufficient potential of lawn cuttings within a 5-km radius. The lawn cuttings must be pre-treated for wet fermentation. Two options were investigated: press juice preparation and wet shredding of the fresh and silage lawn. Batch test was used to determine the biogas potential of the substrates with the following average results: blackwater 500 nl/kg oDM, grease trap residues 1000 nl/kg oDM, lawn cuttings 400 nl/kg oDM and lawn juice 500 nl/kg oDM.

The effects of the composition of the substrate mixture and of the retention time in the reactor on biogas quantity and process stability were studied in semi-continuous operating reactors. Experiments showed that a stable process with an average biogas production of 800 nl/kg oDM is, e.g., possible with a mixture of blackwater, press juice of lawn cuttings and grease trap residues in a fresh mass ratio of 1:1:1.

Furthermore, the N and P contents in digestates were determined. These nutrients are valuable for fertilization.

(Continued on next page)

* Correspondence: saskia.hertel@tuhh.de

Bioconversion and Emission Control Group, Institute of Wastewater Management and Water Protection, Hamburg University of Technology (TUHH), Hamburg, Germany

(Continued from previous page)

Conclusion: It has been shown that blackwater combined with local waste streams can be used for biogas generation and that it has a potential as fertilizer. The experiments have shown that co-digestion has a positive effect on biogas yields and lawn cuttings are suitable as co-substrate. Lawn cuttings can be applied as lawn juice or lawn suspension. Ways of an integral utilization and the potential of nutrient recovery are shown in this work.

Keywords: Blackwater; Grease trap residues; Lawn cuttings; Lawn juice; Anaerobic digestion; Biogas; Bioresources; Inventory; Fertilizer; Nutrient recovery; Digestate utilization; Hamburg; Jenfelder Au

Background

Overview on the Jenfelder Au project

Fossil raw materials for energy production are becoming scarcer. The natural resources for phosphorus (P) fertilizers are declining. The production of the mineral nitrogen (N) fertilizer by the Haber-Bosch process needs a lot of energy. The demands of fertilizers increase. Human toilet waste (blackwater = urine + faeces + toilet paper + flushing water) contains energy and a lot of nutrients. In Hamburg, it is actually treated together with rainwater and greywater for N elimination in a centralized way using nitrification and denitrification processes. Anaerobic treatment, centrifugation, drying and incineration are further cost-intensive follow-up steps for sewage sludge treatment. The whole actual cascade is very energy intensive and the nutrients are lost. Similarly, in urban areas, also other residues are actually often inefficiently used or disposed. Besides blackwater, organic waste from households, gardens, public areas, industry and commerce can be valuable secondary, tertiary or quaternary bioresources for energy and nutrient recovery.

The aim of this paper is it to show that it is possible to produce energy from wastewater and to recover nutrients. These works are incorporated in the KREIS project. The project focuses on the combination of systems for innovative wastewater management, renewable energy generation and waste utilization. The basic system is the "Hamburg Water Cycle®" (HWC) developed by Hamburg Wasser. HWC contains a separate collection of rainwater, greywater and blackwater. Vacuum toilets are used to concentrate blackwater; via a vacuum system, it is transported to a decentralized anaerobic treatment facility where biogas will be produced from blackwater in an anaerobic digestion process together with co-substrates [20].

In the *Jenfelder Au neighbourhood*, a new residential area for about 2000 people is under development. Vacuum toilets with only 1 l water per flush are used to concentrate the blackwater. A daily blackwater generation of 12 m^3 has been calculated. Targets given by the operating company for anaerobic digestion include 900 m^3 fermenter volume and 25 days substrate retention time. Under these conditions, blackwater will fill a third of the reactor and biogas production of blackwater alone would be low. This is the reason why regional co-substrates with a suitable biogas potential are required. For co-substrates, waste streams should be used which have no useful application so far. Blackwater and co-substrates contain N and P, which are valuable nutrients for fertilization purposes, and therefore provide a further utilization option [20].

The Jenfelder Au project is the first project in that dimension. It is realized by the municipal wastewater disposer. Earlier projects using blackwater as biogas substrate were realized in a much smaller scale and mostly carried out by the habitants themselves, e.g. in China and India, family-sized biogas plants exist, where blackwater is treated together with animal and kitchen waste to produce biogas for cooking [37]. One example for a housing estate is in Northern Germany in Lübeck called *Flintenbreite*, where blackwater from about 400 people is collected via vacuum toilets. Investigations of the biogas yield were done by Wendland [37] and found a specific methane generation rate of 14 l CH_4/cap/day. Also in *Sneek*, a Dutch city, an anaerobic blackwater treatment was realized for 32 houses [39].

In order to support and achieve the sustainable implementation of this new wastewater, waste and energy management system in Hamburg's Jenfelder Au neighbourhood with a high bioresource utilization efficiency, an accompanied research programme was carried out and some results are presented in this paper.

Organic waste as a bioresource

"Bioresources are non-fossil biogenic resources which can be used by humans for multiple purposes: to produce food, substantial products and/or energy carriers." That definition of bioresources is given by [17]. Körner also classifies bioresources into primary, secondary, tertiary and quaternary bioresources. The preferable utilization steps depend on the type of bioresources. Multichain and cascade utilization are important aspects for ecological sustainability.

Primary bioresources

Primary bioresources are generated for a specific application-oriented purpose in forestry, agriculture or aquaculture. The most important primary bioresources are not only harvested plants, but also slaughtered animals. The minor part is given by cultivated algae, microorganisms or fungi. Plants which are not specifically cultivated with the aim of food, material or energy production cannot be classified as primary bioresources. In the Jenfelder Au project, primary bioresources shall not be used.

Secondary bioresources

Secondary bioresources are residues that occur during processing of primary bioresources or are generated within landscaping activities. Regarding properties, they have to be available in large quantities and define qualities with low impurity contents. For the Jenfelder Au project, commercial processing residues from nearby industries as well as cuttings from large green areas such as parks, lawns or sport places may be of interest but were not considered in this publication.

Tertiary bioresources

Tertiary bioresources are residues which occur in rather small amounts and/or in undefined fractions. They have a lower value than secondary bioresources. Tertiary bioresources may be generated along the chain in primary bioresource production, harvesting, post-harvesting and storage activities, and during industrial processing, packaging and distribution up to the retail sector. Furthermore, they may occur at consumer level in small commerces (e.g. restaurants, canteens), in private kitchens at household level or in private gardens. In the Jenfelder Au project, kitchen waste was of interest but is not considered in this paper. Considered were lawn cuttings from private gardens, public areas, and green waste collection by landscaping businesses.

Quaternary bioresources

Quaternary bioresources occur after a product was used by the consumer. It can be distinguished between the time frames of their generation into short, mid and long term after start of product use. Short-term-after-use quaternary bioresources are generated in all cases of food consumption in the form of faeces and urine. Such bioresources are generated with short delay after food consumption at a time scale of hours. The blackwater used in Jenfelder Au belongs to that group. Furthermore, the greasy water belongs to that group, too. Fats are used for cooking and accumulate after cooking in grease traps in restaurants and canteens.

Aim of the work

The potential of blackwater, greasy water and lawn cuttings as substrates for energy and nutrient recovery within the integrated Jenfelder Au system shall be evaluated. The objective is the utilization of actually unused or inefficiently used tertiary and quaternary bioresources for the anaerobic digestion with blackwater as the main substrate and the others as the co-substrate. Suitable substrate mixtures should guarantee a stable biogas process and high biogas production. Also, options for digestate treatment to recover nutrients and transform them into mineral fertilizers are part of the investigations. The digestate should not be disposed into the wastewater system, since all the advantages of the previous step would be lost. The digestate would run through the very energy-intensive and nutrient-destroying wastewater treatment complex again.

Methods

Origin of substrates

Blackwater (BW). For experiments, concentrated blackwater was taken from vacuum toilets in the Flintenbreite settlement in Lübeck, Germany. The installed vacuum toilets use 0.7 to 1.0 l of water per flush with a suction pressure of 0.3 to 0.5 bar for the vacuum pumps and deliver the BW into a vacuum tank.

Greasy water (grease trap residue, GW). Grease traps are used, for example, in food processing companies, restaurants and canteens. They are installed for the pretreatment of wastewater containing high amounts of fat before the wastewater is disposed into the sewer. The grease traps should be emptied by regulation once a month (DIN 4040-100). In practice, however, the intervals are often longer. This results in the fact that a major part of the greasy waters disappears in the sewerage. GW used in this experiment was taken from the wastewater treatment plant *Köhlbrandhöft* in Hamburg, Germany. To this facility, Hamburg's GW is actually delivered and processed by anaerobic treatment.

Lawn cuttings (LC). Lawn cuttings for experiments were collected from different places. The main part originated from public areas and was cut by landscapers and delivered to the private waste company BUHCK. The specific original harvesting areas are not known as well as the harvesting type and harvesting history. The sampling was carried out at one of the BUHCK collection places. Samples were taken in May, June, July, September and October and used for substrate characterization. Samples for batch test were collected in June and September. LC for semi-continuous experiments were taken in September and October. Furthermore, lawn cuttings from private gardens in *Hamburg Niendorf* and *Bergedorf* were investigated. The samples were taken from April to July and used for substrate characteristics.

The average values of all samples are given in Table 5. For substrate characterization, the substrates were stored frozen until use. Additional samples were stored in vacuum bags before biogas potential was determined.

Methods for substrate analyses

The analysis of the substrate was carried out at the Institute of Wastewater Management and Water Protection

Table 1 Methods for substrate analysis

Parameter		Guideline/equipment
Dry matter content	DM in % FM	DIN EN 12880 [8]
Organic dry matter content	oDM in % DM	DIN EN 12879 [9]
Chemical oxygen demand	COD in mg/l	Cuvette test by Hach Lange, Germany
Total organic carbon	TOC in mg/l	TOC/TN analyser multi N/C 3000 by Analytika Jena, Germany
Total nitrogen	TN in mg/l	TOC/TN analyser multi N/C 3000 by Analytika Jena, Germany
Ammoniacal nitrogen	NH_4^+/NH_3-N in mg/l	Destillation unit K350 by Büchi, Germany
Total phosphorus	TP in mg/l	Cuvette test by Hach Lange, Germany
pH value	pH	pH meter, model 323 by WTW, Germany
Ratio volatile fatty acids to total inorganic carbonate	FOS/TAC	Automatized titrator model FOS/TAC 2000 by Pronova, Germany

FM fresh matter

of Hamburg University of Technology (TUHH). The used methods are summarized in Table 1.

Methods for substrate pre-treatment and anaerobic digestion

Table 2 shows an overview of the equipments used for pre-treatment of lawn cuttings and the experimental set-up to determine digestibility of substrates and substrate mixtures. For use of LC in wet fermentation, it has to be pumpable and mixable with the liquid substrates. For that reason, a pre-treatment is necessary. For biogas test, the substrate was frozen until use to get values of the fresh material. LC were stored in vacuum bags in the cooling chamber at 8 °C for 3 to 6 month before use in anaerobic fermentation batch tests. The lawn juice (LJ) was prepared from the same fresh or stored LC. LJ preparation for the fresh LC used for batch tests was done by a screw press, and for stored material, it was done with a hydraulic laboratory press. The LJ preparation for semi-continuous experiments was carried out for both types in the same way using a slab press.

Collected BW and GW were homogenized in 60-l barrels and distributed in 10-l canisters which were then stored in the refrigerator until the moment they were used for loading the reactors. Every time a new canister was utilized, a sample of it was analysed.

The batch tests were performed to determine the biogas potential of the pure substrates and mixtures of the substrates. The semi-continuous experiments were carried out for determination of biogas amount of different substrate mixtures according to the retention time in the reactor. In four bench-scale continuous stirred tank reactors (CSTR), various mixtures of substrates were investigated.

Calculation of digestate characterization

Scenarios for substrate mixtures in Jenfelder Au were selected, and the composition of the digestate was calculated with an excel model based on the substrate characteristics (Table 8). The biogas yield was calculated via addition of the biogas yield of the single substrates depending on the amount of input.

$$\dot{V}_{\text{biogas}} = \dot{m}_{\text{feed}} \sum (V_{Si} * x_i * w_{\text{DM}i} * w_{\text{oDM}i})$$

$$\dot{m}_{\text{biogas}} = \dot{V}_{\text{biogas}} * \rho_{\text{biogas}}$$

where

$\dot{V}_{\text{biogas}}$ biogas yield in m^3_{biogas}/day

$\dot{m}_{\text{feed}}$ mass of feed in t_{FM}/day

V_{Si} specific biogas production of substrate in nl/kg organic dry matter (oDM)

Table 2 Methods for substrate pre-treatment and anaerobic digestion

	Guideline/equipment
Technique for substrate pre-treatment	
Maceration of lawn cuttings	Knife mill GRINDOMIX GM 300 by Retsch, Germany
Preparation of lawn juice	Screw press-type CV by Anhydro, Germany[a], slab press LA 180 by Bürkle, Germany[b]
Storage of lawn cuttings	Vacuum bags and store in a cooling chamber (8 °C) until utilization
Technique for anaerobic digestion	
Batch tests for biogas potential	VDI 4630 [35]
Semi-continuously fed biogas experiments	Continuous stirred tank reactors (CSTR) with a volume of 10 l, mesophilic conditions about 37 °C ("Semi-continuous biogas production")

[a]Used at the Leibniz Institute for Agricultural Engineering Potsdam-Bornim (ATB), Germany

[b]Used at the Thünen Institute Hamburg-Bergedorf, Germany

x_i mass ratio of substrate i in the feed

w_{DMi} dry matter content of the substrate i in kg dry matter (DM)/kg fresh matter (FM)

w_{oDMi} dry organic matter content of substrate i in kg oDM/kg DM

$\dot{m}_{biogas}$ mass of biogas in t/day

ρ_{biogas} biogas density depending on methane quantity in t/m^3: $\rho_{biogas} = -0.013\, x_{CH_4} + 1.9933$ [30]

x_{CH_4} content of CH_4 in %

For the estimation of digestate composition, it is assumed that biogas is produced from 85 % organic compounds and 15 % water [30]. The mass of nitrogen and phosphorous is not changing and remains in the digestate. The results are given in Fig. 5 and Table 12.

Based on a literature research [16], a preferred way of digestate utilization was selected and is given in Fig. 6. Separation efficiency of solid–liquid separation is estimated by literature data given by Fuchs and Drosg [12] and Hjorth et al. [15].

For stripping process, an ammoniacal nitrogen recovery of 90 % is supposed. It is assumed that the total stripped ammoniacal nitrogen can be transformed into ammonium sulphate, which contains 21 % nitrogen. The amount of limestone is calculated by an assumed ratio of limestone to ammonium sulphate solution of 0.58, which is calculated from data given by Bauermeister and Wild [5]. The lime content of the ammonium limestone is 70 %.

The expected structure material for composting is woody material with 70 % DM, and the amount is calculated for a supposed DM content of 55 % for structure material and digestate mixture. For composting process, an organic degradation of 20 % is assumed.

Results and discussion

Inventory and substrate analysis

An inventory study determined the potential of various secondary, tertiary and quaternary bioresources in Jenfelder Au and the surrounding district *Wandsbek*, which were expected to deliver suitable biogas amounts. They were partly published in [18, 19] and summarized in the following. The inventory results were completed by substrate analysis data for the substrates investigated in this publication.

Blackwater. BW is a quaternary bioresource and should be treated in Jenfelder Au. A person produces 500 l of urine and 50 kg of excrement per annum [20]. Furthermore, 1 l flush water per flush [20] and 15 kg toilet paper per person and annum are used on average [33]. The specific rate of blackwater from vacuum toilets is given with 6 l per person and day on average [14]. In Jenfelder Au, around 2000 people will generate about 12 m^3 blackwater per day.

Table 3 shows analytical results for BW from our own analysis as well as from the analysis carried out by Wendland [37] and Alp [3]. All analysed BW is from Flintenbreite and in a similar range. Our own analysis and the values of Alp compared with those of Wendland show a lower DM and oDM content due to a change in the flush settings between 2008 and 2010.

Greasy water. GW is a quaternary bioresource and actually treated in the municipal wastewater treatment complex Köhlbrandhöft by anaerobic digestion. It is very inhomogeneous and contains, in addition to fats and oils, food leftovers, detergents and cleaning agents. The amount of greasy water collected in the district Wandsbek was estimated via the amounts generated in the close-by district Bergedorf [25], considering the number of habitants of both districts. This resulted in a daily amount of 4 m^3 greasy water from the district Wandsbek.

Our own characterization of GW compared with literature data is given in Table 4. The table shows that the analysed greasy water consists mainly of water, but the organic matter based on DM is high. Additionally, the high chemical oxygen demand (COD) and total organic carbon (TOC) values probably result in high biogas potentials.

Lawn cuttings. LC from lower quality are tertiary bioresources. Approximately 1–2 kg lawn cuttings per square metre and annum accrue in private gardens and on public green space from March to October. The amount and quality depend mainly on the season and on the type and care of the lawn. Fresh lawn cuttings are humid and contain high shares of easily fermentable organic material. Lawn cuttings are actually either left on the lawn or on a heap, composted in the garden, supplied to the collection station, collected by gardeners or landscapers, or collected in the biowaste and residual waste bin. Lawn cuttings also contains high amount of N and P. The area-specific generation rates of 1.7 kg/m^2 for lawn cuttings from public areas and 1.0 kg/m^2 for lawn cuttings from private areas were given by the literature [1, 6]. An inventory study, using GIS (Geographic Information System), was done with zones of 5, 10, 15 and 20 km around Jenfelder Au. The study shows an annual potential of lawn cuttings of 21,000 t/a in 5 km surrounding Jenfelder Au [10]. Another study resulted in nearly 30,000 t/a of lawn cuttings for the district around Jenfelder Au, called Wandsbek [28].

In literature, limited data are available for fresh lawn. The mostly used substrate is grass silage from farmland. A comparison of our own investigations and literature values for lawn cuttings and grass silage is given in Table 5. The values for grass silage are in the same range of our own analyses, and only for ammoniacal nitrogen, the value is lower, possibility due to ammonia losses during the ensilage process.

Table 3 Characterization of blackwater, our own analysis in comparison with literature values

Parameter	Unit	Own analysis		Literature	
		No. of samples	AWW[a]	Wendland [37]	Alp [3]
			Average ± SD[b]	Average ± SD	Average ± SD
DM	%	10	0.57 ± 0.15	0.65 ± 0.21	0.48 ± 0.03
oDM	% DM	10	55.54 ± 6.35	62.36 ± 12.29	55.93 ± 5.04
COD	mg/l	12	7615 ± 2990	8060 ± 2950	5461 ± 1634
TOC	mg/l	12	2428 ± 878	2410 ± 720	2253 ± 780
TN	mg/l	12	1455 ± 220	1495 ± 244	1382 ± 435
NH_4^+/NH_3-N	mg/l	4	1090 ± 121	1111 ± 137	1002 ± 59
pH	–	9	7.58 ± 0.17	7.7	7.1 ± 0.3

[a]Institute of Wastewater Management and Water Protection
[b]Standard deviation

Kitchen waste. Kitchen waste is a tertiary bioresource and generated in the district. Due to the material properties of kitchen waste, it would be an ideal co-substrate for fermentation. Currently, it is collected in Wandsbek via the biowaste and residual waste bin. As a result of mixings with other waste types, the contents of these bins are not suitable for the wet fermentation in Jenfelder Au. Kitchen waste grinders installed in the sink offer a new collection possibility. These promise a nearly complete and, in addition, very user-friendly possibility of kitchen waste collection, with a homogeneous sludge as a result. In Jenfelder Au, a few kitchen waste grinders are expected to be installed for a test operation. However, an area-covering installation is not possible under the actual situation but may be a promising option for similar undertakings [18, 19]. In these investigations, they are not further considered.

Other bioresources. In gardens and public areas, large amounts of green waste such as weeds, branches and boughs from bushes and trees accumulate. These fractions are not directly suitable for fermentation, as they either contain too high soil shares or woody components. However, after separation of disturbing fractions, they may be handled in similar ways as lawn cuttings. Leaves are, on principle, fermentable. The biogas yields significantly depend on the type and age. Furthermore, residues from a fruit-processing company that accumulate in the proximity of 7 km are very well fermentable. These residues occur as fruity water composed of shredded fruit peeling residues, mixed with process water, that accrue during the industrial fruit processing [18, 19]. These fractions could also not be considered in these investigations. But especially the fruity water should be kept in mind for later inclusion.

Pre-treatment and storage of lawn cuttings

Pre-treatment by pressing and maceration

In Jenfelder Au, a wet fermentation system will be installed. To use the bulky LC together with BW, a pre-treatment is necessary. The dry matter content of the mixture shall not to be higher than approximately 10 %, since the substrate mixture has to be pumpable. Two options are tested to transform LC into a pumpable form: generation of a suspension with macerated lawn particles and of a juice pressed out of lawn cuttings.

- Lawn has long fibres and is not easy to shred. To get good grinding results, the addition of a liquid was necessary. The homogeneous suspension was prepared by maceration of lawn fresh matter with

Table 4 Characterization of greasy water, our own analysis in comparison with literature values

Parameter	Unit	Own analysis		Literature	
		No. of samples	AWW	FNR 2006 [13]	Degener [7]
			Average ± SD	Min–max values	Average ± SD
DM	%	6	2.32 ± 0.22	2–70	4.52 ± 4.03
oDM	% DM	6	81.59 ± 11.31	75–93	88.4 ± 10.5
COD	mg/l	7	63,374 ± 33,582	–	63,900 ± 56,500
TOC	mg/l	7	9158 ± 2805	–	16,100 ± 13,600
TN	mg/l	7	567 ± 230	0.1–3.6 % DM	540 ± 520
NH_4^+/NH_3-N	mg/l	–	–	0.02–1.5 % DM	–
pH	–	4	4.54 ± 0.32	–	4.1 ± 0.7

Table 5 Characterization of lawn cuttings, our own analysis in comparison with literature values

Parameter	Unit	Own analysis		Literature		
		No. of samples	AWW Average ± SD	Green cuttings [13] Min–max values	Lawn cuttings [38] Min–max values	Grass silage [34] Average values
DM	%	19	30.59 ± 13.46	12	18–37	30.66
oDM	% DM	19	82.53 ± 7.54	83–92	77–88	92.46
COD	g/kg	4	275.8 ± 81.1	–	–	396.88
TOC	g/kg	4	80.3 ± 17.4	–	–	–
NH_4^+/NH_3-N	g/kg	6	1.98 ± 1.18	–	–	0.44
TN	% DM	6	3.1 ± 0.6	2–3	1.7	1.6[a]
TC	% DM	4	34.78 ± 6.73	–	34–45	43.04
pH		8	5.3 ± 1.15	–	–	4.3

[a]TKN

water in a ratio of 1:4. In practice, water shall be exchanged by the liquid digestate or liquid substrates like BW and GW. Lawn cuttings may also contain contaminations like branches, stones and sand. Large particles were separated out, but especially sand lead to high wear of cutters and should be taken into account in the planning process for a fermentation plant.

- LJ was prepared by pressing the LC in a screw press or a slab press. The characterization is shown in Table 6. On average, 40 % LJ related to the lawn fresh matter can be obtained from fresh LC; the harvest varies from 20 to 60 %. The amount depends strongly on the type of lawn and decreases with a higher amount of woody, dry or mossy material. The juice is easily mixable with BW and GW.

Previous research works in the field of juice preparation from lawn or grass were done mainly for green biorefineries with the aim of protein and lactic acid extraction. Within the Austrian research project "FABRIK der Zukunft", grass silage was pressed out. Mandl et al. [24] carried out that 85–95 % of total lactic acid and about 55–65 % of crude protein can be transferred into the juice.

Table 6 Characterization of lawn juice

Parameter	Unit	No. of samples	Average ± SD	Min	Max
DM	%	12	4.24 ± 1.50	1.82	8.30
oDM	% DM	12	65.37 ± 11.19	47.30	72.80
COD	mg/l	7	41,581 ± 27,209	14,194	105,520
TOC	mg/l	10	16,135 ± 9336	6438	39,786
TN	mg/l	10	1914 ± 1052	672	4527
NH_4^+/NH_3-N	mg/l	7	386 ± 285	279	980
pH	–	9	5.6 ± 0.89	4.2	7.4

Storage of lawn cuttings

LC only accrue between March and October. To ensure substrate availability for the biogas plant in the whole year, lawn cuttings have to be stored. One possibility is preparing silage, which is a common way used in agricultural biogas plants for maize and grass cuttings. Grass is compacted as bale silo or in a driving silo. Under anaerobic conditions, lactic acid fermentation takes place; due to a production of acids, the pH value decreases and leads to inhabitation of spoilage microorganisms, and a conservation of the material is achieved. [27]

Lab-scale storage was done by using vacuum bags which also imply anaerobic conditions and can be compared with the silage process.

Biogas potential of substrates

Biogas potentials for black and greasy water as well as fresh and stored LC and LJ were determined in mesophilic anaerobic batch tests. Average results for BW and GW as well as results from specific lawn samples are shown in Table 7 and compared with literature sources. LJ and LC correspond to the same original samples. The results of our own substrate analytics was within the same range as literature values.

The specific biogas potential (based on organic dry matter) of BW (500 nl/kg oDM) and GW (1000 nl/kg oDM) has been in the expected range given by the literature. Since BW contains a low amount of organics, the absolute biogas production (based on fresh matter) is low. For lawn, the variation within one group has several reasons. The biogas potential of LC and LJ is highly dependent on the substrate quality. For example, the DM content of LC harvested in June of around 20 % was lower compared to that of LC harvested in September of 30 %, which was dry and more ligneous. Additional leaves which are poorly anaerobically degradable were contained as impurity due to the beginning of autumn. The biogas potential of the stored material was in the same range like the fresh

Table 7 Biogas potential of different substrates, our own analysis (average values + standard deviation) in comparison with literature values

Substrate	No. of samples	DM	oDM	Bulk density	Own investigation			Literature	
		%	% DM	kg/m³	Specific biogas	Absolute biogas	Methane	Specific biogas	Source
					nl[a]/kg oDM	nl/kgFM	%	nl/kg oDM	
Blackwater	6	0.6	60	1000	489 ± 51	2	n.a.	500	[37]
Greasy water	10	3	94	1000	1066 ± 42	30	n.a.	1000	[22]
Lawn cuttings									
Fresh	2	38	52	300	393 ± 4	78	n.a.	300	[13]
								550–680	[22]
								300–600	[38]
								700	[21]
3 months stored	2	23	62	600	329 ± 12	87	71		
6 months stored	2	33	78	600	419 ± 19	60	70	600	[22][b]
Lawn juice									
From fresh cuttings	2	8	72	1000	507 ± 1	29	n.a.	617	[36]
								600[c]	[11]
From 3 months stored cuttings	2	4	65	1000	352 ± 25	4	59		
From 6 months stored cuttings	2	2	55	1000	633 ± 66	17	64		

[a]Norm litre
[b]Grass silage
[c]350–400 l methane/kg oDM with an assumed methane rate of 60 % in the biogas

material. Also, the investigation of [32] showed that ensilage of LC had no mentionable influence on biogas potential. On one side, the lactic acid fermentation process of ensilage leads to degradation of organics which could lower the biogas potential. On the other side, a cell disruption takes place which could improve the biogas potential [31]. Similar to fresh cuttings also in the stored variants, the DM content and degree of lignification of the lawn had a main influence of the specific biogas potential. It decreases with increasing water content and lignification, but differences were low (Fig. 1).

The specific biogas potential (related to organic dry matter content) was higher for LJ than for LC with the highest value for the LJ from June cuttings with approximately 600 nl/kg oDM and 350 nl/kg oDM for September cuttings. It is assumed that the moister the LC, the more water with dissolved degradable substances is pressed out.

Also, various mixtures of BW with GW, LC and LJ were tested regarding their biogas potential [19]. They showed the expected range in comparison to the calculated values considering the biogas potential and share

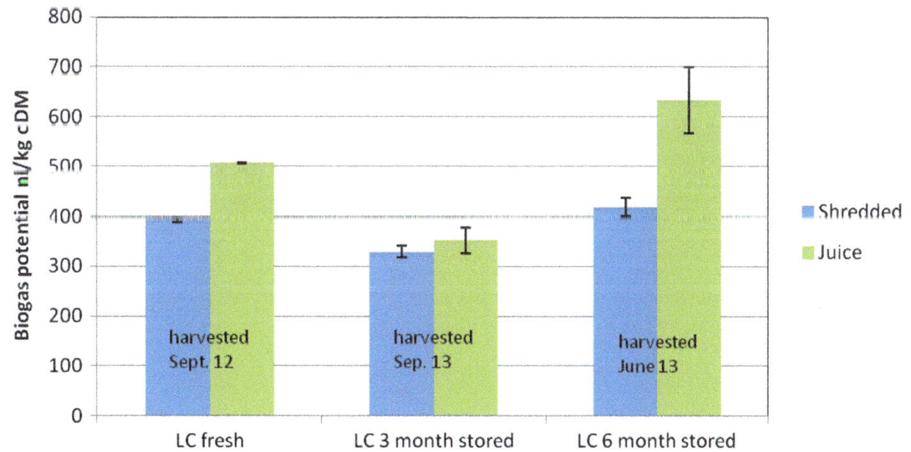

Fig. 1 Biogas production of shredded lawn cuttings and lawn juice from fresh and stored LC with min and max values

of the single substrates. That allows to model biogas potentials of mixtures based on single substrate data.

Basic modelling data

The variation in composition of the bioresources can vary in a wide range and is highly dependent on weather, season, origin and harvest type. Table 8 shows the results of inventory with average values from our own analysis of selected properties, which are used for calculation of expected biogas and nutrient amount in substrate mixtures.

Semi-continuous biogas production

The batch test gives values for biogas production in an ideal case, without adding a substrate to the original during the process. In the real scale of Jenfelder Au, a semi-continuous operating fermenter will be used. Therefore, semi-continuous experiments must be carried out in order to have values closer to the reality. The method for the experiments is given in "Methods for substrate pre-treatment and anaerobic digestion".

In the following, the influence of LJ and GW on the digestion of BW was investigated in semi-continuous experiments at different retention times. The objective was to find a mixture that enables a stable process with low retention time and high biogas production.

Characteristics of substrates used for semi-continuous fermentation

The specific characteristic of substrates used for the semi-continuous biogas experiments is given in Table 9. The specific data are shown, since the inventory and substrate analyses ("Inventory and substrate analysis") showed a certain range of variability within one substrate type and biogas potentials showed a dependency from specific substrate parameters ("Biogas potential of substrates").

BW was collected twice during the experiment. Table 9 shows that the BW composition was quite constant over the whole period of the investigation. A similar variation was observed for the GW; it was collected once before the experiment. Origin and storage are described above ("Origin of substrates" and "Methods for substrate pre-treatment and anaerobic digestion"). However, for the LJ, there was a considerable change in the composition between LJ 1 and LJ 2 (Table 6). LJ 1 was richer in all properties compared to LJ 2 and additionally more acidic. The main reason for the differences in composition presumably resulted from the different harvesting times; LC for LJ 1 were taken in summer while LC for LJ 2 in autumn. Therefore, the latter were very dry, and additionally, they contained a high proportion of branches and leaves. The higher acidity in LJ 1 is probably due to a longer storage time of the juice before use.

Substrates mixtures and feeding period in semi-continuous fermentation

The fresh mass composition of the substrate mixtures investigated in four 10-L-CSTR systems ("Methods for substrate pre-treatment and anaerobic digestion") is shown in Table 10.

The time lapse of the experiments was divided into three periods in which the substrates and retention times were changing (Table 11), but the composition was the same in each reactor over the whole experiment time.

Biogas production in semi-continuous fermentation

Figure 2 shows the absolute biogas production per kilogram of fresh matter, and Fig. 3 shows the specific biogas production per kilogram of organic dry matter (oDM) for the semi-continuously fed anaerobic digestion experiment in the four reactors over a period of 130 days.

Absolute biogas production refers to the fresh mass fed to the reactor as already-prepared substrate (e.g. kilogram of lawn juice and not kilogram of fresh lawn cuttings). The highest absolute biogas production was observed in reactor 4 during periods I and II (start). It was the variant with the highest share of LJ from the average composition. In period II, the value decreases due to, especially in the case of LJ, the composition of

Table 8 Results of inventory and biogas potential of substrates; values rounded for modelling [18, 19]

Bioresource	Bio-resource potential t FM/a	Biogas potential l/kg FM	Biogas potential l/kg oDM	Methane %	Bulk density kg/m^3	DM %	oDM % DM	Nitrogen % DM	Phosphorous %DM
Blackwater	5000[a]	2	500	75	1000	0.5	65	28.0	2.7
Greasy water	1500[b]	13	1000	70	1000	2	85	2.5	0.5
	7830[c]								
Lawn cuttings	21,000[d]	96	400	60	300	30	80	3.0	0.5
Lawn juice	8400[d]	18	500	70	1000	5	70	4.0	2.0

[a]Jenfelder Au
[b]Whole Wandsbek
[c]Whole Hamburg
[d]5 km surrounding within Wandsbek

Table 9 Characterization of substrates used for semi-continuous biogas production (average ± standard deviation) [26]

Parameter	Unit	Blackwater	Greasy water	Lawn juice 1 fed in periods I and II	Lawn juice 2 fed in period III
pH	–	7.4 ± 0.2[a]	4.8 ± 0.2[a]	4.2 ± 0.2[d]	6.2 ± 0.2[e]
DM	%	0.52 ± 0.06[b]	1.95 ± 0.22[c]	4.64 ± 0.12[c]	2.07 ± 0.25[f]
oDM	% DM	59.6 ± 7.69[b]	67.69 ± 10.25[c]	77.59 ± 14.01[c]	68.12 ± 13.04[f]
COD	mg/l	8611 ± 1664[b]	41,783 ± 11,185[c]	64,061 ± 11,325[c]	23,200 ± 3726[f]
TOC	mg/l	2371 ± 477[b]	6544 ± 957[c]	27,450 ± 7,215[c]	8748 ± 2714[f]
TN	mg/l	1753 ± 248[b]	323 ± 136[c]	2,272 ± 713[c]	917 ± 312[f]

[a]60 analysed samples
[b]9 analysed samples
[c]5 analysed samples
[d]38 analysed samples
[e]22 analysed samples
[f]3 analysed samples
[g]7
[h]10

solids which varied highly and thus also the absolute biogas production.

During period III, the biogas production decreased further since the autumn LJ 2 had a low quality and contained much less degradable solids (Table 9). Regarding organic dry matter and water content, it was comparable with GW. The biogas production decrease of reactor 2 in period II was caused by accumulation of acids due to higher loading rates and will be discussed below. Reactor 3 showed a constant value in periods I and II, and only the use of low-potential LJ 2 in period III led to lower biogas production rates.

The *specific biogas productions* are compared in Fig. 3 for the four reactors. Reactor 2, which contained only BW and GW, had the highest specific biogas production presumably due to the high proportion of fat-containing molecules in the mixture. The degradation of fat provides higher yields (1390 nl/kg oDM) than the degradation of carbohydrates (750 nl/kg oDM) and proteins (800 nl/kg oDM) (VDI 4630). However, it is also visible that a change in conditions, namely the retention time, had a big influence on the biogas production of reactor 2. Shorter retention times and consequently higher loading rates in period II (Table 11) resulted in the accumulation of acidic compounds which may be harmful to the methanogens. If concentrations of those become too high, it can lead to reactor breakdown, as what happened in reactor 2 which was only able to recover after feeding only BW for several days until its recovery in period III (Fig. 4).

Table 10 Composition of the substrate mixtures fed into reactors of the semi-continuous anaerobic test system [26]

Reactor	Blackwater (BW)	Greasy water (GW)	Lawn juice (LJ)
	% FM	% FM	1/2 % FM
1	100.0	–	–
2	33.3	66.7	–
3	33.3	33.3	33.3
4	33.3	–	66.7

The *process stability* can be determined by FOS/TAC (volatile fatty acids to total inorganic carbonate) measurement. The FOS/TAC value represents the organic acids in relation to total inorganic carbonate in the substrate. For a stable process, the FOS/TAC value should be around 0.3 and the pH value should be around 7 [29]. Figure 4 shows the development of the FOS/TAC value for the substrate mixture BW with GW in reactor 2. After the increase of substrate loading rate, connected to a decrease of hydraulic retention time (HRT) from 55 to 40 days at the beginning of period II (on day 41), the FOS/TAC value increased from 0.2 to 0.4 and correspondingly the biogas production decreased by half, as shown in Fig. 4. On day 60 within period II, the FOS/TAC value increased further up to 0.7 and the pH dropped below 7. In order to avoid a reactor breakdown, the feeding of GW was stopped and substituted by BW. Until day 80, pH rose above 7 and the FOS/TAC value decreased to 0.2. The process was considered as stable again. The feeding with GW was resumed, and high loading rates with retention times of 40 days were adjusted again (period III). The biogas production improved again and remained stable this time. The specific biogas production related to organic dry matter amount was comparable between low (period I) and high (period III) loading rates. That shows that methanogenic bacteria can tolerate also a higher loading rate but need time to adapt to the new conditions [26].

All other reactors showed pH values ranging between 7 and 7.8 and FOS/TAC values around 0.20 all the time. They were considered as stable and have room for further reductions of retention times. The retention time should be reduced as far as possible to increase the overall process efficiency.

Overall evaluation. The mixture of BW, GW and LJ in reactor 3 showed not only a stable process (FOS/TAC

Table 11 Different timely periods according to changes of hydraulic retention time (HRT) and substrate characteristics performed in the experiments [26]

Period	Relevant changes in the process made in this period	Reactor 1		Reactors 2–4	
		HRT	Loading rate	HRT	Loading rate
		days	kg FM/m^3*day	days	kg FM/m^3*day
Period I	LJ1 in reactors 3 and 4	20	50	55	18.18
Period II	Change of HRT from 55 to 40 days in reactors 2–4; reactor 2 only fed with BW between day 61 and 75[a]	20	50	40	25.00
Period III	Use of LJ 2 with low VS, COD and TOC content (Table 4) in reactors 3 and 4	20	50	40	25.00

[a]In order to recover from organic overfed

around 0.2 ± 0.05), but also a constant and highly specific biogas production over the whole experimental period. It produced an average of 834 ± 87 nl/kg oDM of biogas.

In comparison to the calculated theoretical biogas potentials based on the batch experiments (Table 7), the continuous experiments showed higher biogas yields regarding GW and LJ. This may be because unadapted inoculum was used for potential determination. The mixture of GW with BW in reactor 2 produced an average of 1234 ± 152 nl/kg oDM, while the mixture of LJ with BW in reactor 4 produced 728 ± 90 nl/kg oDM (with LJ 1), in comparison to the respectively theoretically calculated 1000 and 500 nl/kg oDM.

Fertilizer from digestate

Feeding scenarios

After anaerobic digestion, a digestate remains. Three scenarios with different substrate mixtures were chosen to simulate the digestate flows which could be expected in Jenfelder Au. All scenarios are calculated with a hydraulic retention time of 25 days, which is the target time of the operators. The feed mixtures are based on the findings in the previous chapters:

- *Scenario 1* contains only BW (40 % FM) and GW (60 % FM). It illustrates the actual plannings in the Jenfelder Au demonstration project.
- *Scenario 2* contains BW (40 % FM) with GW (30 % FM) and LC (30 % FM) as co-substrate. The mass of LC addition was limited by the dry matter content in the fermenter, which should be not more than 10 % to allow pumping and stirring operations.
- *Scenario 3* contains BW (40 % FM) with GW (13 % FM) and LJ (47 % FM) as co-substrate. The amount of GW was restricted by the potential available in the district Wandsbek where Jenfelder Au is located.

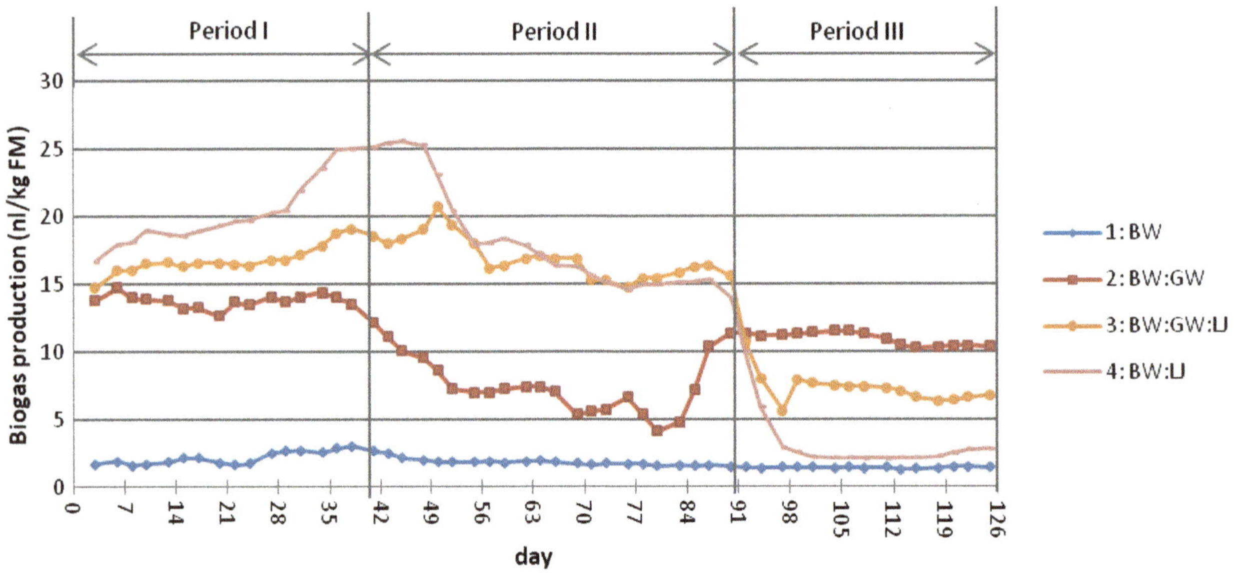

Fig. 2 Absolute biogas production in semi-continuous fermentation in 10-l reactors for four different mixtures with the substrates BW, GW and LJ during three different feeding periods

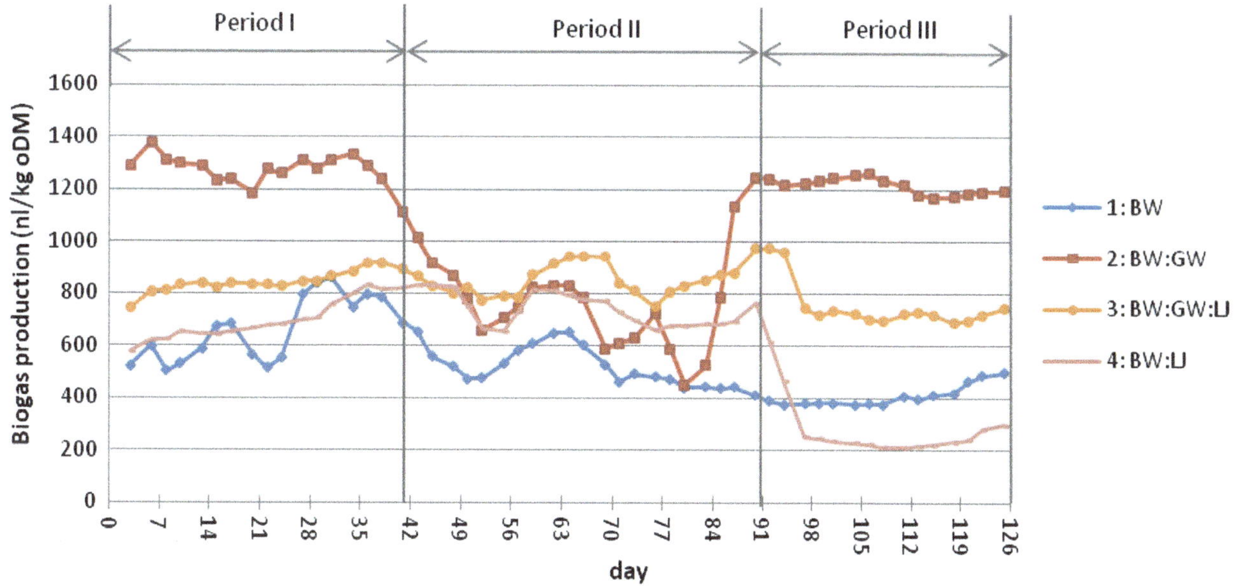

Fig. 3 Specific biogas production in semi-continuous fermentation in 10-l reactors for four different mixtures with the substrates BW, GW, LJ during three different feeding periods

Digestate characterization

The characteristic of digestates which could be expected from the three scenarios is given in Fig. 5. The methodology was described in "Calculation of digestate characterisation".

Figure 5 illustrates that the initial masses are transformed only to a small extent into biogas between 1.2 and 4.4 % of input mass and that digestates remain in amounts comparable to the input. The most important compound in the digestate is water with more than 90 %. The remaining is composed of organics and ash. Digestate contains also nutrients, which may be dissolved in the water fraction and are bound in organics and ashes. The most important nutrients regarding quantity are nitrogen (N) and phosphorous (P). Their contents in the feed are given in Table 8. The concentrations of the relevant nutrients in the digestate were calculated following the procedure in "Calculation of digestate characterisation", and the results for the three scenarios are given in Table 12.

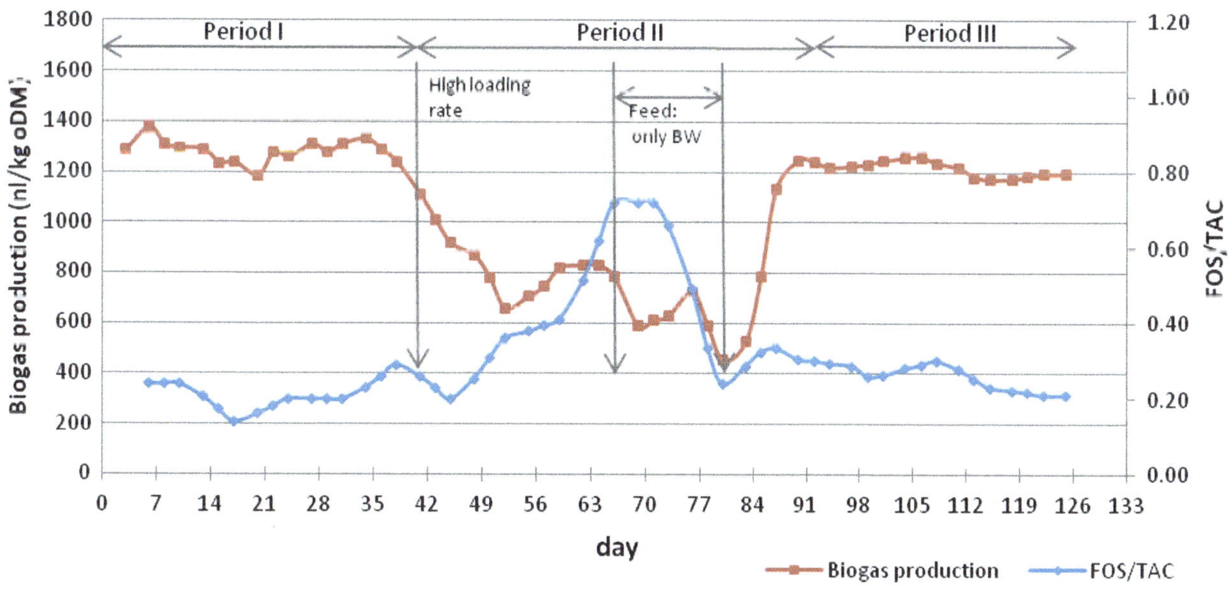

Fig. 4 Biogas production and FOS/TAC during semi-continuous fermentation in 10-l reactors with BW and GW mixture as substrate

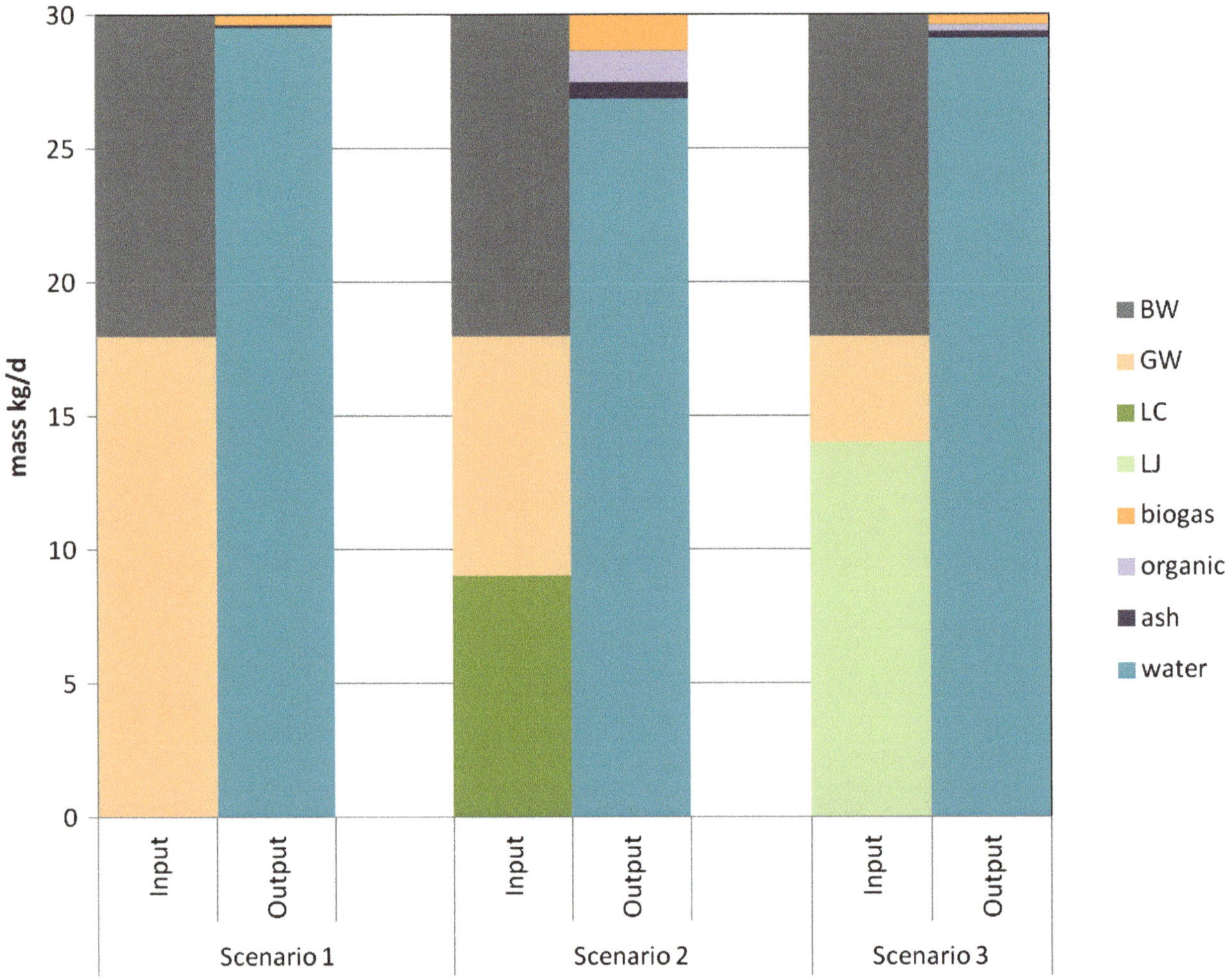

Fig. 5 Mass balances for the three scenarios with different feed mixtures and corresponding outputs for the Jenfelder Au situation. Output: digestate (composed from water, ash and organic) and biogas

The content of ammoniacal nitrogen ranges between 0.6 and 1.2 g/l. The ammoniacal nitrogen is almost completely contained in the liquid phase. It is plant available and therefore has a potential for fertilization. The organic nitrogen remains mainly in the particles, due to a high input of LC in scenario 2 with a high content of organic matter and organic N, and also the N potential of the digestate is high. The P content ranges between 0.2 and 0.5 g/l, which is mostly obtained in scenario 2 due to the high amount of LC. But also scenario 3 with LJ has notable P potential, because P is mainly solved in the liquid phase and a high amount can be recovered in the LJ after pressing.

Utilization of digestate

N and P are important fertilizing elements but have to be recovered to make them usable. A direct application of the digestate within the near surrounding of Jenfelder Au is not possible because of an urban development. Transport to rural areas is expensive and mainly water is transported. Additionally, a hygienization of digestate has to be done with high input of heat. Therefore, different methods for treatment based on literature research were investigated with the objective to generate valuable fertilizer products.

One considered process step is *solid–liquid separation*. A literature research has been carried out for solid–

Table 12 Calculated nutrient concentration in digestate for the three Jenfelder Au scenarios

Scenario		Units	Dry matter	Organic N	NH_4^+/NH_3-N	Total N	Total P
1	BW:GW	mg/l	4120	220	680	900	180
2	BW:GW:LC	mg/l	65770	1700	1160	2860	450
3	BW:GW:LJ	mg/l	18080	950	650	1600	370

liquid separation practices of agricultural digestates, and the theoretic separation efficiency of mass, DM, oDM TN, NH_4^+/NH_3-N and TP was determined [16]. A screw press is assumed with separation efficiency based on data given by [12, 15] (Table 13).

From the *liquid fraction*, ammoniacal nitrogen can be removed from the digestate by stripping with sulphuric acid or limestone and recovered from the gaseous phase via scrubbing with an ammonium sulphate solution [5]. Additionally, the recovery of phosphorus could take place in a stirred loop reactor through the addition of lime [20]. For the *solid fraction*, the prioritized utilization option is composting. In order to do so and to generate compost, mixing with structure-rich materials, for example with wood chippings, is necessary.

Figure 6 shows one possibility of digestate utilization with solid–liquid separation, ammonia stripping with the production of ASL fertilizer,[1] ammonium lime and composting [23]. Ammonium lime is also used as a fertilizer in agriculture, and compost is an important soil conditioner with fertilizing properties. The remaining water could eventually be treated together with greywater of Jenfelder Au [20]. The stripping process takes place with high temperatures and may lead to a hygienization of the digestate.

Based on the exemplarily process cascade in Fig. 6, various amounts of the product for the three different scenarios were calculated. The method and the assumptions are given in "Calculation of digestate characterisation" and table 8. Table 14 shows the expected results. An additional, even more advantageous option may be to align the solid–liquid separation after the stripping.

The comparison of the scenarios shows that the addition of LC increases the amount of nutrients and decreases water content in the digestate. For that reason, scenario 2 produces about twice as much ASL fertilizer compared to scenarios 1 and 3. The amount of compost in scenario 2 is lower since less structure material is necessary due to lower water contents in the solid phase. Additionally, phosphorus precipitation could take place with a supposed recovery rate of 90 % of the phosphorus in the liquid phase. Table 14 shows the amount of pure phosphorus, and also scenario 2 provides the highest amount.

Conclusions

It has been shown that blackwater can be used for biogas generation and that it has a potential as fertilizer. The organic material derived from excrements and toilet paper has an energetic potential. Although the fermentation of blackwater does not result in very high biogas yields, it is energetically advantageous compared to the conventional wastewater treatment which consume energy. The nutrients N and P are of special interest for material recycling. These components can be found mainly in urine and can be used for the production of fertilizers. Blackwater combined with local waste streams is beneficial in order to gain biogas amounts in a considerable range. Inventory studies showed a sufficient potential of secondary, tertiary and quaternary bioresources around Jenfelder Au which have a high biogas potential and also contain nutrients.

The experiments have shown that co-digestion has a positive effect on biogas yields and lawn cuttings are suitable as co-substrate. Lawn cuttings can be applied as lawn juice or lawn suspension. The lawn juice showed a higher specific biogas yield than the used lawn suspension and is easily mixable with blackwater and greasy water. Greasy water showed the highest specific biogas potential related to organic matter but has also high water content; therefore, absolute biogas generation rates are rather low. The high content of greasy water combined with a low retention time of 25 days leads to instability and can easily lead to reactor breakdown due to the high content of volatile acids and is not recommended. In case of no further availability of other co-substrates, a retention time of not less than 40 days is recommended. Process optimizations, e.g. by trace element additions and microbial adaptions, could eventually lead to further reductions of retention time.

It is advised to use greasy water only together with the other lawn-based co-substrates up to a share of one third. Furthermore, the regional available amounts have to be considered which is in Jenfelder Au and even less than 20 % of the required feed mass. In practice, dry matter content of lawn suspension can be increased up to 10 %. That means the biogas production of suspension in relation to fresh matter can be improved and the quality of digestate products can be increased.

Therefore, scenario 2 should be the best scenario for Jenfelder Au with the aim of biogas production. It is combined with a high nutrient recovery and the production of high-quality fertilizer products. For decisions on co-substrates, biogas facility operation scheme and digestate treatment strategy, the whole chains have to be considered and compared with actual strategies. Since the Jenfelder Au project aims at demonstration in small scale, technical feasibility, positive energetic gains and material product possibilities should be especially considered. For instance, if the digestate from the Jenfelder Au facility would be disposed into the sewer and not utilized, it would run through the whole central wastewater treatment

Table 13 Separation efficiency of a screw press, percent by mass of selected components that remains in the solid phase [16]

Separation efficiency	Mass	DM	oDM	TN	NH_4^+/NH_3-N	TP
	%	%	%	%	%	%
Screw press	12	37	40	14	4	26

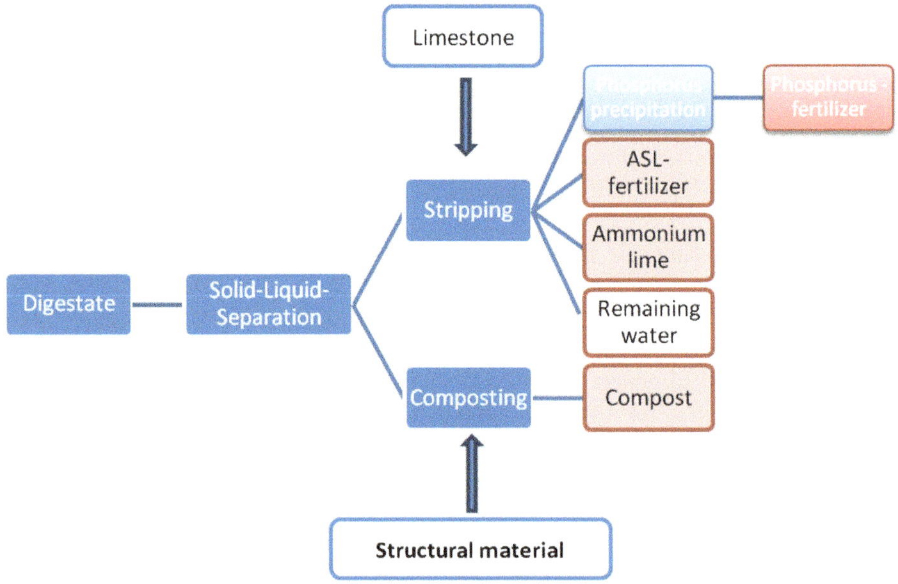

Fig 6 Suggested way of digestate utilization

complex including an anaerobic digestion step later. The new biogas facility would be an unnecessary additional step for blackwater and greasy water. Only for not yet collected lawn cuttings will the fermentation process lead to energetic gains. But in this case, a separate lawn cutting digestion would be preferable.

Several scenarios for complete or partial utilization of digestate were assessed. To calculate the amount of products, which could be generated in Jenfelder Au, a model is under development that includes cost-benefit calculation at all stages. A further step will be an energetic and economical assessment of several scenarios, including transport and pre-treatment of substrates as well as treatment of digestate in order to choose the best option for Jenfelder Au.

Also if coupled biogas and digestate facility would have no positive economical balance considering bioresource collection, pre-treatment, facility operation and investment costs, the overall system could be beneficial compared to the state of the art. The common wastewater treatment process is very energy intensive and leads to the losses of valuable plant nutrients (N and P). Compared to the actual situation of central wastewater treatment in Hamburg (for 150 million (Mio) t wastewater/a: electricity consumption 76.5 Mio MWh/a, heat consumption, 71 Mio MWh/a; [2]), the suggested procedure would be beneficial, since the most energy-intensive steps are avoided. This is on the one hand the nitrification/denitrification step for N elimination; instead, nutrients are recovered in the new process. On the other hand, wastewater can be handled decentrally and does not have to be pumed via long underground pipelines. Both the new process as well as the actual process contain an anaerobic fermentation step. The biogas potential from blackwater from the vacuum toilets from the new process (500 nl/kg oDM) is comparable with the biogas potential of sewage sludge used in the actual process (525 nl/kg oDM, ARCHEA [4]).

The new process is especially advantageous for regions where actually no wastewater treatment exists. Additionally, further actually unused residual streams can be included in the process which gives additional advantages.

Table 14 Examples for products received after digestate utilization processes with solid–liquid-separation, stripping and composting

Scenario	Digestate and its fractions			Additional input		Products			Additional product	Rest
	Digestate	Solid phase	Liquid phase	Structure material	Limestone	Compost	ASL (40 %)	Agricultural lime	Phosphorus	Remaining water
	t/a	t/a	t/a	t/a	t/a	t/a	t/a	t/a	t/a	t/a
1	10,800	1300	9500	3500	44	4800	77	37	1.3	9500
2	10,500	1300	9200	2100	76	3400	130	63	3.3	9100
3	10,800	1300	9500	3200	42	4500	73	35	2.7	9500

Further works shall focus on detailed mass and energy balances from the process and a comparison with other innovative processes for wastewater treatment. If various scenarios are suggested, the various aspects of sustainability—economical, ecological and social—should be compared for a decision [17].

Endnotes

[1]ASL fertilizer, ammonium-sulfat-lösung—ammonium sulphate solution (liquid fertilizer containing 40 % ammonium sulphate)

Abbreviations

ASL: ammonium-sulfat-lösung (ammonium sulphate solution); BW: blackwater; COD: chemical oxygen demand; CSTR: continuous stirred tank reactors; DM: dry matter; FM: fresh matter; FOS/TAC: volatile fatty acids to total inorganic carbonate; GW: greasy water; HWC: Hamburg Water Cycle®; LC: lawn cuttings; LJ: lawn juice; N: nitrogen; nl: norm litre; oDM: organic dry matter; P: phosphorous; TN: total nitrogen; TOC: total organic carbon; TP: total phosphorous; TUHH: Hamburg University of Technology.

Competing interests

The work was founded by the German Federal Ministry of Education and Research (BMBF) in the frame of the KREIS project (Titel, FKZ 033L047C). The academic and intellectual property rights of the authors have to be considered.

Authors' contributions

SH has made substantial contributions to the conception and design, acquisition of data, and analysis and interpretation of data and drafted the manuscript. PN has made substantial contributions in parts of the experimental work, analysis and interpretation of data and drafting of the manuscript.SD has made substantial contributions in parts of the experimental work and interpretation of data. IK has made substantial contributions to the conception and design, was involved in the drafting of the manuscript or revised it critically for important intellectual content, and have given final approval of the version to be published. All authors approved the final manuscript.

Acknowledgements

We would like to acknowledge the support of the Thünen Institute Hamburg-Bergedorf and the Leibniz Institute for Agricultural Engineering Potsdam-Bornim for allowing the use of their presses. Also, we would like to thank all who provided substrates for our analyses, the settlement Lübeck Flintenbreite for the supply of blackwater, the company BUHCK for support and supply of lawn cutting, the caretakers of the TUHH for the provision of lawn cuttings and the waste water treatment plant Köhlbrandhöft, Hamburg for supply of greasy water.
Also, we would like to express our appreciation to the coordinators of the KREIS project, Hamburg Wasser and Bauhaus-Universität Weimar and the funding organizations, Projektträger Jülich (PTJ) and the Federal Ministry of Education and Research (BMBF).
We would like to thank our lab assistants Binnur Kuscu and Susanne Eggers and, last but not the least, all the students who worked for the institute and did their thesis as part of the project

References

1. Adwiraah H, Oldenburg S, Körner I, Stegmann R (2012) Bioresources inventory. GIS-based identification, quantification and characterisation methods. In: H.W (ed) ORBIT 2012-Global assessment of organic resources and waste management. 8th International Conference on Organic Resources and Biological Treatment. ORBIT. ORBIT, Rennes, Frankreich. Weimar, Germany, pp 80–87
2. Agüero Murillo A (2014) A decision support model for the assessment of bioresource management systems. Case study: Jenfelder Au's decentralized system vs. centralized system in Hamburg. Master thesis at Hamburg University of Technology, Institute of Wastewater Management and Water Protection, Bioconversion and Emission Control Group, unpublished
3. Alp Ö (2010) Further treatment of digested blackwater for extraction of valuable components. Dissertation. Technische Universität Hamburg Harburg, Hamburg. Institut für Abwasserwirtschaft und Gewässerschutz. Available online at http://www.tuhh.de/t3resources/aww/publikationen/pdf/diss/Doktorarbeit_Oeznur_Alp.pdf. Accessed 22 Feb 2013
4. ARCHEA Service GmbH: Gaserträge und Nährstoffgehalte - Abfall. Available online at http://www.archea-biogas.de/_mediafiles/9-substrate.pdf. Accessed 25 June 2015
5. Bauermeister U, Wild A (2009) Stickstoffabtrennung mit dem ANAStrip-Verfahren System GNS. In: Fachagentur Nachwachsende Rohstoffe e.V. (FNR) (ed) Gärrestaufbereitung für eine pflanzenbauliche Nutzung - Stand und F + E-Bedarf, vol. 30. With assistance of Fachagentur Nachwachsende Rohstoffe e.V. (FNR). Fachagentur Nachwachsende Rohstoffe (Gülzower Fachgespräche, 30), Gülzow, pp 78–95
6. Bocksch M (2006) Rasenschnittgut - "Fluch oder Segen"? Manuskript: DRG Rasen-Thema Juni 2006. Edited by Deutsche Rasengesellschaft e.V. Deutsche Rasengesellschaft e.V. DRG Hompage. Available online at http://www.rasengesellschaft.de/content/rasenthema/2006/06.php, updated on 6/23/2006. Accessed 26 Feb 2013
7. Deegener S (2012) Halbtechnische Untersuchungen zur Vergärung von Schwarzwasser und Fettabscheiderabwässern. Forschungsbericht. Technische Universität Hamburg-Harburg, Hamburg. Institut für Abwasserwirtschaft und Gewässerschutz, checked on 31 Jan 2013
8. DIN EN 12880, (2000) Charakterisierung von Schlämmen - Bestimmung des Trockenrückstandes und des Wassergehalts.
9. DIN EN 12879, (2001) Charakterisierung von Schlämmen - Bestimmung des Glühverlustes der Trockenmasse.
10. Flerlage JM (2012) GIS-basierte Analyse des Biogaspotentials im Hamburger Osten. Eine Biomassestudie im Rahmen des Projekts "Jenfelder Au". Bachelorarbeit. Technische Universität Hamburg-Harburg, Hamburg. Institut für Abwasserwirtschaft und Gewässerschutz, checked on 31 Jan 2013.
11. Florafuel AG Das florafuel-Verfahren: Bioenergie aus Biomasse. Available online at http://www.florafuel.de/de/florafuel-verfahren/das-florafuel-verfahren/. Accessed 2 Oct 2014
12. Fuchs W, Drosg B (2010) Technologiebewertung von Gärrestbehandlungs- und Verwertungskonzepten. Univ. für Bodenkultur, Wien, Available online at http://www.worldcat.org/oclc/707168218
13. FNR (2006) Fachagentur Nachwachsende Rohstoffe e.V. (Ed.): Handreichung Biogasgewinnung und -nutzung (2006). Available online at http://www.big-east.eu/downloads/FNR_HR_Biogas.pdf. Accessed 25 Feb 2014
14. Hamburg Wasser (2012) Statusbericht-Bemessungswerte SW-Behandlung., E-Mail to TUHH, February 2012
15. Hjorth M, Christensen KV, Christensen ML, Sommer SG (2009) Solid–liquid separation of animal slurry in theory and practice. A review. Agron Sustain Dev 30(1):153–180.
16. Huhn J (2013) Erstellung und Vergleich von Szenarien zur Gärrestverwertung. Dargestellt am Beispiel der anaeroben Schwarzwasserbehandlungsanlage "Jenfelder Au". Masterarbeit. Technische Universität Hamburg-Harburg, Hamburg. Institut für Abwasserwirtscahft und Gewässerschutz.
17. Körner I (2015) Civilisation biorefineries: efficient utilization of residue-based bioresources. In: Pandey A, Hofer R, Taherzadeh M, Nampoothiri K Madhavan, Larroche Christian (eds.) Industrial biorefineries and white biorefinery. Elsevier, pp. 296–340; chapter 7, ISBN 978-0-444-63453-5
18. Körner I, Hertel S, Deegner S (2013) The transformation of blackwater, lawn cuttings and grease trap residues into biogas and fertilisers in neighbourhood of Hamburg—an integrated approach on the territorial scale. In Université de Versailles St-Quentin-en-Yvellines (Ed.): RAMIRAN 2013. Recycling of organic residues for agriculture: from waste management to ecosystem service. 15th International Conference. RAMIRAN. Versailles, France, 03.-05.06.2013. Université de Versailles St-Quentin-en-Yvellines.
19. Körner I, Hertel S, Deegner S, Boltz K (2013) Schwarzwasser von Vakuumtoiletten zur Biogaserzeugung. Aspekte zur Integration in ein Bauprojekt unter besonderer Berücksichtigung von Co-Substraten. In: Nelles M (ed) Tagungsband 7. Rostocker Bioenergieforum. Schriftenreihe Umweltingenieurwesen, vol. 36. Universität Rostock, 20.-21.06.2013. Rostocker Bioenergieforum. Univ., Agrar- und Umweltwiss, Rostock, pp 483–492, Fak. (Schriftenreihe Umweltingenieurwesen, 36)
20. KREIS (2012) Versorgung durch Entsorgung. Kopplung von regenerativer Energiegewinnung mit innovativer Stadtentwässerung. Edited by Hamburg Wasser. Available online at www.kreis-jenfeld.de

21. Kruse A (2011) Einfluss der verschiedenen Lagerugnsverfahren auf das Biogaspotenzial von Rasenschnitt. Bachelorarbeit. Technische Universität Hamburg-Harburg, Hamburg. Institut für Umwelttechnik und Energiewirtschaft.
22. KTBL (2014) Kuratorium für Technik und Bauwesen in der Landwirtschaft; Wirtschaftlichkeitsrechner Biogas. Edited by KTBL. Available online at http://daten.ktbl.de/biogas/startseite.do.
23. Lüssenhop P (2014) Simulation eines Szenarios zur Gärrestaufbereitung im Wohngebiet Jenfelder Au. Projektarbeit. Technische Universität Hamburg-Harburg, Hamburg. Institute für Abwasserwirtschaft und Gewässerschutz.
24. Mandl M, Graf N, Thaller A, Böchzelt H, Schnitzer H, Steinwender M et al. (2006) Grüne Bioraffinerie - Aufbereitung und Verwertung der Grasfaserfraktion. Ein Projektbericht im Rahmen der Programmlinie FABIRK der Zukunft. Edited by Innovation und Technologie Bundesministerium für Verkehr. Wien (Berichte aus Energie- und Umweltforschung, 67/2006), checked on 27 Jan 2015
25. Meinzinger F (2010) BERBION. Ergebnisse der Abwasserinventur. Leipzig, Germany, 2010, checked on 8 Apr 2014
26. Navarro NP (2014) Evaluation of anaerobic digestion of blackwater with different co-substrates in bench-scale continuous stirred tank reactors. Master thesis. Technische Universität Hamburg Harburg, Hamburg. Institut für Abwasserwirtschaft und Gewässerschutz, checked on 10 Apr 2014
27. Oude E, Stefanie JWH, Driehuis F, Gottschal JC, Spoelastra SF (1999) Silage fermentation processes and their manipulation. In: Food and Agriculture Organization of the United Nations (FAO) (ed.) FAO Electronic Conference on Tropical Silage, 01.-15.12.1999. The Grassland and Pasture Group and Feed Resource Group of FAO. Available online at http://www.fao.org/WAICENT/FAOINFO/AGRICULT/AGP/AGPC/gp/SILAGE/PDF/Paper2.pdf. Accessed 9 Jan 2015
28. projects energy gmbh (2009) Energie aus heimischer Biomasse. Studie zum Biomasseptotenial in der Freien Hansestadt Hamburg. Edited by Landwirtschaftskammer Hamburg. Behörde für Stadtentwicklung und Umwelt. Hamburg
29. rec-energy: Renewable energy concepts. Edited by Stefan Gleis. Available online at http://www.rec-energy.com/. Accessed 26 Jan 2015
30. Reinhold G (2005) Masse- und Trockensubstanzbilanz in landwirtschaftlichen Biogasanlagen. Edited by Thüringer Ministerium für Landwirtschaft, Naturschutz und Umwelt. Thüringer Ministerium für Landwirtschaft, Naturschutz und Umwelt, checked on 12 Jan 2015
31. Rösch C, Raab K, Skarka J, Stelzer V (2007) Energie aus dem Grünland – eine nachhaltige Entwicklung? Edited by Forschungszentrum Karlsruhe GmbH. Forschungszentrum Karlsruhe in der Helmoltz-Gemeinschaft; Institut für Technikfolgenab. Karlsruhe. Available online at http://www.itas.kit.edu/pub/v/2007/roua07b.pdf. Accessed 15 Jan 2015
32. Schidler S, Adensam H, Hofman R, Kromus S, Will M (2003) Technikfolgen-Abschätzung der grünen Bioraffinerie. Band II Materialsammlung. Edited by iTA. Institut für Technikfolgen-Abschätzung der Österreichischen Akademie der Wissenschaften. Wien. Available online at http://epub.oeaw.ac.at:8000/ita/ita-projektberichte/d2-2d21-2.pdf. Accessed 24 Nov 2014
33. Statista (2010) Pro-Kopf-Verbrauch von Toilettenpapier in Deutschland und der EU | Statistik. Available online at http://de.statista.com/statistik/daten/studie/157665/umfrage/pro-kopf-verbrauch-von-toilettenpapier-in-deutschland-und-der-eu/. Accessed 7 Apr 2014
34. Thamsiriroj T, Murphy JD (2011) Modelling mono-digestion of grass silage in a 2-stage CSTR anaerobic digester using ADM1. Bioresour Technol 102(2):948–959. doi:10.1016/j.biortech.2010.09.051
35. VDI 4630, (2014) Vergärung organischer Stoffe - Substratcharakterisierung, Probenahme, Stoffdatenerhebung, Gärversuche.
36. Wachendorf, M, Hensgen, F, Bühle, L, Möller, D, Blumenstein, B (2012) Aus Grass Pellets und Biogas erzeugen. In Top Agrar -Energiemagazin (2). Available online at http://www.prograss.eu/fileadmin/PROGRASS/documentation/german_press/topagrar_2_2012.pdf. Accessed 2 Oct 2014
37. Wendland, C (2008) Anaerobic digestion of blackwater and kitchen refuse. Dissertation. Technische Universität Hamburg-Harburg, Hamburg. Institut für Abwasserwirtschaft und Gewässerschutz.
38. Westphal L (2010) Biogaspotenzial von Gras- und Rasenschnitt. Energetische Verwertung von organischen Reststoffen am Beispiel der grasartigen Bioressourcen im Bezirk Bergedorf. Diplomarbeit. Technische Universität Hamburg-Harburg, Hamburg. Institut für Umwelttechnik und Energiewirtschaft.
39. Zeeman G, Kujawa K, de Mes T, Hernandez L, de Graaff M, Abu-Ghunmi L et al (2008) Anaerobic treatment as a core technology for energy, nutrients and water recovery from source-separated domestic waste(water). Water Sci Technol 57(8):1207. doi:10.2166/wst.2008.101

18

The integration of renewable energy sources and electric vehicles into the power system of the Dubrovnik region

Anamarija Šare[2*], Goran Krajačić[1], Tomislav Pukšec[1] and Neven Duić[1]

Abstract

Background: In order to reduce greenhouse gas emissions, governments seek to replace conventional fuels by renewable ones. Nowadays, most attention is paid to electric vehicles in the transport systems and the use of renewable energy in the power systems. The aim of this work is to achieve a 100 % renewable and sustainable system and to examine the impact of electrification in the transport sector on the power curve and the integration of renewable energy into the power systems of the Dubrovnik region up to 2050.

Methods: The analyses of different charging regulation models for the electric vehicles were derived in the EnergyPLAN, which is a computer model for Energy Systems Analysis of the major energy systems and runs on an hourly basis. Calculations were done for selected years—2020, 2030 and 2050. Charging models provided in the EnergyPLAN were dumb charge, flexible demand, smart charge and smart charge with vehicle-to-grid. For each year, two different charging models were selected. Charging regulations according to three tariff models, based on a lower and higher electricity price, with different distributions, were also done for 2050, i.e. tariff model 1, 2 and 3.

Results: The results for the year 2020 showed no difference between the models. In 2030, smart charge gained better results than a flexible demand. In 2050, the flexible demand allowed to achieve better results than the smart charge with vehicle-to-grid and the tariff model 1, while tariff model 3 provided the best results for 2050. It is also shown that the energy systems which include electric vehicles have a greater impact on the reduction of a critical excess electricity production than the systems excluding electric vehicles.

Conclusions: The power system of the Dubrovnik region was set up as an isolated system, and the electric vehicle batteries are the only storage provided. The results showed that each scenario yielded an excess in electricity produced in the system, which means that the available storage was insufficient and there was a need for more storage capacities in order to achieve a 100 % renewable and sustainable power system.

Keywords: Electric vehicles; Vehicles-to-grid; Renewable energy sources; Electric distribution system; Smart charge

Background

Fuel consumption in the transport sector contributes significantly to the overall greenhouse gas emissions. The transport sector in the European Union participates with a share of 31.7 % in the total final energy consumption [1] and with a share of 20 % in the total greenhouse gas emissions [2]. Joining the European Union, Croatia has assumed the obligations of the Directive 2009/28/EC, which among other things require that each member state by 2020 achieves a minimum share of renewable energy in the final energy consumption in transport by 10 % [3]. Croatia has also published the 'Energy Strategy' following the example of the European Union (EU) which commits to achieve a 20 % production from the renewable energy sources (RES) in the total energy production by 2020 [4]. Promoting the electrification in the transport sector as well as an electricity production based on RES would enable the achievement of these goals. Due to the intermitted electricity production of RES, the corresponding energy systems should include significant reserves in a

* Correspondence: anamarija.sare@gmail.com
[2]Maritime Department, University of Dubrovnik, Dubrovnik, Croatia
Full list of author information is available at the end of the article

form of conventional energy sources or significant capacities of energy storages. It leads to an increase in the prices of electricity and the overall energy system and limits a wider distribution of RES. The increased use of electric vehicles (EVs) provides a significant capacity of the distributed battery chargers, being connected to the grid, for creating new possibilities to integrate RES in the power systems. The study presented in [5] analysed the optimal investments and the scheduling of distributed energy resources with uncertainty in the electric vehicle driving schedules. The results of the study have shown that EVs can have a significant impact on a distributed energy resource investment, provided the payback periods are relatively small. On the other hand, RES are proven to reduce electricity prices in the short term as it is shown in the case study carried out for Germany [6]. These two technologies can benefit from each other because EVs can provide storages of a huge capacity for the electricity produced by intermitted RES, with a quick response in a time of peak demand, while RES have proven to be cost effective and a clean way of the EV energy supply [7]. EVs can represent the containers of the electricity that might be connected to the grid in order to supply the power system during the lack of electricity production from RES, what is also known as a Vehicle-to-Grid (V2G) model [8].

Some previous studies have been performed which have considered the integration of RES into the power systems. Most of them indicated a need for future work and improvements. The analysis of the integration of RES into the power system of Macedonia [9] was carried out using the EnergyPLAN program. EnergyPLAN is a computer model for an Energy Systems Analysis of the major energy systems. The program runs on an hourly basis and is suitable for the analysis of energy systems of regions and countries. The results showed an increase in the RES electricity production by 13 %, a decrease in CO_2 emissions as well as a decrease of critical excess of electricity production. Another case study of Macedonia was performed for a 50 and 100 % share of RES in the power system [10]. It is concluded that a share of 50 % is achievable by applying mandatory directives and reducing the electricity consumption. In both cases, a higher percentage of RES would be considered if it were combined with some kinds of electricity storage such as EVs. Higher implementation of RES into the power system of Serbia was also analysed using the EnergyPLAN [11]. These authors suggested a future analysis of the new tariff models for the electricity prices. The case study of the 100 % renewable power system was done for Croatia [12], and the result points to the possibility of achieving a 78.4 % share of RES in the electricity production as well as a reduction of CO_2 emissions by 20 Mt. The authors of the study dealing with a 100 % renewable energy system of Portugal [13] and another study analysing a higher penetration of wind energy into the power system [14] indicated the need of energy storages in order to achieve these goals. The importance of the role of storages in the power systems with high penetration of RES is emphasised in a doctoral thesis [15] in order to maintain the security of electricity supply and to have less negative impact on the environment. It was concluded that a 100 % renewable system could be achieved if it is combined with energy storages and if it is not set up as an isolated system.

Most of the previous studies indicated the need for energy storage, and one of the possible options is the integration of EVs into the power systems. In a doctoral thesis [16], Denmark's power system, with a high penetration of wind energy, was analysed. Using the EnergyPLAN program, including the V2G model, it was concluded that the V2G model provided a better stability of the energy supply than the existing capacity of conventional power plants. The author suggested a future analysis of different charging models for EVs in order to achieve an optimal integration of EVs into the power system. Previous studies [17, 18] demonstrated that the distributed charging stations would provide a more stable and flexible power system, as well as a minor impact on peak loads.

The aim of this paper is to achieve a 100 % renewable energy system with the use of EVs. The system should be regulated by way of optimum EV charging cycles in an order that the planned electricity production meets the needs of the demand in the best possible way. This paper was prepared as part of the i-RESEV project (Information and Communication Technologies (ICT)—aided integration of electric vehicles into the energy systems with a high share of renewable energy sources) which deals with the implementation of EVs into the power system with a high share of RES in the electricity production that is controlled by a widespread use of ICT tools [19]. The paper will not deal with communication technologies between the vehicles and the energy system, but will only use IT tools.

The analyses provided in this work are done for the Dubrovnik region which includes the city of Dubrovnik with its surroundings. It is situated in the southern part of Croatia at 42°38′25″N and 18°06′30″E. Due to the significant potential of RES, such as solar radiation, hydro and wind energy, the computations should be carried out for a wider area of the Dubrovnik region. In order to achieve a 100 % renewable system, analyses had to be carried out according to both the demand and supply side of the power system. The supply side is obtained from RES including hydro, solar and wind potential of the region. Basic hourly data of the electricity demand and RES potential were taken for the year 2010. The aim of this work is to predict the future capacity to meet the consumer needs, by using the known power consumption and the consumption calculations by 2050, as well as the existing installed capacity of power generation. The regulation of the power system is planned to be carried out using EVs as the energy storage. The data on the EVs are based on the gathered information on the number of vehicles in the region and their driving cycles for the year 2010. Year 2010 represented the reference year based on which the future energy plan till 2050 was derived for the Dubrovnik region power system, with the selected scenarios for the years 2020, 2030 and 2050. The EnergyPLAN program is used to optimize all the components of the electric power system so that the production satisfies the demand. The EnergyPLAN analysis is carried out in hourly steps for 1 year, and the consequences are analysed on the basis of different optimisation strategies. The EnergyPLAN manual with the instructions for use is available online [20]. Analyses of some previous studies, abovementioned in the text, were carried out using the EnergyPLAN model.

Previous studies have demonstrated that the energy systems with a high penetration of RES in the electricity production need to contain some kinds of electricity storage to enable an optimal functioning of the system in a way that the planned electricity production meets the needs of the demand in the best possible way. The goal of this work is to achieve a 100 % renewable energy system using the EVs as the electricity storage. The power system of the Dubrovnik region was set up as an isolated system, so that all of the electricity produced from RES had to be used at a time or stored in the EV batteries. Optimisation of the power system is carried out in the EnergyPLAN using the regulation of the system according to the different EV charging models. Beside the regulation by different EV charging models, analyses were carried out by setting different tariff models of the electricity prices in order to regulate EV charging during the lower tariff and low electricity demand. The results showed that the planned electricity storage is not large enough to store all of the electricity planned to be produced from RES, and the power system has to consider the critical excess in electricity production (CEEP). When we compare this work with some previous studies mentioned above, dealing with a similar problem and not realizing 100 % renewable systems, some of them allowed to achieve systems with a high penetration of RES.

This work represents a good basis for future work in energy system planning related to the Dubrovnik region, as well as for other regions, cities and states that can be considered as an independent energy system. The results of the work showed that there is more work to be done with regard to the optimisation of the energy system in order to achieve sustainability of the system, to encourage the production from RES and reduce the negative impact on the environment and society. Some other kinds of storages need to be taken into consideration, and new analyses need to be performed using EV charging regulations strategies.

Methods

The Dubrovnik area is supplied from the electricity produced by the Dubrovnik hydroelectric power plant working within the Croatian power system with a capacity of 108 MW and is owned by the Croatian Electrical Utility (HEP). It is not only supplying the Dubrovnik region, but also the surrounding areas. The Dubrovnik area has a considerable potential of renewable sources and plans to construct new facilities for hydropower, wind power and solar radiation, which has been taken into account in this paper when planning the production for future scenarios. Production from RES presents a challenge in the operation of the power system, as it is explained in the literature [21]. It depends on the weather conditions, which is the reason why often an excess or a shortage of production occurs. That is why it is necessary to provide reservoirs for the electricity produced and the batteries of EVs. The assumption is that batteries serve as a storage for the excess of electricity produced; while, on the other hand, they can also feedback energy to the grid during the lack of electricity production, the so-called vehicle-to-grid. It is necessary to know the number of EVs that will be entering the system by 2050 in order to calculate the capacity of battery storage. It is assumed that EVs are not always available for charging, so it is necessary to determine their availability for charging in order to provide the capacity of the transport electricity demand. Calculations of the capacity are done using driving cycles and the number of vehicles which are available at a certain time. The method used for the optimization of the electric power system was carried out in the EnergyPLAN by use of various models of EV charging. The impact of EV charging models on the electricity demand, as well as their impact on the CEEP production from RES was analysed.

The paper analyses the power system, which includes the production from RES to meet the consumption needs. The flow chart in Fig. 1 shows the methodical flow including electricity supply, demand and electricity storage of the energy system. The EneryPLAN input data were previously calculated using Excel and HOMER. HOMER is a computer program that enables technological and economical analyses of micro grid energy systems powered by RES. In this work, HOMER is used to calculate wind electricity production using the known hourly wind speed data, as it is explained later in the text. The EnergyPLAN input data consist of hourly distribution curves of all the components of the system and the data on their total year capacity. Analyses were carried out for three

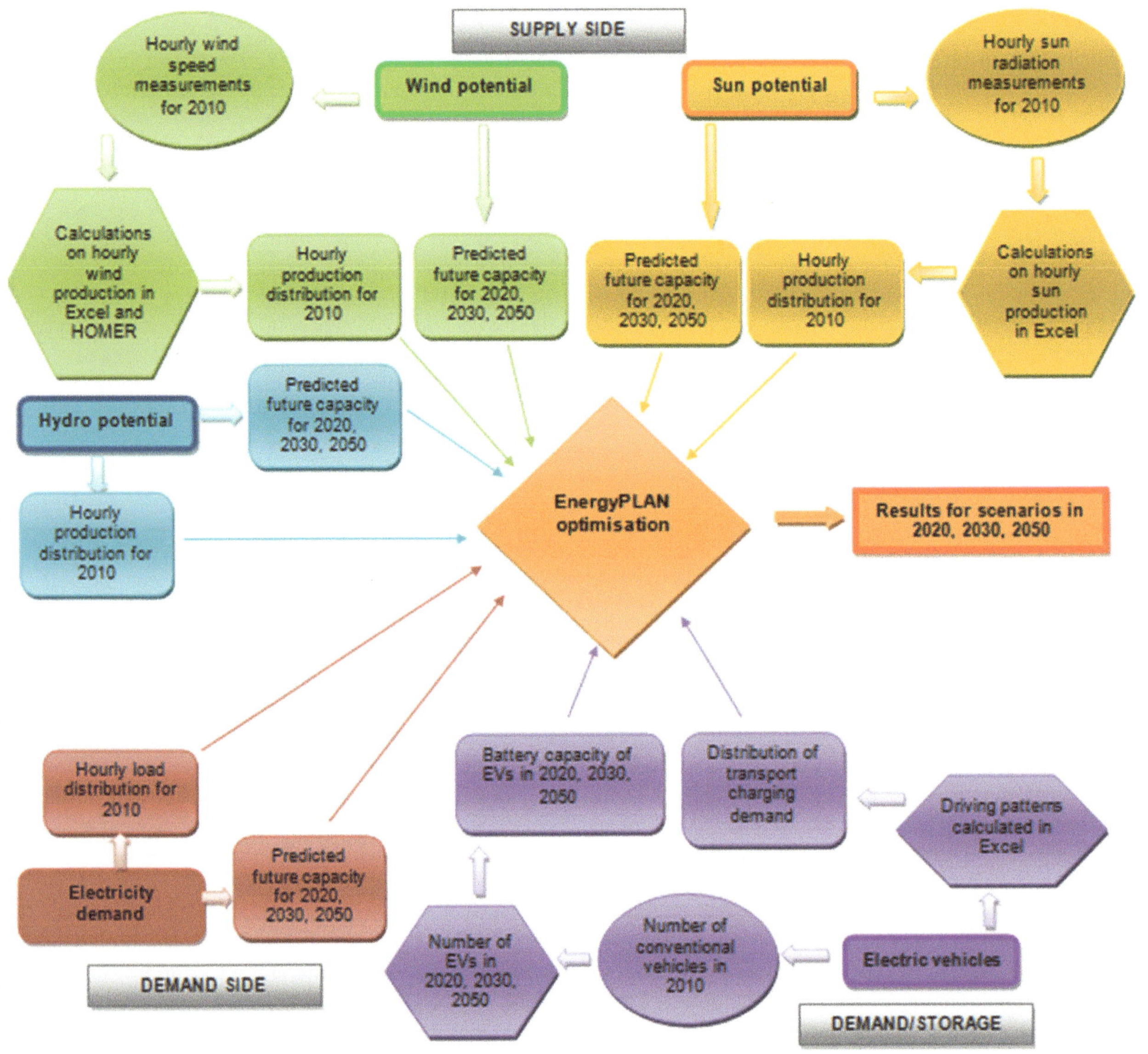

Fig. 1 Flow chart of the paper methodology

selected scenarios—year 2020, 2030 and 2050. The hourly distribution curves were the same for each scenario but the capacities were changed. The next step was the EnergyPLAN optimisation using different regulation models in order to provide the results for scenarios in 2020, 2030 and 2050.

Input data for the EnergyPLAN calculations

The input data for the EnergyPLAN calculations consist of the installed capacity data, the demand capacity data and their hourly distribution curves for a given year. The EnergyPLAN model is used to optimize the energy system of the Dubrovnik region by the given input data on electricity production and electricity demand for selected future scenarios for the years 2020, 2030 and 2050.

Hydroelectric power plant production

The hourly distribution curve of the electricity production of the Dubrovnik hydropower plant was obtained from the data taken for the year 2010. The data were purchased from the HEP Group (HEP d.d.—*Hrvatska elektroprivreda d.d.*), the Croatian national electricity company, for each hour of the year. The description of the power plant and its work is presented in [22]. The projected future capacities of hydropower plant production are shown for selected scenarios in Table 1.

Table 1 Capacity of the production units for all scenarios

Year	2020	2030	2050
Installed capacity of hydropower plant (MW)	70	85	100
Installed capacity of wind turbines (MW)	32	160	320
Installed capacity of photovoltaic systems (MW)	9	18	36

Wind potential

The hourly wind production distribution curve was derived from the measurement data of wind speeds taken from the Croatian Meteorological and Hydrological Service for the year 2010. The data were processed by using the HOMER program, which is the micro power optimization model which details were given in the literature [23]. In this work, HOMER is only used to process the data of wind speeds for a particular wind farm to obtain the production based on wind energy for the Dubrovnik region in the year 2010. The wind speed distribution for the year 2010 was analysed using HOMER as shown in Fig. 2 and the hourly wind production was provided for the installed wind capacity of 6 MW. The maximum wind speed for the year 2010 was 21.4 m/s. The curve of wind production per hour obtained by HOMER was used in further calculations where the higher installed capacities were distributed using the provided curve. Capacities of the future installed wind farms were projected up to 2050. It was carried out by a given projection of the Croatian Ministry of Economy [24]. The projected future capacities of wind power production for selected scenarios are shown in Table 1.

Solar potential

Dubrovnik region has 2623 h of sunshine per year, and it belongs to the areas with the highest irradiation on Croatian territory. Maximum solar irradiation for the year 2010 was 981 W/m^2. The hourly distribution curve of the solar production was derived from the measurement data of the solar radiation for the year 2010, which were taken from the Croatian Meteorological and Hydrological Service. The future installed capacity was projected up to 2050 by a given projection of the Croatian Ministry of Economy [24]. The projected future capacities of solar power production for selected scenarios are presented in Table 1.

Electricity demand

The hourly distribution curve of the electricity demand for the Dubrovnik region was obtained from the measurements of the substation Komolac. The data were provided for the year 2010 by the Croatian Electricity Company—distribution system operator, Elektrojug. The maximum power demand for 2010 reached 64.4 MW, with a total demand of 311 GWh/year. The electricity demand by 2050 was modelled following a study available for Croatia [25]. The projected electricity demand for selected scenarios is shown in Table 5.

Figure 3 shows the relation in distribution between the load curve and the solar and wind potential of the Dubrovnik region in 2010 with the given maximum values of each month. There is a good relationship between solar radiation and energy demand during the summer period (in June, August and September), when both the load and radiation are high. During the winter period, there is a better relationship between demand and wind potential, as the wind potential is higher during October, November and December. The diagram illustrates a good correlation between solar and wind potential. During the summer period, the solar potential is high whereas the wind potential is low, while during the winter period, the opposite occurs. It can be expected that the most excess in electricity production from RES will be available in April, May and June when the load curve is low.

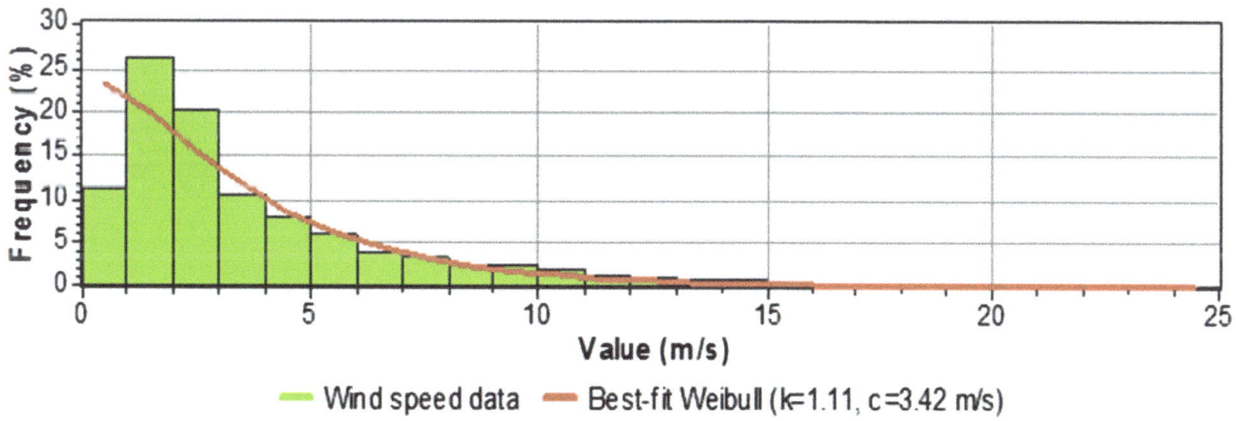

Fig. 2 Weibull distribution of wind speed for the year 2010. *k* (shape parameter), *c* (scale parameter)

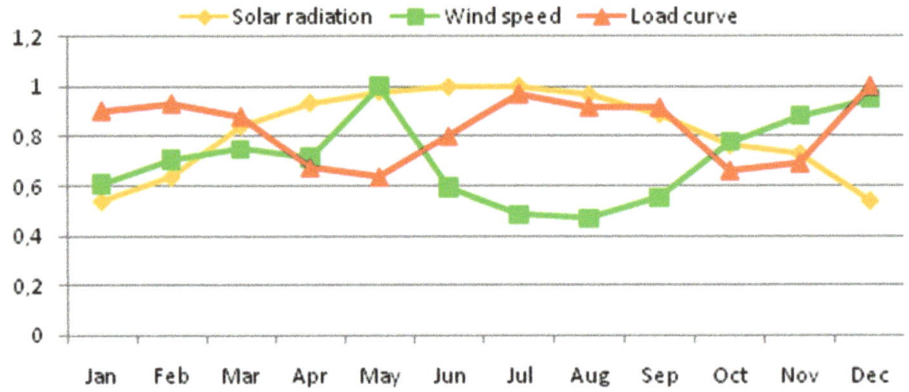

Fig. 3 Monthly variation of the parameters normalized to their maximum value

Transport demand

In order to obtain the transport demand, it was necessary to specify the penetration of electric vehicles by 2050 as well as the EV charging cycles. Many studies focusing on driving cycles and methods were performed for obtaining the transport demand [26–29]. EV penetration in the Dubrovnik region was obtained based on the number of personal vehicles on the territory of the Dubrovnik region in 2010 and a study done for Croatia, in which the EV penetration by 2050 was processed [25], as given in Table 2. It is assumed that in 2050, all of the vehicles will be replaced with EVs [25].

The capacity of the EV batteries is an important data for calculating the transport electricity demand. Some studies have analysed the battery sizes and their impact on the power system as well as their combination with different RES [30, 31]. In this work, this was not done in that way, as only three different battery sizes were chosen for EV fleet according to a test described in [32]. Following their battery size, the vehicles were divided into three groups by the provided number of EVs for all chosen scenarios, as shown in Table 3. Each group has the same number of vehicles and has its own kind of battery with different characteristics.

According to some studies [33, 34], the average daily trip of one particular vehicle is chosen to be 45 km for the entire Dubrovnik region. For the known capacity of each battery (C_B), its range and average daily trip, further calculations can provide the results on how long a full battery will last. The times of charging per day (Ch_i), for each battery type, are gained based on the battery life, which gives us the number of EVs ($EV_{tot,i}$), from each group, that need to be charged during one day, as shown in Table 4. The subscription i represents each group of vehicles with small, medium and large batteries.

$$Ch_i \cdot EV_i = EV_{i,tot} \tag{1}$$

With a known number of EVs that need to be charged per day, as well as their battery capacities per group, the whole daily capacity demand of the entire fleet ($C_{B,tot}$) can be calculated for all scenarios as shown in Table 5.

$$EV_{tot,i} \cdot C_{B,i} = C_{B,i,tot} \tag{2}$$

$$C_{B,tot} = \sum C_{B,i,tot} \tag{3}$$

The total capacity data of the transport demand for each specific year has to be inserted in the EnergyPLAN, along with the hourly distribution curve for the same year, in order to be able to perform the calculations for the EnergyPLAN optimization. Due to the lack of transport data, it is considered that the capacity is going to be the same for each day of the year, so that one specific day is taken into account. The hourly distribution curve of the transport demand was modelled according to the traffic load profile for one specific day gained from the traffic study done for the Dubrovnik region [35] and illustrated in Fig. 4.

The inverted curve (TLC_{inv}) is obtained from the known traffic load curve (TLC) for the Dubrovnik region which represents the vehicles that are available for

Table 2 Number of EVs in Croatia and Dubrovnik region for each scenario

Year	2010	2020	2030	2050
Number of EVs in Croatia	0	10,723	581,802	1,368,462
Number of EVs in Dubrovnik	0	146	7922	18,635

Table 3 Characteristics of selected types of EVs [32]

Battery size	Capacity (kWh)	Range (km)	Consumption (kWh/100 km)	Power required to the grid (kW)
Small	10	100	10	2.2
Medium	20	130	15.38	4.4
Large	35	180	19.44	7.8

Table 4 Number of charging per day and number of each type of EVs to be charged

		Number of vehicles to charge per day		
Size	Number of charging per day	2020	2030	2050
Small	0.450	22	1188	2795
Medium	0.346	17	914	2150
Large	0.250	12	660	1553

charging. The whole capacity for EV charging in one day ($C_{B,tot}$) is then distributed based on the inverted curve.

$$\mathrm{TLC}_{\mathrm{inv},j} = 1\text{–}\mathrm{TLC}_j \tag{4}$$

The sum of the data of the inverted curve is equal to 1 as the data will represent the percentage of the total battery capacity (CLC_j) needed in each hour of the day.

$$\mathrm{CLC}_j = \frac{\mathrm{TLC}_{\mathrm{inv},j}}{\sum_{j=1}^{24} \mathrm{TLC}_{\mathrm{inv},j}} \tag{5}$$

$$\sum_{j=1}^{24} \mathrm{CLC}_j = 1 \tag{6}$$

The battery capacity per hour of the day is calculated by multiplying the percentage of the capacity load curve (CLC_j) with the total battery capacity per each day ($C_{B,tot}$):

$$\mathrm{CLC}_j \cdot C_{B,tot} = C_{B,tot,j} \tag{7}$$

Only one specific day is taken into account due to the lack of traffic data which is repeated throughout the year (Fig. 5).

Calculations using the EnergyPLAN

The calculations and the analysis of the energy system are performed using the EnergyPLAN, and the aim is that the production from RES satisfies the specified demand. The calculations allowed accounting of the values for each hour throughout the year, i.e. 8784 h. The distribution curves of the hourly electricity demand and the production represent the input data for the EnergyPLAN program. The curves are the same for each scenario as it is assumed that the distributions of the production and the demand will remain the same throughout the whole year every year, whereas only the installed production capacities from Table 1 and the annual demand capacities from Table 5 are changed according to each scenario. Production (e_{Total}) represents the sum of all the renewable electricity production in the system:

$$e_{\mathrm{Total}} = e_{\mathrm{Res}} = e_{\mathrm{Res1}} + e_{\mathrm{Res2}} + e_{\mathrm{Res3}} \tag{8}$$

The total demand (d_{Total}) is the sum of electricity demand (d_E) and transport demand which includes the flexible transport demand (d_{FX}), the smart and dumb charging regulation (d_{BEV}) and V2G (d_{V2G}).

$$d_{\mathrm{Total}} = d_E + d_{FX} + d_{BEV} + d_{V2G} \tag{9}$$

The difference between the consumption and the production represents an excess or a deficit in the electricity production, depending on whether less or more energy is produced in a particular time span than it is needed.

$$e_{pp} = d_{\mathrm{Total}} - e_{\mathrm{Total}} \tag{10}$$

In order to reduce an excess or a shortage of production, EVs have been added to the system. EVs are charged in times of an excess electricity production, and the V2G model is replacing the grid in times of shortage of production. When the system produces more energy than what is needed, the energy can be transmitted to the energy transmission lines, i.e. to the neighbouring energy markets. The CEEP represents the production of electricity used in export that goes beyond the capacity values of the energy transmission lines. In this work, the capacity values of the energy transmission lines were not set, so all of the excess of energy produced is observed as critical.

Table 5 Electricity and transport demand for each scenario

Year	2010	2020	2030	2050
Electricity demand (GWh/year)	311	346	408	474
Transport demand (GWh/year)	0	0.4	21.86	51.43

EV charging models

This paper considers four models of EV charging that are given by the EnergyPLAN:

(a) Dumb charge
(b) Flexible demand
(c) Smart charge
(d) Smart charge including V2G

Apart from these four models of EV charging, we analysed also the regulation of EV charging according to the different tariff models in electricity prices.

Dumb charge

The existing infrastructure of the power distribution sets certain limits on the integration of a large number of EVs into the power system. If the system does not include regulation, EV charging can influence the increase of a peak demand even in the case when the annual transport demand is low, for example, several thousand kilowatts [36]. The model of EV charging without any

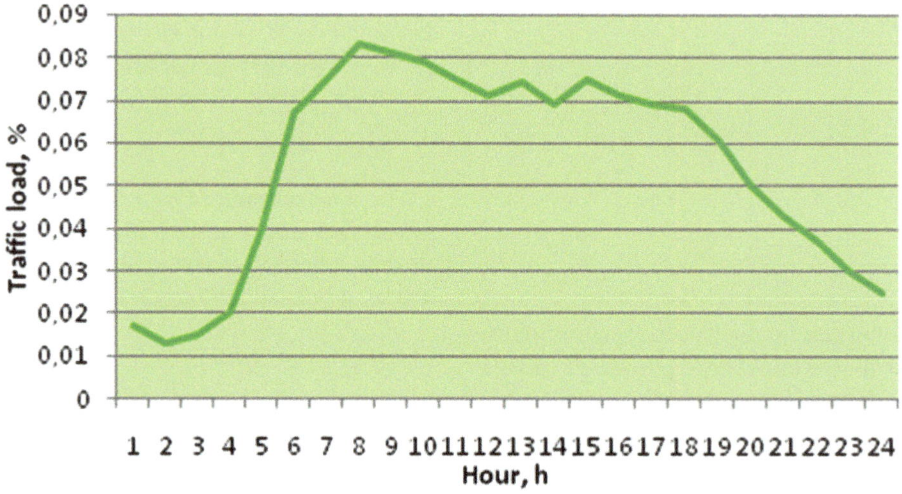

Fig. 4 Traffic load measured for the Dubrovnik region for one specific day

regulation in the system is called the dumb charge, and it is based solely on the needs and the habits of the drivers. Vehicles are loaded at the time when parked, depending on the demands of the consumers. A distribution curve of a dumb charge demand is gained by processing the transport demand data in Excel, as it was explained in the previous text focusing on the transport demand input data, and using an input distribution curve for the EnergyPLAN calculations. The dumb charge distribution represents the drivers' needs, as it can be assumed that all the vehicles parked are available for charging, since drivers do not use them at the time.

Flexible demand

Other additional technologies, including the controlled models of charging, have also been considered, for example the flexible electricity demand of transport demand. A flexible demand due to the charging EVs will ensure that the vehicles are full during a low electricity demand, mostly in the night hours. At the same time, it also regulates the EV charging during a high electricity production from RES, in order to reduce the production peak loads. The input distribution curve of the transport demand for the flexible demand model is the same as the one for the dumb charge. The difference is that the flexible demand regulation using the EnergyPLAN distributes the curve by changing its shape according to the charge of EVs, in order to charge the vehicles in the time of low electricity demand and during a large production from RES [37, 38]. A flexible demand cannot function without using ICT tools. The application of a flexible demand regulation requires the knowledge on weathercast

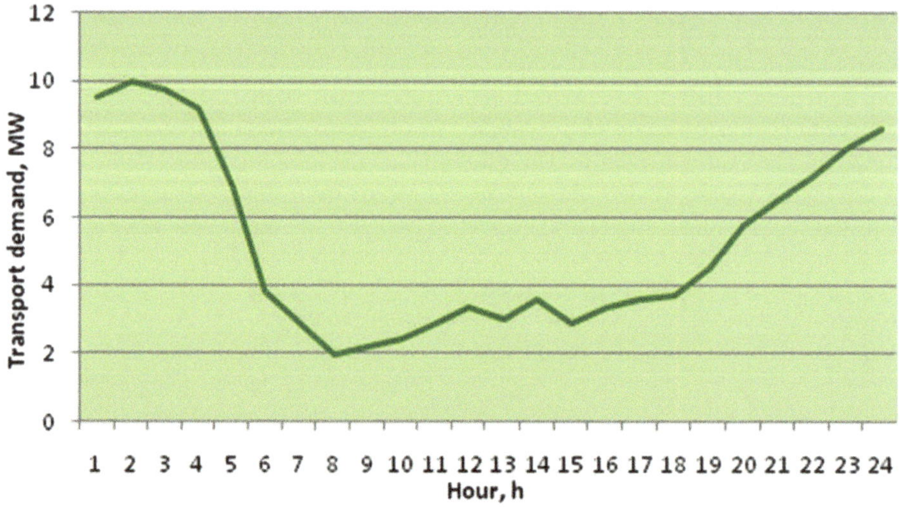

Fig. 5 Hourly distribution of electricity transport demand for one specific day in 2050

in advance, in order to predict the production from RES and be able to estimate the needs for EV charging, so that the latter might not affect the electricity peak load demand and that they might reduce the peak load of electricity production.

Smart charge

Developing communication technologies between EV and the infrastructure of the power system might switch the EV charging during low electricity demand [39]. The smart charge model works in a way that it ensures that the battery of each vehicle is fully charged before switching off. Generally, the battery is full in the case of CEEP. However, when there is no CEEP, the model must ensure that the battery is charged before the driving cycle. Through smart charging, EVs can effectively contribute to the grid due to the rapid response to the needs of the energy system and the expected growth of the number of EVs in the future. The smart charge model provides the charging of EVs during low-power demand by controlling the charging of each vehicle in order to meet the needs of the drivers to recharge the vehicle at a certain time, as well as to control it for the purposes of the grid in order to provide a regulation and avoid an overloading of the grid [40]. The input curve for the smart charge regulation model in the EnergyPLAN is the distributed curve of transport demand for 2010 obtained in Excel based on the traffic data. It represents the transport electricity demand distribution based on the drivers' needs, and it is redistributed in the EnergyPLAN according to the smart charge regulation model.

Smart charge using V2G

V2G is one of the many technologies in the energy sector that goes in favour of storing electricity. The vehicles use the power from the grid for driving and have also the ability to feedback electricity to the grid. They are partly controlled to meet the needs of the power system. The batteries of V2G vehicles are charged during a low demand and discharged during a shortage of a power generation in the system [41]. Analysis and testing for V2G were carried out using various models [42–44]. The smart charge model that includes V2G in the production system, allows charging V2G vehicles at the time of available CEEP and available battery capacity.

Tariff models in electricity prices

The regulation of EV charging is provided by setting three different tariff models for electricity prices in the EnergyPLAN. Each model is based on two tariffs, a lower tariff with a price of 0.45 kn/kWh and a higher tariff with 0.91 kn/kWh [45]. During day times, we have a higher tariff and during night times a lower one. The EVs are charged during the lower tariff period. Each tariff model has the same price for lower and higher tariffs but with a different distribution. In order to estimate the efficiency of the V2G technology in the integration with the production of energy from RES, detailed simulations are carried out for each hour [46]. This type of analysis is made by using the EnergyPLAN for the Dubrovnik region.

Different combinations of charging models were considered for each scenario, based on some future assumptions, as it is described in the results. Their impact on the system optimization was observed as we analysed the differences between the models according to their charging patterns, their impact on the maximum electricity demand and their correlation with the production from RES. The charging models and the formulae for all of the charging models are detailed, as described in the EnergyPLAN manual [20, 47].

Results and discussion

There have been some studies so far dealing with charging models for EVs. They compared regulated and unregulated charging and their effect on the electrical system [48–50]. Something similar is shown in this work. Considering the four existing models of the EV charging that have been explained in the previous text, some combinations of different charging models were selected for each of the scenarios by 2050, according to their characteristics. Their analysis determined the impact of each model on the electricity demand and the impact on the production from RES. Comparisons were carried out for each scenario. The impact of EVs on the hourly electricity demand was analysed by using the increase of the maximum annual peak demand due to charges of EVs. The impact of the model on the electricity production from RES is investigated with regard to the increase in the CEEP, which indicates an excess of energy produced used in export. In the calculation, the values of the energy transmission capacity which might be provided to the neighbouring energy markets were not regarded. The system was set up as an isolated one.

Scenario results for 2020

Capacity input data of the consumption from Table 5 and the production of electricity from Table 1 for 2020 are entered into the EnergyPLAN, along with annual hourly distribution curves, as it was described in the previous text. The simulation in the EnergyPLAN is carried out on an hourly basis for the year 2020 and rested on two assumptions, scenario (A) and (B). Figure 6 presents the results for 1 week in January. Scenario (A) includes the model of a dumb charge and a flexible demand in the transport sector for year 2020. It is considered that in 2020, 50 % of the consumption will be flexibly regulated, while the other 50 % of the vehicle

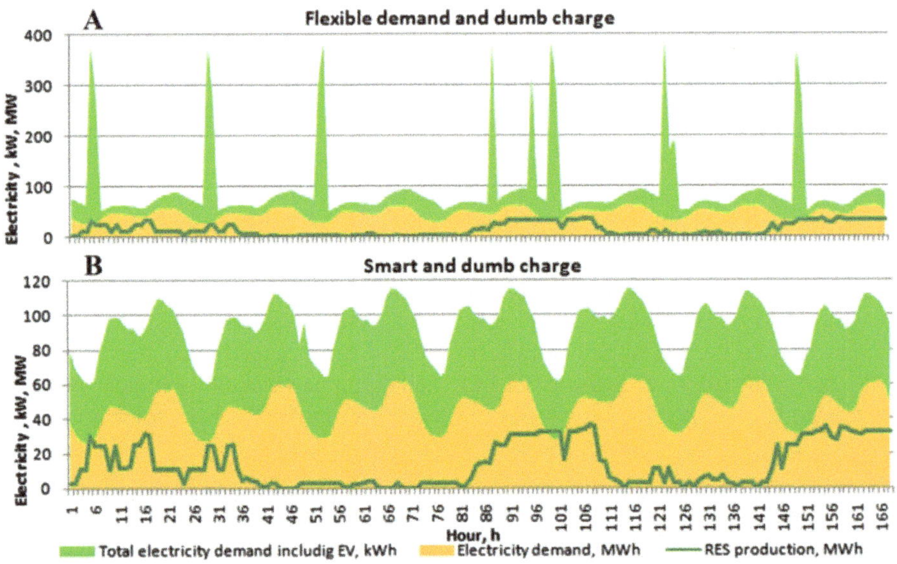

Fig. 6 Distribution of transport demand for two assumed EV charging models in 2020. The green field in the diagram represents the distribution of EV charging for scenario (A) and (B) in kWh as the capacity of the EV batteries is thousand times lower as compared to the electricity demand and RES production. There are only 146 EVs in 2020 and the capacity of their batteries is too small

fleet will be charged by dumb charge since the regulation technology will not be fully represented in the transport sector in the near future. In scenario (B), the flexible demand is replaced with a smart charge, in the same way.

The results of two scenarios indicated a difference in the charging patterns due to the different regulation sets of EV charging models. The EV charging regulation in scenario (A), based on a flexible demand, provided some peeks in the transport demand distribution during the week. The reason for the peeks can be found in the fact that a flexible demand ensures to charge the EV batteries in the time of low electricity load and high production from RES, as it has been explained before in (b). Half of the EV fleet is charged according to the dumb charge which is based on the consumer needs, as has been explained before in (a). The hourly dumb charge curve is based on one specific day distributed through the whole year, which is the reason why the rest of the transport distribution is the same for each day of the week.

The charging regulation in scenario (B) provided a distribution with almost the same charging patterns for each day of the week. The reason is that half of the EV fleet is charged according to the dumb charge. Smart charge regulation did not provide any peeks as it controls EV charging in order to avoid overloads of the grid. Transport demand distribution is the same for each day because the regulation goes in favour of the drivers, which basically means that the smart charging regulation might partly ensure a distribution according to the dumb charge.

In order to compare scenarios (A) and (B), we analysed their impacts on the electricity load and CEEP, as shown in Table 6. The results showed no difference between the scenarios due to the small numbers of EVs in 2020, which points to a low capacity storage for the excess electricity produced. The increase in the maximum peak demand is the same in both scenarios as the EVs are partly charged according to the dumb charge. The number of the EVs projected for the year 2020 is not enough to store all of the electricity produced from RES as 33 MW more storage capacity is needed to store all of the excess of electricity produced.

Scenario results for 2030

For 2030, it could be assumed that the EV charging regulation technology will be fully represented in the power system and is based on the fact that the two scenarios were determined and simulated in the EnergyPLAN. The results are illustrated in Fig. 7 for 1 week in January. For scenario (A), it is assumed that the whole EV fleet will be charged according to a flexible demand control, while for scenario (B), it is assumed that the EV

Table 6 Impact of assumed EV charging models on the electricity demand curve and the critical excess electricity production in 2020

2020	Increase in CEEP (GWh/year)	Increase of the maximum peak demand (kW)
Smart and dumb charge	288.88	21
Flexible demand and dumb charge	288.88	21

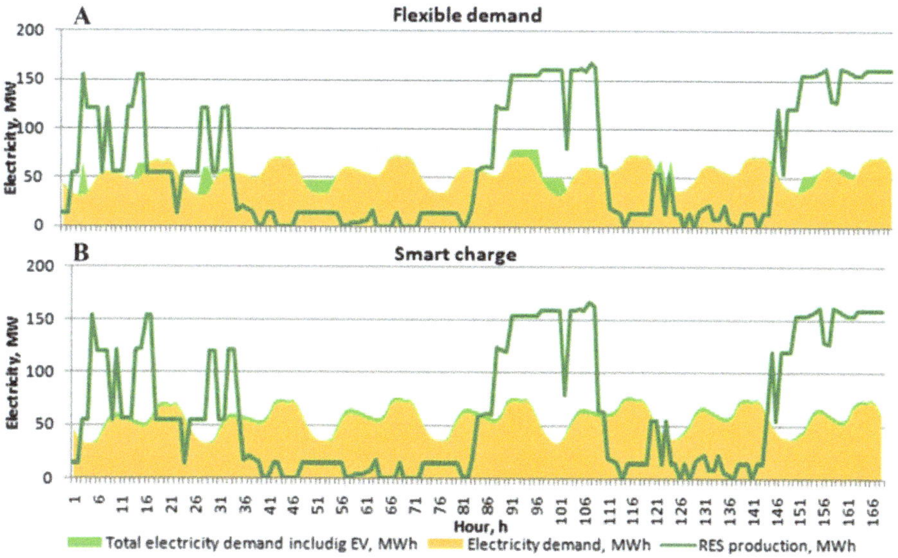

Fig. 7 Distribution of transport demand for two assumed EV charging models in 2030

charging will be regulated by using a smart charge control.

The results in the diagram for scenario (A) demonstrated that the flexible demand regulation distributes the transport electricity demand in a way that the EVs are charged during low electricity demand and the time of high electricity produced from RES (b). The smart charge control in scenario (B) provided almost the same EV charging patterns for each day of the week. The reason is that the smart charge controls charging in a way that it meets the consumer needs, such as it was explained for scenario (B) for the year 2020 (c). When the results on impacts of electricity load and CEEP are compared as shown in Table 7, it might be concluded that smart charge control gives better results than the flexible demand control for 2030. There is no big difference between scenarios (A) and (B) in their impact on the increase in CEEP, although smart charge provides a storage with a slightly larger capacity of 47 kW, and has a significantly lower impact on the increase of the maximum peak demand in the amount of 12.4 MW. This means that smart charge control distributes EV charging in order to provide more capacity storage for the excess electricity produced and avoids overloading of the grid. The capacity of EV batteries projected for 2030 was not large enough to store all of the excess of produced electricity forms of RES. To store all of the CEEP obtained in the system in 2030, 74 MW more storage capacity is needed.

Table 7 Impact of the assumed EV charge models on the electricity demand curve and the critical excess electricity production in 2030

2030	Increase in CEEP (GWh/year)	Increase of the maximum peak demand (kW)
Smart charge	646.45	9075
Flexible demand	646.86	21496

Scenario results for 2050

In the scenario for 2050, it is considered that a smart charge technology combined with V2G vehicles will be fully developed, and it is taken into account as a first assumption for EV charge control. V2G vehicles are able to feedback energy to the grid in the case of a lack of production from RES. According to the previous work [51], new models of charging regulation are analysed in this work, dealing with charging regulated by setting three different tariff models of the electricity prices. Tariff models are set for two different electricity prices, lower and higher tariff, with a different distribution shown in Table 8. Tariff model 1 is the one that is in use in the Dubrovnik region. Only certain sets of tariff for the final consumers were analysed in this work, and they are not formatted according to the prices on the wholesale market. The reason is that Croatia is not yet connected to any stock market, but in the future, it probably will be connected to the regional wholesale electricity market of Central Easter Europe along with Czech Republic, Hungary, Poland, Romania, Slovakia

Table 8 Tariff models in electricity prices

Tariff models	Lower tariff (0,45 kn/kWh)	Higher tariff (0,91 kn/kWh)
Tariff model 1	9 pm–7 am	7 am–9 pm
Tariff model 2	11 pm–9 am	9 am–11 pm
Tariff model 3	2 pm–6 pm; 12 pm–6 am	6 am–2 pm; 6 pm–12 pm

and Slovenia [52]. In scenario (A), it is assumed that the EV charging is regulated according to the flexible demand. For scenario (B), the regulation of EV charging is set up for a smart charge with both the V2G and the tariff model 1. Figure 8 illustrates the results for 1 week in January.

The results for scenario (A), presented in the diagram of Fig. 8, demonstrate that the flexible demand model for charging the regulation goes in favour of high electricity production and low electricity demand (b). The charging demand provides peeks in the time of high electricity production by RES, and in some parts of the week, it flattens the load curve by the regulation of EV charging during low electricity demand. Scenario (B) shows the same charging and discharging patterns of V2G vehicles according to the regulation sets for tariff model 1. The vehicles are charged in a time of a higher tariff shown in Table 8 and discharged in a time of a lower tariff. The charging patterns are the same for each day of the week as the hourly curve of the transport demand, used as an input data for the EnergyPLAN, is distributed according to one specific day, as it was explained in the paragraph on the input data for the transport demand. The hourly curve of the transport demand represents the consumer needs where as the smart charge regulates the charging in order to satisfy those needs, which provide almost the same charging patterns for each day.

According to the results listed in Table 9, it can be concluded that scenario (A) allows to obtain better results in the case of CEEP because it provides an electricity storage of a larger capacity of up to 532 kW. Scenario (B), however, gives better results according to the increase of the maximum peak demand. In comparison to the results for scenario (A), it decreases the maximum peak demand by 14 MW. In 2050, all of the conventional vehicles are replaced with EVs, but the installed capacity of RES in the system, projected for 2050 (see Table 1), still produces CEEP and needed a 125 MW storage capacity to store all of the CEEP produced from RES.

Two more scenarios were analysed for 2050, according to the charging regulation controlled by different tariff models presented in Table 8, i.e. tariff model 2 and 3. It is assumed that a charging regulation will be provided by a smart charge and the V2G model as well as two different sets for tariff models. The plotted results are presented in Fig. 9 for both scenarios, whereas the results obtained for the impact of EV charging on both the increase in CEEP and the maximum peak demand are given in Table 10.

According to the results in Fig. 9, V2G are charged in a time of a lower tariff and discharged in a time of a higher tariff. The charging distribution is the same for each day of the week according to the regulation sets, as it was previously explained for scenario (B) for 2050.

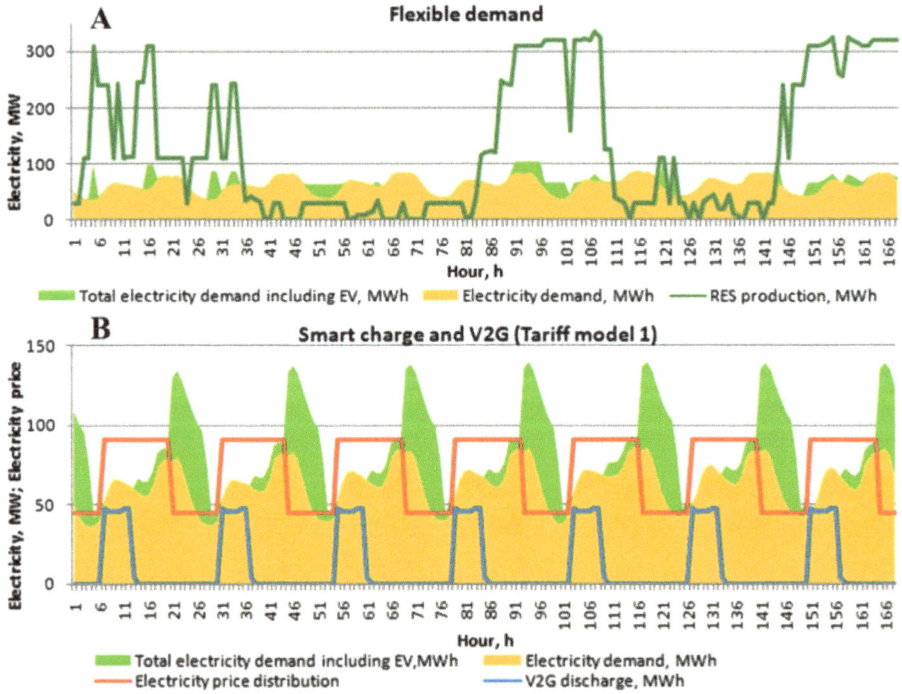

Fig. 8 Distribution of transport demand for two assumed EV charging models in 2050. The *red line* in the diagram (B) represents the electricity price distribution of the tariff model 1 explained in Table 8

Table 9 Impact of assumed EV charge models on the electricity demand curve and the critical excess electricity production in 2050

2050	Increase in CEEP (GWh/year)	Increase of the maximum peak demand (kW)
Smart charge and V2G (Tariff model 1)	1092.9	52,518
Flexible demand	1088.24	66,468

The results presented in Table 10 illustrate a comparison between three different tariff models. It is shown that the regulation sets of tariff model 3 give best results compared to the other tariff models. Tariff model 3 provides a 1.3 MW larger storage capacity and has an 11 MW less increase in the maximum peak demand compared to tariff model 1. Compared to tariff model 2, it provides a 1 MW larger storage capacity and a 7.5 MW less increase in the maximum peak demand.

The impact of EVs and RES penetration on CEEP

In the EnergyPLAN, an additional analysis of the impact of EV penetration on the critical excess of annual electricity production is carried out for each scenario, which has resulted from an increase in installed capacity of renewable energy. Smart charge models for the analysis were selected from all scenarios.

The diagrams in Fig. 10 represent the relationship between the CEEP and the installed capacity of wind turbines (A) and photovoltaic systems (B), based on the assumption that EVs participate in the system and for the case that they do not participate in the system, for each scenario. From the diagram, it can be concluded that the CEEP is higher in the system without EVs, and the difference is most apparent in 2050. The reason is the larger number of EVs that represent higher storage capacity for the produced electricity. In 2020, there is a small number of EVs, of 146, so they have low storage capacity, which gives no big difference for the system if it includes EVs or not. That is why the lines for 2020 with and without EVs are overlapping. It can be seen from the diagram that in the case (A) CEEP is rising more rapidly with an increase of installed capacity for wind production in the power system than in the case (B). In the case (B), the CEEP is high, but does not increase much with an increase in the installed capacity of the photovoltaic systems, as it is true for the case (A), because its installed capacity is lower. However, a significant increase in CEEP from one scenario to another is noted. We would now draw our attention to a case where with the penetration of wind turbines, the implementation of EVs in the energy system reduces the CEEP, which reflects the expected situation in 2050 because of a larger storage capacity provided by a higher number of EV batteries. The EV storage capacities for 2020, 2030 and 2050 are listed in Table 5.

Conclusions

In this paper, three scenarios for the years 2020, 2030 and 2050 are presented, and an analysis was carried out

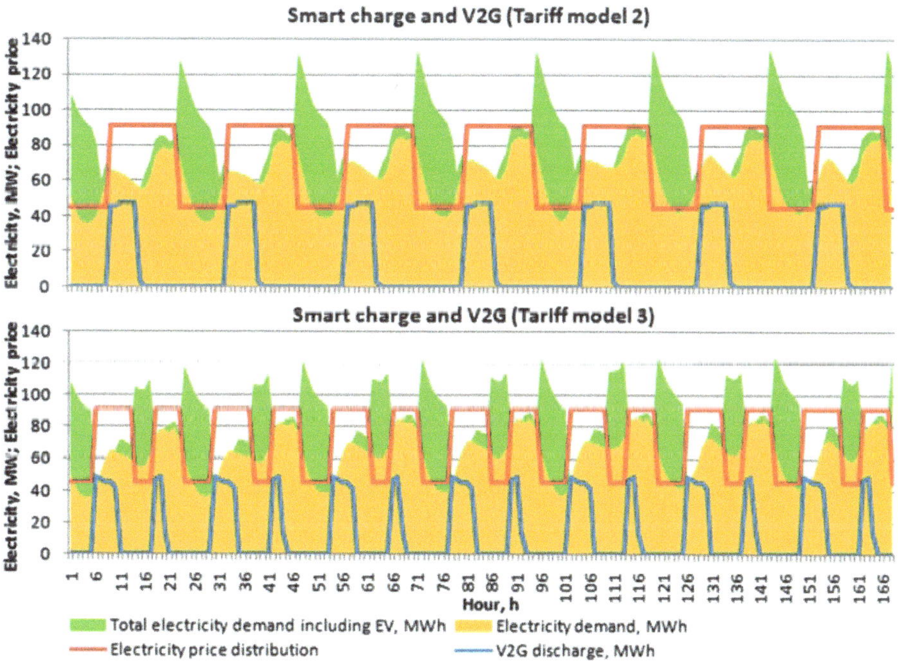

Fig. 9 Distribution of transport demand for the smart charge model including V2G with tariff model regulation. The *red line* in diagrams represents the electricity price distribution of the tariff model 2 and 3 explained in Table 8

Table 10 Impact of tariff model regulation of EV charging on the electricity demand curve and the critical excess electricity production in 2050

2050	Increase in CEEP (GWh/year)	Increase of the maximum peak demand (kW)
Tariff model 1	1092.9	52,518
Tariff model 2	1090.29	49,151
Tariff model 3	1081.62	41,620

to determine the influence of the EVs on both the hourly distribution demand curve and the production of energy from RES.

In 2020, both models have the same effect on the increase in the peak demand, reaching an amount of 21 kW, as each model represented the same percentage of dumb charge. Both models generate CEEP in an amount of 288.88 GWh/year. The charging demand distributions for the two assumed scenarios for the year 2020 are different, but their impact on both the maximum peak demand and the CEEP is the same due to the low number of EVs present in the system. The capacity of EV batteries is not large enough, and 33 MW more storage capacity is needed in order to store the excess of electricity produced by RES.

Smart charge control is generally considered more favourable than flexible demand as it satisfies the needs of the drivers and avoids overloading of the grid. Flexible demand regulation, on the other hand, is aiming to reduce peak production and to recharge the vehicles at the time of low demand, but regulation is not implemented in order to satisfy the needs of drivers. In the scenario for 2030, a flexible demand regulation increases the CEEP by 0.41 GWh/year and the peak demand is 12.4 MW larger than in the case of smart charge. Smart charge provides better results, but there is still a need for a 74 MW larger storage in the system for 2030.

The model of smart charge with both the V2G and tariff model 1 charging regulation, for the year 2050, is compared with a model of a flexible demand. A flexible demand control gave better results in reducing the CEEP in an amount of 4.66 GWh/year less than for the V2G

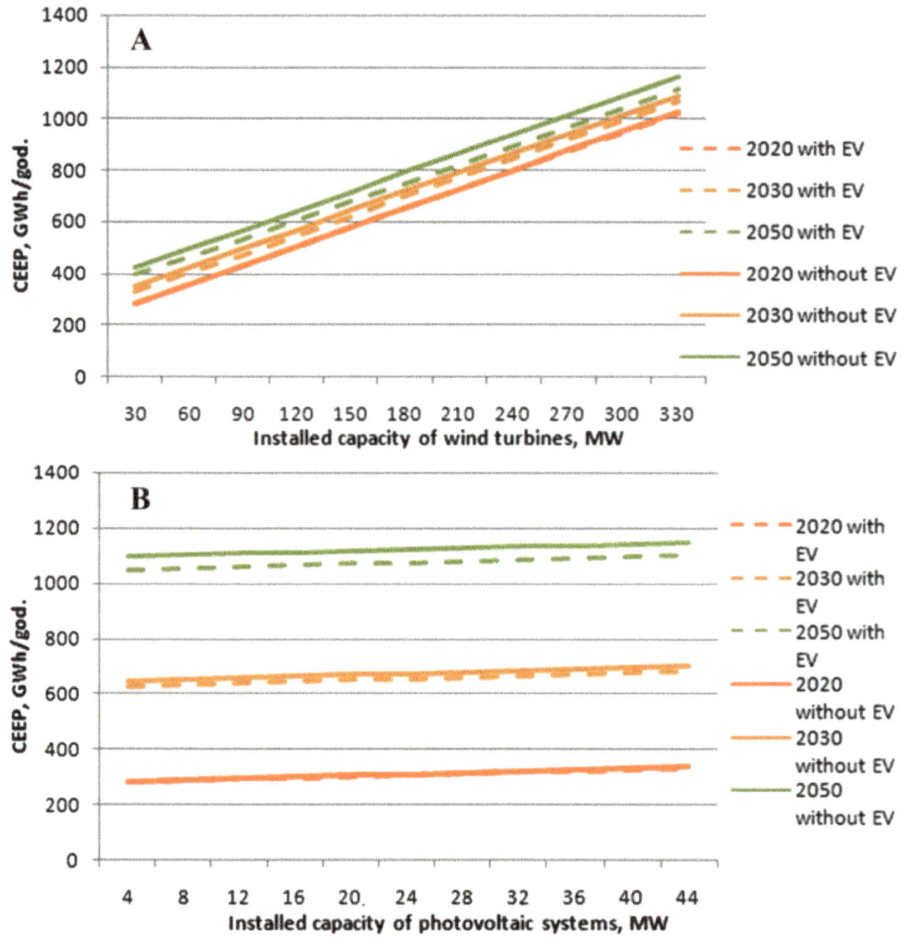

Fig. 10 Increase of the critical excess electricity production with penetration of RES in the energy system, the comparison of scenarios with the EVs in the system and without them

model, but increased the peak demand by 14 MW more than that of the V2G model. The reason is that in the model, which includes smart charge together with V2G vehicles, an additional regulation was introduced in order to meet the minimum cost, whereas the model, apart from the requirements of the driver and the grid, adjusts the electricity price in order to achieve a minimum charge cost. Vehicles are charged in a time of lower tariff and discharged in a time of higher tariff. There is still a need for 125 MW of storage capacities in order to store all of the CEEP produced in the system.

Two additional scenarios were analysed for 2050 according to the different sets for the tariff models 2 and 3. Tariff model 3 gave the best results according to the impact on the CEEP and the maximum peak demand, as compared to other scenarios presented for 2050. This leads to the conclusion that the optimal sets of tariff models could provide a better functioning of the power system in order to reduce the costs and to satisfy the needs of the system in the best possible way. Higher tariffs could be set up in a time of high electricity load and low production from RES and in the reverse case, we can do the opposite and set tariff to a lower price. The system could have more than two tariffs. Charging regulation according to the tariff models is something that could be part of the future work in system optimization. In order to improve an optimization of the electricity market, we need to consider an application of ICT tools which would provide the communication between consumer and supplier.

All of the scenarios have been set up as isolated systems in the EnergyPLAN optimization, and therefore, the CEEP and also a lack of production is appearing because the transmission line capacities for import and export are not specified. The paper also provides an overview of the CEEP increase accompanied by an increase in the installed capacity of RES. From this, it will be evident that the power system, which includes the EVs in the regulation of the relationship between production and consumption of electricity, produces less CEEP than the power system that does not include the EVs in the system. The EVs store the excess of electricity produced from RES and provide less CEEP in the system. This leads to the conclusion that a power system that includes RES cannot function without the electricity storage. Although EVs are present in the Dubrovnik region power system up to 2050, it appears that the EV battery capacities are not enough to store all of the electricity produced from RES. Some future work should consider more installed capacity of new storages for the region or maybe set up capacities of transmission lines for import and export of electricity in order to provide good functioning of the power system.

Although this work did not provide an optimal regulation of the power system by the chosen scenarios to achieve a 100 % renewable electricity production, it derived guidelines in which direction future work should go and provided results that could be useful for further work. Due to the lack of data, the transport demand had to be distributed according to one specific day, and thus, some new information and new data should be considered in future investigations in order to provide more realistic results. This work can be helpful for the optimization of both the power systems in regard to the power produced by RES and the EV charging technologies developed for cities, regions or entire countries. In order to provide sustainability of such systems, they should contain enough storage capacities. There is more work necessary to be carried out to reach the sustainability of these systems and to apply alternative technologies with a more positive impact on the environment.

Competing interests
The authors declare that they have no competing interests.

Authors' contributions
The article was jointly prepared by all authors. All authors read and approved the final manuscript.

Acknowledgements
A financial support from the Croatian Science Foundation through the *"ICT-aided integration of Electric Vehicles into the Energy Systems with a high share of Renewable Energy Sources"* project is gratefully acknowledged.

Author details
[1]Faculty of Mechanical Engineering and Naval Architecture, University of Zagreb, Zagreb, Croatia. [2]Maritime Department, University of Dubrovnik, Dubrovnik, Croatia.

References
1. European Commission (2010) Eurostat., http://epp.eurostat.ec.europa.eu/statistics_explained/index.php/Consumption_of_energy. Accessed 23 Apr 2015.
2. EEA (2009) European Environment Agency., http://www.eea.europa.eu/. Accessed 23 Apr 2015.
3. Direktiva 2009/30/EZ Europskog parlamenta i vijeća, (2009). http://eur lex.europa.eu/legal-content/HR/TXT/PDF/?uri=CELEX:32009L0030&from=HR. Accessed 23 Apr 2015.
4. "Strategija energetskog razvoja Republike Hrvatske" (2009) »Narodne novine«, br. 68/01., 177/04., 76/07. i 152/08. http://narodne-novine.nn.hr/clanci/sluzbeni/2009_10_130_3192.html. Accessed 23 Apr 2015.
5. Cardoso G, Stadler M, Bozchalui MC, Sharma R, Marnay C, Barbosa-Póvoa A, Ferrão P (2014) Optimal investment and scheduling of distributed energy resources with uncertainty in electric vehicle driving schedules. Energy 64:17–30.
6. Sensfus F, Ragwitz M, Genoese M (2008) The merit-order effect: a detailed analysis of the price effect of renewable electricity generation on spot market prices in Germany. Energy Policy 36(8):3086–3094. doi:10.1016/j.enpol.2008.03.035.
7. Kempton W, Tomic J (2005) Vehicle to grid fundamentals: calculating capacity and net revenue. J Power Sources 144:268–279. doi:10.1016/j.jpowsour.2004.12.025.
8. Kempton W, Tomić J (2005) Vehicle-to-grid power implementation: from stabilizing the grid to supporting large-scale renewable energy. J Power Sources 144:280–294
9. Ćosić B, Markovska N, Taseska V, Krajačić G, Duić N (2013) Increasing the renewable energy sources absorption capacity of the Macedonian energy system. J Renewable Sustainable Energy 5:4, 041805-1-041805-8

10. Ćosić B, Krajačić G, Duić N (2012) A 100 % renewable energy system in the year 2050: the case of Macedonia. Energy (Oxford) 48:80–87.
11. Batas Bjelić I, Rajaković N, Ćosić B, Duić N (2013) Increasing wind power penetration into the existing Serbian energy system. Energy 57:30–37.
12. Krajačić G, Duić N, Zmijarević Z, Mathiesen B, Anić Vučinić A, Carvalho M (2011) Planning for a 100 % independent energy system based on smart energy storage for integration of renewables and CO2 emissions reduction. Appl Therm Eng 31:2073–2083.
13. Krajačić G, Duić N, Carvalho M (2011) How to achieve a 100 % RES electricity supply for Portugal? Appl Energy 88(2):508–517.
14. Segurado R, Krajačić G, Duić N, Alves LM (2011) Increasing the penetration of renewable energy resources in S. Vicente, Cape Verde. Appl Energy 88(2):466–472.
15. Krajačić G (2012) The role of energy storage in planning of 100 % renewable energy systems. University of Zagreb, Dissertation.
16. Pillai R, Jayakrishnan (2010) Electric vehicle based battery storages for large scale wind power integration in Denmark. Aalborg University.
17. de Gennaro M, Paffumi E, Martini G (2015) Customer-driven design of the recharge infrastructure and vehicle-to-grid in urban areas: a large-scale application for electric vehicles deployment. Energy 82:294–311.
18. Neaimeh M, Wardle R, Jenkins AM, Yi J, Hill G, Lyons PF, Hübner Y, Blythe PT, Taylor PC (2015). A probabilistic approach to combining smart meter and electric vehicle charging data to investigate distribution network impacts. In: Applied Energy. Elsevier (in press).
19. i-RESEV project (2012) ICT-aided integration of electric vehicles into the energy systems with a high share o renewable energy sources., http://powerlab.fsb.hr/iresev/. Accessed 23 Apr 2015.
20. Lund H (2012) EnergyPLAN—advanced energy systems analysis computer model, documentation version 10.0., http://energy.plan.aau.dk/EnergyPLAN-Version10-August2012.pdf. Accessed 23 Apr 2015.
21. Johansson B (2013) Security aspects of future renewable energy systems—a short overview. Energy 61:598–605.
22. Hidroelektrana Dubrovnik (2012) HEP Proizvodnja d.o.o., http://www.hep.hr/proizvodnja/en/basicdata/hydro/dubrovnik/default.aspx. Accessed 23 Apr 2015.
23. National renewable energy laboratory (2005) getting started guide for HOMER version 2.1. USA. http://www.homerenergy.com/pdf/HOMER GettingStarted210.pdf. Accessed 23 Apr 2015.
24. Ministarstvo gospodarstva Republike Hrvatske (2012) http://oie-aplikacije.mingo.hr/pregledi/Popuplzvjestaj.aspx?ReportId=5b47346e-67aa-4df2-9603-fa83c47061e3#P32ed9c04a3d843c08f17e75a12c7aad8_2_179iT1R0x0. Accessed 23 Apr 2015.
25. Pukšec T, Krajačić G, Lulić Z, Van Mathiesen B, Duić N (2013) Forecasting long-term energy demand of Croatian transport sector. Energy 57:169–176.
26. Metz M, Doetsch C (2012) Electric vehicles as flexible loads—a simulation approach using empirical mobility data. Energy 48:369–374.
27. Shah N, Cho B, Geth F, Clement K, Tant P, Driesen J (2011) Electric vehicle impact assessment study based on data-logged vehicle and driver behavior. Vehicle Power and Propulsion Conference (VPPC). IEEE. 1–6. doi:10.1109/VPPC.2011.6043146.
28. He Y, Vankatesh B, Guan L (2012) Optimal scheduling for charging and discharging of electric vehicles. IEEE, TRANSACTIONS ON SMART GRID.3:095 – 1105. doi: 10.1109/TSG.2011.2173507.
29. Grahn P, Alvehag K, Söder L (2012) Plug-in-vehicle mobility and charging flexibility Markov model based on driving behavior. 9th International Conference on the European Energy Market (EEM). IEEE. 1–8. doi:10.1109/EEM.2012.6254711.
30. Juul N (2012) Battery prices and capacity sensitivity: electric drive vehicles. Energy 47:403–410.
31. Fernández IJ, Calvillo CF, Sánchez-Miralles A, Boal J (2013) Capacity fade and aging models for electric batteries and optimal charging strategy for electric vehicles. Energy 60:35–43.
32. Perujo A, Ciuffo B (2010) The introduction of electric vehicles in the private fleet: potential impact on the electric supply system and on the environment. A case study for the Province of Milan, Italy. Energy Policy 38:4549–4561.
33. Kang DJ, Park S (2011) A case study on the grid impact of PHEV to distribution power system. Yonsei University. 44th Hawaii International Conference on System Sciences (HICSS), Kauai, HI, 4–7 Jan. 2011.IEEE. 1–6. doi:10.1109/HICSS.2011.2.
34. Clement K, Van Reusel K, Driesen K (2007) The consumption of electrical energy of plug-in hybrid electric vehicles in Belgium. Proceedings of the European Ele-Drive Conference. Brussels, Belgium, In.
35. Bando M, Bukvić I, Cvjetković I, Miličić N, Peran V, Vuković Z, Kosmat L, Landek S, Majstorović I, Njegovec M, Pološki D, Rigo A, Stepan Ž (2012) Prometna studija Grada Dubrovnika.
36. Lassila J, Kaipia T, Haakana J, Partanen J, Ja¨rventausta P, Rautiainen A (2009) Electric cars—challenge or opportunity for the electricity distribution infrastructure? European conference on smart grids and mobility, 53851 Lappeenranta, Finland.
37. Lund H, Münster E (2006) Integrated transportation and energy sector CO2 emission control strategies. Transp Policy 13(5):426–433.
38. Mathiesen BV, Lund H, Nørgaard P (2008) Integrated transport and renewable energy systems. Util Policy 16(2):107–116.
39. Markel T, Kuss M, Denholm P (2009) Communication and control of electric vehicles supporting renewables. Vehicle Power and Propulsion Conference (VPPC). IEEE 27–34. doi:10.1109/VPPC.2009.5289874.
40. Bracale, A., Caramia, P., Proto, D. (2011) Optimal operation of smart grids including distributed generation units and plug in vehicles. International Conference on Renewable Energies and Power Quality (ICREPQ'11), Las Palmas de Gran Canaria (Spain), 13th to 15th April, 2011.
41. Brooks A, Lu E, Reicher D, Spirakis C, Weihl B (2010) Demand dispatch: using real-time control of demand to help balance generation and load. IEEE Power & Energy Magazine., pp 20–29.
42. Kempton W, Kubo T (2000) Electric-drive vehicles for peak power in Japan. Energy Policy 28(1):9–18.
43. Kempton W, Letendre SE (1997) Electric vehicles as a new power source for electric utilities. Transp Res Part D: Transp Environ 2(3):157–175.
44. Miloš P (2011) Stochastic optimal charging of electric-drive vehicles with renewable energy. Energy 36(11):6567–6576. doi:10.1016/j.energy.2011.09.006.
45. HEP (Hrvatska elektroprivreda)—Operator distribucijskog sustava d.o.o.(2010) http://www.hep.hr/ods/kupci/kucanstvo.aspx. Accessed 23 Apr 2015.
46. Lund H, Kempton W (2008) Integration of renewable energy into the transport and electricity sectors through V2G. Energy Policy 36:3578–3587.
47. Kempton W, Tomic J, Letendre S, Brooks A, Lipman T (2001) Vehicle-to-grid power: battery, hybrid, and fuel cell vehicles as resources for distributed electric power in California. UCD-ITS-RR-01-03, 2001.
48. Kiviluoma J, Meibom P (2011) Methodology for modelling plug-in electric vehicles in the power system and cost estimates for a system with either smart or dumb electric vehicles. Energy 36:1758–1767.
49. Madzharov D, Delarue E, D'haeseleer W (2014) Integrating electric vehicles as flexible load in unit commitment modelling. Energy 65:285–294.
50. Honarmand M, Zakariazadeh A, Jadid S (2014) Optimal scheduling of electric vehicles in an intelligent parking lot considering vehicle-to-grid concept and battery condition. Energy 65:572–579.
51. Šare A, Krajačić G, Pukšec T, Duić N (2013) The impact of electrification in the transport sector on the power curve, and the integration of renewable energy sources in to the power systems of the Dubrovnik region. 8th SDEWES Conference, Dubrovnik, Croatia, 2013.
52. European Commission (2015) Quarterly report on European electricity markets. Market Observatory for Energy. Energy, Volume 8, issue 1.

19

Design, fabrication and performance evaluation of solar dryer for banana

Vinay Narayan Hegde*, Viraj Shrikanth Hosur, Samyukthkumar K Rathod, Puneet A Harsoor and K Badari Narayana

Abstract

Background: An indirect, active-type, environmentally friendly, low-cost solar dryer was designed to dry various agricultural products. The dryer was built by locally available, biologically degradable, low-cost materials. The dryer consists of solar flat plate air heater with three layers of insulation, drying chamber and a fan with a regulator to induce required air flow in the system. Banana is the chosen crop for the experimentation since it is high in production and also has substantial loss in India. Also, dried bananas are having good nutritive value which makes it as essential diet.

Methods: The experiments were conducted to dry banana slices and to study its drying characteristics like rate of drying and quality of dried banana in terms of taste, colour and shape. The dryer has the following features: two different air flow configurations (air flow between glass cover and absorber plate called as the top flow and air flow between absorber plate and the bottom insulation of solar collector called as the bottom flow), forced flow with variable flow rates from 0–3 m/s and two different mounting schemes (conventional trays and wooden skewers).

Results: In the top and bottom flow experiments, the bottom flow provided about 2.5 °C higher chamber temperatures than the top flow for the same solar energy input. The efficiency of top flow configuration was found to be 27.5 %, whereas the efficiency of bottom flow configuration was found to be higher at 38.21 %. The results also agree well with the theoretical calculations performed as 60 W of energy can be saved for the same energy input.

Conclusions: The drying rate was found to increase when wooden skewers were used instead of conventional trays. At the end of the day, the total difference in moisture content is found to be 3.1 % which is considerable knowing that the rate of drying drastically decreases with time. Banana dried at 1 m/s air flow rate was of the best quality in terms of colour, taste and shape when compared to drying at 0.5 and 2 m/s air flow rate while the weather condition and ambient conditions were almost the same for all the cases with negligible difference.

Keywords: Solar drying; Design of solar dryer; Thermal analysis; Experimentation; Top flow and bottom flow; Conventional trays and wooden skewers; Varying air flow rate

Background

In many parts of the world, awareness is growing about renewable energy which has an important role to play in extending technology to the farmer in developing countries like India to increase their productivity. Poor infrastructure for storage, processing and marketing in many countries of the Asia-Pacific region results to a high proportion of waste, which average between 10 and 40 % [1]. Although India is a major producer of horticultural crops, many Indians are unable to obtain their daily requirement of fruits and vegetables and the Human Development Index (HDI) is very low. Considerable quantities of fruits and vegetables produced in India go to waste owing to improper postharvest operations and the lack of processing [1]. This results in a considerable gap between gross food production and net availability [1]. Reduction of postharvest losses is essential in increasing food availability from existing production [2]. Traditional techniques used in food preservation are drying, refrigeration, freezing, salting (curing), sugaring, smoking, pickling, canning and bottling. Among these, drying is especially suited for developing countries with poorly established low-temperature and thermal processing facilities. It offers a highly effective and practical means of preservation to reduce postharvest losses and offset the shortages in supply.

* Correspondence: vinumanu07@gmail.com
Department of Mechanical Engineering, RV College of Engineering, Bangalore, Karnataka 560059, India

Drying is a method of dehydration of food products which means reducing the moisture content from the food to improve its shelf life by preventing bacterial growth [3]. It is still used in domestic up to small commercial size drying of crops, agricultural products and foodstuff such as fruits, vegetables, aromatic herbs, wood etc. contributing thus significantly to the economy of small agricultural communities and farms [4–6].

Hii et al. [7] have shown that sun drying (laying the crops under direct sunlight) is economical, but the product obtained by it is of lower quality due to contamination by dust, insects, birds, pets and rain. Also, loss of vitamins, nutrients and unacceptable colour changes due to direct exposure to ultraviolet rays, and it takes long time to dry. Solar dryers are specialized devices that control the drying process and protect agricultural products from damage by insect pests, dust and rain. Umogbai et al. [8] made a comparison between sun drying and solar drying and obtained that solar dryers generate higher temperatures, lower relative humidity, lower product moisture content and reduced spoilage during the drying process than sun drying. Rajeshwari and Ramalingam [9] have demonstrated that the drying time in case of solar dryers compared to open air drying reduced by about 20 % and produces better quality dried products. Solar dryers are available in a range of size and design such as tunnel dryers, hybrid dryers, horizontal- and vertical-type dryers, multi-pass dryers and active and passive dryers [10–17]. Hii et al. [7] classified solar dryer according to their heating modes and the manner in which the solar heat is utilized, namely forced air circulation or active solar dryers and natural air circulation or passive solar dryers. Three distinct sub-classes of either the active or passive solar drying system can be identified depending upon the design or working principle of the dryer, mode of drying and type of product to be dried, namely integral or direct mode, distributed or indirect mode and mixed mode solar dryers. It should be noted that sunlight may affect certain essential components in the product, e.g. chlorophyll is quickly decomposed. If available places are scarce, indirect mode types of dryers are preferred for drying larger quantities. In such case of indirect mode, nutritive value of the food product and colour is retained.

Mohanraj and Chandrasekar [18] and Banout and Ehl [19] concluded that forced convection solar dryer is more efficient than natural convection dryers. Also, products can be dried faster in the case of forced convection solar dryer than in the case of natural convection solar dryer, and end products obtained from forced convection drying have a superior quality.

From the literature survey, it is evident that though there are many dryer designs which involve the flow of air between glass cover and absorber plate in the collector [20] and also in some other designs, the flow is maintained between absorber plate and bottom insulation [21]. However, there is no comparison of the performance done between these two cases in a single setup. Hence, there is a requirement for comparative study to address the relative performance of the above-mentioned cases and hence to arrive at a better and efficient flow configuration.

Indian Horticulture Database 2013 [22] shows that banana is the most important fruit crop in India, accounting for 32.6 % of the total fruit production. Almost the entire production is used fresh, and hence, the entire production is subjected to the postharvest losses of 17.87 %. Banana is the chosen crop for the experimentation since it is high in production and also has substantial loss in India. Also, dried bananas are having good nutritive value which makes it as essential diet [23–25].

Most of the thin-layer drying of fruits is carried on using stainless steel meshed trays [26]. In practice, trays have many disadvantages which include sticking of the dried products to the trays, difficulty in loading and unloading, hygiene among others. Hence, an innovative way of placing the bananas in the trays is devised in which wooden skewers are used to hold the fruits. Though the various literatures on drying banana included the studies on the optimum slice thickness, solar-assisted dryer for bananas and effect of various pre-treatments and temperatures on banana among others [26–29], the effect of the air velocities on the moisture removal rate and quality of the dried bananas obtained has not been studied. As air velocity also plays an important role in drying of food products [30–33], there is a need to address the effect of air velocity on drying of banana.

In the present study, low-cost indirect-type solar dryer was designed and constructed using locally available environmentally friendly materials to compare the performance of the flat plate collector for the top flow and the bottom flow of air both theoretically and experimentally, to compare the performance in the form of moisture removal rate and dryer efficiency for different banana mounting methods such as wooden skewers and on conventional trays and to dry banana using different air velocities 0.5, 1 and 2 m/s and to compare the quality of the end product in terms of taste, texture, colour and final moisture content.

Methods

The design process of the dryer first involved the collection of the climatic data of the study location, i.e. Bengaluru. Further, the other important data such as insolation was studied and calculated as per the collector configuration. For the initial phase of dryer design, many existing designs were studied and some of the design parameters were determined. The performance of the dryer was then analysed [34–37]. Once the dimensions of the dryer were fixed, an appropriate axial fan was selected to obtain the required flow rates.

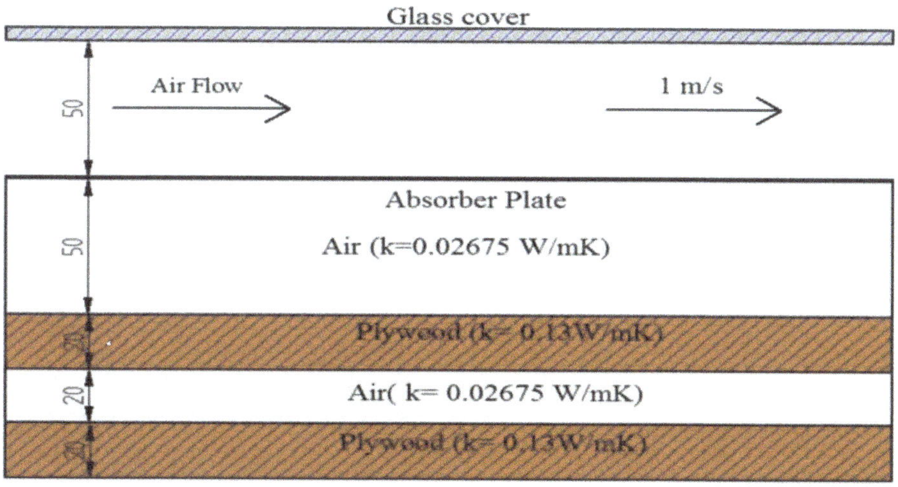

Fig. 1 Top flow

Climatic data collection

Bengaluru is located in Karnataka, India, at a latitude of 12° 58′ North and longitude of 77° 34′ East. Solar radiation over the year on horizontal surface in Bengaluru is found to be 666.635 W/m^2 [38]. Total solar radiation on a 13° tilted surface is calculated as 676.367 W/m^2.

Design consideration

The dryer was constructed using plywood, stainless steel mesh, wooden skewers, clear glass, galvanized iron sheet and axial fan for operation of the dryer which are locally available with low cost.

The thickness of banana slices was selected to be 4–5 mm [17, 27, 29]. An indirect type of solar dryer was considered as it does not affect the colour and nutrient content of the produce as in the case with a direct type. Also, the drying is uniform without any localized heating. Flat plate collector is used since it is easy to fabricate and also economical. The collector is made up of GI sheet of 0.6 mm thick as it is a good conductor and economical. It was painted black to increase the absorption of heat [39, 40]. The recommended glass thickness for collector is 5 mm [37]. Air gap of 5 cm is recommended for a tropical climate [37]. The insulating material was selected to be plywood as it is a good insulator as well as environmentally friendly. It also does not have any carcinogenic effects which other popular insulating materials like glass wool have. To reduce the heat loss, a layer of air sand witched between two plywood sheets (Figs. 1 and 2). To further reduce the heat loss by radiation and to avoid moisture absorption by wood, aluminium foil is wrapped on the inside of the chamber [41]. Food grade stainless steel mesh for the trays and food grade wooden skewers were selected for placing of

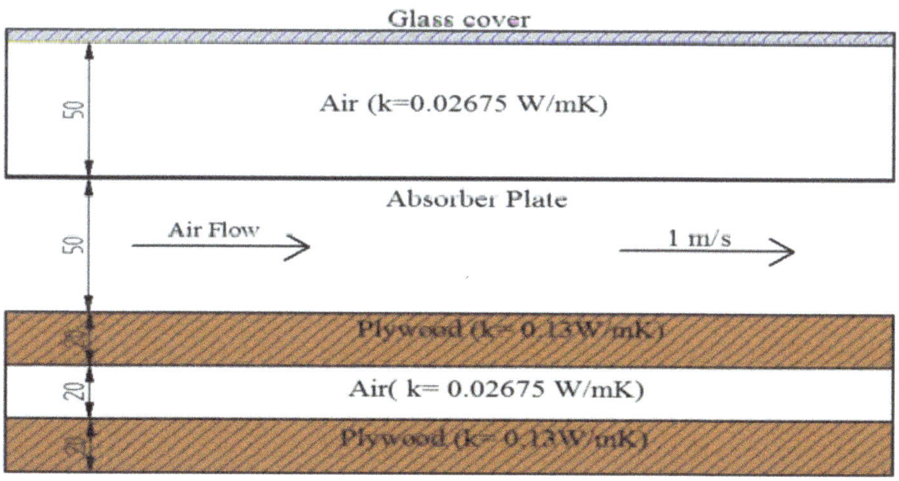

Fig. 2 Bottom flow

Table 1 Heat losses

	Top loss (W)	Side loss (W)	Bottom loss (W)	Total (W)
Top flow	259.2	7.058	25.879	292.137
Bottom flow	215.657	6.931	1.818	224.406

banana. To ensure the constant flow rate of air during the experimentation, an axial flow fan was selected based on the calculations of pressure drop in the system and the required flow rate limit of air at 3 m/s. For the purpose of experimentation, 1.5 kg of banana of Poovan variety which is locally available is used.

Thermal analysis

In the design, a flat plate collector with an area of 1.6 × 0.6 m^2 is considered. The performance of the collector is described by an energy balance that indicates the conversion of solar radiation into useful energy gain and losses. The thermal analysis was done to calculate the heat gain and losses for flow of air between glass cover and absorber plate which is the top flow and flow of air between absorber plate and bottom insulation which is the bottom flow [3, 36, 41]. Figures 1 and 2 show the typical configuration of top flow and bottom flow, respectively.

Table 1 consolidates the results from the thermal analysis. It is seen that heat loss from the top, side and bottom of the collector is more for top flow configuration compared to bottom flow configuration. This is due to the reduced temperature difference between the collector and the ambient.

Specification of the dryer

Table 2 gives the specification of the dryer.

Experimental procedure

The solar dryer was placed over the roof top of a building based on the design (Fig. 3). Axial flow fan was fixed at the top of the drying chamber and tested. The experiments were conducted in the month of March, from daily

Fig. 3 Solar dryer setup

9 am to 5 pm. The solar radiation was measured using pyranometer. The K-type thermocouples were used for the measurement of temperature in the collector assembly. The temperature was measured for each hour from 9 am to 5 pm at three points, namely entry, middle and exit of the glass cover, absorber plate and bottom insulation as it can be seen in Fig. 4. The temperature of the air in the drying chamber and the atmosphere were measured by the thermometer. A vane-type anemometer is used to measure the air velocity. The weight of the banana is

Table 2 Specification of the dryer

Overall length	2.04 m
Overall height	1.38 m
Absorber plate dimension	1.6 × 0.6 m
Glass cover thickness	0.005 m
Insulation total thickness (bottom)	0.06 m
Gap between absorber plate and glass cover	0.05 m
Gap between absorber plate and insulation	0.05 m
Number of trays	3
Tray dimension	0.3 × 0.6 m
Distance between trays	0.15 m
Tilt angle of the collector	13° due south

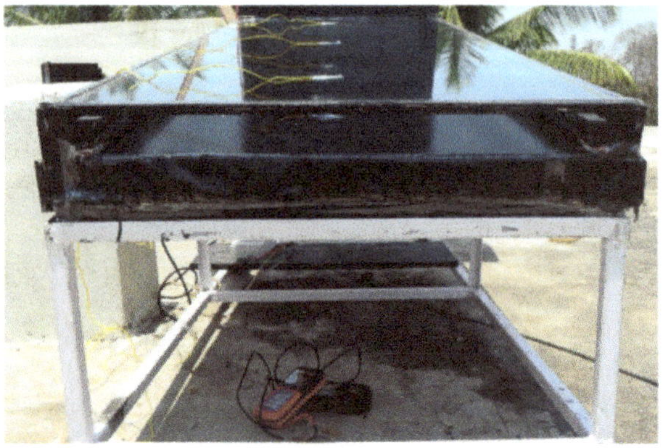

Fig. 4 Top flow

Fig. 5 Bottom flow

measured using a digital weighing pan. All the experiments were repeated to confirm the repeatability of the data obtained. The following experiments are carried out:

Flow over the absorber plate (top flow) and flow beneath the absorber plate (bottom flow)

Case A: Air is allowed to pass between the absorber plate and glass cover (Fig. 4). The air passage between absorber plate and bottom insulation is blocked using a cardboard with adhesive tape and glue.

Case B: Air is allowed between the absorber plate and the bottom insulation (Fig. 5). The air passage between glass cover and absorber plate is blocked using a cardboard with adhesive tape and glue.

The temperature readings were taken, and the losses and gain are calculated and compared.

Conventional trays and wooden skewers

In this case, the banana was placed on conventional trays (Fig. 6) and wooden skewers (Fig. 7) and allowed to dry for 8 h from 9 am to 5 pm. The air flow velocity is maintained as 1 m/s in the collector. The moisture content removed from the banana slices was compared.

To calculate the final moisture content, the following formula is used:

$$\text{Percentage moisture content} = \frac{77.2 \times w_i - 100 \times w_l}{w_i - w_l} \tag{1}$$

where 77.2 % is the initial moisture content of banana variety selected.

Efficiency for the dryer system is given by [41].

Fig. 6 Banana kept on trays

Fig. 7 Wooden skewers

$$\eta = \frac{mL_w}{AIt}$$

Varying air flow rate

The air flow though the dryer was varied using a speed regulator for the axial flow fan. The experiments are conducted, and repeatability tests are also made. As the experiments have been done on consecutive days, there is a very little change in atmospheric temperature and solar radiation.

Based on the test results of drying banana slices kept in trays and wooden skewers, varying velocity tests were conducted only on banana slices attached to skewers. The velocity of air flow is maintained as 0.5, 1 and 2 m/s in the collector region (0.0169, 0.0338 and 0.0676 m^3/s volume flow rate) for a duration of 16 h of drying time for consecutive days with each day 8 h from 9 am to 5 pm. Every hour the weight of the banana slices was measured, and the moisture content and the efficiency of the dryer were calculated. At the end of the day, banana slices were stored in air-tight bags. After drying for 16 h, the dried banana samples obtained from using 0.5, 1 and 2 m/s were compared in terms of taste, texture, colour and final moisture content.

Results and discussions

The average solar insolation in Bengaluru in the month of March as measured with a pyranometer is shown in Fig. 8. It can be noticed that the insolation increases by the day from 9 am to 1 pm and then starts to drop. The maximum radiation received was 1033 W/m^2 at 1 pm, whereas it was 293 W/m^2 at 5 pm.

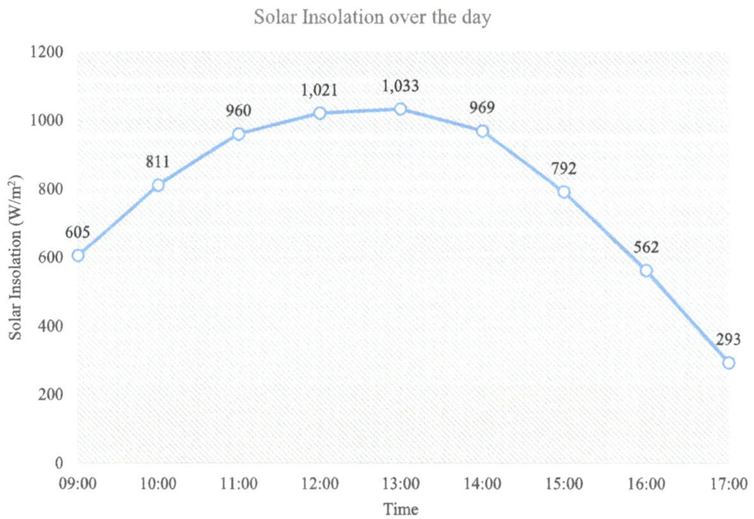

Fig. 8 Solar insolation over the day

Comparison of the top flow and bottom flow

Figure 9 shows the temperatures of the ambient air and chamber temperatures over the entire day from 9 am to 4 pm. The experiments were conducted on consecutive days and observed that the ambient temperatures were quite similar. Hence, the average ambient temperatures of 2 days were used to plot the comparison. In Fig. 9, for the case of the bottom flow, it can be seen that when the air inlet velocity is maintained at 1 m/s, the maximum temperature of the air outlet is 45 °C at an ambient temperature of 34 °C which is 11 °C above the ambient temperature. In case of the top flow experiment conducted at the same air inlet velocity, the maximum air outlet temperature reached is 42.5 °C at an ambient temperature of 34.5 °C which is about 8 °C above the ambient temperature. It can also be seen from the figure that the air temperature rise and fall closely follow the insolation curve except at the end of the day for the case of the bottom flow. This is due to the heat storage effect of the insulation which helps to maintain the air temperatures in the evening even though the insolation drastically drops. So it is evident that the configuration with the air flow below the absorber plate gives higher air outlet temperature, i.e. about 2.5 °C more than the one with the air flow above the absorber plate for same solar energy input at the peak insolation. A maximum difference of 3.5 °C was obtained at 11 am and 4 pm between these two configurations. Another point to be noted is that though the solar insolation drops quite drastically in the post noon period, i.e. from 12 pm, the ambient air temperature does not drop that drastically as it retains the heat being a good insulator. Also, it absorbs the heat radiated by the earth.

The collector efficiency indicates the utilized heat against the heat input in the form of solar insolation. Figure 10 compares the efficiency for the two configurations. Lower efficiency at 9 am is due to the fact that the experiment was started at 9 am and the setup had not yet stabilized. The increase in efficiency during the evening in the case of the bottom flow may be attributed to the heat storage by the insulation. When insolation drops, the stored heat is retrieved, thus maintaining higher air temperature and hence higher efficiency. The efficiency of the system was least at the peak insolation hour of 12 pm. This is because the plate temperature rises rapidly in the noon with higher insolation, but the heat removal capacity of the air does not meet this additional load due to its fixed velocity. Thus, the air does not take away the heat which stays in the collector chamber and is hence lost to the surroundings in the form of various losses leading to lower efficiencies of the system. If the air velocity is increased in order to improve the efficiency, the air outlet temperature from the collector would decrease. Hence, a balance has to be maintained between collector efficiency and air outlet temperatures. The total loss for the top flow is 201.9 W, whereas it is 139 W for the bottom flow. This saves nearly 62.9 W in total if bottom flow configuration is used. The mean efficiency for top flow configuration is 27.5 %, whereas it is 38.21 % for bottom flow for a day.

Thus, it can be concluded from this experiment that the bottom flow configuration is more efficient than the top flow configuration.

Comparison of conventional trays and wooden skewers

Figure 11 shows the variation of moisture content with time for the two mounting configurations, i.e. one with conventional trays and the other with wooden skewers.

It is clear from Fig. 11 that the rate of moisture removal is better with the skewers at every stage of time.

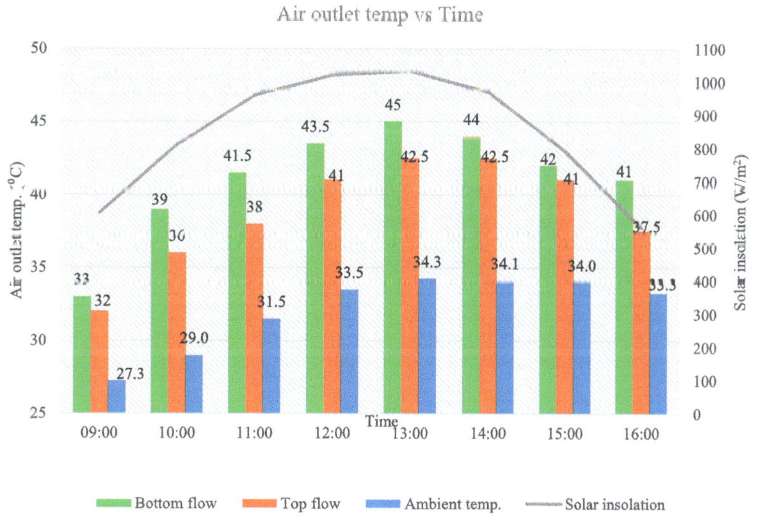

Fig. 9 Comparison of air outlet temperature for top and bottom flow over the day

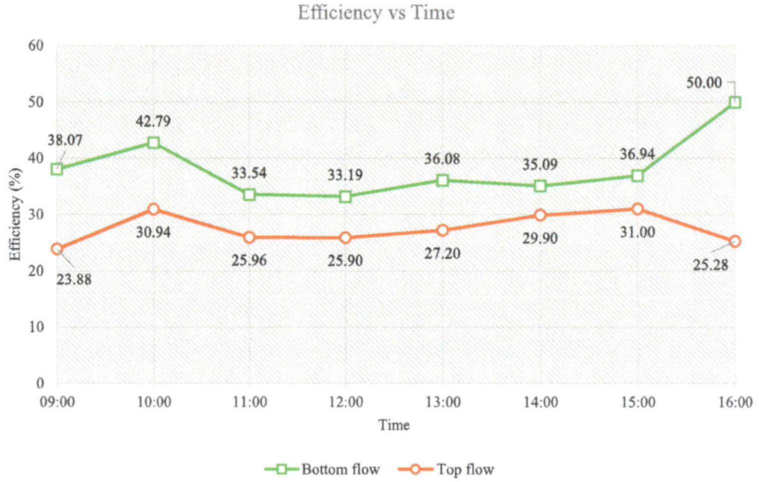

Fig. 10 Efficiency for top and bottom flow configuration

At the end of the day, the total difference in moisture content is found to be 3.1 % which is quite considerable knowing that the rate of drying drastically decreases with time. In terms of the weight of moisture removed, it is 825 g for trays while it is 864 g for skewers, the difference being 39 g.

Figure 12 shows the cumulative dryer efficiency for the above two mounting configurations. The cumulative efficiency is the ratio of the total moisture removed in the form of latent heat to the total energy supplied calculated up to specified time. It can be noted that initially, the difference in efficiencies is higher compared to the later periods, as initially, it is the unbound moisture that is being removed and just depends on the surface area. As the time goes by, the difference between the two curves reduces as the rate of moisture removal becomes much less dependent on the surface area and due to the start of the falling rate phase of drying.

The cumulative efficiency continuously drops as the rate of moisture removal drops even though the input energy is the same because of this falling rate period. Starting off with better efficiency, the skewer configuration maintains the higher efficiency throughout the day over the tray-type configuration. At the end of the day, the efficiencies are 8.45 and 8.06 % for skewer and tray type, respectively.

Flow rate for drying of banana

Figure 13 shows the comparison of moisture content for different flow rates (0.5, 1 and 2 m/s) over drying time.

The maximum temperatures achieved with 0.5, 1 and 2 m/s are 49.5, 45 and 41 °C, respectively, with almost similar ambient air temperature for all the velocities. At the end of 2 days of drying, i.e. 16 h, the moisture content in the bananas is 34.98, 29.63 and 36.04 % for 0.5, 1 and 2 m/s, respectively. If the absolute moisture removal

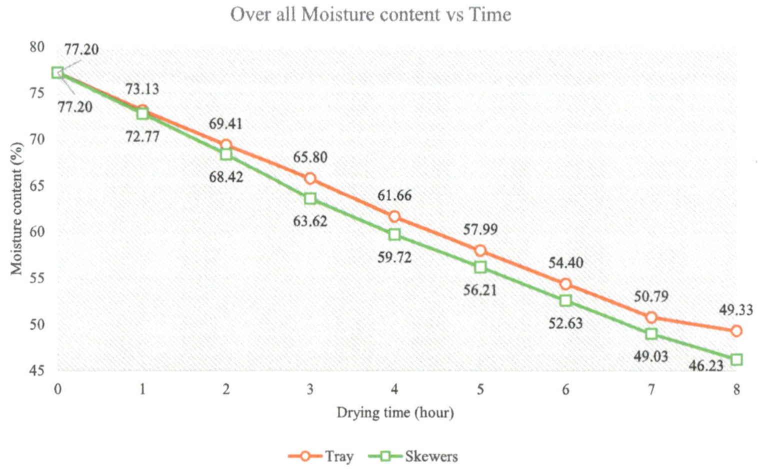

Fig. 11 Moisture content in trays and skewers over a day

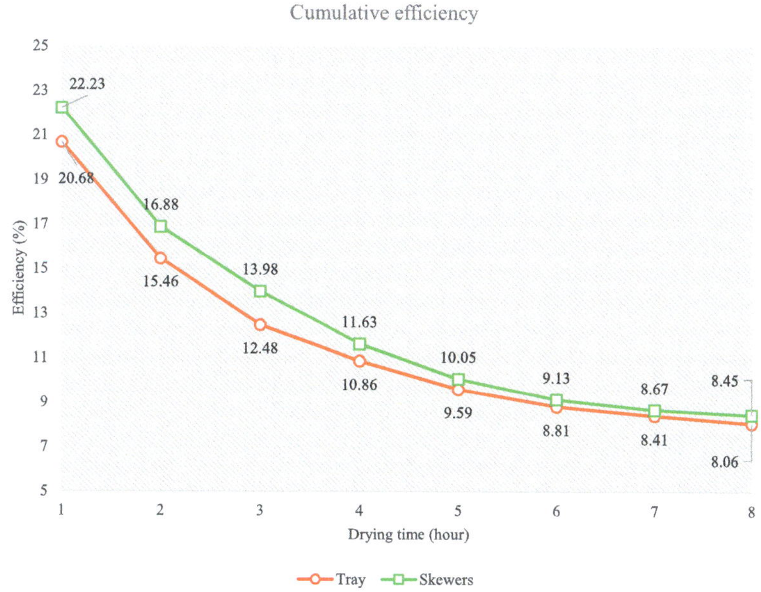

Fig. 12 Cumulative dryer efficiency for trays and skewers

rate is considered then the moisture removal rate is fastest with the velocity of 1 m/s, followed by 0.5 and 2 m/s as seen in Fig. 13.

The dryer efficiency at the end of 16 h of drying is 5.75, 4.96 and 5.05 % for 0.5, 1 and 2 m/s, respectively. The reason for slight increase in dryer efficiency for 2 m/s over 1 m/s is that the bananas dried at 1 m/s have reached the falling rate stage of drying for the given time which has impacted the dryer efficiency for the time frame considered. It is seen in Fig. 14 that the drying efficiency is clearly higher with air inlet velocity of 0.5 m/s, followed by 2 and 1 m/s. This indicates that higher air temperatures are much more effective in increasing the drying rate with the air velocity playing a minor role. But it is also to be noted that it is just not the drying rate that is important, the quality of the products obtained is more important. It is noted that with velocity of 0.5 m/s, the dried banana obtained has cardboard-like structure, hard outer surface, too light and looks like not ripened which is unacceptable. Because of the faster rate of moisture removal with 0.5 m/s, the rate of evaporation increased which resulted in hardening of the surface. The dried banana samples are shown in

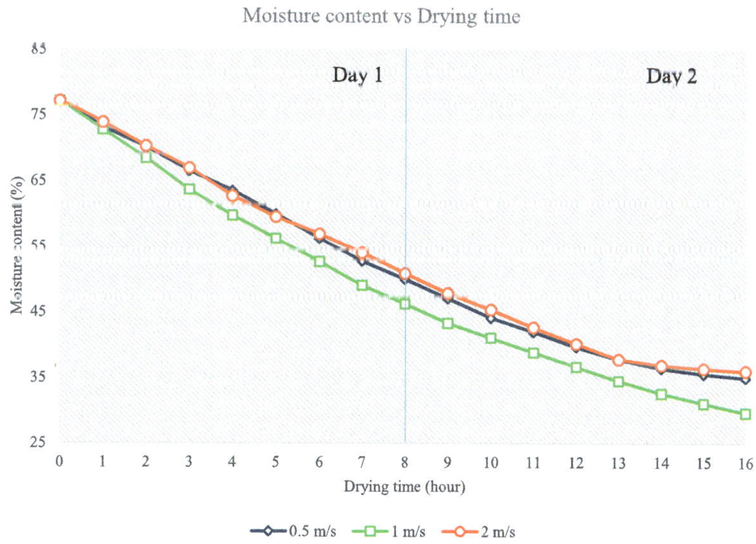

Fig. 13 Comparison of moisture content for different flow rates over drying time

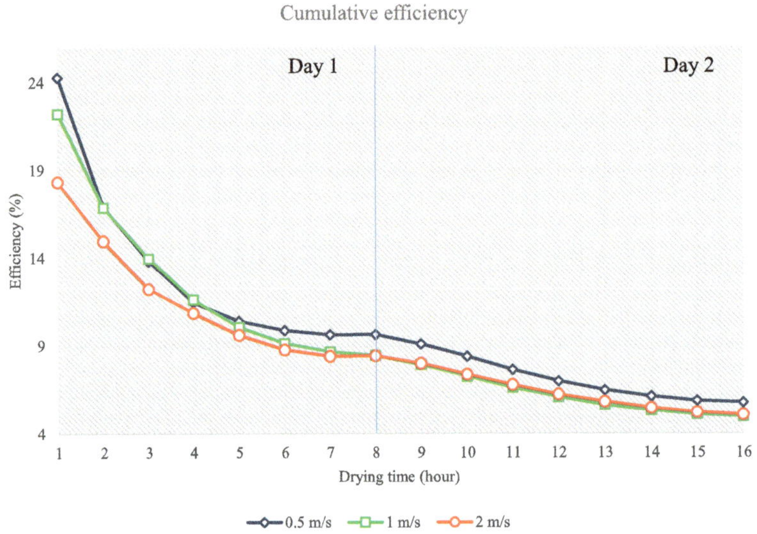

Fig. 14 Cumulative efficiency for different flow rates

Fig. 15. On the other hand, the rapid cooling of the surface of banana slices due to faster air velocity with the air inlet velocity of 2 m/s resulted in dark colour with blackening of the surface, and the surface became hard. Due which the banana obtained is of unacceptable quality. This also impacted the rate of drying as can be seen in Fig. 13. But the banana samples obtained by drying at a velocity of 1 m/s at the collector inlet are having more consistent quality of dried banana with good colour, texture, no dusty appearance, chewy and natural aroma. So with 1 m/s, good quality of banana can be obtained with quite high drying rates. The dried banana samples obtained with different air velocities were also compared with the dried banana samples obtained by Wakjira et al. [29] and Brett et al. [42]. These also confirm that the dried banana samples obtained by a velocity of 1 m/s are acceptable.

Conclusions

In the top and bottom flow experiments, the bottom flow provided about 2.5 °C higher chamber temperatures than the top flow for the same solar energy input. The efficiency of top flow configuration is found to be 27.5 % and the total heat loss or the case is found to be 201.9 W, whereas the efficiency of bottom flow is found to be higher at 38.21 % and the total heat loss is found to be 139 W. The experimental results are in excellent agreement with the theoretical values with the savings of 62 W energy. Hence, the bottom flow configuration is

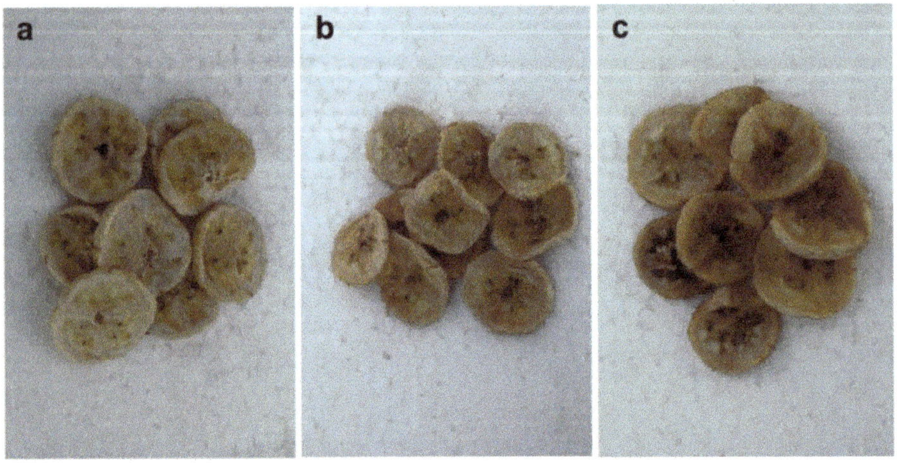

Fig. 15 Dried banana sample. **a** 0.5 m/s. **b** 1 m/s. **c** 2 m/s

more efficient. The drying rate is found to be increased when skewers are used instead of conventional trays with ease of loading and unloading of banana in the case of skewers. At the end of 16 h of drying, about 3.1 % difference in moisture content is obtained between the two configurations which is significant. The result also shows that the banana dried at 0.0338 m^3/s volume flow rate (velocity of 1 m/s over the collector) is of the best quality in terms of colour, taste and shape when compared to drying at 0.5 and 2 m/s flow rate for the same solar energy input and atmospheric conditions.

Competing interests
The authors declare that they have no competing interests.

Authors' contributions
This article is based on the student project done at the RV College of Engineering Bangalore as a part of curriculum for final semester under the guidance of Dr. KBN, Professor in the Mechanical Engineering Department at the RV College of Engineering Bangalore. Vinay, Viraj, Samyukth and Puneet are the students involved in this project. VNH contributed to the conception and design, fabrication of the model and interpretation of the data collected from the experiments. VSH contributed to the data analysis, calculations and interpretation of the data. SKR contributed to the acquisition of the data and data interpretation and fabrication of the model. PAH contributed to the experimentation and data collection and fabrication of the model. Dr. KBN involved in drafting the manuscript and revised it properly for important technical content. He is also involved in the data interpretation. All authors read and approved the final manuscript.

Acknowledgements
We are thankful to the Principal, RV College of Engineering, Head of the Department, Mechanical Engineering for their support throughout the work. We are thankful to Professor CS Prasad for sharing with us the technical knowledge and helping us to achieve the results. We are also thankful to Professors M S Krupashankara and Dr. J R Nataraja for their invaluable support and encouragement.

References
1. Rosa Rolle S. (2006) Postharvest management of fruit and vegetables in the Asia-Pacific region. Asian Productivity Organization, Tokyo, Japan
2. Abdullahi Y, Momoh M, Garba MM, Musa M (2013) Design and construction of an adjustable and collapsible natural convection solar food dryer. Int J Comput Eng Res 3(6):1–8
3. Garg H P, Prakash J. (1997) Solar energy fundamentals and applications. Tata McGraw Hill, New Delhi, India
4. Emad Almuhanna A (2012) Utilization of a solar greenhouse as a solar dryer for drying dates under the climatic conditions of the eastern province of Saudi Arabia. J Agric Sci 4(3):237–246
5. Ghaly AE, MacDonald KN (2012) An effective passive solar dryer for thin layer drying of poultry manure. Am J Eng Appl Sci 5(2):136–150
6. Manoj M, Manivannan A (2013) Simulation of solar dryer utilizing greenhouse effect for cocoa bean drying. Int J Advanc Eng Technol 4(2):24–27
7. Hii C L, Jangam S V, Ong S P, Mujumdar A S (eds) (2012) Solar drying: Fundamentals, applications and innovations. TPR Group Publication, Singapore
8. Umogbai VI, Iorter HA (2013) Design, construction and performance evaluation of a passive solar dryer for maize cobs. Afr J Food Sci Technol 4(5):110–115
9. Rajeshwari N, Ramalingam A (2012) Low cost material used to construct effective box type solar dryer. Arch Appl Sci Res 4(3):1476–1482
10. Tiwari G, Katiyar VK, Dwivedi V, Katiyar AK, Pandey CK (2013) A comparative study of commonly used solar dryers in India. Int J Current Eng Tech 3(3):1–6
11. Madhlopa A, Jones SA, Kalenga Saka JD (2002) A solar air heater with composite absorber systems for food dehydration. J Renewable Energy 27:27–30
12. Brett A, Cox DR, Simmons R, Anstee G (1996) A solar tunnel dryer for natural convection drying of vegetables and other commodities in Cameroon. J Am Med Assoc 35(2):31–35
13. Isiaka M, El-Okene AMI, Muhammed US (2012) Effect of selected factors on drying process of tomato in forced convection solar energy dryer. Res J Appl Sci Eng Technol 4(19):1–4
14. Arinze EA, Sokhansanj S, Schoenau GJ, Trauttmans Dorff FG (1996) Experimental evaluation, simulation and optimization of a commercial heated-air batch hay drier. J Agric Eng Res 63:301–314
15. Afriyie JK, Rajakaruna H, Nazha MAA, Forson FK (2011) Simulation and optimization of the ventilation in a chimney-dependent solar crop dryer. Sol Energy 85:1560–1573
16. UmeshToshniwal KSR (2013) A review paper on solar dryer. Int J Eng Res Appl 3(2):896–902
17. Amer BMA, Hossain MA, Gottschalk K (2010) Design and performance evaluation of a new hybrid solar dryer for banana. Energy Convers Manag 51:813–820
18. Mohanraj M, Chandrasekar P (2009) Performance of a forced convection solar drier integrated with gravel as heat storage material for chilly drying. J Eng Sci Technol 4(3):305–314
19. Banout J, Ehl P (2010) Using a double-pass solar drier for drying of bamboo shoots. J Agric Rural Dev Trop Subtrop 111(2):119–127
20. Folaranmi J (2008) Design, construction and testing of simple solar maize dryer. Leonardo Electronic J Practicals Technol 7(13):122–130
21. Adelaja AO, Babatope BI (2013) Analysis and testing of a natural convection solar dryer for the tropics. J Energy 2013:1–8
22. Rajendra Kumar Tiwari, Mistry N C, Brajendra Singh, Chander Gandhi P. (2014) Indian Horticulture Database 2013. National Horticulture Board, Ministry of Agriculture, Government of India
23. Thilagavathi T (2013) Nutrient content of banana varieties dehydrated by various methods. Int J Innovative Res Stud 2(11):225–231
24. Kostaropoulos AE, Saravacos GD (2006) Microwave pre-treatment for sun-dried raisins. J Food Sci 60(2):344–347
25. Schirmer P, Janjai S, Esper A, Smitabhindu R, Mühlbauer W (1996) Experimental investigation of the performance of the solar dryer for drying bananas. Renew Energy 7(2):119–129
26. Hassnain AA (2009) Simple solar drying system for banana fruit. World J Agric Sci 5(4):446–455
27. Nguyen M-H, William Price E (2007) Air-drying of banana: Influence of experimental parameters, slab thickness, banana maturity and harvesting season. J Food Eng 79(1):200–207
28. Abano EE, Sam-Amoah LK (2011) Effects of different pretreatments on drying characteristics of banana slices. ARPN J Eng Appl Sci 6(3):121–129
29. Wakjira M, Adugna D, Berecha G (2011) Determining slice thickness of banana (Musa spp.) for enclosed solar drying using solar cabinet dryer under Ethiopian condition. Am J Food Technol 6(7):568–580
30. Robert Foster, Majid Ghassemi, Amla Cota. (2009) Solar energy: renewable energy and the environment. CRC Press, Traylor & Francis Group, Boca Reton, Florida
31. Manjarres-Pinzon K, Cortes Rodriguez M, Rodriguez Sandaval E (2013) Effect of drying conditions on the physical properties of impregnated orange peel. Braz J Chem Eng 30(3):667–676
32. Jokic S, Velic D, Bilic M, Lukinac J, Planinic M, Kojic AB (2009) Influence of process parameters and pre-treatments on quality and drying kinetics of apple samples. Czech J Food Sci 27(2):88–94
33. Bulent Koc A, Toy M, Hayoglu I, Vardın H (2007) Solar drying of red peppers: effects of air velocity and product size. J Appl Sci 7(11):1490–1496
34. Vlachos NA, Karapantsios TD, Balouktsis AI, Chassapis D (2002) Design and testing of a new solar tray dryer. Dry Technol 20(6):1243–1271
35. Jain D, Jain RK (2004) Performance evaluation of an inclined multi-pass solar air heater with in-built thermal storage on deep-bed drying application. J Food Eng 65:497–509
36. Pangavhane DR, Sawhney RL (2002) Review of research and development work on solar driers for grape drying. Energy Conversion Manage 43(1):45–61
37. Suhas P Sukhatme. (2000) Solar energy, principles of thermal conduction and storage. New Delhi, India
38. Solar Radiation Handbook (2008) Solar Energy Centre. MNRE Indian Metrological Department, New Delhi, India

39. Alexandre Queiroz JM, Hermeval Dantas J, Rossana Figueiredo MF, Karla dos Melo S (2011) Solar drying of jack fruit almonds. Post-Harvest Sci Technol 31(6):1–7
40. Ogunkoya AK, Ukoba KO, Olunlade BA (2011) Development of a low cost solar dryer. Pacific J Sci Technol 12(1):98–101
41. Gatea AA (2011) Performance evaluation of a mixed-mode solar dryer for evaporating moisture in beans. J Agricultural Biotechnol Sustainable Dev 3(4):65–71
42. Brett A, Cox DRS, Trim DS, Simmons R, Anstee G (1996) Producing fruit and vegetables for micro-and small-scale rural enterprise development, handbook 2: dryer construction. University of Greenwich, UK

Gathering and processing energy consumption data from public educational buildings over IPv6

Kostas Koumoutsos[1,2], Aristotelis Kretsis[1,2*], Panagiotis Kokkinos[1,2], Emmanouel A Varvarigos[1,2], Vassilis Nikolopoulos[3], Eirini Gkioxi[3] and Anastatios Zafeiropoulos[4]

Abstract

Background: Reducing energy consumption and CO2 emissions in order to address climate change requires behavioral changes by the citizens, who will have to adopt more environmentally friendly and energy-saving practices.

Methods: We present a system and corresponding practices for gathering energy consumption data from public school buildings over the Internet, processing them to identify hidden correlations and produce actionable advice and presenting the results in real-time to its occupants over the Web. We describe the metering infrastructure installed at schools for energy consumption monitoring and the related actions carried out in order to motivate local school communities towards an environmentally friendly behavior. The introduction of IPv6 was found to be a key enabling technology for setting up such a system in a simple, secure, and efficient way.

Results: The proposed system and processes are put under test in a pilot installation composed of about 50 IPv6-enabled schools of the Greek School Network (GSN). These are also combined with appropriate educational and social engagement tools. In this way, energy consumption, operating costs, and greenhouse effects can be reduced in the area of educational and public administration buildings in general. Thus, people become almost instantly aware of the energy and environmental implications of their actions, which motivates them towards behavioral changes and the adoption of environmentally friendly practices.

Conclusions: The pilot provides a good example of collaboration between the ICT sector, the smart building and automation vendors, and the public authorities.

Keywords: Smart energy meters; Energy information system; Advance energy analytics; IPv6; School network

Background

Reducing energy consumption and CO_2 emissions in order to address climate change requires behavioral changes by the citizens, who will have to adopt more environmentally friendly and energy-saving practices. As buildings account for about 40 % of total energy consumption and emissions [1, 2], improving the energy behavior of their occupants is a key factor for achieving the desired climate and energy objectives. The first, but necessary and important, step in this direction is raising the energy awareness of the buildings' occupants by providing them immediate feedback on the energy implications of their actions (e.g., when turning on the lights or some equipment), in the same way that the speed measurement equipment present in a car gives the driver immediate feedback on the implications of his actions while driving (e.g., when stepping on the gas pedal).

Energy awareness is even more important when it refers to pupils, as behavioral changes in them are more long lasting, easier to accomplish, and have larger cumulative impact. The application of behavioral change theories in the field of energy consumption supports criticism of a too narrow focus on individual behavior and the need to broadening to include social interaction [3, 4]. So, intervention programs aimed at the change of energy consumption patterns need to take into account that behavioral change is best achieved and maintained if supported by tailored information and changes in context, for example, through supportive social networks.

It is evident that energy monitoring systems are important for energy awareness (e.g., in manufacturing companies [5], schools [6], households [7, 8], universities [9], etc.). Studies have shown that by simply monitoring one's energy consumption, the savings on the electric bills can be 5 to 15 %. However, Van Dam et al. [8]

* Correspondence: akretsis@ceid.upatras.gr
[1]Department of Computer Engineering and Informatics, University of Patras, Patras, Greece
[2]Computer Technology Institute "Diophantus", Patras, Greece
Full list of author information is available at the end of the article

found that the introduction of energy monitors in buildings results in only a temporary decrease in the level of energy savings made by occupants, which could not be sustained after a period of only a few months. This is attributed to a lack of habit formation as well as poor design with overly complex interfaces. The authors showed that where people adopt a regular habit of looking at energy monitors on a daily basis, they exhibit larger savings over time compared with others. An implication is to question the notion of mass-produced "one-size-fits-all" home energy monitors and whether solely technological solutions (such as energy monitors) actually achieve the desired results. It also suggests that a deeper understanding of the relationship between the user and these systems is needed. Darby [10] also discusses the level of "affordance" that smart meters offer, i.e., their usability and effectiveness and points out that: "Taking control away from the customer cannot be relied upon to improve the situation: it may actually entrench and legitimize high-demand practices, disengaging customers from any need to consider and question them". Darby argues that effective forms of interface, feedback, narrative, and support need to be developed to reach more diverse populations and to reduce actual consumption. The challenge is for the smart meter to prove itself in terms of developing a useful relationship with the user, through new forms of user engagement (understanding, expectation, and behavior) enabled by new design and media approaches.

Today, the many recent advances in energy metering technology and the design of more energy-efficient equipment have only partially been adopted in public and residential buildings. One of the reasons limiting their adoption is that their internetworking is often difficult to implement due to the lack of public IPv4 addresses (and the use of Network Address Translation (NAT) technologies, which complicate things) and the difficulties encountered in configuring the network of devices. IPv6 [11] is a technology that promises to meet the needs of the huge Internet growth (and the "Internet of things" vision), eliminating some of the restrictions caused by IPv4 technology. Thus, the IPv6 technology takes the role of an enabler and introduces various additional benefits such as security, effective access control, reduced management, and maintenance effort.

Most related works focus either on the behavioral and social aspects of energy monitoring [6] or on the pure technical details of related energy monitoring solutions [12, 13]. In our work, we present a system and set of practices for gathering energy consumption data from public educational buildings over the Internet. We concentrate more on the technical aspects of developing, installing, and operating such a system and on the visual interactions with the users, while the presented energy consumption data are indicative of the value and the benefits of the system. In particular, the proposed system consists of a set of IPv6 enabled smart meters, an energy information system, a Web platform with social engagement tools for the pupils of the schools, and an IPv6 enabled network that interconnects them. The proposed system and processes are put under test in a pilot installation composed of about 50 schools of the Greek School Network (GSN) [14], implemented in the context of the Governments Enabled with IPv6 (GEN6) project [15]. IPv6 technology has already been deployed in GSN, providing a leveraging technology for enhancing existing services or providing new services to the end users, including the advanced energy services described in the present paper. The infrastructure installed makes use of cloud computing technologies, as the smart meters measure, and push the recorded energy consumption data to a scalable cloud aggregation system via a secure communication channel over the IPv6 protocol. The collected data is being stored, aggregated (if necessary), and analyzed using a meter data management (MDM). The services providing energy key performance indicators (KPIs), such as the kilowatt hours consumed and the corresponding volume of CO_2 emissions, will be offered according to Software as a Service (SaaS) model by exposing Web services to third authorized partners. Data is made available through an API in javascript object notation (JSON) format.

An interactive IPv6 Web platform (http://gen6.sch.gr) is in charge of monitoring the processed energy consumption data coming from MDM. Via a real-time intuitive interface, the school community will be taught the correlation between the actions they undertake and the energy consumption and CO_2 emissions of their schools, providing in this way significant motivation for behavioral changes. Moreover, the IPv6 Web-based platform will become an educational and social engagement tool for teachers and students, stimulating discussions and enabling actions within schools in order to better understand complex energy consumption data, through their interactive display using animations, graphical statistics, historical data, comparative schools' energy data, average factor, best performing school competition, etc. Results from the competition among the participating school communities will be disseminated to the public through various communication channels including the Web portal and social networks. At the final stage, the students will be given the opportunity to participate in a series of interactive workshops designed to develop their knowledge on environmental and energy issues, which will encourage them to take actions in the direction of sustainable living, energy consumption minimization, and climate change and motivate them to adopt a long-term environmentally friendly behavior, inside and outside the school.

The proposed system and processes are implemented in the context of the GEN6 project [15] through the cooperation of the Computer Technology Institute (CTI) [16], Intelen [17], and the Greek Research and Technology Network (GRNET) [18] on the GSN [14]. The GEN6 project has an ambitious mission in setting up complementary and interoperable national and cross-border pilots across different EU States, in order to validate and catalyze the deployment of IPv6 in the existing EU government infrastructures, services, and applications. The GEN6 Greek Pilot presented in this work has already installed a set of IPv6-enabled smart meters at 50 schools for the real-time monitoring of their energy consumption over the IPv6-enabled network and has kicked off other dissemination actions aiming at motivating students to change their behavior towards an energy-efficient and environmentally friendly direction. Our initial tests have shown energy savings of more than 10 % and reductions in the carbon footprint by at least 10 % at the participating school buildings. Our target is that the developed system for monitoring, analyzing, and reporting energy consumption will also become an educational and social engagement tool for students.

The remaining of this paper is organized as follows. In the "Public education buildings" section, we describe the basic characteristics of the Greek public education buildings. The architecture of the system responsible for collecting and analyzing energy data from these buildings is presented in the "Proposed architecture" section. Details for the smart meters and the IPv6 connectivity, the data management system, and the Web platform are presented in the "Smart power meters and IPv6," "Meter data management," and "Web platform and social engagement" sections, respectively. The "Results and discussion" section presents the collected energy consumption data from a number of schools. In the "Conclusions" section, we conclude the paper.

Methods

Public education buildings

One of the reasons we focused our efforts on public educational buildings as opposed to general public building is that in the former case the buildings' architecture and topology are more suitable for running energy efficiency case studies. Usually, educational buildings are more complex and have a more distributed architecture with many individual facilities inside the campus (classroom space, library/resource space, indoor and external sports facilities), so that the energy consumed is much more than that of common public building. Moreover, the occupants of the educational buildings are students, whose energy consumption behavior should be easier to modify when compared to that of adults. Indeed, the goal of education is behavioral change, and the use of ICT in the formulation of effective teaching methods that tap into the mechanisms of behavioral change can be very efficient. In fact, some of the behavioral change theories, like the Social Learning Theory and Theory of Planned Behavior, were developed as attempts to improve behavior through education in schools. Several works that consider energy consumption issues consider schools as the primary use case [6, 19].

In numbers, today in Greece, there are around 3600 secondary education units and 12,000 primary education units, hosting more than 800,000 students and 100,000 teachers [20]. Public education buildings in Greece are distributed over the whole country, with varying student population sizes per building, varying floor area, and varying weather conditions. The mere energy consumption of a building is not a good measure of energy efficiency, as it has to be adjusted to account for these factors. One of the most objective KPIs used to quantify energy efficiency in schools is the "energy utilization index (EUI)" of a school which is measured in kWh /m^2/student. Using objective and easy to compare KPIs is important in our case where we are also interested in initiating an energy efficiency improvements "competition" among pupils of different schools, and the competition will of course have to be fair to all participants.

GRNET [18] is the National Research and Education Network (NREN) provider, operating the Greek Academic network, that connects local universities and research institutions via dark fiber at speeds up to 10 Gbps, and offering to the Greek R&E community access to the pan-European GEANT network through 4 × 10 Gbps links. The Greek Schools' Network (GSN) [14] is the educational intranet of the Greek Ministry of Education that interlinks all primary and secondary education schools and respective administrative units, providing basic and advanced telematics services. GSN is a nationwide network that spans to all 51 prefectures of Greece, using GRNET as its backbone network (Fig. 1.) The majority of schools in Greece are connected to GSN using ADSL and optical technology.

GRNET and GSN are in strong cooperation in order to complete the transition from an IPv4 only network to a dual stack (IPv4/IPv6) network. The Greek School Network already supports IPv6 interconnection services and other basic services over IPv6 [21]. Such services are gradually extended in the intranets within schools (approximately 15,000 sites). The backbone network of GSN is fully IPv6-enabled including IPv6 support. On the access network, IPv6 interconnection has been activated for the ADSL users, i.e., for 95 % of the schools. This has been achieved by enabling IPv6 on the radius profiles of each ADSL router connected to GSN. All

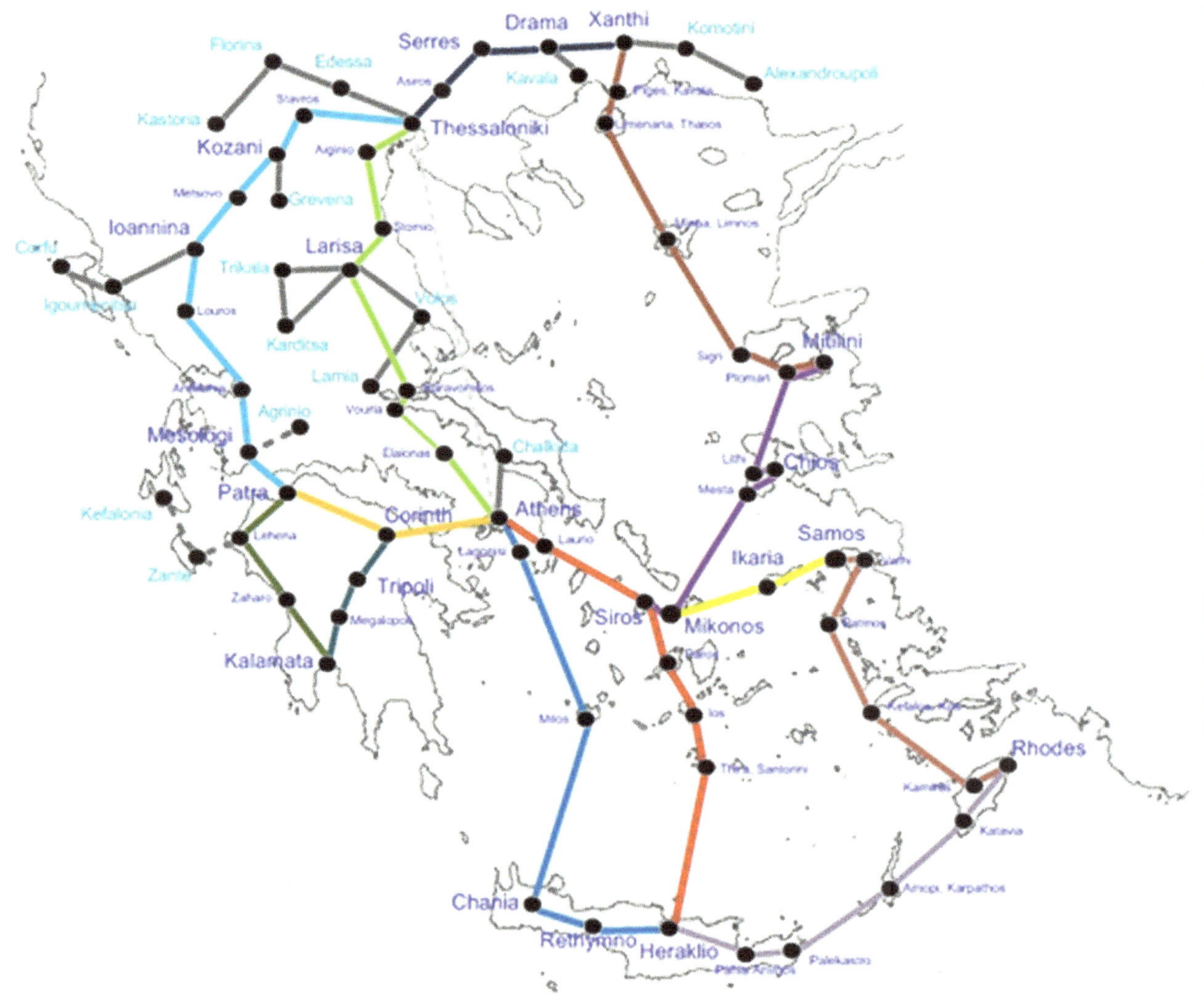

Fig. 1 GRNET and GSN network topology [24]

public education buildings are connected to the Public Power Corporation (PPC).

Developing and deploying an energy monitoring and analysis system in all public education buildings are huge tasks, considering the number of buildings and responsible personnel, the possible electrical and network peculiarities of each educational unit, and the huge amount of information that needs to be collected and analyzed. The advantages to be obtained by accomplishing this task are however also important, with the most obvious one being the energy savings and corresponding environmental benefits that could be obtained and the elimination of great costs that are spent annually to meet the schools' energy needs. The money savings can also be translated into the ability to invest in additional intervention actions for additional high performance facilities, hiring more teachers and purchasing more textbooks and computers. Beyond these bottom line benefits, schools also can realize better overall student health and better serve as centers of community life.

Proposed architecture

Advanced metering infrastructure (AMI) systems are systems that measure, collect and analyze energy usage, and communicate with metering devices such as electricity meters either on request or on a schedule. These systems include hardware, software, communications, consumer energy displays and controllers, and a MDM. AMI differs from traditional automatic meter reading (AMR) in that it enables two-way communication with the meters.

In what follows, we present such an AMI system designed for public education buildings in Greece consisting of components provided by Computer Technology Institute (CTI) [16] of the GSN [14], Intelen [17], and the GRNET [18]. In particular, the proposed system consists of the following components:

- energy meter
- network
- aggregator and analysis system
- user interface

Figure 2 presents the logical topology of the GEN6 Greek Pilot in schools, which consists of the aforementioned components.

The proposed system (Fig. 3) utilizes Intelen's [17] smart metering infrastructure consisting of a consumption metering device (CMD) along with its current transformer (CT), which does not measure voltage, a transmitter, and the so-called i-box. The i-box is a smart network device that acts as a data bridge between the power meter and the local building router. Both i-box and the GSN/GRNET local router can have IPv6 global address using stateless auto-configuration method (prefix delegation). The primary internet connection in education buildings is broadband (xDSL) connection with IPv4 or IPv6 characteristics. The installed power meters collect energy consumption data that are sent over the GSN/GRNET network aggregated to Intelen's Head-End System (HES) and then to Intelen's MDM. The MDM is responsible for data acquisition and is capable for a series of services, including energy analysis, financial analysis, and demand response (signal notification, manual and automated response, response analysis, and quantification). Finally, the MDM feeds the processed information to Intelen's intercompany data exchange (IDE) system, which exposes the Web services used by the Web platform ("Web platform and social engagement" section).

Smart power meters and IPv6

The smart metering infrastructure is capable of offering a series of services, including the following:

- It enables real-time or near real-time registration of electricity use.
- It provides local and remote access to the meter (on demand).
- It performs basic calculations in order to translate the raw data that are collected by the specific CMD, to information that can be handled by centralized (cloud) infrastructure.

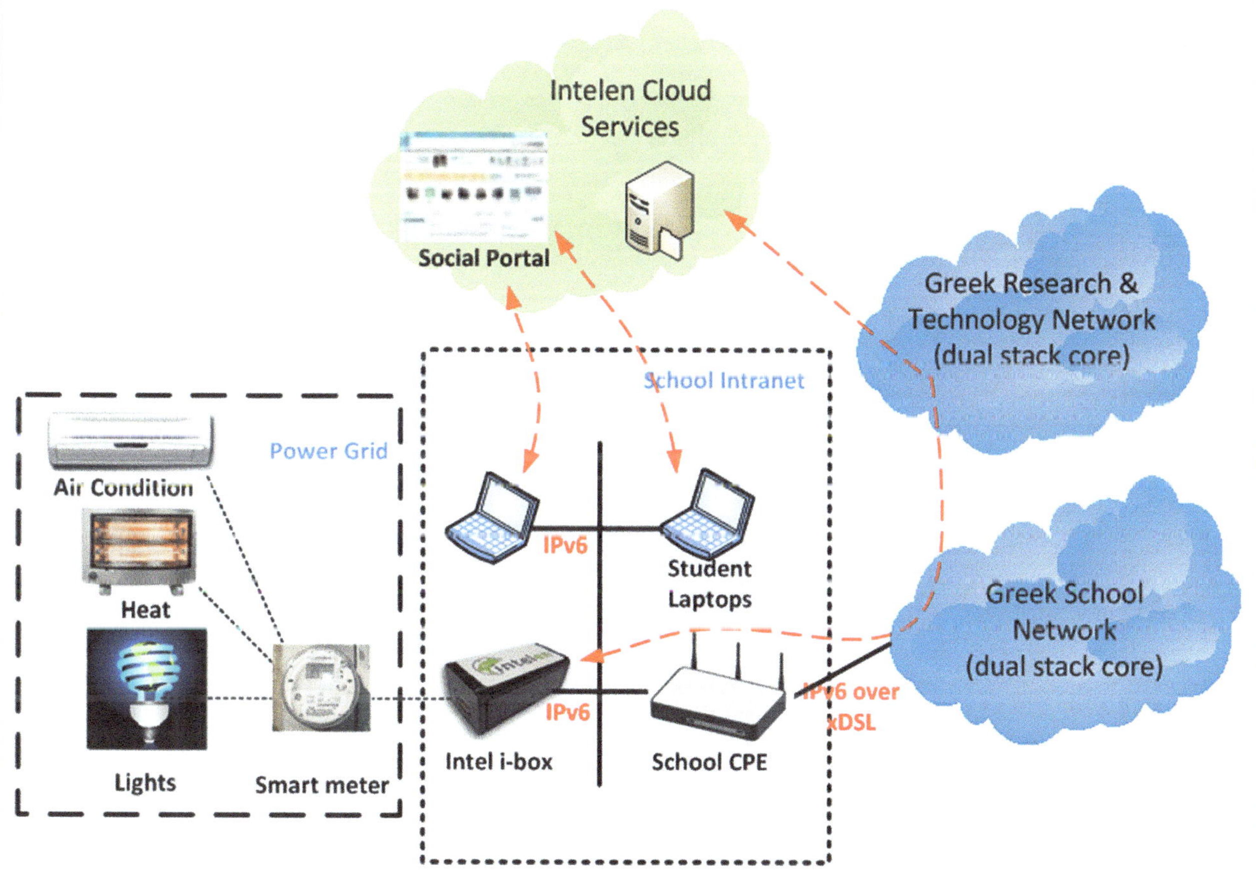

Fig. 2 Logical topology of the GEN6 Greek Pilot in schools

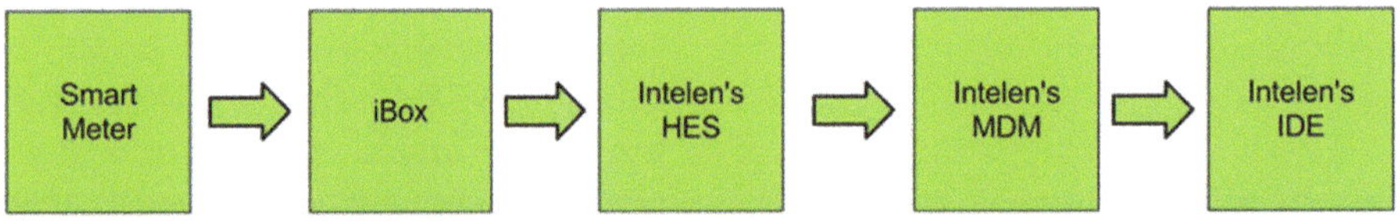

Fig. 3 The data flow from a smart meter to Intelen's intercompany data exchange (IDE) system

- It acts as memory storage (buffer) so that data are not lost in case of a network or other error but are stored (up to the size of the respective embedded storage device) for transmission when network connectivity is re-established.
- It extracts the appropriate KPIs from the raw data received by the CMD.
- It performs error handling concerning the CMD, the connectivity of the system.
- It receives instructions for extra KPIs from Intelen's MDM.

The proposed system requires not only the installation of smart energy meters but also of additional sensor nodes (e.g., temperature and humidity sensors, meteorological stations, etc.). These nodes have to operate in a reliable way, be able to recover from failures and instant networking problems, and operate for a large time period (minimum energy consumption). The deployed nodes are based on the 6LoWPAN standard (IPv6 over low power wireless personal area networks) that allows IPv6 packets to be carried efficiently within small link layer frames such as those used by IEEE 802.15.4 achieving in this way lower energy consumption. The energy efficiency of 6LoWPAN is also combined with the support of advanced ad-hoc routing, autonomic networking, and infrastructure bootstrapping characteristics. Self-* functionalities (e.g., self-configuration, self-healing, self-management) can be designed and implemented over IPv6 permitting the deployment of advanced communication schemes among the sensor networking devices and reducing the administration overhead for handling faults or alarms. Based on the addressing scheme applied, each device is able to recognize its neighboring devices and an ad-hoc network may be established.

Intelen's smart meters are fully IPv6 enabled and can be assigned a global IPv6 address for their connection to the GSN/GRNET network. IPv6 is as a necessary technology for the reliable, efficient, and scalable implementation of the proposed energy metering system. Even if the major force for the transition to IPv6 is related to the absence of public addresses in IPv4, there are several IPv6 protocol characteristics that facilitate the deployment of advanced networking services, especially in the sensor networking world, and were found to be particularly useful in our proposed system. In particular, the IPv6 technology takes the role of a "green" enabler and raises various benefits such as the following:

- Reduced management and maintenance effort due to end-to-end communication and management of the smart sensors. The end-to-end communication scheme facilitates the proper management/configuration of the devices, the installation of software updates, and the monitoring of the infrastructure.
- Better policy schemes over heterogeneous networks are made possible; for example, nodes of various on-site communications systems can be provided with unique public IPv6 addresses to avoid conflicting use of private IPv4 addressing; centralized communication initiation and management can be provided and services can be directly connected as all nodes across different on-site networks can have globally unique public IP addresses.
- Advanced security features to the schools intranets are enabled; for example, transparent end-to-end security without complex NAT traversal mechanisms, fine-grained security policies, and filtering rules can be applied based on unique end system addressing scheme.
- Advanced auto-configuration features (e.g., IPv6 stateless auto-configuration, bootstrapping of the infrastructure) and ad-hoc routing are made possible.
- Quality of Service (QoS) can be supported in local and global network environment.
- Multicast transmission features are enabled.
- Facilities are provided to deploy new services without NAT-related limitations and requirements for application gateway implementation (e.g., exchange of sensor data, services for situation monitoring, etc.).

Meter data management

A meter data management (MDM) system stores, analyzes, and displays building energy data. At a minimum, MDM provides hourly whole-building electric data that are Web accessible, with analytical and graphical capabilities. The MDM system was extended in order to meet the envisioned system's requirements and is used from the project Web portal in order to retrieve data from electrical energy meters. Energy consumption data from energy meters are being collected through the i-box and, every 15 min (default value), are sent to the MDM system.

The MDM system comprises of the following:

- Specific middleware for data acquisition, data aggregation, data structuring, and transformation, so as to feed all functionalities and services in a generic way.
- Functionalities and services for data validation, integration/storage, consolidation, and access.
- Generic applications that can be realized in a modular way through the invocation/combination/orchestration of functionalities and/or services. These applications are still generic but form the basis for those offered to clients, either through parameterization or through client specific instantiation.

In the developed MDM system, energy data are processed and enriched with other school characteristics and data, such as school area, school population, etc. Data analysis is one of the most important parts of the MDM system and is also the system responsible for all calculations and data-mining algorithms, either in batch mode or in real-time. Data mining algorithms and techniques that are used by the MDM system allow users to analyze data from many different angles, categorize it, and summarize the relationships identified. In essence, data mining is the process of finding the correlations or patterns among dozens of fields in the existing large relational energy database. A recursive algorithm is executed in order to analyze the results through time and correlate them with other external variables (temp, humidity, etc.). The process of finding hidden information from data has different definitions, including knowledge extraction, information discovery, data mining, and the more proper knowledge discovering in energy database. In particular, data mining includes other activities, such as data selection, checking, cleaning, preparation, pattern presentation, and knowledge refinement and visualization. In the MDM system, a load profiling method is being applied and a specific data-mining technique on agent grid clustering is used for the processing of the data. The objective of this design is to determine the best techniques to be used, in order to reach the energy-saving target that has been set.

For effective load profiling and data mining, the following steps were followed:

- Set specific energy key performance indicators (KPI) from the metering grid and the load curve mathematics.
- Categorize and compute specific objective functions, comprised of specific KPIs.
- Perform a weighted version of the recursive analytics algorithms, based on a distributed agent structure, routed on the cloud.
- Store the numerical values in matrices, called energy relevance matrices (ERM), and use numerical linear algebra to find correlations and hidden patterns.
- Use the results of the energy relevance matrix analysis together with the output of the agent results vector in order to identify cross-hidden correlations, periodicity, and peak trends on the energy load curves.
- The proposed approach, regarding the application of game mechanics and behavioral engagement to the students, will enhance their awareness and push them towards a change in their behavior that will save energy. The IPv6-based system will analyze the real-time energy consumption and various dynamic statistical indices and metrics. Special algorithms and game mechanics/scenarios will engage the student in a continuous game by offering comparison tables and personalized incentives to save energy.

Energy data become available through APIs in JSON, XML, plain text, HTML, and CSV formats. The most appropriate format is JSON due to its flexibility and the smaller size of its files. The available API methods include daily data in 1-hour intervals, monthly data in 1-day intervals, and yearly data in 1-month intervals.

Web platform and social engagement

Citizens' behavior is an important factor that determines the overall energy consumption of buildings and may be improved through energy awareness. An energy consumer who is "energy-aware" and is able to know in real-time the energy consumption that his/her behavior is causing and compare it to that of other similar users is more likely to initiate changes and adopt a more environmentally friendly behavior. The term "energy awareness" implies also the training of a user in terms of understanding the contribution to the total energy consumption caused by his/her use of appliances and helping him/her define specific actions that could lower the energy consumption in a quantifiable way.

We have developed an interactive Web platform [22] that is in charge of presenting the processed energy consumption data coming from the MDM system. In order to achieve this and enable the students' community better understand complex energy data, the display of the data in the students' portal has been enriched

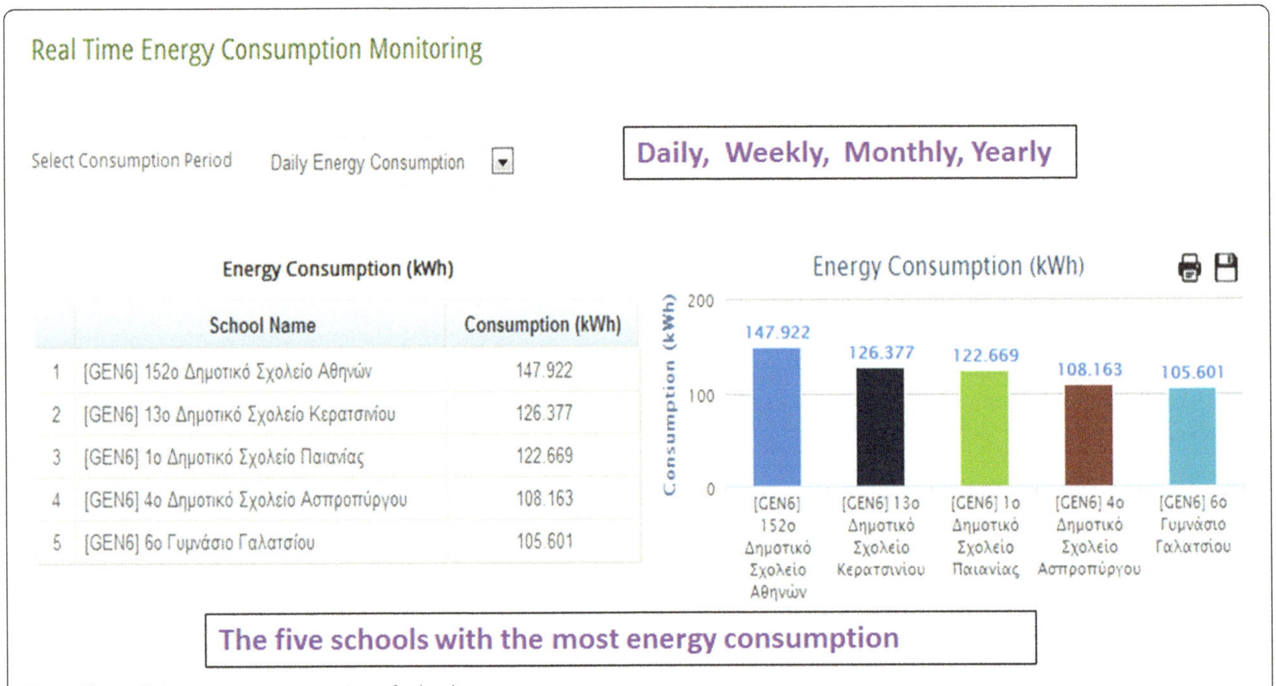

Energy Consumption (kWh)

	School Name	Consumption (kWh)
1	[GEN6] 152o Δημοτικό Σχολείο Αθηνών	147.922
2	[GEN6] 13o Δημοτικό Σχολείο Κερατσινίου	126.377
3	[GEN6] 1o Δημοτικό Σχολείο Παιανίας	122.669
4	[GEN6] 4o Δημοτικό Σχολείο Ασπροπύργου	108.163
5	[GEN6] 6o Γυμνάσιο Γαλατσίου	105.601

Fig. 4 The real-time energy consumption of schools

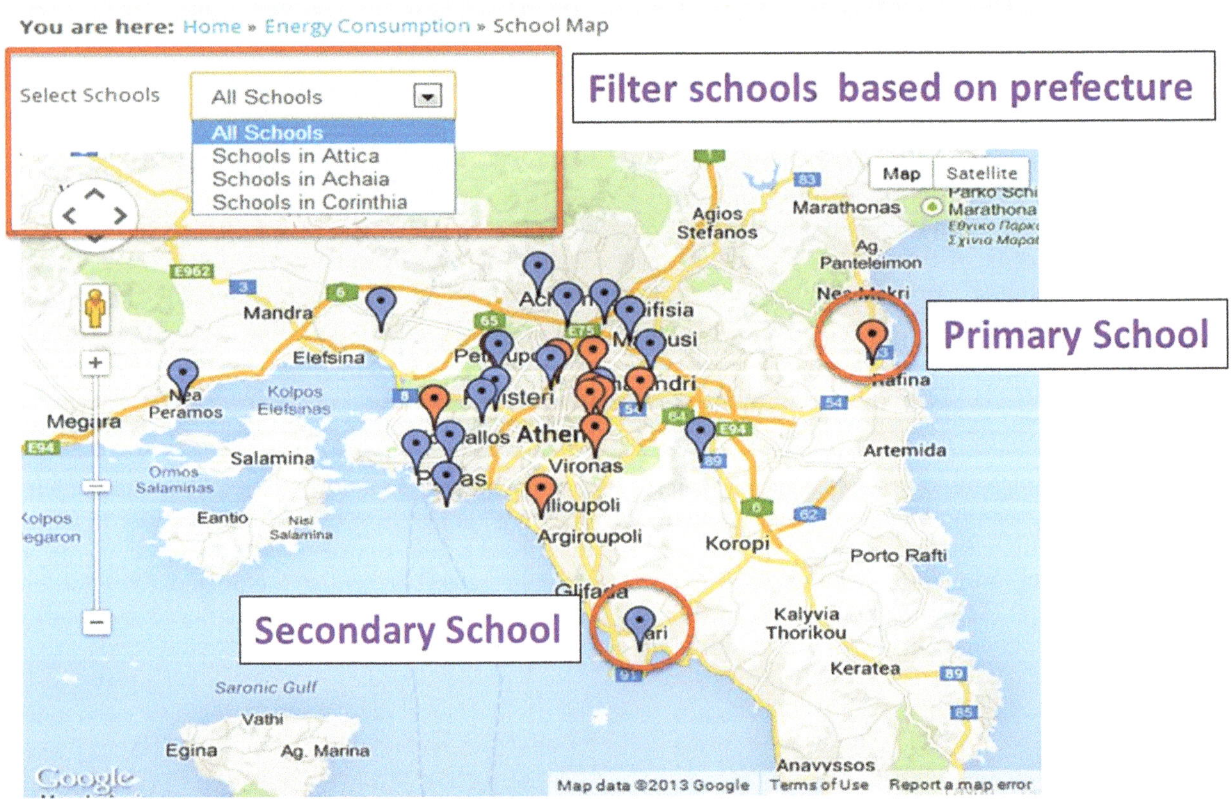

Fig. 5 Geospatial map of the measured schools

Table 1 Energy savings

School name	Energy saving (%)
Attiki-14th Primary School Acharnon	22.79
Attiki-1o Primary School Varkizas	16.98
Attiki-152nd Primary School of Athens	19.42
Attiki-7th Primary School Transformation	27.00
Attiki-6o Primary School Upper Liosion	22.66
Attiki-1st Gymnasium of Alimos	35.00
Attiki-6th Gymnasium Galatsi	35.57
Attiki-49th High School of Athens	39.18
Attiki-41o Primary School of Peristeri	19.50
Attiki-1st Primary School Peanias	33.49

with animations, graphical statistics, historical data, comparative school energy data, average factor, information on the best performing school, etc.

For instance, Fig. 4 presents the real-time energy consumption of schools as it is shown in the portal [22]. The schools' energy consumption information is made more directly comparable by accounting for the number of students and the area of the school. Historical data regarding the energy consumption of schools are also available. The Web platform also associates geo-information with schools energy profile (Fig. 5).

In general, the Web platform illustrates to the students the correlation between their actions and energy consumption/CO_2 emissions of their school, targeting to become an educational and social engagement tool for students. This platform will be used to motivate a kind of competition among schools regarding which of them will achieve the most significant energy savings in a given time period. It can also be used to stimulate discussions and actions within schools related to energy consumption and its environmental implications.

We also plan to use social media to promote energy efficiency and sustainability campaigns not only to the school community but also to the Greek public sector in general. By using social competitions, social group benchmarking, and social interfacing, the students will interact with the service in real-time (as playing a social game with real energy data), and the service will offer adaptive incentives to the student, according to their social profile and their habits. The social media campaign program will be reinforced by hands-on workshops that will engage students on issues of sustainability and empower them to make further behavioral changes. School energy competitions will create interaction among students and create additional "social media traffic."

Table 2 Special characteristics

No. of Buildings	10
Av. school area (m^2)	~1800 m^2
Av. student population	~257 students
Av. energy per student/week consumption before smart energy meters	2.5 kWh/ student
Av. energy per student/week consumption after smart energy meters	1.6 kWh/ student
Av. energy savings/week	27.16 %

Results and discussion

Pilot setup

An important initial decision that had to be made and would influence the success and the impact of the Greek pilot was the selection of the participating schools. The selection was based on the location of the schools and their detailed characteristics:

- Location of the schools: we decided that the selected schools will be located across three adjacent prefectures (namely, Achaia, Korinthia, and Attiki) within the Greek territory. The selected prefectures present sufficiently diverse geological and climate conditions [23]. Variations in energy consumption can be monitored without, however, affecting the impact on the achieved energy savings in the pilot, which will be based on a percentage basis.
- Detailed school characteristics: GSN organized a survey to take place among the schools. The survey that was completed at the end of March 2012 collected information regarding the characteristics of the schools and the school buildings, the students' and teachers' populations, and their degree of interest in environmental issues, including existing or past activities on energy efficiency and related issues. Some important parameters of interest are the following: (1) the willingness of the school to participate in the pilot, (2) the existence or lack of broadband connectivity, (3) type of school (primary or secondary education), and the specific grades it includes; students at the last grades of primary education and the first grades of secondary education would be preferable as participants in the pilot's activities, (4) the number of students in the school (we prefer schools with a relatively high student population), (5) the availability or lack of local technical support staff, (6) any student activities related to environmental and energy efficiency issues, (7) the total area of the school, (8) the daily electricity load, (9) the ease of accessibility of the electrical panel, (10) the proximity of the electrical panel to the school computer room, (11) the number of high consumption (over 500 W)

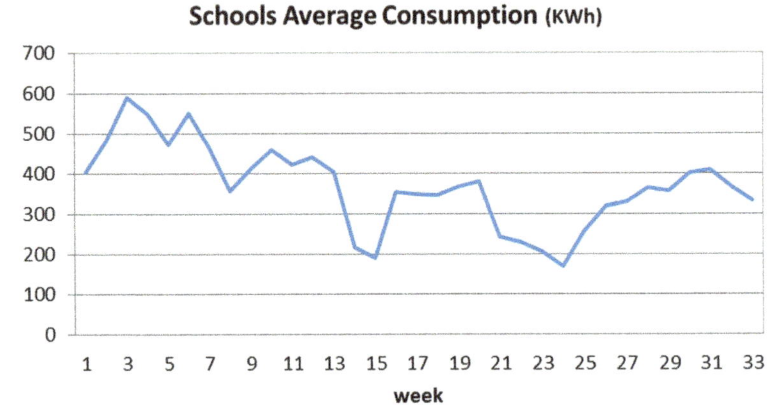

Fig. 6 Schools' average energy consumption (kWh) per week

energy devices or appliances that operate at the school, (12) whether single-phase or three-phase power supply is available at the school, (13) the existence or not of a person responsible for environmental activities in the school, (14) the availability of sufficient contact information for the schools, and (15) the nominal amperes of the building's general switch.

The previous factors are actually soft requirements that have been used for the selection, and some of the schools that finally participated may not have all the desired characteristics. The final selection of the schools in the pilot has been done based on the analysis of the questionnaire's responses, interviews with interested teachers, and on-site visits that were also performed.

Pilot results from 25 school buildings

In this section, we present our initial results from the deployment of smart meters at 25 (out of the 50 schools). Actually, energy meters are currently installed in all 50 schools; however, not sufficient data from all these schools are yet available. The total energy savings for these 25 school units over a period of 33 weeks was 2945 kWh. The calculated monetary benefit is 353 €, when the electricity school price for energy unit is 0, 12€ per kWh.

The energy saving results in kilowatt hours from 10 schools after the period of 33 weeks is presented in Table 1.

These energy saving values were calculated for the 10 most "energy-efficient" schools, based on their average weekly energy consumption (over the 33-week period) in relation to the energy consumption in the first week of the evaluation. No adjustments have been made regarding other factors that may have affected the schools' consumption, such as weather (because we did not have such data available in a useable manner). Nevertheless, we believe that these results are indicative of the presented system's value and the benefits that can be obtained, even though the lack of such adjustments decrease the ability to make definite conclusions regarding the effect of the system on the students' energy behavior.

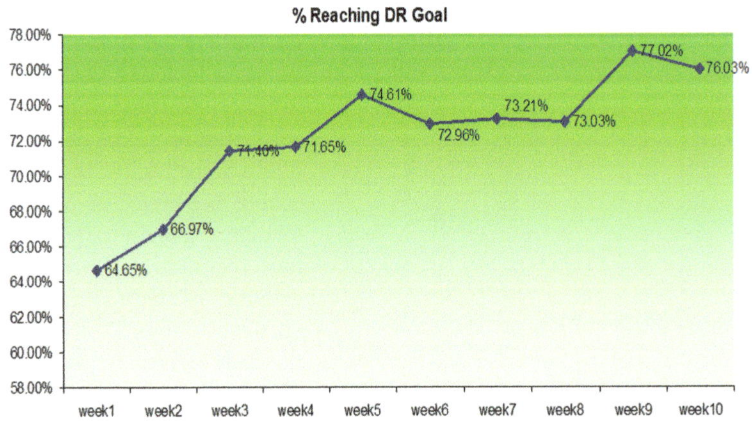

Fig. 7 Engagement efficiency in percentage to the demand response signals that were sent to the student teams

Table 2 depicts useful details and school characteristics such as school density, school population, cumulative average energy consumption before and after the pilot installation at 10, highly "energy-efficient" schools, and finally, the cumulative average energy saving per week.

Figure 6 shows the 25 schools' average energy consumption per week. We observe that the schools consumption is decreasing, though not steadily due to variations in energy needs based weather and other reasons. The decrease cannot be with certainty attributed to the platform, as mentioned previously.

Figure 7 represents some initial results we had, following a simple demand response (DR) pilot campaign over a 10-week period, for the engagement of the students. The graph in Fig. 7 represents that the engagement efficiency as a percentage of the demand response signals that were sent to the student teams, in order to reduce their demand consumptions by specific targets in a specific timeframe (i.e., drop by 1.4 kW in the next 30 min). The DR signals were sent using mobile SMS and email to the teams responsible. During the first weeks, the DR efficiency rose from 65 % up to 74 % and then was stable, fluctuating over a median. This means that from the initial DR targets, the schools were able to respond to the DR signals with some delays (~15–20 min) and to meet their initial DR targets (drop in kW) approximately by 75 %. Thus, the average DR efficiency reaching goal is calculated for every school for each week, as the efficiency of the transition speed (min) and the efficiency of the target demand (kW). The dynamic average during the 10 weeks can be seen in the figure, indicating the amazing engagement progress that the students and schools have shown, towards demand response and energy efficiency targets.

Conclusions

The lack of consumers' awareness regarding their own electricity usage is a significant problem, but it also provides an opportunity for public and private stakeholders to save money and reduce CO_2 emissions. The Greek School Network energy pilot aims at reducing the schools' carbon footprint by at least 10 % and offer advanced real-time energy efficiency services over IPv6-enabled networks. It focuses on positively affecting the students' behavior and raising energy awareness within the school communities. The installed energy meters demonstrate to participating students in real-time the energy consumption and environmental consequences of their actions, providing in this way significant motivation for behavioral changes. Given the long lifespan of most governmental buildings (including schools), their energy efficiency influences energy consumption for many years.

Our work concentrated more and presented the technical aspects of developing, installing, and deploying such an energy monitoring system in schools and on the visual interactions with the users. In this context, we recognized the importance of using the IPv6 technology for the interconnection of the smart meters with the rest of the infrastructure. Though initial results indicated an actual reduce of the energy consumption in several schools participating in the pilot, these results are not conclusive or global. Regarding overall energy consumption and reduction, it is evident that a system like the one presented in this work is only a part of the equation towards changing the energy related behavior and habits of students and people in general. Education, engagement in social interactions, and gamification are also means for achieving a sustainable energy aware and friendly attitude for all members of the society.

Abbreviations

6LoWPAN: IPv6 over low power wireless personal area networks; AMI: advanced metering infrastructure; AMR: automatic meter reading; CMD: consumption metering device; CT: current transformer; CTI: computer technology institute; DR: demand response; ERM: energy relevance matrices; EUI: energy utilization index; GEN6: governments enabled with IPv6; GRNET: Greek Research and Technology Network; GSN: Greek School Network; HES: head-end system; IDE: intercompany data exchange; KPI: key performance indicators; MDM: meter data management; NAT: network address translation; NREN: National Research and Education Network; PPC: Public Power Corporation; QoS: quality of service; SaaS: software as a service.

Competing interests

The authors declare that they have no competing interests.

Authors' contributions

KK organized the school selection procedure and drove the communication with each school authorities. Also, he was responsible for the configuration of the GSN network. KK had a lead role in the communication with the other partners for the setup of the pilot. Also, KK contributed in the writing of this paper. AK was the developer of the Web portal and contributed in the writing of this paper. PK had a role in the communication with the other partners for the setup of the pilot and in its maintenance. Also, PK contributed in the writing of this paper and in the collection of the measurements. VN participated in the development of the used advanced metering infrastructure (AMI) and also had a lead role in the installation and deployment of its components. VN provided information for the AMI that was included in this paper. EG participated in the development of the used AMI and also had a role in the installation and deployment of its components. EG provided information for the AMI that was included in this paper. AZ was responsible for the GRNET's network configuration and provided related information for the paper. EAV critically revised the paper and gave final approval for the paper to be published. All authors read and approved the final manuscript.

Authors' information

KK is a senior researcher in the Computer Technology Institute "Diophantus."
AK is a senior researcher in the Computer Technology Institute "Diophantus."
PK is a senior researcher in the Computer Technology Institute "Diophantus."
EAV is a scientific director in the Computer Technology Institute "Diophantus" and professor in the Department of Computer Engineering and Informatics, University of Patras, Greece.
VN is a CEO and cofounder of Intelen.
EG is a senior manager in Intelen.
AZ is a senior researcher in the Greek Research and Technology Network, Greece.

Acknowledgements
This work was funded by the European Commission through the GEN6 project [15].

Author details
[1]Department of Computer Engineering and Informatics, University of Patras, Patras, Greece. [2]Computer Technology Institute "Diophantus", Patras, Greece. [3]Intelen Advanced Engineering Division, Nicosia, Cyprus. [4]Greek Research and Technology Network, Athens, Greece.

References

1. Energy-efficient Buildings (2015) (EeB), ec.europa.eu/research/industrial_technologies/energy-efficient-buildings_en.html, last seen May 2015
2. US energy information administration, www.eia.gov/tools/faqs/faq.cfm?id = 86&t = 1, last seen May 2015.
3. Gill T, Pegg and Allan (2010) Low-energy dwellings: the contribution of behaviours to actual performance. Build Res Inf 38:5
4. Schultz W, Nolan M, Cialdini B, Goldstein J, Griskevicius V (2007) The constructive, destructive, and reconstructive power of social norms". Psychological Science 18(5):429–434
5. Bogdanski G et al (2012) Energy monitoring in manufacturing companies—generating energy awareness through feedback. Leveraging technology for a sustainable world, Springer Berlin Heidelberg, pp 539–544
6. R. Kermish-Allen, K. Peterman, S. MacDonald, R. Thompson, B. Winner, "Student and teacher teams using high resolution electricity monitoring to create local change", Journal of Sustainability Education, Vol. 8, 2015.
7. Hargreaves T, Michael N, Jacquelin B (2013) Keeping energy visible? Exploring how householders interact with feedback from smart energy monitors in the longer term. Energy Policy 52:126–134
8. Van Dam SS, Bakker CA, Van Hal JDM (2010) Home energy monitors: impact over the medium-term. Build Res Inf 38(5):458–469
9. Trejo-Perea M et al (2013) Development of a real time energy monitoring platform user-friendly for buildings. Procedia Technology 7:238–247
10. Darby S (2006) The effectiveness of feedback on energy consumption. A review for DEFRA of the literature on metering, billing and direct displays. Environmental Change Institute, University of Oxford, Oxford
11. Popoviciu CP, Levy-Abegnoli E, Grossetete P (2006) Deploying IPv6 networks
12. Zhao L, Ji-li Z, Ruo-bing L (2013) Development of an energy monitoring system for large public buildings. Energy Build 66:41–48
13. A. Al-Daraiseh, S. Nazaraf, and E. Eyas "An intelligent energy management system for educational buildings", International Journal of Distributed Sensor Networks, 2013. Volume 2013 (2013), Article ID 209803, 13 pages http://dx.doi.org/10.1155/2013/209803
14. Greek School Network (GSN): www.sch.gr
15. GEN6—governments enabled with IPv6, www.gen6-project.eu/.
16. CTI, Computer Technology Institute-"Diophantus", http://www.cti.gr/en/
17. Intelen: http://www.intelen.com
18. Greek Research and Technology Network—GRNET, https://www.grnet.gr/en
19. Dascalaki EG, Sermpetzoglou VG (2011) Energy performance and indoor environmental quality in Hellenic schools. Energy Build 43:718–727
20. Greek ministry of education and religious affairs archives: archive.minedu.gov.gr/el_ec_category127.htm
21. Kalogeras D, Friaças C, Ferreira JN (2005) Cookbook on deploying IPv6 in school networks
22. GEN6 Greek pilot: gen6.sch.gr/index.php/en
23. GeoKlima: http://www.geoclima.eu/node/91
24. GRNET topology: www.noc.grnet.gr/en/node/25

21

Sustainable systems - game theory as a tool for preserving energy resources

Martin Dolinsky

Abstract

Background: Mankind is highly dependent upon fossil fuels for its ever-increasing energy demands. The future supply is unreliable mainly due to the fact that it is a finite resource which is becoming more and more expensive to produce. As a reaction to this current threat and uncertainty, a mathematical model is developed in this study in order to identify more sustainable quantities. The main objective of this study was to provide companies with a tool to assist in the prevention of overproduction, in other words, to minimize unsalable production quantities and to preserve natural resources.

Method: Regression analysis, game theory - Cournot game.

Results: The study was carried out in cooperation with the Italian company PintInox, S. p. A., located in the Italian region of Lombardia. After the implementation of sustainability measurement metrics into the company's processes and the discovery of a profitable market niche (market principles applying value to sustainable development) for PintInox, S. p. A., a natural grouping (oligopoly) of Italian cutlery producers was identified. Entrance barriers posed by this group (oligopoly) are all kinds of inabilities to prove a verifiable concept of their Corporate Social Responsibility practices. Companies that failed to communicate their commitment regarding sustainable development to their business-to-business (B2B) partners cannot enter this oligopoly and are, consequently, excluded from markets placing value on sustainability principles. Based on economic effectiveness, the marketable output of every oligopoly member was calculated.

Conclusions: The desire to preserve natural resources is driven by incentives within the marketplace. These incentives were identified through the game theory. The companies firstly concentrate themselves on social responsibility and environmental safety and then, as oligopoly members, on economic effectiveness. The study reconfirms that also the economic dimension plays an important role in a 'triple bottom line' philosophy.

Keywords: Oligopoly; Cournot game; Game theory; Corporate Social Responsibility

Background

Introduction

This paper has an ambition to fill the gap in research activities in regard to sustainable development. Many studies in this field contain an economic dimension. Just to mention some of the institutions and their contributions: the Institute for Environmental Decisions from Zurich dealt with the topic 'Economic and environmental benefits of dry discharge of bottom ash and acid washing of fly ash from municipal solid waste incineration in Switzerland', the Department of Energy Engineering from The University of Florence worked on the 'Economic Assessment of Innovative Processes for Biomethane Production', the Faculty of Civil Engineering and Geosciences from The Delft University of Technology on the 'Economic and Carbon Footprint Evaluation of Thin Stillage Treatment', CIRSEE Suez Environnement on the 'Technical and Economic Assessment of Micro-cogeneration Technologies for Sewage Biogas', and the Deutsches Biomasseforschungszentrum (gemeinnützige GmbH) from Leipzig on the 'Environmental Impacts and Economic Feasibility of Producing Bioenergy under Various Frame Conditions' [1]. All the mentioned studies contained

Correspondence: martin.dolinsky@sustainability.sk
Department of Management and Marketing, Faculty of Economics and Business, Pan-European University, Tematinska 10, 85105 Bratislava, Slovakia

integrative assessments. The aim of the economic parts of the integrative assessments was the selection of the most feasible investment strategy, whilst classical financial indicators were used to determine the economic feasibility (e.g., simple cost analyses, net present value analyses, payback period anticipation, capital expenditure analyses). However, none of them paid scientific attention to a rigorous analysis of the real market situation. Economic issues were considered only from the company's perspective, and the market perspective was neglected. In spite of this, we can find only a few contributions studying the market situation and solving different problems in a manner which is attempted in this paper. The paper 'Environmental innovation under Cournot competition' from June 2007 contains a unique market approach. According to the comments of the authors who presented this paper, contrary to the existing literature, a market approach was used in modeling the interaction between the output market, where companies play á la Cournot, and the emission permit market [2]. The authors make mention that besides their contribution, only Montero's article takes into account a strategic interaction in the output market [3]. Sanin and Zanaj correlate output market revenues and revenues or costs from trading in the permit market. This paper, in contrary, examines the total costs of the production and revenues from selling in a market evaluating the triple bottom line of sustainability. In this paper, a concrete case of the potential success/failure of a business strategy oriented on sustainable development is investigated under market conditions.

The role of a game theory in sustainability

Mathematical models can help preserve the Earth's resources. A game theory (i.e., an interactive decision theory) can play an important role in the preservation of energy resources - through its ability to identify the salable quantities and consequently minimize energy consumption. Today, fossil resources are not considered to be a perspective natural resource for a sustainable future. On the one hand, there are countries that are prominent users of oil, such as the USA, becoming almost net exporters of oil; but what is the price paid for the self-sufficiency in strategic commodities? Energy Return on Investment (EROI) would be a partial answer to this question. During the 1950s in the USA, every barrel of oil invested in exploration had led to the discovery of about 50 or more, and in 1990, a ratio of one to five [4]. In other words, different deposits are characterized by different cost levels of production. We are depleting the most accessible supplies first and then investing more and more energy into the extraction of less accessible resources. This paper offers a solution of how to circumvent the problem of the scarcity of resources by providing companies with guidance on how to better match the production setup to the customer's needs - this is precisely the role that the game theory can play in sustainable development. An application of the game theory (Cournot game) would lead to less energy resources being consumed since it would help a company to better prevent excessive (not needed) production. Our solution is economically driven - based on the presumption that companies will always want to lower production costs by consuming less raw material. 'One (opinion) group argues from an economic point of view with the goal of breaking the dependency on a feedstock that is becoming scarcer and more contentious'. This economic view is presented by Sandholzer and Narodoslawsky in the literature [5]. In general, it is also cost-effectiveness (not solely altruism) which drives companies to optimize their setup of production under the requirements of sustainable development. Environmental economists are now solving the question - how large, in its physical dimensions, should the economic system be in relation to the ecosystem that sustains it [4]? To answer this question, one has to link together economic and ecological indicators, and that is also the focus of this paper.

The role of an ecological footprint in the relationship between economics and ecology

In order to be able to create a link between economic activities and their environmental impact, we need an appropriate indicator. Our research outcomes were tested along with the management of an Italian small- and medium-sized enterprise (SME) PintInox, S. p. A. For assessment purposes, we were searching for an indicator which would have a user friendly format and which would enable us to implement a life cycle assessment. After consultation with the management of PintInox, S. p. A., we decided to use the ecological footprint. The latter is also available in the form of an easily accessible web-based tool, and an Austrian version of the ecological footprint was available through the university, which has made the tool reliable. The ecological footprint relates economic/ business activities to the ecosystem by offering the possibility to convert economic dimensions into environmental dimensions. The maximum capacity of our ecosystem is represented by the surface area of our planet Earth. The anthropogenic impact (represented by economic activities) on the environment which is measured and transferred into square meter of earth surface cannot exceed the area of planet Earth. Otherwise, the anthropogenic activities cannot be considered sustainable. Obviously, a single small- or medium-sized enterprise cannot track the ecological footprint of all companies in the world and, based on this information, derive its own allowable output in square meter. Today, there is no option to calculate something like that (see Equation 1).

$$S_1 + \sum_{2}^{n} S \leq S_{\text{planet}} \tag{1}$$

where S_1 is the earth surface occupied by activities of selected SME, $\sum_{2}^{n} S$ is the surface occupied by other entities in the economy, and S_{planet} is the total surface of planet Earth.

Consequently, the only strategic move a company can effectively make on its own initiative is to declare a real (measurable) commitment to the environment and society. One of the best solutions is to work out a policy or a strategy which will effectively reduce the negative environmental impact.

The uniqueness of the proposed method

The main goal of the presented initiative is to help SMEs to apply sustainability measurement metrics in a manner that will not jeopardize their ability to successfully compete in the market (it should even boost their ability to compete rather than jeopardize it). The paper presents a concept of sustainability implementation by respecting the business limitations (e.g., the economic effectiveness). Companies can accept a philosophy of sustainable development only to an extent which will not endanger their ability to compete in the marketplace. The principle is not to insist on compulsory levels of rescued square meter of earth surface but to take into account an environmental dimension in a way that managers would voluntarily lower the environmental impact in cases when it is economically feasible (i.e., the preference of a market-based approach to a command-and-control approach). In other words, this paper offers a methodology that companies can employ to boost their competitiveness by realizing sustainability-friendly decisions. There must be a number of market incentives which support the evolution of verifiable Corporate Social Responsibility as a functional concept. If we want to make our planet habitable in the future, we must find approaches to realize sustainable development. The aim is to implement new principles within the boundaries of those impacting the environment most - the SMEs (the SMEs contribute an amount of at least 70% of industry-related environmental damage due to the fact that the damage is caused by 99% of all companies in Europe [6]). However, this would not be possible without respecting their primary objective, which is in general to remain competitive. These seem to be contradictory goals - to help SMEs to implement climate-friendly processes whilst not jeopardizing their ability to remain competitive. The resolution of this situation lies in the successful implementation of sustainable development principles in the form of a well-targeted and well-performed business strategy. In other words, the implementation of an innovation lowering environmental impact will come at a cost, but if customers are willing to pay for that, then it would be a win-win situation for a company and society. The key to a successful implementation of verifiable and trustworthy sustainable measurement metrics is the consecutive successful commercialization of the results of this activity.

Markets placing value on sustainable development

One of the main characteristics of every business is the natural willingness of every company to remain competitive. Another important feature is that an innovation must be marketable, as seen from a business perspective. Otherwise, it would be easily forgotten. Taking these into consideration, this paper strongly concerns itself with the new thriving market which places value on sustainability principles - customers expect responsible behavior from companies and buyers, whereas the latter require responsible behavior from their suppliers and push in this way the responsible entities directly into the acceptance of the sustainable development philosophy. For example, a large company such as the ACCOR hotel chain could have decided to choose only the threshold suppliers. It would therefore not be possible for the company to exceed this threshold with any other products. The threshold could be measured using ecological footprints. Products exceeding the maximum footprint would not be procured by such a large company. This could be one of the ways in which barriers to entering a market principle, applying value to sustainable development for SMEs characterized by their low environmental performance, could be overcome. Other authors also mention the role of markets in the dissemination of sustainable development principles:

- Local markets for combustible waste in Europe are gradually opening up, and a liberalization of trade in products, such as paper and plastics, has created a thriving international market that valorizes secondary resources and improves the treatment efficiency [7];
- Life cycle assessment has become an important issue for products and services, since environmental concerns become increasingly decisive for the admission to markets as well as customer's decisions [5].

Implementation in SMEs

This paper was prepared in cooperation with PintInox, S. p. A., an Italian SME. Small- and medium-sized enterprises are the key element in decarbonizing the economy. Those SMEs employ 66.9% of the European labor force and form 58.4% of the total Gross Value Added produced by private businesses in the EU in 2010 [8]. The cooperation with small- and medium-sized enterprises from Lombardy dates back to 2011, when the initial parts of the assessments were carried out. The selected geographical field of research was Northern Italy. This paper focusses on the implementation phase of an innovative and sustainable product. Figure 1 depicts the life cycle of this product. In a time period from 2011 to 2012, we performed pioneering sustainability measurements of the internal processes in the company. We were then able to create a knowledge base used for the elaboration of a pioneering Corporate Social Responsibility concept based on the existing efforts leading towards sustainability. In our work, we discovered eco-innovations that had already been realized and that had not yet been labeled with an ecological label, because the company management did not have the relevant metrics which provided the evidence of sustainability. By the identification of already elaborated environmental innovation, we realized that one of the traditional managerial beliefs, stating that an orientation on sustainable development is undoubtedly a money-consuming activity requiring additional investments, did not hold true. Figure 2 illustrates an example of our assessment results. The company produced their model Sirio in 1987 with successive innovations in the design realized in 1991 and 2004. Those innovations were responsible for savings in the raw material used, and therefore, calculated by using the ecological footprint, 9 m^2 of earth surface was rescued per one item produced. This is a result that could be compared with the threshold set up by a large company (i.e., a potential buyer of an 'eco-innovated product'). Provided the savings in footprint were enough/satisfactory for a buyer, then the 'sustainability market entrance barrier' could be overcome.

The aim of this paper is not only to describe the pioneering sustainability assessment processes for a selected company but also to fill the gap in sustainability assessments in terms of integrative assessment processes. An integration of technical (e.g., footprint) and economic assessments (e.g., game theory) is needed but has been sporadic until now. In other words, companies need to think holistically in realizing eco-innovation. The holistic view has to be incorporated in the assessment of the ecological and social performance of a concrete sustainable idea (to prove and to verify that it is really a sustainable solution) along with the assessment of the economic feasibility of this sustainable idea under real market conditions. Sustainability initiatives strive for commercialization; otherwise, the philosophy of 'being sustainable' and 'behaving responsibly' can hardly be widespread. This study is devoted to an analysis of the market because selling to the end customer is the most critical phase of a product's life cycle, determining its success or failure. In Figure 1, the most critical point in a life cycle (the market penetration), determining the success or failure of a concept, is presented.

The aims of the study

The broader aim of this study was to describe the steps necessary to analyze whether the sustainable idea (e.g., environmental innovation) is environmentally compatible, socially acceptable, and economically viable. In simple words, we have searched for a viable business solution reducing the negative anthropogenic impact. In our opinion, the most important steps are the specification of the sustainability measurement metrics to be used drawing attention to the fact that the implementation of these metrics cannot endanger the competitiveness of companies applying them. The specific research aims of the study were as follows:

(1) Within the life cycle of a product, the specific aim in terms of the market penetration phase is to analyze the potential success of a product at markets assigning value to principles of sustainable development (markets based on sustainability);

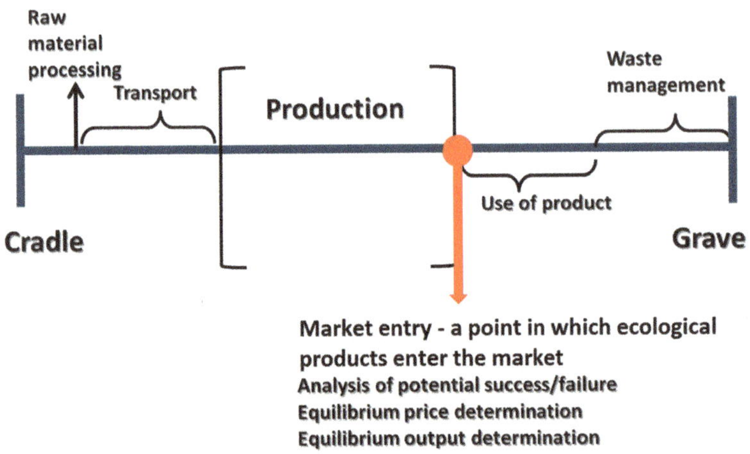

Figure 1 Critical point of life cycle - analysis of market penetration.

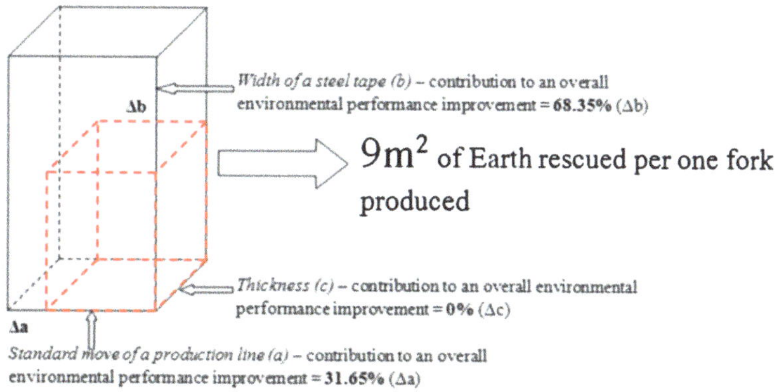

Figure 2 Environmental innovation discovered for a production plant.

(2) to provide an example to be followed, showing companies how they can commercialize their newly introduced concepts of Corporate Social Responsibility; and

(3) to present a methodology that other companies can use in order to predict their potential success using market principles based on sustainable development.

Methods

Game theory in general

The game theory (interactive decision theory) is a key tool used in this study as a part of the implementation of sustainability measures in business. It is a rational interaction of counterparts (labeled by the game as players). In business-to-business (B2B) relations, there are numerous interactive situations that can be analyzed by the game theory. In our case, the game theory is utilizable in the identification of the behavior of smaller business entities in relation to large companies (B2B customers) that already have a significant interest in sustainability. The interest of large companies in sustainability is derived from their understanding of this concept as a marketing advantage (when applied as a differentiation strategy) or, in some cases, as an incentive to preserve material and energy resources (economy and ecology are sometimes in symbiosis - they both want to 'consume less'). Large companies, after establishing their own internal environmental certification, 'prescribe' social responsibility to their suppliers (e.g., the obligation posed to supplier SME to deliver life cycle assessment information about supplied products, the obligation to prove a functional Corporate Social Responsibility policy using verifiable metrics). It is generally agreed that the first formal theorem in the theory of games was proven by E. Zermelo in an article on chess that appeared in German in 1913[a]. However, a game theory contains several of topics (monopolies, law of demand, competition) that existed independently of the game theory[a] itself. Today's Cournot game principles were brought to the forefront by Antoine Augustin Cournot in 1838 in the book 'Recherches sur les principles mathématiques de la théorie des richesses' (who studied the mathematical principles of the theory of wealth) but have not referred to the game theory rules since that time[b]. This means that the principles did not exist in the form that we know them today. To summarize briefly, the very first time when an interactive decision theory was professionally used and labeled as a game was in the relation to saloon games and is dated back to the year 1913. Today, the game theory is being used in economics, political science, military strategic and tactical decisions, computer science, evolutionary biology, and a more recent field of application - the sustainable development. The game theory analyzes behavior and interaction of economic agents. To date, 15 scientists have been awarded The Sveriges Riksbank Prize in Economic Sciences in Memory of Alfred Nobel for their work, which either directly or indirectly contained game theory principles (very recent laureate dealing with the game theory is Jean Tirole, for the year 2014). The 2012 laureates, Alvin E. Roth and Lloyd S. Shapley, have also studied how societies allocate resources[c].

Game theory application

Promising markets placing value on sustainability already exist. Such markets are not being created by governmental regulations but by large companies themselves. This situation is being partially triggered by ISO 14 040, which requires companies to proceed with life cycle assessment. Large companies are therefore requiring their suppliers to deliver information about the previous phases of the life cycle. This interaction between large companies and their suppliers is responsible for the creation of a new market applying value to sustainability. The following sections of the paper contain the interactions of competitors of these types of promising markets. The Cournot game is being used for the analysis of this interaction. The following sections also describe the entrance barriers that this new market type typically lays down for those not pursuing a Corporate Social Responsibility policy. The game theory in this paper analyzes the market situation being created by a supply chain successor - the Accor hotel chain (a traditional B2B customer of cutlery producers) and the competitive response of cutlery producers willing to penetrate the market created by the Accor hotel chain by meeting the sustainability criteria being set by the Accor hotel chain to be met by its suppliers. The game theory is here in a position to drive the SMEs towards sustainability.

Orientation on sustainability as a source of competitive advantage

Due to disparities between Italian labor standards and Asian labor standards, an Italian company will never be able to match Asian cost-effectiveness. In the opinion of Italian managers, the scarce concern for health and safety rules and precarious employment makes some Asian companies more cost-effective in comparison to Italian companies. Italian companies have optimized their production automation and employ the minimum workforce required. However, with other factors (e.g., material effectiveness, pace of production, high-innovation potential), the director of marketing considers their ability to realize a cost leadership strategy as very limited. After cooperation in a field of implementation of sustainable practices, he believes that a sustainable development philosophy is one of the very few possibilities to successfully remain competitive with cheaper Asian companies. The emergence of thriving markets placing value on triple bottom line principles means that there is a market space which cannot be entered by companies with low social and environmental performance. This paper introduces the Cournot game theory as a tool enabling companies to analyze their potential to succeed in these new markets. Large companies with their environmental obligations posed to their suppliers are direct creators of such markets. Those suppliers unable to meet the prescribed environmental standards are automatically deterred from this market assigning value to principles of sustainable development. In other words, environmental standards are considered as entrance barriers for competitors with low social and environmental performance. For example, a footprint threshold set up for supplier's products could be one of the measurable and clear entrance barriers.

Introduction to a Cournot game

In terms of comparability of environmental and social performance of entities (e.g., companies, municipalities, departments), standardized indicators are necessary. The negative side of standardization is the reductionism, the loss of certain dimension. In our opinion, there is a unique value mind-set behind all sustainability measurement metrics, and this mind-set is expressed in the willingness to prioritize things and to compare them. The game theory applied within the sustainable development field cannot have an ambition to measure environmental impact, but its role is to help identify the best economic decisions that will lead companies to the new type of market, giving them information about their potential success/failure in the market. Economic advice given by the game theory is as follows:

- How costly, in terms of economic effectiveness, should the pioneering sustainable steps be;
- how much should a company ideally invest into eco-innovations;
- how much the ideal technology, ensuring the presence in the new type of market, could cost; and
- what is the ideal bulk of production?

In this paper, we decided to use the game theory as one of the assessment tools that are usually categorized as sustainability measurement metrics. Using the Cournot game theory, we are capable of thinking in another dimension - the dimension of competitiveness. Using a system of equations (polynomial equations of the second degree), we can shift from the cost dimension to the competitiveness dimension. In the course of the identification of the competitiveness of a product, which went through an eco-innovation process, we firstly observe the total costs of a production (including labor costs and overheads). In general, we need to make this observation in every company which would like to enter a market value on principles of sustainable development. The other data needed represent the market space to be served. To derive a demand function, we need to know the required amount of products and prices a buyer will accept. Using linear regression splines, we derive at the cost function of every competitor willing to enter the market. Profit functions, derived from demand and cost functions, represent payoff functions. First-order conditions for the maximization of the profit of each company lead to a system of linear equations. Solving the system of equations will underscore companies with positive and negative outputs (produced amounts of eco-friendly products). An economic interpretation of a negative output is zero production - such a company will have failed to deliver its eco-products to the market. To correctly complete this mathematical analysis, additional verifications are inevitable - after a second-order condition test, we define intervals of feasible outputs because we exclude intervals of outputs for which cost functions do not have an economic interpretation.

Demand function derivation

The main purpose of the Cournot game was in our case to identify how successful a company would be under real market conditions. The market actors are as follows:

- PintInox, S. p. A.;
- Other SME competitors of PintInox, S. p. A., with an adequately elaborated Corporate Social Responsibility (CSR) policy; and
- Large companies in the role of buyers (B2B partners, market creators) from PintInox, S. p. A., and other competing SMEs.

This new type of market is only accessible for companies with verifiable economic, social, and ecological performance and with a reliable orientation towards sustainable development. As we have already mentioned, the decision as to which competitors are admitted to the market should be made by the demand creator - the large company in this case. In the game á la Cournot, the company's strategies are quantified. The companies simultaneously choose their respective output levels q_i from feasible sets $Q_i = [0,x)$ [9]. They sell their output at a market-clearing price $p(q)$, where $q = q_1 + q_2 + \ldots + qn$. The company's total profit is then generally expressed by an Equation 2 and Cournot reaction functions specifying each company's optimal output for each fixed output level of every opponent in the game ((3) and (4)).

$$u_i(q_1, q_2) = q_i p(q) - c_i(q_i) \tag{2}$$

$$r_i : Q_{-i} \rightarrow Q_i \tag{3}$$

$$r_j : Q_{-j} \rightarrow Q_j \tag{4}$$

where $c_i : [0, \infty) \rightarrow [0, \infty)$ is company i's cost function and Q_{-i} is the sum of the outputs of all the members of the oligopoly except competitor i. PintInox, S. p. A., is in our game with its competitors from Italy. Asian low-cost competitors were not able to enter this market due to their inability to come up with a functional and verifiable Corporate Social Responsibility policy - which enables companies in Italy to create a natural oligopoly composed of Italian companies. Large companies (e.g., the ACCOR hotel chain, COOP Italia) have set environmental rules for their supplier SMEs. Such rules are in the form of an internal environmental policy and set requirements for SMEs (SMEs are considered as potential business partners):

- The ACCOR hotel chain ('Planet 21 certification') - involves partners in the sustainable development strategy and integrates them into a responsibility process. Purchasing contracts have to be in compliance with 'Procurement charter 21';
- COOP Italia ('10 principles of COOP environmental policy') - requires a reduction in transportation distance, a better utilization of raw materials, and the tracking of the environmental impact of the production process happening on the suppliers' production site and involves the education of business partners.

In our investigated case, non-compliance with the sustainability requirements posed by large companies automatically repelled the SMEs (suppliers) from business deals with a large company. Decisions regarding who would be admitted and which cooperation would be terminated were based on the actual measured environmental and social performance. Based on the delivery of the required sustainability measurement metrics results, the first pre-selection of SMEs is carried out. A natural oligopoly of SMEs fulfilling the requirements in the application of sustainability measurement metrics is created during this step. Since the triple bottom line of sustainability is composed of three equal dimensions, in the second step of the market selection, the economic dimension plays its role. In this paper, the Cournot game forms the economic part of sustainability measurement metrics and can be considered as a complex economic indicator. Based on economic effectiveness, the Cournot game defines a market space every competitor in a formed natural oligopoly is able to fulfill. The Cournot game is in this paper presented as an internal part of sustainability measurement metrics. For the explanation of natural oligopoly, this paper uses a definition provided by M. Horniaček, as it allows companies in the industry to have different cost functions - which will likely be true for our case. The definition of a natural oligopoly in a single product industry is as follows ((5) and (6)) [10]:

Let m be the number of types of products considered in the model, $J^{(k)}$ the set of companies capable of producing type $k \in \{1, \ldots, m\}$ of product, $Y_j = [0, \chi_j]$ the set of feasible output levels of the company $j \in J^{(k)}$, and χ_j the maximal feasible output of the company j; we can define a natural oligopoly as follows:

Considering $k \in \{1, \ldots, m\}$ and output $Q_k \in (0, \Sigma_{j \in J(k)} X_j]$, a set of companies $J^{(k)*} \subseteq J^{(k)}$ with $\#(J^{(k)*}) \geq 2$ is a natural oligopoly in the industry producing type k of a product with an output level Q_k provided there exists an output vector $y^{(k)*} \in \Pi_{j \in J}(k) * Y_j$ such that

$$\sum\nolimits_{j \in J}(k) * Yj^{(k)*} = Q_k \tag{5}$$

and

$$\sum\nolimits_{j \in J}(k) * c_j\left(Y_j{}^{(k)*}\right) \leq \sum\nolimits_{j \in J}(k)_{1Yj} > 0 C_j(Y_j), \forall y \in \prod\nolimits_{j \in J}(k) Y_j \text{with} \sum\nolimits_{j \in J}(k) Y_j = Q_k \tag{6}$$

The market demand is being represented by the Accor hotel chain and is geographically limited to Italy. Hotel chains are typical B2B customer of PintInox, S. p. A., and its competitors. The demand is being represented by 65 hotels residing in Italy and belonging to the Accor chain (Mercure, Sofitel, Ibis, Novotel). We assume one typical product of PintInox, S. p. A., and its competitors (a fork in this case) per one hotel guest. We have presumed the full utilization of the hotel capacities, i.e., the maximum allowed amount of guests in a hotel room (standard or family rooms). The total demand for Italy is then 8,766 forks. We have considered only hotels with internal certification 'Planet 21' because this form of standardization is the key element in setting up sustainability boundaries. A hotel which possesses this certification (assigned by Accor headquarters) should track the origin of raw and auxiliary materials being used during the previous phases of a life cycle of products and being purchased by the hotel management. As we have already mentioned, hotel chains are one of the traditional B2B customers of cutlery producers, and in this case, the Accor is a direct creator of a market based on sustainability principles. This market is closed for competitors with weak environmental performance, as cost-effectiveness itself does not represent a competitive advantage here. After the selection of a price from the catalogue and the discussion with the management about the potential demand referring to the modeled price, we set up an inverse demand function (8) from the demand function (7). The coefficients of the function were obtained using regression analysis.

$$Q_d = 8,787–104P \quad (7)$$

$$P = 84.5–0.009Q = 84.5–0.009(q_1 + q_2 \cdots \quad + q_n) \quad (8)$$

Cost function derivation

We started our game with 11 eco-friendly and responsible players. We identified 10 direct competitors of PintInox, S. p. A. in Italy; however, we did not go into further specifications of each of them, since we did not have direct access to their facilities. We assumed their levels of production costs as similar to those of PintInox, S. p. A. In this industry, every company has detailed knowledge about technology variations typical for cutlery production and their related costs. In the sample of 10 companies, we compare various cost levels of the production of an identical product (a model Sirio produced by PintInox, S. p. A., and related substitutes made by different companies). Small differences in the technology used were expressed by the pace of production. The higher the time donation to produce one product (a fork), the less innovative technology was used. The cost functions we derived for every company in a game are based on 30 observations occurring during 1 week. In PintInox, S. p. A., we observed a production process of a model Sirio (fork) during 1 week. We took into consideration fixed and variable costs.

- Fixed costs: Total production costs of a shearing form used in the cutting process to cut the exact shape of a fork;
- Variable costs: Salaries and wages, material costs, and other types of costs influenced by a change in the amount of production.

There are many more obligations that PintInox, S. p. A., has to fulfill on a periodic basis in comparison with the cheap Asian producers (e.g., prescribed probation period, equal opportunities, compulsory content of labor agreement, classification of work types in the metal industry and related remuneration). The variability of costs in our game was influenced by the employee operating the machine during our observation. During 1 week, four employees belonging to a different category each (based on the character of duties, difficulty, and work experience) were rotated at the observed workplace. The monthly salary levels in effect from the 1st of January 2012 are depicted in Table 1. Using linear regression splines, we derived a cost function for PintInox, S. p. A., and for the 10 Italian competitors. In the dataset, there are three samples for every competitor - total costs (TC), amounts produced (q_n), and amounts produced squared (q_n^2). The size of each sample is represented by the number of observations - 30. We employed a technique known as 'curvilinear regression analysis'. The data was fitted to the model using polynomial regression, and a polynomial equation of the second degree was derived. In order to explain this situation, we have to delve deeper into this analysis and to firstly explain the use of degrees of freedom, as it is important for the t-statistic. The degrees of freedom are given by the number of observations minus the number of coefficients in the fitted regression model [11]. According to the lecture material of the University of Iowa, for a quadratic curve model, this gives n-3 since we are fitting two slopes and one intercept[d]. Having 27 degrees of freedom, the critical value for t_.025 one-tailed and t_0.05 two-tailed is 2.052. Using the critical value approach, we computed $t = -1.495$ [(coefficient − 1)/standard error]. So, for a

Table 1 Prescribed minimum monthly salary levels

Category	Monthly salary levels in effect from 1st of January 2012 (EUR)
1st	1,206.23
2nd	1,327.19
3rd	1,466.17
4th	1,528.32
5th	1,634.56
5th superior	1,748.28
6th	1,876.27
7th	2,038.21

linear coefficient, we do not reject the null hypothesis at a level of .05 since $t = |-1.495| < 2.052$. For a quadratic coefficient, we do not reject a null hypothesis at a level .05 since $t = |-0.99| < 2.052$. From the statistical point of view, we are not able to confirm causality between the level of output and total costs in PintInox, S. p. A. In the case of the competitor's data, there are two competitors of 10, where we are not able to reject the H_0 hypothesis. This does not mean, we have to accept H_0. The data from the regression analysis are used in the Cournot game, which represents the sustainability decision support tool. The data feed will be realized by companies, and the decision will be theirs as to what kind of 'typology' of competitors they will choose; they may create various scenarios based on different cost levels of production. The data in our case are from the actual production plant and were confirmed by the management of the cooperating company. The competitor's data were also confirmed as one possible scenario. Therefore, we consider them relevant and trustworthy. The computed cost functions based on regression analysis are as follows:

$$0.0055q_1^2 - 3.5306_{q1} + 5,562 \text{ PintInox} \quad (9)$$

$$-0.0408q_2^2 - 19.78_{q2} + 4,184 \text{ Competitor } x_2 \quad (10)$$

$$0.0007q_3^2 - 1.08q_3 + 5,573 \text{ Competitor } x_3 \quad (11)$$

$$-0.0017q_4^2 + 4.02q_4 + 5,096 \text{ Competitor } x_4 \quad (12)$$

$$-0.0130q_5^2 + 7.43q_5 + 4,567 \text{ Competitor } x_5 \quad (13)$$

$$-0.0056q_6^2 + 8.42q_6 + 4,402 \text{ Competitor } x_6 \quad (14)$$

$$-0.0075q_7^2 + 10.49q_7 + 4,540 \text{ Competitor } x_7 \quad (15)$$

$$-0.0241q_8^2 + 22.7q_8 + 3,951 \text{ Competitor } x_8 \quad (16)$$

$$-0.0031q_9^2 + 6.55q_9 + 4,818 \text{ Competitor } x_9 \quad (17)$$

$$0.0021q_{10}^2 + 0.61q_{10} + 5,681 \text{ Competitor } x_{10} \quad (18)$$

$$0.0088q_{11}^2 + 0.002q_{11} + 4,930 \text{ Competitor } x_{11} \quad (19)$$

Regression analysis based on the data from PintInox, S. p. A. is available in the Additional file 1; regression analysis based on the data from competitors is available in the Additional file 2. Initial demand function is available in the Additional file 3, and residual demand function is available in the Additional file 4. Additional file 5 contains geographical distribution of hotels in Italy that were considered for derivation of demand functions. The detailed calculation description is available in the Additional file 6.

Results and discussion

This paper has reconfirmed that for companies excluded from the Cournot game due to zero production levels or those which have not conformed to the second-order condition, it is not in their interest to reenter the Cournot game. The Cournot game's results have actually shown the optimal output for each fixed output level of every opponent in the game of companies.

In Figure 3, we can see the overall satisfaction of the market demand, represented by sold quantities of every competitor. For a better recognition, the sold quantities are available in Table 2. From 11 competitors that adhere to the principles of sustainable development, only six were able to enter the market based on the triple bottom line of sustainability. The reason for not entering was at this time the insufficient economic effectiveness (despite the fact that the environmental performance of these competitors was acceptable). The Cournot game helps companies to obtain a picture of their competitiveness and abilities to place a certain amount of environmentally friendly production within the market. The game should be in a reality based on the collections of data (production costs) mutually exchanged by economic departments of competing companies. The English word 'competition' is 'competere' in Latin and does not mean to act against one another but means to search for something together. Having competitors exchanging information about their production costs is surely a thought-provoking piece of information worth thinking about. The Cournot game is a tool with an ability to draft the potential situation of the market that is typically evaluated using the triple bottom approach - especially in that case where the sustainable development philosophy is one of the very few possibilities of how to contend with low cost competitors. This paper offers clear guidance on the determination of

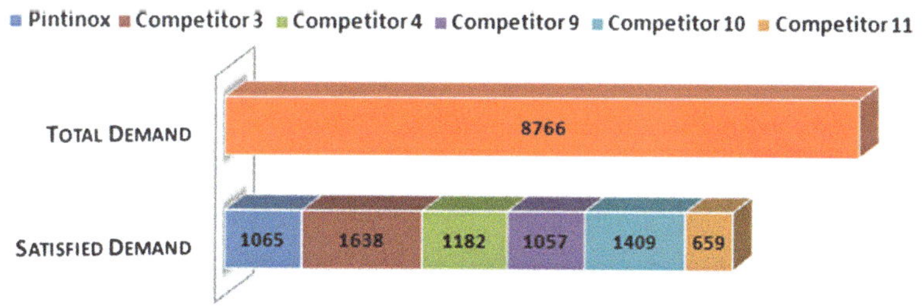

Figure 3 Total initial demand versus real satisfied demand.

Table 2 Final subgame perfect equilibrium

$q_1 = 1{,}065$	$q_3 = 1{,}638$	$q_4 = 1{,}182$
$q_9 = 1{,}057$	$q_{10} = 1{,}409$	$q_{11} = 659$

Italics refers to Pintlnox, S. p. A. data.

the potential success or failure of Corporate Social Responsibility policy concepts. The guidance given assists company managements in the determination of the salable quantities of their production. In other words, the Cournot game represents a method to preserve the natural resources (amounts that are not needed would not be produced). In Figure 4, the sold quantity is compared to a weekly production program. From this figure, it can be concluded that a company's management could reconsider their production program. During a 1-week observation of a production line, the company manufactured 11,840 pieces (forks) whilst the sold amount, according to the Cournot game results, totalled 1,638 pieces. Which means that 10,202 pieces would not be sold, i.e., considered for the market. Stainless steel used to produce this unsold quantity caused an environmental impact of 4,682,718 m^2 of consumed earth surface. If there are no additional markets where the excess amounts could be sold or if there is no potential to discover new markets for this surplus production, then (based on the Cournot game information about saleable quantities) management should reassess the volumes produced daily. Management should assume a smaller number of daily production quantities and to 'play' the Cournot game once again with lowered amounts of production. Using various scenarios, management would be able to prepare more alternatives (each with different input data) in order to determine a more sustainable production program. A simulation of the market situation by 'playing' the Cournot game would assist management in preparing a more sustainable production program. This means that a decrease of unsold quantities would be accompanied by a lower footprint. Excessive amounts of production are common phenomena, as witnessed during the post-Christmas period ('Day-After-Christmas-Sales') when huge amounts of unsold products are sold undervalue. The Cournot game represents the economic side of the triple bottom line (TBL) reporting. The Cournot game is therefore an integral part of sustainability measurement metrics.

Some large companies have already discovered situations in which economy and ecology strive to reach the same goal - to consume fewer resources. It is cheaper if it preserves natural resources (e.g., the hotels encourage customers to reuse their towels). There are more basic incentives that drive companies towards the application of sustainability measurement metrics as the main component of their marketing strategy:

- Differentiation strategy - a need to compete against low-cost competitors;
- The image - creating an image of a responsible company, especially when serving the high-end customer segment;
- A need to fulfill requirements valid at new emerging markets placing value on sustainability; and
- Cases when the economy and ecology have the same goal.

In summary, the bonus for companies that offer eco-innovations is a gained ability to utilize an opportunity to enter new prospective markets based on principles of sustainable development. This paper did not have an ambition to present a detailed way of identifying eco-innovations by using sustainability measurement metrics. We had concentrated on the consecutive steps in the life cycle assessment - the identification of success/failure of an eco-innovation by using the Cournot game. The Cournot

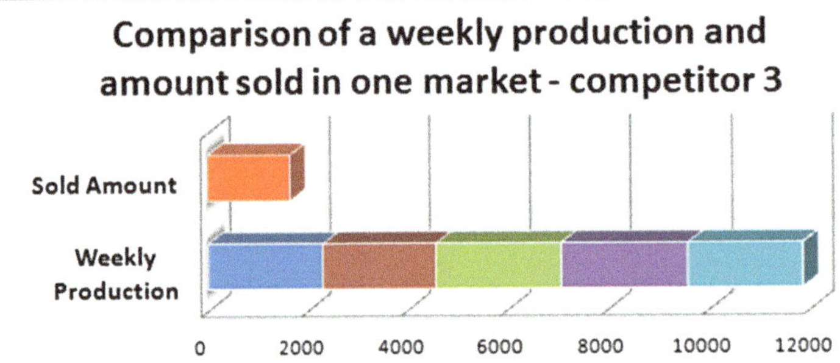

	Weekly Production	Sold Amount
Monday	2300	1638
Tuesday	2270	
Wednesday	2500	
Thursday	2510	
Friday	2260	

Figure 4 Production versus ability to sell the production.

game represents the economic part of the TBL reporting. The Cournot game measures the competitiveness of companies using their CSR policy concepts and the salable quantities of their eco-innovated products.

Conclusions

In this study, the application of the game theory is a reaction to the current needs of companies, which have already applied sustainability measurement metrics. The Cournot game has a strong economic dimension, as this type of economic analysis not only derives the quadratic cost functions of competitors (using regression analysis) but also calculates the market demand and the residual market demand. In addition, it mathematically verifies the allowable production outputs and tests to find out which of the competitors cannot enter or reenter the market. The study expects the existence of market principles applying value to sustainable development. In the presented example, the SMEs had to present their commitment to society and the environment in order to access the market created by the Accor hotel chain. After admission to this market (based upon the presentation of a verifiable sustainability measurement metrics using life cycle assessment), the admitted companies form a natural oligopoly with entrance barriers for low-cost competitors and in this case become competitive with one another. In this oligopoly, economic feasibility plays a decisive role as has been evidenced by the Cournot game example. This fact reconfirms the correctness of the triple bottom line concept - environmental safety, social responsibility, and economic effectiveness play equally important roles. We do not only draw the attention of managers to the fact that Corporate Social Responsibility does have a business potential but also introduce the very complex economic part of TBL (enabling companies to fully embrace the opportunities being offered by sustainable development).

The Cournot game enables the company to think in competitiveness dimensions instead of in simple cost dimensions and delivers results utilizable in, for example, break-even point analysis. The final outcome from using the Cournot game is the ability to identify a sustainable system - which is in this case an amount of production fitted to the market needs (i.e., the salable quantities). Company managements receive a tool which protects them from overproduction. The Cournot game helps to diminish the difference between both the produced/overproduced and sold amounts. Hence, this would be the best application of the Cournot game in order to protect natural resources. From an ecological point of view, the preservation of resources means less deviation from natural cycles and from an economic point of view; it means less variable costs, less storage costs and in some cases less transportation costs. As ecological economics is a cross-discipline, we have combined the economic tool (game theory) for the evaluation of economic issues with the ecological footprint used for the evaluation of environmental dimensions in this paper. This study was prepared as a reaction to a wide gap perceived in sustainable development research in a phase of commercialization. An integration of technical and economical assessment is needed but rare. From those few studies containing an integrative assessment, the economic dimension of integrative studies is generally aimed at feasibility studies, analysis of world prices, cost benefit analysis (CBA), payback period, capital and operating cost analysis, or capital expenditures (CAPEX) analysis. However, rigorous studies of the implementation phase accompanied by market situation analyses are critically missing. The implementation and commercialization is in fact a decisive step in the life cycle of a product. The sustainable development initiatives strive for commercialization; otherwise, the philosophy of 'being sustainable' and 'behaving responsibly' can hardly be widespread. The presented paper is devoted to an analysis of the market, as selling to the end customer is in our opinion the most critical phase of the life cycle of a product which determines its success or failure. The use of a game theory is greatly underestimated by the current sustainability science. In this study, the Cournot game has a role to preserve material and energy resources. Companies limiting overproduction will directly contribute to a better future for generations to come. In this paper, it has been stated that companies cooperate within a competitive oligopoly. As mentioned, to compete ('competere') in Latin means to work on something together. A real market interaction within an oligopoly is based on economic effectiveness which is represented by the production costs. More accurate information about shared costs can be more precisely predicted by regarding the reality (the market space acquired by every competitor) in their practice. In other words, cooperation may assist them to better predict the real market situation and to prepare themselves for this goal.

Endnotes

[a]www.math.harvard.edu/~elkies/FS23j.03/zermelo.pdf, accessed 12. Feb. 2015.

[b]http://cruel.org/econthought/profiles/cournot.html, accessed 12. Feb. 2015.

[c]www.nobelprize.org/nobel_prizes/economic-sciences/laureates/2012/popular-economicsciences2012.pdf, accessed 12. Feb. 2015.

[d]http://homepage.stat.uiowa.edu/~jcryer/22s008/Lecture%20Notes/chap8.PDF, accessed 12. Feb. 2015.

Additional files

Additional file 1: Regression analysis - Pintlnox, S. p. A. Results of regression analysis based on the data from the Pintlnox, S. p. A., production plant.

Additional file 2: Regression analysis - other competitors. The results of the regression analysis based on the developed scenarios.

Additional file 3: Initial demand function. The demand derived from the total capacity of the considered hotels in Italy - the Accor hotel chain.

Additional file 4: Residual demand function. The residual demand is calculated without the upper bounds of two competitors.

Additional file 5: Hotels - geography. Summary of all considered hotels in Italy forming a demand in our game.

Additional file 6: Calculation part of the Cournot game. Additional proofs required from the formal mathematical point of view.

Abbreviations

S. p. A.: La societá per azioni - joint stock company; B2B: Business to business; SMEs: Small- and medium-sized enterprises; CSR: Corporate Social Responsibility; TBL: Triple bottom line; CBA: Cost benefit analysis; CAPEX: Capital expenditures.

Competing interests

The author declares that he has no competing interests.

Authors' information

MD is currently employed as a research assistant at the Faculty of Economics and Business of the Pan-European University. Here, he also defended his PhD thesis in November 2013. The PhD research was being supervised by Assoc. Prof. Pavol Molnár from the Pan-European University (the first half of the PhD study was devoted to environmental innovations) and then supervised by Assoc. Prof Milan Horniaček from the Comenius University in Bratislava (the second half of the PhD study was devoted to the game theory). MD is a graduate of the University of Economics.

Acknowledgements

The author of the article, MD, started work in PintInox, S. p. A., as a master's student in 2009 doing his research fellowship funded by the Erasmus work placement program. MD has renewed his relationship with the management of PintInox, S. p. A., in June 2011 with the proposal to apply the research outcomes of this company. The approval was granted by the co-owner, Dr. Silvia Pinti. The data sources were provided by the director of the technical office, Ing. Roberto Pinti; the director of the financial office, Ing. Pinardi; and the director of the marketing department, Dr. Mauro Romani. These persons were also involved in discussions and a continued information exchange with MD.

References

1. 4th International Symposium on Energy from Biomass and Waste (2012) International Waste Working Group. Venice
2. Sanin ME, Zanaj S (2007) Environmental innovation under Cournot competition. CORE 2007/50 33:1–3
3. Montero JP (2002) Permits, standards, and technology innovation. J Environ Econom Manage 44:23–44
4. Daly H, Farly J (2004) Ecological economics: principles and applications. Island Press 449:33–75
5. Sandholzer D, Narodoslawsky M (2007) Sustainable? Processes from renewable resources, vol 15. Graz University of Technology, Institute for Resource Efficient and Sustainable Systems, Graz, Austria, pp 1–15
6. Constantinos C, Sørensen SY, Larsen BP, Alexopoulou S (2010) SMEs and the environment in the European Union. PLANET SA and Danish. European Commission, vol 238. DG Enterprise and Industry, p ii
7. Dubois M (2012) Towards a coherent European strategy for taxation of combustible waste. In: International Symposium on Energy from Biomass and Waste, vol 8. International Waste Working Group, Venice, pp 1–8
8. Wymenga P, Derbyshire J (2011) Are EU SMEs recovering? Annual Report on EU SMEs for European Commission, vol 58. ECORYS Nederland BV, Rotterdam, Cambridge, p 8
9. Fudenberg D, Tirole J (1991) Game theory, vol 579. The MIT Press, Boston, pp 80–100
10. Horniaček M (2011) Cooperation and efficiency in markets. Lecture notes in economics and mathematical systems, vol 649th. Springer, Heidelberg, doi:10.1007/978-3-642-19763-5
11. Schlotzhauer SD (2009) Elementary statistics using SAS. SaS Institute Inc, New York

Efficient growth of *Kluyveromyces marxianus* biomass used as a biocatalyst in the sustainable production of ethyl acetate

Christian Löser[1*], Thanet Urit[1,2], Erik Gruner[1] and Thomas Bley[1]

Abstract

Background: Whey is just turning from a waste of milk processing to a renewable raw material in biotechnology for producing single-cell protein, bio-ethanol, or ethyl acetate as an economic alternative. Conversion of whey-borne sugar into ethyl acetate requires yeast biomass as a biocatalyst. A high cell concentration results in a quick ester synthesis, but biomass growth means consumption of sugar at the expense of ester production. Efficient and cost-saving biomass production is thus a practical requirement. Whey is poor in nitrogen and has therefore to be supplemented with a bioavailable N source.

Methods: Several aerobic growth tests were performed with *Kluyveromyces marxianus* DSM 5422 as a potent producer of ethyl acetate in whey-borne media supplemented with various N sources. Preliminary tests were done in shake flasks while detailed studies were performed in a stirred bioreactor.

Results: Ammonium sulfate resulted in strong acidification due to remaining sulfate, but costly pH control increases the salt load, being inhibitory to yeasts and causing environmental impacts. Ammonium carbonate lessened acidification, but its supplement increased the initial pH to 7.5 and delayed growth. Urea as an alternative N source was easily assimilated by the studied yeast and avoided strong acidification (much less base was required for pH control). Urea was assimilated intracellularly rather than hydrolyzed extracellularly by urease. Conversion of urea to ammonium and usage of formed ammonium for biomass production occurred with a similar rate so that the amount of excreted ammonium was small. Ammonium hydroxide as another N source was successfully added by the pH controller during the growth of *K. marxianus* DSM 5422, but the medium had to be supplemented with some ammonium sulfate to avoid sulfur limitation and to initiate acidification. Non-limited growth resulted in 82 mg N per g of biomass, but N-limited growth diminished the N content.

Conclusions: *K. marxianus* could be efficiently produced by supplementing the whey with nitrogen. Urea and ammonia were the favored N sources due to the proton neutrality at assimilation which lessened the salt load and reduced the supply of alkali for pH control or made this even needless.

Keywords: *Kluyveromyces marxianus*; Whey; Growth; Nitrogen; Ammonium; Ammonia; Urea

* Correspondence: christian.loeser@tu-dresden.de
[1]Institute of Food Technology and Bioprocess Engineering, TU Dresden, 01062 Dresden, Germany
Full list of author information is available at the end of the article

Background

Sustainable and environmentally compatible development requires successive substitution of fossil resources by renewable raw materials. This applies to the energy sector but is also true for production of industrial bulk materials. Such a bulk chemical is ethyl acetate with an annual world production of 1.7 million tons [1]. Ethyl acetate is an organic solvent of moderate polarity with versatile industrial applications. Another prospective application is the biodiesel production from vegetable oil; here, triglycerides are transformed to fatty acid ethyl esters in a lipase-catalyzed transesterification reaction with ethyl acetate as an acyl acceptor instead of methanol [2-5]. Although being an irritant and intoxicant at higher concentrations, ethyl acetate is less toxic to humans compared to many other solvents. Ethyl acetate is an environmentally friendly compound since the ester is easily degraded by bacteria [6-8] and is regarded as a non-persistent pollutant of the atmosphere [9].

Synthesis of ethyl acetate currently proceeds by petrochemical processes, which are based on crude oil constituents or natural gas, run at elevated temperature and pressure and commonly require catalysts [9]. The reactions are often incomplete, and the recovery of ethyl acetate and residual precursors needs a high input of energy.

Microbial production of ethyl acetate from renewables could become an interesting alternative. Various yeast species can synthesize ethyl acetate (reviewed by Löser et al. [9]), but only *Pichia anomala*, *Candida utilis*, and *Kluyveromyces marxianus* produce this ester in larger amounts. *K. marxianus* is the most promising candidate for large-scale ester production since this dairy yeast with GRAS status grows quickly, converts sugar directly into ethyl acetate without ethanol as an essential intermediate, and produces the ester with a high rate and yield [9-17]. The ester synthesis in *K. marxianus* is easy to control by the level of iron [11,13,16-18]. *K. marxianus* exhibits a distinct thermal tolerance which allows cultivation at an elevated temperature [14,19,20] which in turn accelerates the ester stripping and advances its process-integrated recovery. The outstanding capability of *K. marxianus* for lactose utilization offers the chance for using whey as a resource of ester synthesis. *K. marxianus* converts whey-borne lactose with a high yield into ethyl acetate [9], but its ability for sucrose and glucose assimilation [19-21] enables production of ethyl acetate with this yeast from renewables like sugarcane, grain, and corn.

Ethanol is another product of microbial sugar conversion, but several factors favor microbial ester over ethanol production [22]: the higher market price of the ester, a reduced number of process stages, a faster process, and a cost-saving product recovery.

Conversion of sugar into ethyl acetate requires yeast biomass as a biocatalyst. A high biomass concentration results in a quick process, but the production of this biomass, on the other hand, is connected with sugar consumption which reduces the portion of sugar available for ester production (as demonstrated in pilot-scale experiments [13]). The right balance between yeast growth and ester synthesis or, in other words, a compromise between a quick process and a high ester yield is of practical importance. This balance can be controlled by the available iron [13,16,23].

An efficient production of the required yeast biomass is an important factor for the economy of the total process of ester and ethanol production. At whey-based bio-ethanol production, the biomass is often considered as a gift and sugar utilization for biomass production is ignored. Such an 'out-sourcing' of biomass processing only seemingly improves the economy.

Much research was done in the field of whey-based *K. marxianus* cultivation for single-cell protein production [24-28]. The results refer to some problems at cultivation of *K. marxianus* in whey-based media; whey is poor in bioavailable nitrogen so that nitrogen can limit yeast growth [26-29]. Nitrogen-limited growth can even deregulate yeast metabolism and provoke ethanol formation at aerobic conditions [30].

Whey was often supplemented with ammonium as a source of nitrogen to stimulate growth of *K. marxianus* [10-14,16,27,29,31-37]. Added ammonium increased the yield of biomass [27,32,35] or was without significant effect [29,31,36]. Ammonium is usually supplemented in the form of ammonium sulfate where ammonium is intensively consumed while most of the sulfate remains in the medium and causes an ionic imbalance and acidification [15,38-41]. Such acidification can be inhibitory to *K. marxianus* since its growth rate is distinctly reduced at pH ≤3.5 [22,28,42,43]. Supplementation of whey with $(NH_4)_2SO_4$ or NH_4Cl requires cost-intensive pH correction with NaOH or KOH which in turn increases the salt load and thus inhibits yeast growth [31,42] and creates waste-water problems.

Urea is an alternative N source since growth with urea exhibits proton neutrality without significant pH changes [38-40]. Assimilation of urea by yeasts is not an exception but the rule; 122 of 123 tested yeasts were able to metabolize urea [44]. There are several potential advantages of urea [25,28,39]: a high amount of nitrogen per unit weight, a low price, and a reduced or even omitted supply of pH correctives. *K. marxianus* definitely metabolizes urea, and whey was repeatedly supplemented with urea as an N source for this yeast [28,31,32,37,45-47]. However, the effect of added urea on growth was often not described [37,45,46], or the published results were contradictory. Yadav et al. [28] observed an increased biomass yield, Kar and Misra [32] found no positive effect, Mahmoud and Kosikowski [31] described slight inhibition by added urea, and Rech et al. [47] even detected strong inhibition of growth which had been attributed to alkalinization. These inconsistent results require clarification by more detailed studies on this subject.

The first objective of this work was testing the effect of $(NH_4)_2SO_4$, $(NH_4)_2CO_3$, or urea as sources of nitrogen at aerobic growth of *K. marxianus* DSM 5422 in whey-borne medium with special attention on yeast growth and acidification. When using $(NH_4)_2SO_4$, sulfate remains in the medium and causes proton imbalance and acidification but, when using $(NH_4)_2CO_3$, carbonate disappears in form of CO_2 during cultivation which possibly avoids acidification. The second objective was studying the growth of *K. marxianus* DSM 5422 in whey-borne medium with urea in detail at defined conditions to get a deeper insight into the urea metabolism. The third objective was testing the pH-controlled feed of ammonia during aerobic growth of *K. marxianus* DSM 5422 in whey-borne medium. Ammonia as the cheapest source of nitrogen is interesting for large-scale processes.

Methods

Microorganism

K. marxianus DSM 5422 from the Deutsche Sammlung von Mikroorganismen und Zellkulturen GmbH (Braunschweig, Germany) was maintained with the Cryoinstant preservation system (Lomb Scientific Pty Ltd, Vienna, Austria), cultivated on yeast-glucose-chloramphenicol agar (Roth GmbH, Karlsruhe, Germany) for 2 days at 32°C, and then used as an inoculum.

Non-supplemented DW medium

All media originate from concentrated and partially demineralized sweet whey. Sweet whey was ultrafiltrated, concentrated by reverse osmosis, and then demineralized by slight alkalinization and moderate heating to yield the used whey permeate (thus processed in the Sachsenmilch Leppersdorf GmbH, Leppersdorf, Germany). Non-supplemented DW medium was prepared in batches of 1 L by mixing 0.5 L whey permeate with the same volume water. This mixture was autoclaved in a sealed 1-L Schott bottle for 15 min at 121°C. Precipitated minerals were allowed to settle overnight before the upper phase was withdrawn for cultivation experiments. Non supplemented DW medium should not be confused with DW basic medium (as used in [11-14,16]) which contains 10 g/L $(NH_4)_2SO_4$.

Shake-flask cultivation

In a first series of shake-flask experiments, various media were prepared like the just-described non-supplemented DW medium with the modification that various sources of nitrogen (10 g/L $(NH_4)_2SO_4$, 10 g/L $(NH_4)_2CO_3$, or 5 g/L urea) were added before medium sterilization. The media handling (autoclaving, settlement, withdrawal) occurred as in the above-described manner. These media were supplemented with 2 mL/L autoclaved trace-element solution (preparation as described in [10]) and diluted with water to 1/20 for reducing the content of sugar to 3.9 g/L and, thus, to avoid oxygen-limited growth at a low oxygen-transfer rate in shake flasks. Several 500-mL conical flasks with cotton plugs were filled each with 50 mL medium, inoculated with an agar-plate culture as described in [14], and shaken with 250 rpm at 32°C. After 14 h of pre-cultivation, the process was followed by regular analysis of the optical density (OD) and the pH value over a period of at least 10 h. After a total of 40 h of cultivation, the OD and pH were measured and the residual sugar and formed biomass were analyzed.

A second series of shake-flask experiments was based on autoclaved non-supplemented DW medium. Portions of this medium were supplemented with 2 mL/L sterile trace-element solution and separately autoclaved aqueous stock solutions of $(NH_4)_2SO_4$, $(NH_4)_2CO_3$, or urea. The pH was then adjusted to 6.5 with 0.1 M HCl or KOH. These media were diluted with water to reduce the sugar content to 3.9 g/L like in the first series. Discrete sterilization of whey and N sources in this second series avoided unwanted interaction between whey constituents and N compounds such as the Maillard reaction. Inoculation, cultivation, and analyses occurred as in the first series.

Bioreactor cultivation

One reference experiment was done with DW basic medium (DW medium with 10 g/L $(NH_4)_2SO_4$) and described in detail in [11] while all other presented bioreactor processes were based on non-supplemented DW medium. The sterile 1-L stirred bioreactor with 0.6 mL Antifoam A (Fluka, Sigma Aldrich, St. Louis, USA) was filled with 0.6 L DW medium and 1.2 mL trace-element solution (the latter described in [10]). Further supplements such as $(NH_4)_2SO_4$, urea, and Na_2SO_4 were put into 100-mL Schott bottles, autoclaved for 15 min at 121°C, and then transferred to the bioreactor as follows: the bottle was connected with the bioreactor, some culture medium was pumped to the bottle, and, after complete dissolution of the substance, transported back to the reactor. Separate autoclaving avoided unwanted interaction such as the Maillard reaction and sulfate precipitation.

The cultivation occurred as usual [10-12,14]: the medium was inoculated with five loops of biomass, and the reactor was operated at 1,200 rpm and 32°C and gassed with 50 L/h dry and CO_2-free air (given for 0°C and 101,325 Pa). The pH was controlled to ≥5 with 2 M KOH or 2 M ammonia solution. All processes lasted at least 36 h. Sampling and sample preparation were performed as usual [10].

Test for urea assimilation

K. marxianus DSM 5422 was cultivated in diluted DW medium with urea (3.9 g/L sugars, 0.25 g/L urea) as described for the second series of shake-flask experiments. After depletion of the sugar, the cell suspension was separated into biomass and supernatant by centrifugation. The biomass was suspended in 50-mM phosphate buffer at pH 6 containing 0.25 g/L urea to yield an initial biomass concentration of 2 g/L, and the supernatant was supplemented with urea to give a concentration of 0.25 g/L. Both mixtures were then shaken with 250 rpm at 32°C. Sampling occurred at the beginning and after 24 h of incubation. The samples were analyzed regarding urea and ammonium.

Analyses

The optical density of cell suspensions was measured photometrically at 600 nm (after pre-dilution to OD <0.4 if required). The biomass dry weight was determined by separating the yeasts via centrifugation, washing the pellet twice, and drying at 103°C. Sugar was quantified by a modified 3,5-dinitrosalicylic-acid method [48] with lactose as a standard. Ammonium, nitrate, nitrite, sulfate, and phosphate were measured by the LCK303, LCK339, LCK342, LCK153, and LCK049 cuvette tests (Hach Lange GmbH, Düsseldorf, Germany). The total dissolved nitrogen was analyzed by the Kjeldahl method (German DIN 38409 H11). Ethanol was quantified by gas chromatography [10].

Urea was measured in the style of Rahmatullah and Boyde [49] with some modifications. This method had been approved for quantifying urea in wine [50] being a matrix similar to culture liquids. Two reagents were prepared: a mixture of 300 mL sulfuric acid (95% to 98%, ρ = 1,840 g/L), 100 mL phosphoric acid (85%, ρ = 1,670 g/L), and 100 mL water; and a solution of 500 mg butane-2,3-dione monoxime (diacetyl monoxime) and 10 mg thiosemicarbazide in 100 mL water. The analysis reagent was prepared from 40 mL acid mixture and 20 mL of the second solution and used immediately. A volume of 0.1 mL sample was mixed with 3 mL analysis reagent. In the case of colored samples, this mixture was measured photometrically at 525 nm as a blind. Then, the mixture was put in a 25-mL test tube with plastic cap and incubated for 20 min in boiling water. After cooling and short mixing, the absorbance was measured at 525 nm. The blind was subtracted, and the obtained ΔA_{525nm} value was interpreted as a urea concentration by a non-linear calibration curve prepared for several solutions of urea in water (from 0 to 500 mg/L): $C_{Urea,L} = a \cdot (\Delta A_{525nm})^b$ with a = 520 mg/L and b = 1.37.

Data analyses

The O_2 consumption, CO_2 formation, and the respiratory quotient (RQ) were calculated as usual [11]. The time-dependent specific growth rate was calculated from masses of formed CO_2 assuming a correlation between yeast growth and CO_2 formation which only applies to respiratory processes without significant maintenance: $\mu(t) \approx \Delta \ln(m_{CO2}(t))/\Delta t$. The overall biomass yield ($Y_{X/S}$) is the ratio between the mass of yeasts grown and the mass of sugar consumed; determination of these masses took losses by sampling and changes of the liquid volume into account. The pH-controlled feeding of 2 M KOH is given as a specific volume, related to the initial liquid volume of the culture. The mass of fed KOH was related to the formed biomass (given as g_{KOH}/g_X). The fed 2 M ammonia solution is also expressed as a specific volume and as the mass of supplied ammonia-N. The mass of bioavailable nitrogen was calculated as the sum of initial ammonium-N and urea-N plus the mass of ammonia-N supplied till a given time: $m_N(t) = V_L(t_0) \cdot (C_{NH_4\text{-}N}(t_0) + C_{Urea\text{-}N}(t_0)) + m_{NH_4OH\text{-}N}(t)$. The consumed nitrogen is this $m_N(t)$ value minus the nitrogen not yet used or lost by sampling: $\Delta m_N(t) = m_N(t) - V_L(t) \cdot (C_{NH_4\text{-}N}(t) + C_{Urea\text{-}N}(t)) - \Sigma m_N(\text{sampling})$. These $\Delta m_N(t)$ values were used for calculating biomass-specific consumption rates (r_N as $mg_N/g_X/h$). The overall biomass yield for nitrogen ($Y_{X/N}$) is the ratio between the mass of yeasts totally formed and the mass of N altogether consumed. The final N content of the biomass (x_N) is the reciprocal of this overall $Y_{X/N}$ value.

Results and discussion

Nitrogen in non-supplemented DW medium

The composition of whey depends on many factors such as origin of milk (cow, goat, or sheep), technology of curd production, and whey processing (e.g., [51,52]). Casein protein is coagulated by acidification (mineral or organic acids directly added or lactic acid produced in the processed milk by bacteria) and/or by using chymosin. Whey processing modifies the composition of whey as well: whey protein is separated by ultrafiltration, solutes are concentrated by reverse osmosis, and/or minerals are partially removed by alkalinization. Preparation of whey-borne culture media also changes the composition by dilution, adding supplements, and heat sterilization. This explains why published compositions of whey-borne media highly fluctuate.

Several batches of non-supplemented DW medium were analyzed regarding potential sources of nitrogen and some other parameters (Table 1). The medium is rich in \sugar and thus exhibits a high potential for biomass growth. An amount of 78 g/L sugar allows formation of 28 g/L *K. marxianus* biomass assuming $Y_{X/S}$ = 0.36 g/g (as observed at aerobic cultivation in DW basic medium with trace elements [11]). These 28 g/L biomass can only develop when yeast growth is not limited by

Table 1 Composition of non-supplemented DW medium

Parameter	Value
Sugars (88% lactose, 12% galactose)	78 g/L
Total N content (Kjeldahl analysis)	403 mg/L
Ammonium-N	106 mg/L
Urea-N	43 mg/L
Nitrate-N	4 mg/L
Nitrite-N	<1 mg/L
Organic nitrogen (except urea-N)	250 mg/L
Protein[a]	1,560 mg/L
Sulfate-S	78 mg/L
Phosphate-P	410 mg/L
Potassium	2,600 mg/L
pH value	5.7 to 6.0

Concentrated and partially demineralized sweet whey permeate was diluted with the same volume of water and autoclaved for 15 min at 121°C; precipitates were settled, and the upper clear phase was analyzed.
[a]According to the EC Nutrition Labelling Rules Directive 90/496/EEC, organic N multiplied by 6.25.

any other resource than sugar. *K. marxianus* DSM 5422 grown under such conditions exhibited the following content of minerals (in milligrams of the addressed element per gram dry biomass): $x_N = 78...79$ mg/g [10,11], $x_P = 10$ mg/g, $x_S = 4$ mg/g, and $x_K = 2$ mg/g (unpublished results). These data are similar to published compositions of *K. marxianus* [45,53-55] and *Saccharomyces cerevisiae* [40]. Multiplying these *x* values with the cell concentration of 28 g/L gives the required concentration of the respective element in the culture medium to allow non-limited growth: 2,200 mg/L N, 280 mg/L P, 112 mg/L S, and 56 mg/L K. Nitrogen and sulfur are lacking in non-supplemented DW medium while the phosphorous and potassium content covers the demand (Table 1).

DW medium owns 403 mg/L Kjeldahl-N (Table 1). Detailed analysis of five batches of non-supplemented DW medium gave 89 to 130 mg/L ammonium-N (106 mg/L on an average, $\sigma_{n-1} = 15$ mg/L), 38.4 to 45.9 mg/L urea-N (43 mg/L on an average, $\sigma_{n-1} = 2.6$ mg/L), 4 mg/L nitrate-N, and <1 mg/L nitrite-N. The Kjeldahl-N minus the urea-N and inorganic N gives a proteinogenic N of 250 mg/L which corresponds to ca. 1.5 g/L proteins (Table 1).

Utilization of whey proteins by *K. marxianus* is discussed controversially. Raw cheese whey owns 7 g/L protein comprising 50% β-lactoglobulin, 20% α-lactalbumin, 15% glycomacropeptide, and 15% minor protein/peptide components [51]. Their microbial hydrolysis requires excretion of proteases. Decomposition of whey proteins by *K. marxianus* has been repeatedly studied; the results varied from absent hydrolysis [26], over 20% to 33% [28,29,56], up to 80% hydrolysis [33]. Recent studies confirmed an extracellular serine protease for *K. marxianus* [57], and Yadav et al. [28] proved modification of whey proteins by *K. marxianus* via electrophoresis. Indigenous proteases in milk [58] and the proteolytic activity of lactobacilli [56] could also contribute some to protein modification during milk processing. Pre-treatment of whey protein with added proteases resulted in peptides <1 kDa which were efficiently assimilated by *K. marxianus* [59].

Some yeasts and fungi assimilate nitrate by intracellular reduction to ammonium [60]. *K. marxianus* seems to be unable for nitrate assimilation; at least several tested strains were negative [61]. This explains why nitrate added to whey did not improve growth of *K. marxianus* [32]. Some growth of *K. marxianus* in $NaNO_3$-supplemented medium was possibly caused by the added yeast extract [62]. *K. marxianus* DSM 5422 proved to be unable to assimilate nitrate.

Due to the uncertainty of whey-protein assimilation by *K. marxianus*, it is assumed that whey proteins do not contribute to assimilable nitrogen. The utilizable nitrogen in non-supplemented DW medium of ca. 150 mg/L (ammonium-N plus urea-N) is thus much smaller than the calculated demand. Supplementation of DW medium with nitrogen is essentially required.

K. marxianus DSM 5422 grows well with ammonium [10,11], but its growth with urea has not yet been tested. Yeasts generally assimilate urea [44] which should also apply to the studied strain.

A low amount of iron, zinc, and copper in DW medium limits growth of *K. marxianus* DSM 5422 [11]. Here, the medium was ever supplemented with trace-element solution to avoid such limitation.

Test of several sources of nitrogen without previous pH adjustment

This preliminary test for assimilation of diverse sources of nitrogen by *K. marxianus* DSM 5422 was performed in shake flasks. Non-supplemented DW medium was spiked with several N sources (Table 2): no supplement as a reference, $(NH_4)_2SO_4$ as the usual N supplement, $(NH_4)_2CO_3$ as an alternative ammonium resource, and urea as another N compound. The DW medium contributed ca. 7 mg/L N (5 mg/L NH_4-N and 2 mg/L urea-N) to the assimilable N in each culture. The initial pH value was influenced by the added N source: $(NH_4)_2SO_4$ did not change the pH while urea and $(NH_4)_2CO_3$ alkalinized the medium (Table 2). All shake flasks were inoculated and pre-cultivated for 14 h before the process was followed by repeated OD and pH measurements (Figure 1).

The process without a supplement was at first similar to the process with $(NH_4)_2SO_4$ (Figure 1), but later, nitrogen became a limiting factor so that growth and medium acidification slowed down and nearly stopped after 24 h. The N-limited growth resulted in some residual sugar (Table 2).

Table 2 Parameters at aerobic growth of *K. marxianus* DSM 5422 in media with various sources of nitrogen without previous pH adjustment

Parameter		Added source of nitrogen			
		No supplement	$(NH_4)_2SO_4$	$(NH_4)_2CO_3$	Urea
Supplied N	[mg/L]	0	106	146	117
NH_4-N + urea-N	[mg/L]	7	113	153	124
Initial pH value	[–]	5.96	5.94	7.50	7.18
μ at t = 14 h	[h^{-1}]	0.51	0.57	0.29	0.57
Final sugar	[g/L]	0.92	0.00	0.15	0.00
Final biomass	[g/L]	0.47	1.23	1.06	1.29
Final pH value	[–]	3.99	2.47	3.45	7.10

The whey and the sources of nitrogen were autoclaved together; the data belong to the experiment shown in Figure 1; the final parameters were measured after a total of 40 h of cultivation.

With a supplement of $(NH_4)_2SO_4$, *K. marxianus* DSM 5422 grew at first with $\mu = 0.57\ h^{-1}$ as usual in diluted DW medium [12,14], but then, the growth slowed down due to medium acidification (Figure 1). A pH <3.5 impairs growth of *K. marxianus* [28,42,43]. This acidification was caused by consumption of ammonium without an equivalent uptake of sulfate. The use of ammonium sulfate as an N source calls for buffered medium (in shake-flask experiments [12,14]) or for pH control (in bioreactor experiments [10,11,13,16,22]).

The supplement of $(NH_4)_2CO_3$ at first alkalinized the medium. The initial pH of 7.5 impacted growth of *K. marxianus* DSM 5422 and caused a highly retarded process (Figure 1). Vivier et al. [42] and Antoce et al. [43] found that a pH >7 is adverse for growth of *K. marxianus*. Continued growth gradually reduced the pH and accelerated yeast growth so that all sugar had been consumed after 40 h. Carbonate as an exchangeable anion can leave the medium in the form of carbon dioxide which counteracted acidification (medium with $(NH_4)_2CO_3$) while sulfate as a permanent ion remains and resulted in unfavorable acidification (medium with $(NH_4)_2SO_4$).

The supplement of urea also alkalinized the medium to some degree (initial pH = 7.18) which is, at the first view, surprising since urea reacts neutral in aqueous solution. Autoclaving the medium together with urea maybe caused some hydrolysis of urea to form alkaline ammonium carbonate. This slight alkalinization slowed down yeast growth and retarded the process a little at the beginning, but the yeast metabolism lowered the pH so that the growth rate approached a normal value during pre-cultivation (Table 2). After 21 h, the medium acidification stopped and then the pH increased to a final pH of 7.1 (i.e., only temporary decrease in pH). Absence of enduring pH changes confirms the earlier postulated proton neutrality at growth with urea [38-40].

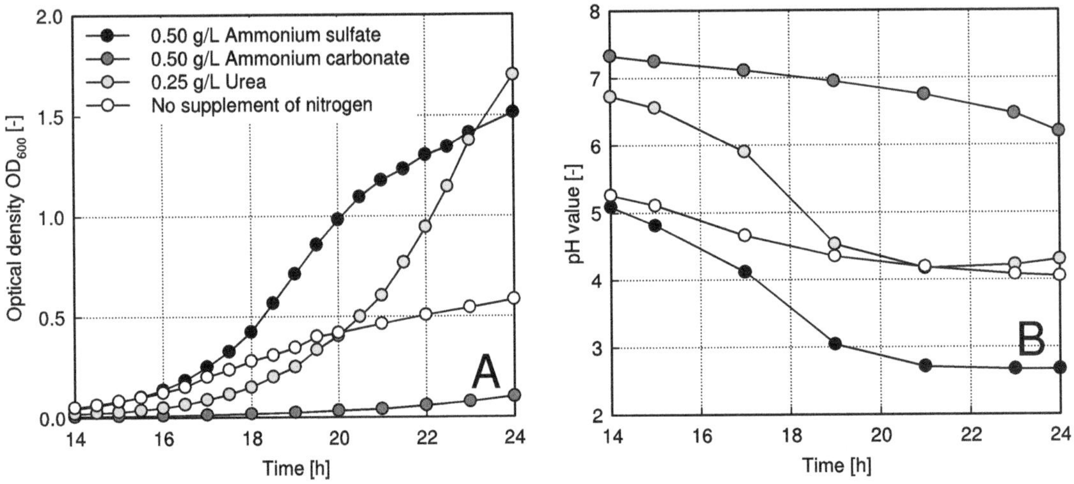

Figure 1 Aerobic batch cultivation of *K. marxianus* DSM 5422 without pH adjustment. Aerobic batch cultivation of *K. marxianus* DSM 5422 in media with various sources of nitrogen without previous pH adjustment. Concentrated and partially demineralized sweet whey was diluted with the same volume of water, supplemented with various sources of nitrogen, and autoclaved; the clear upper phase was supplemented with trace elements and diluted again with water resulting in media with 3.9 g/L sugar; the media were inoculated and cultivated in conical flasks at 250 rpm and 32°C.

Test of several sources of nitrogen with previous pH adjustment

The just-described shake-flask experiments were repeated with two modifications: all sources of nitrogen were autoclaved separately to eliminate unwanted interaction between the N sources and medium constituents, and the initial pH value was adjusted to 6.5 for avoiding alkaline conditions.

The convenient initial pH of 6.5 produced a uniformly high specific growth rate of $\mu = 0.59\ h^{-1}$ (Table 3) resulting in very similar cell densities at the beginning of the observation period (Figure 2A). This μ value was identical with the growth rate of *K. marxianus* DSM 5422 in highly diluted and phosphate-buffered DW basic medium [12,14].

The culture without a supplement of nitrogen exhibited N-limited growth (Figure 2A), only moderate acidification owing to restricted growth (Figure 2B), a low biomass formation, and some residual sugar after 40 h of cultivation (Table 3). Supplementation with $(NH_4)_2SO_4$ or $(NH_4)_2CO_3$ resulted in a nearly identical growth behavior and similar final pH values and biomass yields (Table 2); the pH was suboptimal in both cases which caused impaired growth compared to the process in phosphate-buffered medium [12,14]. A reduced biomass yield for *K. marxianus* in whey at a low pH has also been described by Yadav et al. [28] and interpreted as diversion of lactose from anabolism (growth) toward catabolism (maintenance). The use of $(NH_4)_2CO_3$ instead of $(NH_4)_2SO_4$ was without advantage since the initial pH adjustment of $(NH_4)_2CO_3$-supplemented medium with hydrochloric acid caused substitution of exchangeable by permanent ions (carbonate replaced by chloride). $(NH_4)_2CO_3$ is thus not a useful alternative since a high initial pH inhibits yeast growth while preceding pH adjustment with acids eliminates the buffering effect of carbonate.

Separate autoclaving of the N source and initial pH adjustment caused good growth in the urea-supplemented medium from the beginning; then, the growth became somewhat retarded (at 18 to 22 h; Figure 2A), but afterward, the growth accelerated again (at t >22 h) and even exceeded growth with $(NH_4)_2SO_4$ (Table 3). Quick growth in the first period was possibly based on whey-borne ammonium while the temporal slowdown of growth in the second period perhaps came about through the adaptation of yeast metabolism to urea assimilation after NH_4-N depletion. The quite high biomass yield with urea is certainly attributed to an appropriate pH value over the whole growth period (compare [39,62]). Urea is thus a promising source of nitrogen for *K. marxianus.* The acidification with urea was only of temporal nature; the pH increased again to give a final value being nearly identical with the initial pH (due to proton neutrality at assimilation of urea [38-40]).

Assimilation of urea by *K. marxianus* DSM 5422

Ammonium is utilized by all common yeasts directly while urea is either hydrolyzed by urease to form ammonium or it is assimilated via the urea amydolyase pathway [25,38]. Urease acts extracellularly while the amydolyase pathway works intracellularly. *K. marxianus* is regarded as a urease-negative yeast [40] and should metabolize urea only in the latter way being a two-step process [44,60,63]; urea reacts with hydrocarbonate in an energy-consuming process to form allophanate which, in turn, is hydrolyzed to release ammonium:

$$NH_2\text{-}CO\text{-}NH_2 + HCO_3^- + ATP \rightarrow NH_2\text{-}CO\text{-}NH\text{-}COO^- + H_2O + ADP + P_i \quad (1)$$

$$NH_2\text{-}CO\text{-}NH\text{-}COO^- + 3\,H_2O + H^+ \rightarrow 2\,NH_4^+ + 2\,HCO_3^- \quad (2)$$

These two reactions are catalyzed by urea carboxylase and allophanate hydrolase [44,60,63]. The produced ammonium is then metabolized in the same manner as ammonium taken up directly.

Table 3 Parameters at aerobic growth of *K. marxianus* DSM 5422 in media with various sources of nitrogen with initial pH adjustment

Parameter		Added source of nitrogen			
		No supplement	$(NH_4)_2SO_4$	$(NH_4)_2CO_3$	Urea
Supplied N	[mg/L]	0	106	146	117
NH_4-N + urea-N	[mg/L]	7	113	153	124
Initial pH value	[–]	6.50	6.50	6.50	6.50
μ at t = 14 h	[h^{-1}]	0.59	0.59	0.59	0.59
Final sugar	[g/L]	1.03	0.00	0.00	0.00
Final biomass	[g/L]	0.47	1.29	1.25	1.44
Final pH value	[–]	4.29	2.43	2.54	6.45

The whey and the sources of nitrogen were autoclaved separately; the data belong to the experiment shown in Figure 2; the final parameters were measured after a total of 40 h of cultivation.

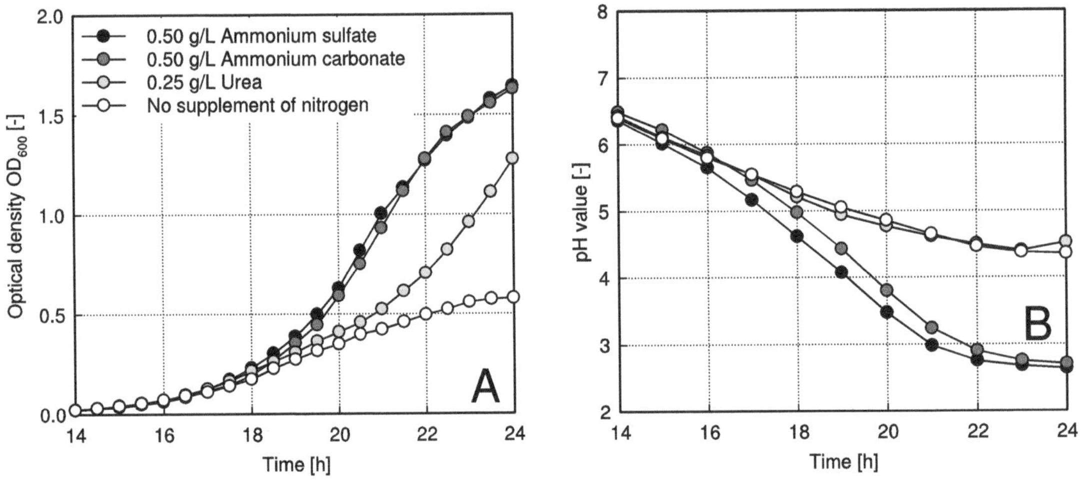

Figure 2 Aerobic batch cultivation of *K. marxianus* DSM 5422 with initial pH adjustment. Aerobic batch cultivation of *K. marxianus* DSM 5422 in media with various sources of nitrogen with initial pH adjustment. Non-supplemented DW medium was supplemented with various sources of nitrogen and trace elements, the pH was adjusted to 6.5 by HCl or KOH, and these media were diluted with water resulting in 3.9 g/L sugar; the media were inoculated and cultivated in conical flasks at 250 rpm and 32°C.

An experiment was done for clarifying the way of urea assimilation in *K. marxianus* DSM 5422. The yeast was cultivated in diluted DW medium with urea, and the obtained culture was then used in an above-described test. *K. marxianus* DSM 5422 did not excrete urease into the medium since urea added to the cell-free aqueous fraction of this culture was not hydrolyzed at all. This is in accordance with Nahvi and Moeini [61] who found all tested *K. marxianus* and *K. lactis* strains being urease negative. These findings argue for assimilation of urea via the amydolyase pathway. The grown yeast biomass was incubated with urea in a phosphate buffer but only a bit urea was reacted to ammonium (average reaction rate 0.25 mg urea/g_X/h). Absent sugar obviously suppressed transfer of urea into ammonium which refers to an amydolyase pathway being under transcriptive control.

Bioreactor cultivation with the addition of urea

The shake-flask experiments clearly demonstrated the capability of *K. marxianus* DSM 5422 for urea assimilation. Urea is a promising source of nitrogen due to the proton neutrality at its consumption during yeast growth [38-40], avoiding strong acidification as happening at growth with $(NH_4)_2SO_4$. Urea has been repeatedly used at growth of *K. marxianus*, but the obtained results were inconsistent [28,31,41,47,62]. Hensing et al. [39] referred to the potential risk of an imbalance between release and assimilation of ammonium; the medium could alkalinize when ammonium is quicker released from urea than is incorporated into biomass. Such an alkalinization was observed during cultivation of *K. marxianus* in urea-supplemented whey [47]. In the above-described experiments, the pH temporally decreased rather than increased. This phenomenon has not yet been understood and requires clarification in a bioreactor experiment.

K. marxianus DSM 5422 was cultivated in a stirred reactor at well-defined conditions (32°C, pO_2 ≥30% air saturation, pH ≥5) in DW medium which was supplemented with urea, Na_2SO_4, and trace-element solution to avoid limitation of growth by nitrogen, sulfur or microelements. Another bioreactor experiment performed with DW medium containing 10 g/L $(NH_4)_2SO_4$ and trace elements was taken from [11] and used here as a reference. These two processes are depicted in Figure 3, and characteristic parameters are summarized in Table 4.

The courses of yeast growth, sugar consumption, and marginal ethanol formation were very similar for $(NH_4)_2SO_4$ or urea as the added N sources (Figure 3A). The only marked difference was the amount of formed biomass which was apparently higher with urea (Table 4). This observation is in accordance with Hensing et al. [39] and Rajoka et al. [62]. From the energetic point of view, growth with urea should be less effective compared to growth with ammonium since assimilation of urea via the amydolyase pathway requires ATP [44,60,63]. Cultivation with urea should therefore result in a lower rather than a higher biomass yield. In case of Hensing et al. [39] and Rajoka et al. [62], the observed low growth with ammonium was possibly caused by inhibitory acidification due to absent pH control at shake-flask cultivation. Such an inhibitory acidification was prevented by controlling the pH during bioreactor cultivation (Figure 3B); here, the diverging biomass yields were presumably caused by slightly different amounts of bioavailable nitrogen (1.49 g N in medium with 5 g/L urea, and 1.30 g N in medium with 10 g/L $(NH_4)_2SO_4$). Nearly all bioavailable nitrogen was assimilated in both processes (only 20 or 30 mg/L residual NH_4-N in the culture broth; Table 4) which refers to a

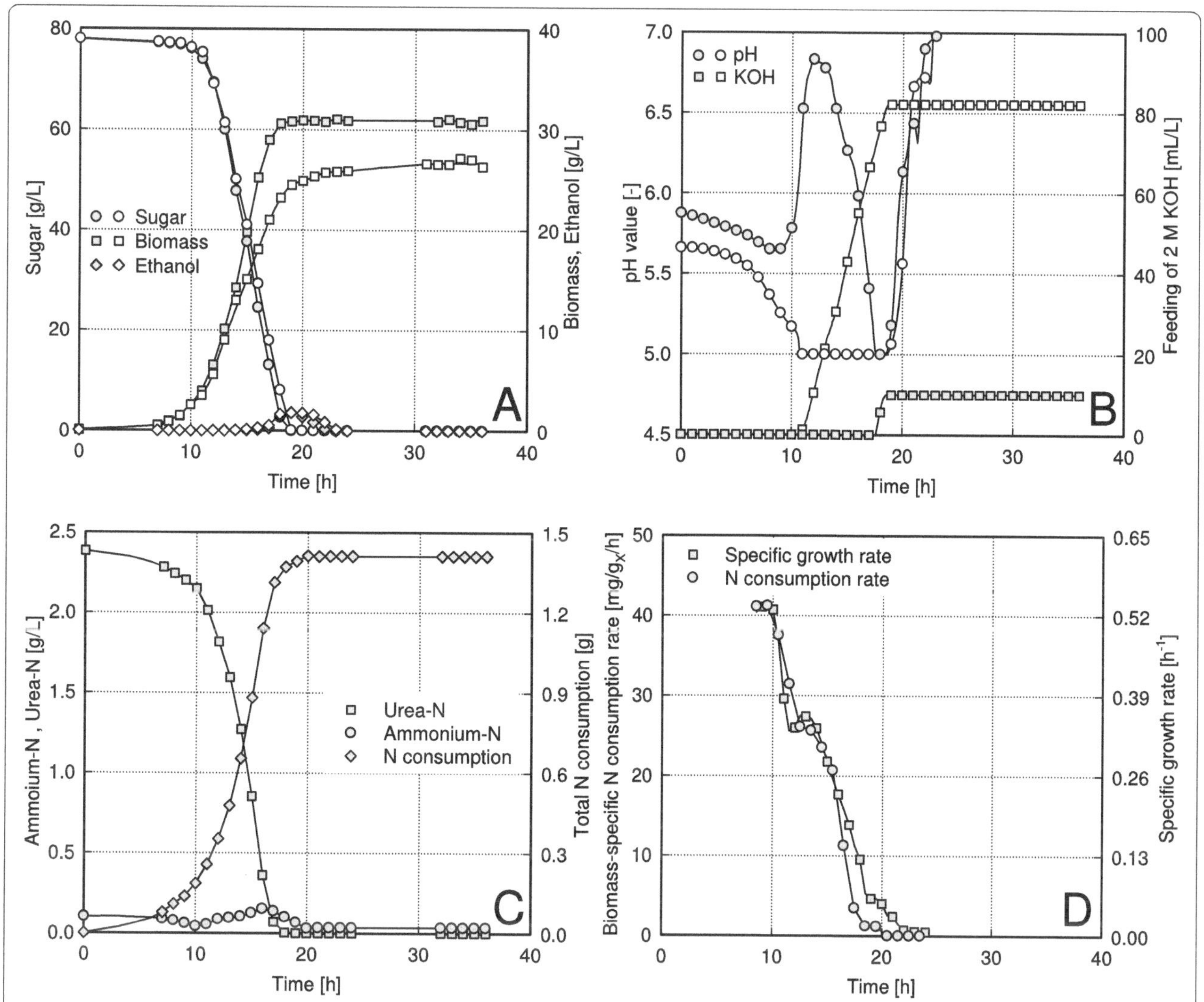

Figure 3 Aerobic batch cultivation of *K. marxianus* DSM 5422 in stirred bioreactors using ammonium or urea. Aerobic batch cultivation of *K. marxianus* DSM 5422 in a stirred bioreactor using ammonium (white symbols) or urea (grey symbols) as a source of nitrogen. DW medium with 2 mL/L trace-element solution was supplemented with $(NH_4)_2SO_4$ or urea plus Na_2SO_4 (white symbols = 10 g/L $(NH_4)_2SO_4$; grey symbols = 5 g/L urea and 0.4 g/L Na_2SO_4); the cultivation occurred in an 1-L stirred reactor at 1,200 rpm, 32°C, and aeration with 50 L/h; the pH was controlled to ≥5 with 2 M KOH; the given growth rate was derived from measured CO_2 data.

slight deficit of nitrogen, and this deficit was more striking with $(NH_4)_2SO_4$. This argumentation is supported by a higher N content in the biomass grown with urea (Table 4).

In urea-supplemented medium, *K. marxianus* DSM 5422 grew at first quickly; but later, the growth slowed down more and more (Figure 3D), although sufficient O, N, P, S, K, and trace elements should allow non-limited growth over an extended period. This behavior could be explained with a lack of vitamins in whey [27,29] or with inhibition by whey-borne minerals [31]. Supplementing whey with yeast extract [27,29] or vitamins [29,32,47] stimulated the growth of *K. marxianus*.

The source of nitrogen distinctly influenced the pH(*t*) course (Figure 3B). With $(NH_4)_2SO_4$, the pH decreased due to ammonium consumption until the pH was controlled to pH 5; 82 mL/L 2 M KOH were supplied which corresponds to a specific dosage of 0.34 g KOH per g of produced biomass. The strong acidification with $(NH_4)_2SO_4$ is explained by a proton imbalance [38-40]: ammonium was consumed while most of the sulfate remained in the medium. With urea, the pH at first a little decreased, then sharply rose to pH 6.8 and reduced again to pH 5 where the pH controller avoided further acidification; the KOH dosage was in fact much smaller (only 10 mL/L 2 M KOH or 0.04 g KOH per g of produced biomass; Table 4). Hensing et al. [39] cultivated *K. lactis* in galactose medium with $(NH_4)_2SO_4$ or urea; the acidification was strong and permanent with $(NH_4)_2SO_4$, while the acidification was moderate and only temporary with urea.

During cultivation of *K. marxianus* DSM 5422 in urea-supplemented DW medium, the dissolved ammonium-N

Table 4 Parameters at aerobic growth of *K. marxianus* DSM 5422 in media with various sources of nitrogen in a stirred bioreactor

Parameter		Added sources of nitrogen…				
		$(NH_4)_2SO_4$ [a]	Urea	NH_4OH	$(NH_4)_2SO_4 + NH_4OH$	$(NH_4)_2SO_4 + NH_4OH$
Added $(NH_4)_2SO_4$	[g/L]	10	0	0	1	2
Added urea	[g/L]	0	5	0	0	0
Added Na_2SO_4	[g/L]	0	0.4	0.4	0	0
Initial NH_4-N content	[g/L]	2.12	0.10	0.13	0.30	0.53
Initial urea-N content	[g/L]	0.04	2.38	0.04	0.04	0.04
Maximum growth rate	[h^{-1}]	0.56	0.54	0.54	0.54	0.56
Consumed sugar	[g]	45.2	44.7	44.3	44.8	44.9
Final cell concentration	[g/L]	26.7	30.8	7.49	22.1	27.0
Formed biomass	[g]	16.3	17.3	4.2	13.0	16.2
Overall $Y_{X/S}$	[g/g]	0.362	0.388	0.095	0.289	0.361
Final NH_4-N content	[g/L]	0.02	0.03	0.00	0.04	0.17
Final urea-N content	[g/L]	0.00	0.00	0.00	0.00	0.00
Supplied 2 M NH_4OH	[mL]	0.0	0.0	0.0	24.8	39.8
Supplied ammonia-N	[g]	0.000	0.000	0.000	0.694	1.114
Consumed NH_4-N	[g]	1.212	0.039	0.077	0.849	1.317
Consumed urea-N	[g]	0.024	1.371	0.025	0.026	0.023
Consumed N	[g]	1.236	1.410	0.102	0.875	1.340
Overall $Y_{X/N}$	[g/g]	13.2	12.3	41.2	14.9	12.1
Final N content, x_N	[mg/g]	76	81	24	67	83
Supplied 2 M KOH	[mL/L]	82	10	0	0	0
Supplied KOH	[g_{KOH}/g_X]	0.34	0.04	0.00	0.00	0.00
Consumed oxygen	[g]	25.3	26.4	28.6	30.3	27.9
Formed CO_2	[g]	35.1	37.2	41.2	42.5	39.0
Average RQ	[mol/mol]	1.01	1.02	1.05	1.02	1.02
Maximum $C_{EtOH,L}$	[g/L]	1.8	1.8	<0.1	0.2	<0.1

Final parameters are valid for 36 h of cultivation with exception of the process with NH_4OH as the only added N source where the final parameters were determined for 40 h of cultivation; cultivation at conditions as stated in Figures 3 and 4.
[a]Data in this column was in part taken from Urit et al. [10,11,14]; small deviations from the formerly published data are due to re-evaluation of basic data taking liquid sampling into account.

and urea-N were repeatedly measured and used for calculating the N consumption (Figure 3C). Intracellular conversion of urea to ammonium and usage of this ammonium for biomass growth occurred with nearly the same rate since ammonium excretion was only marginal (some NH_4-N originated from the used whey). The N consumption correlated well with the yeast growth (compare Figure 3A and C) and, thus, the courses of the N consumption rate and the specific growth rate were similar (Figure 3D). The quotient of these rates represents a momentary $Y_{X/N}$ value.

The small transient NH_4-N accumulation (Figure 3C) partially correlated with the temporary increase in pH. Hensing et al. [39] already referred to the danger of alkalinization when ammonium release exceeds ammonium assimilation. Such an alkalization to pH 8.5 was observed by Rech et al. [47] at cultivation of *K. marxianus* in urea-supplemented whey causing severe growth inhibition. Here, such an inhibition did not occur (only moderate rise of pH to 6.8).

Bioreactor cultivation at a pH-controlled feed of ammonia

Feeding the required nitrogen in form of ammonia could be a cost-saving alternative. Ammonia was repeatedly used as an N source at cultivation of *K. marxianus* in whey or other media [34,64-67] but dissolved ammonium or N consumption has not been paid much attention, with exception of Hack and Marchant [65] who depicted the time-dependent supply of ammonia.

In another series of bioreactor experiments, *K. marxianus* DSM 5422 was cultivated in DW medium as before but 2 M NH_4OH was used as the predominating N source which was supplied by the pH controller at pH <5. Three experiments were performed with a varied mass of

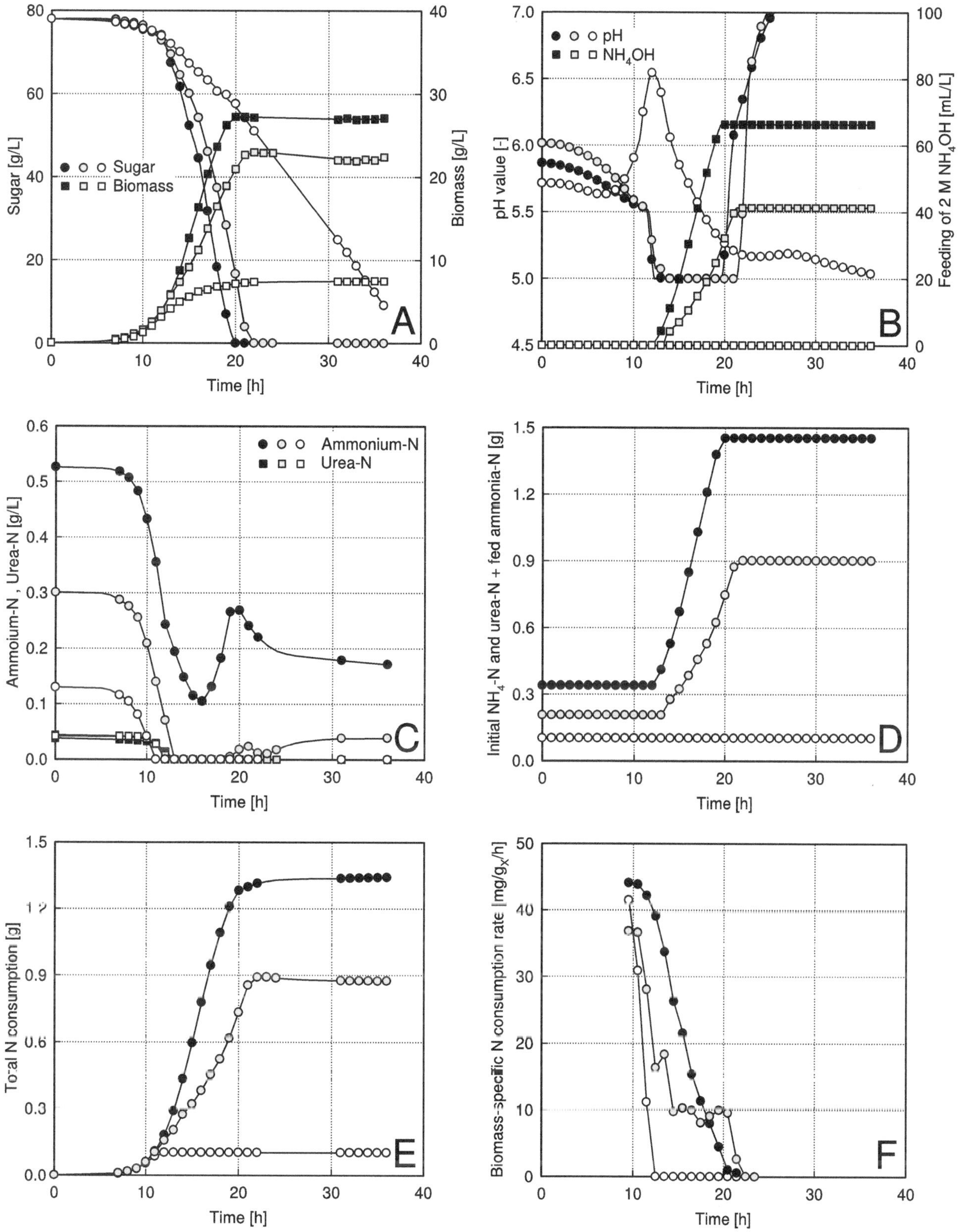

Figure 4 (See legend on next page.)

(See figure on previous page.)
Figure 4 Aerobic batch cultivation of *K. marxianus* DSM 5422 in stirred bioreactors using $(NH_4)_2SO_4$ and NH_4OH. Aerobic batch cultivation of *K. marxianus* DSM 5422 in a stirred bioreactor using various amounts of $(NH_4)_2SO_4$ and NH_4OH as sources of nitrogen. DW medium with 2 mL/L trace-element solution was supplemented with $(NH_4)_2SO_4$ or Na_2SO_4 (white symbols = 0.4 g/L Na_2SO_4; grey symbols = 1 g/L $(NH_4)_2SO_4$; black symbols = 2 g/L $(NH_4)_2SO_4$); the cultivation occurred in an 1-L stirred reactor at 1,200 rpm, 32°C, and aeration with 50 L/h; the pH was controlled to ≥5 with 2 M NH_4OH; the given growth rates were derived from measured CO_2 data.

$(NH_4)_2SO_4$ which was added as a pure substance to the autoclaved DW medium (0, 0.6, or 1.2 g). These processes were limited neither by oxygen (pO_2 always >10%) nor by sulfur (proven by residual sulfate).

In DW medium without an $(NH_4)_2SO_4$ supplement (Figure 4; white symbols), *K. marxianus* DSM 5422 grew on whey-borne nitrogen (0.13 g/L NH_4-N and 0.04 g/L urea-N), but this nitrogen was quickly depleted (Figure 4C) and the growth became N limited (Figure 4A). The pH temporally rose (Figure 4B) which seemingly correlated with urea consumption (Figure 4C). After depletion of all bioavailable N, the pH decreased only slowly due to a low metabolic activity (look at the sugar concentration in Figure 4A). Later, the pH stagnated above pH 5 and ammonia was thus not dosed (Figure 4B). The low availability of nitrogen (Figure 4D) caused restricted yeast growth (Figure 4A). The total N consumption and the N consumption rate were accordingly low (Figure 4E,F).

A supplement of 1 g/L $(NH_4)_2SO_4$ in the second experiment (Figure 4; grey symbols) increased the bioavailable N (Figure 4D) and allowed better yeast growth due to the higher initial NH_4-N which let the pH quickly decrease (Figure 4B). After depletion of this nitrogen, the pH stagnated at 5.05 for a while and then dropped below 5 where the pH controller started dosage of ammonia (Figure 4B). The added ammonia was assimilated immediately, and no ammonium accumulated in the medium (Figure 4C). That is, the yeast growth continued but at N-limited conditions as becoming visible from the low rates of ammonia dosage and N consumption (Figure 4D,F). This deficit of nitrogen slowed down growth and diminished the formed biomass (Figure 4A, Table 4). In the stationary period, some ammonium was released into the medium (Figure 4C) which was also observed by Ghaly and Kamal [26] at the cultivation of *K. marxianus* in whey and interpreted as decomposition of yeast biomass with release of NH_4-N into the medium.

A supplement of 2 g/L $(NH_4)_2SO_4$ in the third experiment (Figure 4; black symbols) increased the bioavailable N most (Figure 4D) and resulted in fast yeast growth and quick acidification. The feed of ammonia at pH <5 started before the initially added NH_4-N had been depleted (Figure 4C). The early start of ammonia dosage prevented limitation of yeast growth by nitrogen (NH_4-N always >100 mg/L), and the supply and uptake of nitrogen were well balanced (Figure 4C). The high rate of ammonia dosage corresponded with an accordingly fast growth and intensive N consumption (Figure 4D,E). The rate of N consumption became gradually smaller which is explained by the gently declining growth rate (possible reasons for this fading growth were discussed above). Urea was assimilated co-metabolically with the ammonium (Figure 4C).

The process with a supplement of 2 g/L $(NH_4)_2SO_4$ and ammonia dosage ran very similar to the process with 10 g/L $(NH_4)_2SO_4$; the final cell concentrations, formed biomasses, and the overall $Y_{X/S}$ values were nearly identical in both processes (Table 4) which demonstrates effective cultivation of *K. marxianus* with a pH-controlled feed of ammonia. Supplementing the medium with some $(NH_4)_2SO_4$ was however required for a quick acidification and for initiation of ammonia dosage. The amount of added $(NH_4)_2SO_4$ could be reduced by changing the setpoint of the pH controller (e.g., to pH 5.5) so that dosage of ammonia starts earlier and avoids N-limited conditions even at a reduced $(NH_4)_2SO_4$ supplement, but some $(NH_4)_2SO_4$ is needed to cover the requirement for sulfur.

Nitrogen in biomass

The content of nitrogen in biomass grown at cultivation in the stirred bioreactor (Figures 3 and 4) is the inverse of the overall biomass yield for nitrogen: $x_N = 1/Y_{X/N}$. The overall $Y_{X/N}$ values were calculated from the produced biomass and the consumed nitrogen, assuming that only ammonium and urea were assimilated (Table 4). This calculation ignores that *K. marxianus* possibly hydrolyzes some whey-borne proteins and assimilates thus-formed peptides and amino acids. The N content of biomass depended on the extent of N limitation (Table 4): cultivation with enough nitrogen (process with 5 g/L urea and process with 2 g/L $(NH_4)_2SO_4$ plus dosed ammonia) gave x_N values of 81 and 83 mg/g, a slight deficit of nitrogen during the late growth stage (process with 10 g/L $(NH_4)_2SO_4$) resulted in x_N = 76 mg/g, distinct N limitation (process with 1 g/L $(NH_4)_2SO_4$ plus ammonia) yielded x_N = 67 mg/g, while severe N limitation (process without any N supplement) produced an x_N value of only 24 mg/g. A diminished N content of *K. marxianus* was also observed at limitation of growth by trace elements [10,11]. A decreased N content can be explained by a lowered portion of active biomass owing to intracellular storage of polysaccharides (details in [10,11]).

The N content can also be derived from the elemental composition of biomass. Several authors measured the cell composition for *K. marxianus* by elemental analyzers and transformed these data into biomass formulae: $CH_{1.78}O_{0.75}N_{0.16}$ [45], $CH_{1.776}O_{0.575}N_{0.159}$ [53], $CH_{1.63}O_{0.54}N_{0.16}$ [54], $CH_{1.94}O_{0.76}N_{0.17}$ [55]. These formulae represent an N content of 88, 80, 91, or 83 mg/g. The fluctuations originate from measuring errors and from a variable cell composition depending on growth conditions [68].

Stoichiometry of yeast growth

Stoichiometric equations for describing the growth of *K. marxianus* has been derived here by using the method of Hensing et al. [39] and Mazutti et al. [69]. Such balancing requires a sum formula for biomass. The above-given formulae for *K. marxianus* biomass are restricted to C, H, O, and N as the predominating elements (derived from elemental analyses [45,53-55]). Here, the elements P, S, and K are included for more precision. The C, H, and O content was taken from the above-given biomass formulae (as averages), the N content of 82 mg/g was taken from own measurements at non-limited yeast growth, and the P, S, and K content was assumed with 10, 4, and 2 mg/g (own assimilation measurements). Combination of these data gives $CH_{1.78}O_{0.66}N_{0.158}P_{0.009}S_{0.0035}K_{0.0015}$ (yielding a molar mass of 27.027 g/mol).

Individual stoichiometric balance equations were derived for ammonium, urea, or ammonia as an N source, assuming respiratory growth (without formation of ethanol or ethyl acetate) of *K. marxianus* DSM 5422 with lactose as a substrate. The included stoichiometry coefficients were determined by balancing each element: seven balance equations were obtained containing nine unknown stoichiometric coefficients. This uncertain algebraic system was dissolved following Hensing et al. [39] by adding a proton balance and introducing the yield coefficient ($Y_{X/S}$ informs about the relation of assimilatory to dissimilatory substrate utilization and allows to establish the mass ratio between formed biomass and consumed lactose). $Y_{X/S}$ = 0.36 g/g was used here uniformly for all balances as found at non-limited growth with ammonium or ammonia (Table 4). Phosphate and sulfate were consumed in form of HPO_4^{2-} and SO_4^{2-} at the prevailing pH. Three equations were obtained for ammonium, ammonium hydroxide, or urea as an N source:

Protons are only formed during growth with ammonium which explains the observed substantial consumption of KOH by the pH controller with ammonium sulfate; the proton release equates to 0.28 g consumed KOH per g grown biomass and hence somewhat deviates from the measured KOH consumption (0.34 g_{KOH}/g_X). With NH_4OH or urea (Equations 4 and 5), the balances predict a slight alkalinization since OH^- ions are formed. The consumption of some KOH with urea as an N source is contradictory to this finding, but it should be kept in mind that the final pH was higher than the initial pH (Figure 3B); synthesis of organic acids (acetate, pyruvate, 2-oxoglutarate, and succinate were by-products of aerobic sugar metabolism of *K. marxianus* [16,68,70,71]) presumably caused KOH consumption, and consumption of these acidic metabolites after depletion of sugar alkalinized the medium. But such temporary metabolite accumulation was not considered at balancing. Another interfering effect originates from whey-borne lactate (ca. 4 g/L in DW medium [10]) whose microbial utilization also causes some alkalinization.

The balance equations allow to compare calculated with measured masses of consumed oxygen and formed carbon dioxide. The expected masses were calculated from the masses of utilized sugar. The measured masses (Table 4) were 1% to 20% smaller than predicted for unknown reason, but the ratio between formed CO_2 and consumed oxygen (the average RQ values) was ca. 1.02 mol/mol (Table 4) and agreed well with the predicted values (1.01 to 1.03 mol/mol).

Conclusions

Whey is poor in nitrogen and requires supplementation with an N source for effective production of yeast biomass. Ammonium sulfate, as usually applied for this reason, causes medium acidification by residual sulfate which requires pH control by alkaline substances to avoid growth inhibition. Application of ammonium carbonate instead of ammonium sulfate is not helpful since added $(NH_4)_2CO_3$ elevates the pH to inhibitory levels. *K. marxianus* DSM 5422 assimilates urea as an

$$C_{12}H_{22}O_{11} + 0.0410\ HPO_4^{2-} + 0.0160\ SO_4^{2-} + 0.0068K^+ + 7.3790\ O_2 + 0.7205\ NH_4^+ \rightarrow 4.5600\ CH_{1.78}O_{0.66}N_{0.158}P_{0.009}S_{0.0035}K_{0.0015} + 7.4400\ CO_2 + 8.0963\ H_2O + 0.6133H^+ \quad (3)$$

$$C_{12}H_{22}O_{11} + 0.0410\ HPO_4^{2-} + 0.0160\ SO_4^{2-} + 0.0068K^+ + 7.3790\ O_2 + 0.7205\ NH_4OH \rightarrow 4.5600\ CH_{1.78}O_{0.66}N_{0.158}P_{0.009}S_{0.0035}K_{0.0015} + 7.4400\ CO_2 + 8.7095\ H_2O + 0.1072\ OH^- \quad (4)$$

$$C_{12}H_{22}O_{11} + 0.0410\ HPO_4^{2-} + 0.0160\ SO_4^{2-} + 0.0068K^+ + 7.9189\ O_2 + 0.7205\ NH_2CONH_2 \rightarrow 4.5600\ CH_{1.78}O_{0.66}N_{0.158}P_{0.009}S_{0.0035}K_{0.0015} + 8.1600\ CO_2 + 8.3495\ H_2O + 0.1072\ OH^- \quad (5)$$

alternative N source. Consumption of urea means proton neutrality, medium acidification is minor, and only a little pH corrective is required. Moreover, the use of urea reduces the salt load (less inhibition, diminished environmental impact). Dosage of ammonia by the pH controller is a cost-saving alternative, but a suitable supplement of $(NH_4)_2SO_4$ is needed as a source of sulfur and for initiating dosage of ammonia.

Competing interests
The authors declare that they have no competing interests.

Authors' contributions
CL and TU conceived of the study. EG, CL, and TU explored relevant literature. CL and TU designed the experiments. EG and TU conducted the experiments. CL performed data analysis. CL and TB drafted the manuscript. All authors read and approved the final manuscript.

Acknowledgements
Thanet Urit would like to express his thanks to the Nakhon Sawan Rajabhat University (Muang Nakhon Sawan, Thailand) for financial support. We are grateful to Mrs. E. Kneschke for technical assistance, to M. Heller from the Sachsenmilch Leppersdorf GmbH (Germany) for providing whey permeate, and to A. Stukert for performing two bioreactor experiments.

Dedication
This publication is dedicated to Prof. Dr. Andreas Zehnsdorf on the occasion of his 50th birthday.

Author details
[1]Institute of Food Technology and Bioprocess Engineering, TU Dresden, 01062 Dresden, Germany. [2]Department of Biology and Biotechnology, Faculty of Science and Technology, Nakhon Sawan Rajabhat University, 60000 Nakhon Sawan, Thailand.

References
1. Posada JA, Patel AD, Roes A, Blok K, Faaij APC, Patel MK (2013) Potential of bioethanol as a chemical building block for biorefineries: preliminary sustainability assessment of 12 bioethanol-based products. Bioresour Technol 135:490–499
2. Kim S-J, Jung S-M, Park Y-C, Park K (2007) Lipase catalyzed transesterification of soybean oil using ethyl acetate, an alternative acyl acceptor. Biotechnol Bioprocess Eng 12:441–445
3. Modi MK, Reddy JRC, Rao BVSK, Prasad RBN (2007) Lipase-mediated conversion of vegetable oils into biodiesel using ethyl acetate as acyl acceptor. Bioresour Technol 98:1260–1264
4. Uthoff S, Bröker D, Steinbüchel A (2009) Current state and perspectives of producing biodiesel-like compounds by biotechnology. Microb Biotechnol 2:551–565
5. Röttig A, Wenning L, Bröker D, Steinbüchel A (2010) Fatty acid alkyl esters: perspectives for production of alternative biofuels. Appl Microbiol Biotechnol 85:1713–1733
6. Hwang S-CJ, Lee C-M, Lee H-C, Pua HF (2003) Biofiltration of waste gases containing both ethyl acetate and toluene using different combinations of bacterial cultures. J Biotechnol 105:83–94
7. Kam S-K, Kang K-H, Lee M-G (2005) Removal characteristics of ethyl acetate and 2-butanol by a biofilter packed with jeju scoria. J Microbiol Biotechnol 15:977–983
8. Chan W-C, Su M-Q (2008) Biofiltration of ethyl acetate and amyl acetate using a composite bead biofilter. Bioresour Technol 99:8016–8021
9. Löser C, Urit T, Bley T (2014) Perspectives for the biotechnological production of ethyl acetate by yeasts. Appl Microbiol Biotechnol 98:5397–5415
10. Urit T, Löser C, Wunderlich M, Bley T (2011) Formation of ethyl acetate by *Kluyveromyces marxianus* on whey: studies of the ester stripping. Bioprocess Biosyst Eng 34:547–559
11. Urit T, Löser C, Stukert A, Bley T (2012) Formation of ethyl acetate by *Kluyveromyces marxianus* on whey during aerobic batch cultivation at specific trace-element limitation. Appl Microbiol Biotechnol 96:1313–1323
12. Urit T, Manthey R, Bley T, Löser C (2013) Formation of ethyl acetate by *Kluyveromyces marxianus* on whey: influence of aeration and inhibition of yeast growth by ethyl acetate. Eng Life Sci 13:247–260
13. Löser C, Urit T, Stukert A, Bley T (2013) Formation of ethyl acetate from whey by *Kluyveromyces marxianus* on a pilot scale. J Biotechnol 163:17–23
14. Urit T, Li M, Bley T, Löser C (2013) Growth of *Kluyveromyces marxianus* and formation of ethyl acetate depending on temperature. Appl Microbiol Biotechnol 97:10359–10371
15. Löser C, Urit T, Nehl F, Bley T (2011) Screening of *Kluyveromyces* strains for the production of ethyl acetate: design and evaluation of a cultivation system. Eng Life Sci 11:369–381
16. Löser C, Urit T, Förster S, Stukert A, Bley T (2012) Formation of ethyl acetate by *Kluyveromyces marxianus* on whey during aerobic batch and chemostat cultivation at iron limitation. Appl Microbiol Biotechnol 96:685–696
17. Kallel-Mhiri H, Engasser J-M, Miclo A (1993) Continuous ethyl acetate production by *Kluyveromyces fragilis* on whey permeate. Appl Microbiol Biotechnol 40:201–205
18. Willetts A (1989) Ester formation from ethanol by *Candida pseudotropicalis*. Antonie Van Leeuwenhoek 56:175–180
19. Aziz S, Memon HUR, Shah FA, Rajoka MI, Soomro SA (2009) Production of ethanol by indigenous wild and mutant strain of thermotolerant *Kluyveromyces marxianus* under optimized fermentation conditions. Pak J Anal Environ Chem 10(1+2):25–33
20. Lertwattanasakul N, Rodrussamee N, Suprayogi LS, Thanonkeo P, Kosaka T, Yamada M (2011) Utilization capability of sucrose, raffinose and inulin and its less-sensitiveness to glucose repression in thermotolerant yeast *Kluyveromyces marxianus* DMKU 3–1042. AMB Express 1:20
21. Fonseca GG, de Carvalho NMB, Gompert AK (2013) Growth of the yeast *Kluyveromyces marxianus* CBS 6556 on different sugar combinations as sole carbon and energy source. Appl Microbiol Biotechnol 97:5055–5067
22. Löser C, Urit T, Keil P, Bley T (2014) Studies on the mechanism of synthesis of ethyl acetate in *Kluyveromyces marxianus* DSM 5422. Appl Microbiol Biotechnol. Accepted, doi:10.1007/s00253-014-6098-4
23. Armstrong DW, Yamazaki H (1984) Effect of iron and EDTA on ethyl acetate accumulation in *Candida utilis*. Biotechnol Lett 6:819–824
24. Fonseca GG, Heinzle E, Wittmann C, Gombert AK (2008) The yeast *Kluyveromyces marxianus* and its biotechnological potential. Appl Microbiol Biotechnol 79:339–354
25. Ugalde OU, Castrillo IJ (2002) Single cell proteins from fungi and yeasts. In: Dilip KA, George GK (eds) Applied mycology and biotechnology, vol 2, Agriculture and food production. Elsevier, Amsterdam, pp 123–149
26. Ghaly AE, Kamal MA (2004) Submerged yeast fermentation of acid cheese whey for protein production and pollution potential reduction. Water Res 38:631–644
27. Schultz N, Chang L, Hauck A, Reuss M, Syldatk C (2006) Microbial production of single-cell protein from deproteinized whey concentrates. Appl Microbiol Biotechnol 69:515–520
28. Yadav JSS, Bezawada J, Elharche S, Yan S, Tyagi RD, Surampalli RY (2014) Simultaneous single-cell protein production and COD removal with characterization of residual protein and intermediate metabolites during whey fermentation by *K. marxianus*. Bioprocess Biosyst Eng 37:1017–1029
29. Parrondo J, García LA, Díaz M (2009) Nutrient balance and metabolic analysis in a *Kluyveromyces marxianus* fermentation with lactose-added whey. Brazil J Chem Eng 26:445–456
30. De Nicola R, Hazelwood LA, De Hulster EAF, Walsh MC, Knijnenburg TA, Reinders MJT, Walker GM, Pronk JT, Daran J-M, Daran-Lapujade P (2007) Physiological and transcriptional responses of *Saccharomyces cerevisiae* to zinc limitation in chemostat cultures. Appl Environ Microbiol 73:7680–7692
31. Mahmoud MM, Kosikowski FV (1982) Alcohol and single cell protein production by *Kluyveromyces* in concentrated whey permeates with reduced ash. J Dairy Sci 65:2082–2087
32. Kar T, Misra AK (1998) Effect of fortification of concentrated whey on growth of *Kluyveromyces sp.* Rev Argent Microbiol 30(4):163–169
33. Belem MAF, Lee BH (1999) Fed-batch fermentation to produce oligonucleotides from *Kluyveromyces marxianus* grown on whey. Process Biochem 34:501–509
34. Domingues L, Lima N, Teixeira JA (2001) Alcohol production from cheese whey permeate using genetically modified flocculent yeast cells. Biotechnol Bioeng 72:507–514

35. Moeini H, Nahvi I, Tavassoli M (2004) Improvement of SCP production and BOD removal of whey with mixed yeast culture. Electronic J Biotechnol 7:249–255
36. Aktaş N, Boyacı İH, Mutlu M, Tanyolaç A (2006) Optimization of lactose utilization in deproteinated whey by *Kluyveromyces marxianus* using response surface methodology (RSM). Bioresour Technol 97:2252–2259
37. Gupte AM, Nair AS (2010) β-galactosidase production and ethanol fermentation from whey using *Kluyveromyces marxianus* NCIM 3551. J Sci Ind Res 69:855–859
38. Castrillo JI, de Miguel I, Ugalde UO (1995) Proton production and consumption pathways in yeast metabolism. A chemostat culture analysis. Yeast 11:1353–1365
39. Hensing MCM, Bangma KA, Raamsdonk LM, de Hulster E, van Dijken JP, Pronk JT (1995) Effects of cultivation conditions on the production of heterologous β-galactosidase by *Kluyveromyces lactis*. Appl Microbiol Biotechnol 43:58–64
40. Vicente A, Castrillo JI, Teixeira JA, Ugalde U (1998) On-line estimation of biomass through pH control analysis in aerobic yeast fermentation systems. Biotechnol Bioeng 58:445–450
41. Tovar-Castro L, García-Garibay M, Saucedo-Castañeda G (2008) Lactase production by solid-state cultivation of *Kluyveromyces marxianus* CDBBL 278 on an inert support: effect of inoculum, buffer, and nitrogen source. Appl Biochem Biotechnol 151:610–617
42. Vivier D, Ratomahenina R, Moulin G, Galzy P (1993) Study of physicochemical factors limiting the growth of *Kluyveromyces marxianus*. J Ind Microbiol 11:157–161
43. Antoce O-A, Antoce V, Takahashi K (1997) Calorimetric study of yeast growth and its inhibition by added ethanol at various pHs and temperatures. Netsu Sokutei 24(4):206–213
44. Large PJ (1986) Degradation of organic nitrogen compounds by yeasts. Yeast 2:1–34
45. Castrillo JI, Ugalde UO (1992) Energy metabolism of *Kluyveromyces marxianus* in deproteinated whey. Chemostat studies. Modelling. J Biotechnol 22:145–152
46. Castrillo JI, Ugalde UO (1993) Patterns of energy metabolism and growth kinetics of *Kluyveromyces marxianus* in whey chemostat culture. Appl Microbiol Biotechnol 40:386–393
47. Rech R, Cassini CF, Secchi A, Ayub MAZ (1999) Utilization of protein-hydrolyzed cheese whey for production of β-galactosidase by *Kluyveromyces marxianus*. J Ind Microbiol Biotechnol 23:91–96
48. Hortsch R, Löser C, Bley T (2008) A two-stage CSTR cascade for studying the effect of inhibitory and toxic substances in bioprocesses. Eng Life Sci 8:650–657
49. Rahmatullah M, Boyde TRC (1980) Improvements in the determination of urea using diacetyl monoxime; methods with and without deproteinisation. Clin Chim Acta 107:3–9
50. Francis PS (2006) The determination of urea in wine - a review. Aust J Grape Wine Res 12:97–106
51. Smithers GW (2008) Whey and whey proteins - from 'gutter-to-gold'. Int Dairy J 18:695–704
52. Prazeres AR, Carvalho F, Rivas J (2012) Cheese whey management: a review. J Environ Manage 110:48–68
53. Cordier J-L, Butsch BM, Birou B, von Stockar U (1987) The relationship between elemental composition and heat of combustion of microbial biomass. Appl Microbiol Biotechnol 25:305–312
54. Krzystek L, Ledakowicz S (2000) Stoichiometric analysis of *Kluyveromyces fragilis* growth on lactose. J Chem Technol Biotechnol 75:1110–1118
55. Silva-Santisteban BOY, Converti A, Filho FM (2006) Intrinsic activity of inulinase from *Kluyveromyces marxianus* ATCC 16045 and carbon and nitrogen balances. Food Technol Biotechnol 44:479–483
56. Hamme V, Sannier F, Piot J-M, Didelot S, Bordenave-Juchereau S (2009) Crude goat whey fermentation by *Kluyveromyces marxianus* and *Lactobacillus rhamnosus*: contribution to proteolysis and ACE inhibitory activity. J Dairy Res 76:152–157
57. Foukis A, Stergiou P-Y, Theodorou LG, Papagianni M, Papamichael EM (2012) Purification, kinetic characterization and properties of a novel thermo-tolerant extracellular protease from *Kluyveromyces marxianus* IFO 0288 with potential biotechnological interest. Bioresour Technol 123:214–220
58. Fox PF (2003) Significance of indigenous enzymes in milk and dairy products. In: Whitaker JR, Voragen AGJ, Wong DWS (eds) Handbook of food enzymology. Marcel Dekker Inc, New York, Basel, pp 255–277
59. Perea A, Ugalde U, Rodriguez I, Serra JL (1993) Preparation and characterization of whey protein hydrolysates: applications in industrial whey bioconversion processes. Enzyme Microb Technol 15:418–423
60. Walker GM, White NA (2005) Introduction to fungal physiology. In: Kavanagh K (ed) Fungi: biology and applications. John Wiley & Sons Ltd, Chichester, pp 1–34
61. Nahvi I, Moeini H (2004) Isolation and identification of yeast strains with high beta-galactosidase activity from dairy products. Biotechnol 3(1):35–40
62. Rajoka MI, Khan S, Latif F, Shahid R (2004) Influence of carbon and nitrogen sources and temperature on hyperproduction of a thermotolerant β-glucosidase from synthetic medium by *Kluyveromyces marxianus*. Appl Biochem Biotechnol 117(2):75–92
63. Whitney PA, Cooper TG (1972) Urea carboxylase and allophanate hydrolase. Two components of adenosine triphosphate: urea amido-lyase in *Saccharomyces cerevisiae*. J Biol Chem 247:1349–1353
64. Shay LK, Hunt HR, Wegner GH (1987) High-productivity fermentation process for cultivating industrial microorganisms. J Ind Microbiol 2:79–85
65. Hack CJ, Marchant R (1998) Characterisation of a novel thermotolerant yeast, *Kluyveromyces marxianus* var *marxianus*: development of an ethanol fermentation process. J Ind Microbiol Biotechnol 20:323–327
66. Nor ZM, Tamer MI, Scharer JM, Moo-Young M, Jervis EJ (2001) Automated fed-batch culture of *Kluyveromyces fragilis* based on a novel method for on-line estimation of cell specific growth rate. Biochem Eng J 9:221–231
67. Gélinas P, Barrette J (2007) Protein enrichment of potato processing waste through yeast fermentation. Bioresour Technol 98:1138–1143
68. Fonseca GG, Gombert AK, Heinzle E, Wittmann C (2007) Physiology of the yeast *Kluyveromyces marxianus* during batch and chemostat cultures with glucose as the sole carbon source. FEMS Yeast Res 7:422–435
69. Mazutti MA, Zabot G, Boni G, Skovronski A, de Oliveira D, Di Luccio M, Rodrigues MI, Maugeri F, Treichel H (2010) Mathematical modeling of *Kluyveromyces marxianus* growth in solid-state fermentation using a packed-bed bioreactor. J Ind Microbiol Biotechnol 37:391–400
70. Silva-Santisteban BOY, Converti A, Filho FM (2009) Effects of carbon and nitrogen sources and oxygenation on the production of inulinase by *Kluyveromyces marxianus*. Appl Biochem Biotechnol 152:249–261
71. Rocha SN, Abrahão-Neto J, Gombert AK (2011) Physiological diversity within the *Kluyveromyces marxianus* species. Antonie Van Leeuwenhoek 100:619–630

Permissions

All chapters in this book were first published in ESS, by Springer; hereby published with permission under the Creative Commons Attribution License or equivalent. Every chapter published in this book has been scrutinized by our experts. Their significance has been extensively debated. The topics covered herein carry significant findings which will fuel the growth of the discipline. They may even be implemented as practical applications or may be referred to as a beginning point for another development.

The contributors of this book come from diverse backgrounds, making this book a truly international effort. This book will bring forth new frontiers with its revolutionizing research information and detailed analysis of the nascent developments around the world.

We would like to thank all the contributing authors for lending their expertise to make the book truly unique. They have played a crucial role in the development of this book. Without their invaluable contributions this book wouldn't have been possible. They have made vital efforts to compile up to date information on the varied aspects of this subject to make this book a valuable addition to the collection of many professionals and students.

This book was conceptualized with the vision of imparting up-to-date information and advanced data in this field. To ensure the same, a matchless editorial board was set up. Every individual on the board went through rigorous rounds of assessment to prove their worth. After which they invested a large part of their time researching and compiling the most relevant data for our readers.

The editorial board has been involved in producing this book since its inception. They have spent rigorous hours researching and exploring the diverse topics which have resulted in the successful publishing of this book. They have passed on their knowledge of decades through this book. To expedite this challenging task, the publisher supported the team at every step. A small team of assistant editors was also appointed to further simplify the editing procedure and attain best results for the readers.

Apart from the editorial board, the designing team has also invested a significant amount of their time in understanding the subject and creating the most relevant covers. They scrutinized every image to scout for the most suitable representation of the subject and create an appropriate cover for the book.

The publishing team has been an ardent support to the editorial, designing and production team. Their endless efforts to recruit the best for this project, has resulted in the accomplishment of this book. They are a veteran in the field of academics and their pool of knowledge is as vast as their experience in printing. Their expertise and guidance has proved useful at every step. Their uncompromising quality standards have made this book an exceptional effort. Their encouragement from time to time has been an inspiration for everyone.

The publisher and the editorial board hope that this book will prove to be a valuable piece of knowledge for researchers, students, practitioners and scholars across the globe.

List of Contributors

Susanne Arndt
Institut für Verkehrssicherheit und Automatisierungstechnik, Technische Universität Braunschweig, Hermann-Blenk-Straße 42, 38108 Braunschweig,Germany

Tatyana Sheveleva
Physikalisch-Technische Bundesanstalt, Bundesallee 100, 38116 Braunschweig, Germany

Christoph Goeker
AG Rechnernetze und verteilte Systeme, Universität Bielefeld, Universitätsstraße 25, 33615 Bielefeld, Germany

Rolando A. Rodriguez and Martin Greiner
Department of Mathematics, Aarhus University, Ny Munkegade 118, 8000 Aarhus C, Denmark

Magnus Dahl
Department of Physics and Astronomy, Aarhus University, Ny Munkegade 120, 8000 Aarhus C, Denmark

Sarah Becker
Frankfurt Institute for Advanced Studies, Frankfurt University, Ruth-Moufang-Str.1, 60438 Frankfurt am Main, Germany

Martin Greiner
Department of Engineering, Aarhus University, Inge Lehmanns Gade 10, 8000 Aarhus C, Denmark

J. A. M Hufen
Research and Consultancy in Leiden, Galileïweg 8, 2333BD, Leiden, The Netherlands

J. F. M. Koppenjan
Department of Public Administration, Faculty of Social Sciences, Erasmus University Rotterdam, P.O. Box 17383000DR, Rotterdam, The Netherlands

Richard P van Leeuwen and Jan B de Wit
Sustainable Energy Group, Saxion University of Applied Sciences, Maarten Harpertsz Tromplaan 28, P.O. Box 70.000, 7500 KB Enschede, the Netherlands

Richard P van Leeuwen, Jirka Fink and Gerard JM Smit
Department of Computer Science, Mathematics and Electrical, Engineering, University of Twente, Drienerlolaan 5, P.O. Box 217, 7500 AE Enschede, the Netherlands

María Elena Arce
Defense University Center at Spanish Naval Academy (Marin), Plaza de España 2, 36920 Marín, Spain

Alexandra Purkus and Erik Gawel
Department of Economics, Helmholtz Centre for Environmental Research - UFZ, Permoser Str. 15, Leipzig 04318, Germany

Erik Gawel
Institute for Infrastructure and Resources Management, University of Leipzig, Grimmaische Str. 12, Leipzig 04109, Germany

Marc Deissenroth and Kristina Nienhaus
Department of Systems Analysis and Technology Assessment, Institute of Engineering Thermodynamics, German Aerospace Center (DLR), Wankelstr. 5, Stuttgart 70563, Germany

Sandra Wassermann
Stuttgart Research Center for Interdisciplinary Risk and Innovation Studies (ZIRIUS), University of Stuttgart, Seidenstr. 36, Stuttgart 70174, Germany

Jörg Raupach-Sumiya
College of Business Administration, Ritsumeikan University, Kyoto, Japan

Hironao Matsubara
Institute for Sustainable Energy Policies (ISEP), Tokyo, Japan

Andreas Prahl
EcologicInstitute, Berlin, Germany

Astrid Aretz and Steven Salecki
Institute for Ecological Economy Research (IÖW), Berlin, Germany

Baboo Lesh Gowreesunker and Savvas A Tassou
RCUK National Centre for Sustainable Energy Use in Food Chains, Brunel University, UB8 3PH, Uxbridge, Middlesex, UK

Birte Viétor
Department: Nature and Environment, Bonner Str. 100, 42697 Solingen, Germany

Thomas Hoppe and Joy Clancy
Department of Governance and Technology for Sustainability (CSTM), Institute for Governance and Innovation Studies (IGS), University of Twente, Drienerlolaan 5, 7522, NB Enschede, The Netherlands

Jaqueline Daniel-Gromke, Jan Liebetrau, Velina Denysenko and Christian Krebs
Department Biochemical Conversion, DBFZ Deutsches Biomasseforschungszentrum gGmbH, Torgauer Straße 116, 04347 Leipzig, Germany

Richard P. van Leeuwen
Saxion University of Applied Sciences, Sustainable Energy Group, P.O. Box 70.000, 7500 KB, Enschede, The Netherlands

Jiří Fink, Richard P. van Leeuwen, Johann L. Hurink and Gerard J.M. Smit
University of Twente, Department of Computer Science, Mathematics and Electrical Engineering, P.O. Box 217, 7500 AE, Enschede, The Netherlands

Christian Büscher and Patrick Sumpf
Karlsruhe Institute of Technology (KIT), Institute for Technology Assessment and Systems Analysis (ITAS), Karlstr. 11, Karlsruhe 76133, Germany

Judi Pajo
Sociology/Anthropology Department, Dyson College, Pace University, 41 Park Row, Room 725, New York, NY 10038, USA

Leonard C. Kibet, Humberto Blanco-Canqui and Walter H. Schacht
Department of Agronomy and Horticulture, University of Nebraska-Lincoln Lincoln, NE 68583, USA

Robert B. Mitchell
Grain, Forage, and Bioenergy Research Unit USDA-ARS, University of Nebraska-Lincoln, Lincoln, NE 68583, USA

Mehmet Senbayram, Christian Wenthe and Annika Lingner
Institute of Applied Plant Nutrition, University of Gottingen, Carl-Sprengel-Weg 1, Gottingen 37075, Germany

Mehmet Senbayram and Cengiz Kaya
Present Address: Department of Plant Nutrition and Soil Science, University of Harran, SanliUrfa TR-63000, Turkey

Johannes Isselstein
Department Crop Science, Grassland Science, University of Gottingen, Gottingen 37075, Germany

Horst Steinmann
Department Biodiversity and Sustainable Land Use, University of Gottingen, Gottingen 37077, Germany

Sarah Köbke
Department Crop Science, Plant Nutrition and Soil Science, University of Gottingen, Gottingen 37075, Germany

Cheryl de Boer and Hans Bressers
CSTM/ITC, University of Twente, Drienerlolaan 5, 7522 NB Enschede, the Netherlands

Richard Hewitt, Patricia Martinez Alonso, Verónica Hernández Jiménez, Jaime Díaz Pacheco and Lara Román Bermejo
Observatory for a Culture of the Territory (OCT), C/ Duque Fernán Nuñez, 2 - 1º, 28012 Madrid, Spain

S. Hertel, P. Navarro, S. Deegener and I. Körner
Bioconversion and Emission Control Group, Institute of Wastewater Management and Water Protection, Hamburg University of Technology (TUHH), Hamburg, Germany

Goran Krajačić, Tomislav Pukšec and Neven Duić
Faculty of Mechanical Engineering and Naval Architecture, University of Zagreb, Zagreb, Croatia

Anamarija Šare
Maritime Department, University of Dubrovnik, Dubrovnik, Croatia

Vinay Narayan Hegde, Viraj Shrikanth Hosur, Samyukthkumar K Rathod, Puneet A Harsoor and K Badari Narayana
Department of Mechanical Engineering, RV College of Engineering, Bangalore, Karnataka 560059, India

Kostas Koumoutsos, Aristotelis Kretsis, Panagiotis Kokkinos and Emmanouel A Varvarigos
Department of Computer Engineering and Informatics, University of Patras, Patras, Greece. 2Computer Technology Institute "Diophantus", Patras, Greece

Vassilis Nikolopoulos and Eirini Gkioxi
Intelen Advanced Engineering Division, Nicosia, Cyprus

Anastatios Zafeiropoulos
Greek Research and Technology Network, Athens, Greece

Martin Dolinsky
Department of Management and Marketing, Faculty of Economics and Business, Pan-European University, Tematinska 10, 85105 Bratislava, Slovakia

Christian Löser, Thanet Urit, Erik Gruner and Thomas Bley
Institute of Food Technology and Bioprocess Engineering, TU Dresden, 01062 Dresden, Germany.

Thanet Urit
Department of Biology and Biotechnology, Faculty of Science and Technology, Nakhon Sawan Rajabhat University, 60000 Nakhon Sawan, Thailand